T4-AHW-002

REVIEWS in MINERALOGY

(Formerly: "Short Course Notes")

Volume 3

OXIDE MINERALS

DOUGLAS RUMBLE, III, Editor

The Authors:

Ahmed El Goresy
Max-Planck-Institut für Kernphysik
Heidelberg, West Germany

Stephen E. Haggerty
Department of Geology
University of Massachusetts
Amherst, Massachusetts 01003

J. Stephen Huebner
959 National Center
United States Geological Survey
Reston, Virginia 22092

Donald H. Lindsley
Department of Earth and Space Sciences
State University of New York
Stony Brook, New York 11794

Douglas Rumble, III
Geophysical Laboratory
2801 Upton St., N.W.
Washington, D.C. 20008

Series Editor:

Paul H. Ribbe
Department of Geological Sciences
Virginia Polytechnic Institute and State University
Blacksburg, Virginia 24061

MINERALOGICAL SOCIETY OF AMERICA

(Second printing, 1981)

PRINTED BY

BookCrafters, Inc.
Chelsea, Michigan 48118

REVIEWS IN MINERALOGY

(Formerly: SHORT COURSE NOTES)
ISSN 0275-0279

Volume 3: OXIDE MINERALS
ISBN 0-939950-03-0

Additional copies of this volume as well as those listed below may be obtained at moderate cost from

Mineralogical Society of America
2000 Florida Avenue, NW
Washington, D.C. 20009

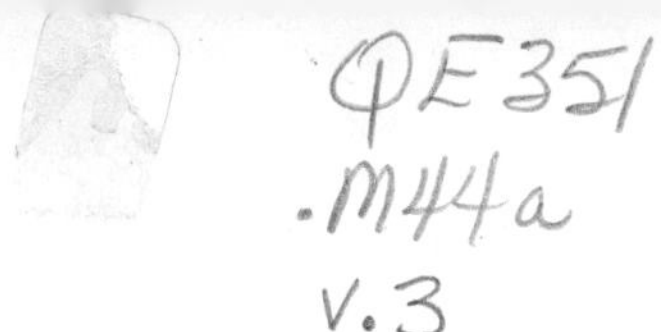

FOREWORD

Oxide Minerals was first printed in 1976 as Volume 3 of the Mineralogical Society of America's "SHORT COURSE NOTES." That series was renamed "REVIEWS IN MINERALOGY" in 1980, and for that reason this, the second printing of *Oxide Minerals*, has been reissued under the new banner. Only minor corrections have been made in this printing.

Paul H. Ribbe
Series Editor
Blacksburg, VA
October 1981

EDITOR'S PREFACE and ACKNOWLEDGEMENTS

The purpose of this volume is to provide, in a rapidly-printed, inexpensive format, an up-to-date review of the mineralogy and petrology of rock-forming opaque oxide minerals. It was the textbook for the short course on rock-forming oxide minerals sponsored by the Mineralogical Society of America at the Colorado School of Mines, November 5-7, 1976. The contributors hope that the work will be valuable not only to participants in the short course, but also to others desiring a modern review of the subject.

The editor is grateful to Don Bloss, Paul Ribbe, and Southern Printing Co. for invaluable assistance in preparing the notes for publication. Especial thanks are due Mrs. Margie Strickler for her outstanding work in typing the entire text. Elsevier Scientific Publishing Co. granted permission to reprint figures from their journal *Earth and Planetary Sciences Letters*.

J.J. Finney, George Fisher, J.F. Hays, Jim Munoz, Paul Ribbe, S.B. Romberger, and E-An Zen contributed vital advice and assistance in organizing the short course. E. Leitz, Inc., C. Reichert (American Optical Corp.), Vickers Instruments, Inc., and Carl Zeiss, Inc., provided demonstration models of their microscopes for use in the ore microscopy workshop.

Finally, I wish to acknowledge the assistance provided by the resources of the Carnegie Institution of Washington through the good offices of H.S. Yoder, Jr., Director of the Geophysical Laboratory, in organizing the short course and publishing this volume.

Douglas Rumble, III
Washington, D.C.
November 1976

TABLE OF CONTENTS

Chapter 8. OPAQUE MINERAL OXIDES in TERRESTRIAL IGNEOUS ROCKS

Stephen E. Haggerty

USEFUL REFERENCES

General

Deer, W. A., R. A. Howie, and J. Zussman (1962) *Rock Forming Minerals*, Vol. 5, Non-silicates. John Wiley, New York.

Elsdon, R. (1975) Iron-titanium oxide minerals in igneous and metamorphic rocks. *Minerals Sci. Eng. 7*, 48-70.

Freund, H. (Ed.) (1966) *Applied Ore Microscopy, Theory and Technique*. Macmillan Co., New York.

Ramdohr, P. (1969) *The Ore Minerals and Their Intergrowths*. Pergamon Press, Oxford.

Meteorites

Bunch, T. E., K. Keil, and K. G. Snetsinger (1967) Chromite composition in relation to chemistry and texture of ordinary chrondites. *Geochim. Cosmochim. Acta 31*, 1569.

Bunch, T. E. and K. Keil (1971) Chromite and ilmenite in nonchronditic meteorites. *Am. Mineral. 56*, 146-157.

Mason, B. (1962) *Meteorites*. John Wiley and Sons, New York.

Ramdohr, P. (1973) *The Opaque Minerals in Stony Meteorites*. Elsevier, Amsterdam.

Wasson, J. T. (1974) *Meteorites*. Springer-Verlag, New York.

The CRYSTAL CHEMISTRY and STRUCTURE of OXIDE MINERALS as EXEMPLIFIED by the Fe-Ti OXIDES

Donald H. Lindsley

Chapter I

INTRODUCTION

Crystal structures play a vital role in the interpretation of chemical reactions and magnetic properties of the oxide minerals. For most purposes it is useful to treat the oxide minerals as ionic crystals that consist of oxygen frameworks (nearly cubic or hexagonal close-packed) with cations occupying octahedral or tetrahedral interstices. Both iron and titanium as well as manganese are members of the first transition metal series; each can therefore exist in more than one valence state. Furthermore, the existence of unpaired $3d$ electrons in Ti^{3+}, Fe^{2+}, Fe^{3+}, Mn^{2+}, and Mn^{3+} imparts net magnetic moments to these ions. Thus a complete characterization of the structure of an oxide mineral must include the determination of valence states and magnetic orientation as well as the position of each atom in the unit cell. In some cases it is necessary to chose a magnetic unit cell that is a multiple of the crystallographic cell, or, alternatively, to view the magnetic cell as having lower symmetry than the crystallographic cell of the same size. No one technique can uniquely characterize the structure and chemistry of an iron-bearing oxide mineral, but a combination of methods has yielded detailed information on the most important structures.

Techniques

The primary method of determining the structures is of course x-ray diffraction. Single-crystal x-ray studies yield the (non-magnetic) symmetry of the structure, the positions of the metal ions, and with lesser precision, the positions of the oxygen ions. Thus in 1915, only three years after the discovery of x-ray diffraction, Bragg and Nishikawa were able independently to determine the main details of the magnetite (spinel) structure. However, they were not able to assign Fe^{2+} and Fe^{3+} to specific sites in the structure (the assignment they assumed is incorrect), nor were they able to determine the oxygen positions with great precision. Nevertheless, the basic structure provided by x-ray diffraction makes possible the utilization of other techniques, for these merely provide refinements of the structure rather than independent determinations of it.

Neutron diffraction is an important technique in determining the structures of the oxide minerals; three characteristics make it especially useful. It can yield information on (1) electron spin orientation and thus on magnetic structures; (2) oxygen parameters, since oxygen has a relatively large scattering cross-section for neutrons; and (3) the distribution of iron and titanium in the structure. (The electron densities of iron and titanium are sufficiently similar that x-ray diffraction cannot always distinguish between them, whereas the scattering cross-section of iron for neutrons is three times that of titanium in magnitude and of the opposite sign.) Most determinations of magnetic structure and precise oxygen parameters utilize neutron diffraction.

Measurement of saturation magnetization at low temperature can test models of magnetic structures, and is particularly useful in studying solid solution series. Electrical conductivity measurements have also been used to predict magnetic symmetry and cation distribution; this technique has been superseded for the most part.

Naturally occurring ^{57}Fe in the iron-titanium oxides permits application of the Mössbauer effect, which has been particularly useful in determining cation valencies, and, to a lesser extent, magnetic structure. Determination of the isomer shift for ^{57}Fe in the sample relative to that in the source provides a quantitative (or at least semiquantitative) measure of the valence of the iron--and, by difference, that of other cations.

Magnetic properties

The ensuing discussion of crystal structures must refer to several concepts of magnetism of the solid state. These concepts are briefly reviewed here; a more extensive review is given by Nagata (1961, p. 1-39) and by Stacey and Banerjee (1974).

An electron in an atom (or ion) has a magnetic moment that results from its spin, from its orbital motion, or from a combination of both. In iron and titanium, the moment results mainly from spin rather than orbital motion. The moments of electrons with opposite spins but with otherwise identical quantum states are self-cancelling. Atoms and ions in which all the electrons are thus paired have no net magnetic moment and are called *diamagnetic*. Atoms or ions with one or more unpaired electrons, on the other hand, have a net magnetic moment and are termed *paramagnetic*. The magnetic moment of an unpaired electron is one Bohr magneton (μ_B) which equals 0.9274×10^{-20} emu. Most paramagnetic particles have but one unpaired electron and therefore have moments of 1 μ_B. However, in the first transition-metal series each of the five states in the $3d$ level tends to be filled first by a single electron, and to become doubly occupied only after all five have been filled singly. (This is known as the high-spin condition. When as many $3d$ electrons as possible are paired, the atom or ion is said to be in the low-spin condition.) The spins of electrons occupying the same state must be opposed, in accord with the Pauli exclusion principle. Mn^{2+} and Fe^{3+}, each having five unpaired $3d$ electrons with parallel spins, have magnetic saturation moments of $5\mu_B$. Fe^{2+} has six $3d$ electrons, two of which are paired and therefore self-cancelling, so the resultant saturation moment due to the four unpaired electrons is $4\mu_B$. Fe^{o} likewise has six $3d$ electrons (as well as two paired $4s$ electrons) and also has a saturation moment of $4\mu_B$. Ti^{3+} with one $3d$ electron has 1 μ_B, but Ti^{4+} with no $3d$ electrons has no net moment.

It might be noted here that the *saturation* moments differ slightly from the *susceptibility* moments of the same ions. Susceptibility moments are somewhat larger, and tend not to be integral numbers of Bohr magnetons.

If paramagnetic atoms or ions are incorporated into a crystal structure with their spins (and hence their magnetic moments) randomly oriented, the resulting solid has a low magnetic susceptibility and no net moment, and is termed *paramagnetic*. But in crystals of α-iron, for example, *exchange interaction* between neighboring iron atoms results in a parallel alignment of moments throughout each of several small volumes (domains) of the crystal. Such crystals that have high magnetic susceptibilities and can acquire remanent (permanent) magnetic moments, are termed *ferromagnetic*. The theory of ferromagnetism is reviewed by Bozorth (1951).

Exchange interactions are strongly effective only between nearest-neighbor atoms. In oxide structures, however, the nearest neighbors of the metals are always oxygen ions, and exchange coupling cannot be invoked. Kramers (1934) proposed that the spins of next-nearest neighbor metal ions are coupled through the intermediate oxygens, a mechanism that he called *superexchange*. Anderson (1950) provided a firm theoretical basis for superexchange interactions, which now play a central role in most interpretations of the magnetic properties of oxides. Superexchange becomes more effective as the metal-oxygen-metal angle approaches 180°, and in general the interaction is negative: the moments of two paramagnetic ions coupled by superexchange through an oxygen are antiparallel and hence self-cancelling. Crystals in which superexchange interactions produce two equal magnetic substructures with opposite spin directions are termed *antiferromagnetic* (Hulthén, 1936). A perfect antiferromagnetic crystal has no magnetic moment, but if impurities, defects, or cation deficiencies are concentrated in one substructure there may be a net moment, as was postulated, for example, by Néel (1949,1953) to explain the "parasitic ferromagnetism" of hematite. Similarly, if the spin directions of the magnetic substructures are not precisely antiparallel, there can be a resultant moment perpendicular to the spin axes; this phenomenon is termed "spin-canting."

For every antiferromagnetic substance there is a characteristic temperature--the Néel point--at and above which the antiferromagnetic structure is destroyed through thermal vibration and the material becomes paramagnetic. Some substances, such as ilmenite, are paramagnetic at room temperature but become antiferromagnetic at lower temperatures. Magnetic susceptibility and specific heat of an antiferromagnetic material usually reach a maximum at or near the Néel point.

For some oxide structures superexchange interaction results in the formation of two magnetic substructures with opposite but unequal moments. Such structures have a net magnetic moment and were called *ferrimagnetic* by Néel (1948), the term being derived from the ferrites (spinels) in which ferrimagnetism is best displayed. Ferrimagnetic substances, like ferromagnetic ones, each have a characteristic temperature--the Curie point--at and above which they become paramagnetic; there is a close analogy between the Curie point and the Néel point of antiferromagnetic materials. Many macroscopic properties of ferrimagnetism are closely similar to those of ferromagnetism, and the former is sometimes conveniently if inelegantly considered a subclass of the latter. Thus it is common to speak of the "ferromagnetic properties" of a rock or mineral, even though those properties are almost certainly due to ferrimagnetism. Both ferrimagnetism and antiferromagnetism with parasitic ferromagnetism result from two opposite but unequal magnetic substructures, but in ferrimagnetism the inequality is a result of the crystal structure and thus would exist even in a perfect crystal.

THE CUBIC OXIDE MINERALS

Many oxide minerals have a cubic structure; others are perhaps best viewed as slight distortions or modifications of isometric structures. In general the oxygens approach cubic close packing. A cubic-close-packed framework of 32 oxygens contains a total of 64 tetrahedral sites and 32 octahedral sites. If only octahedral sites are occupied by divalent cations, as in periclase (MgO), the crystal has the NaCl structure (Fig. L-1). In a very important group of oxides--the spinels--both octahedral sites and tetrahedral sites

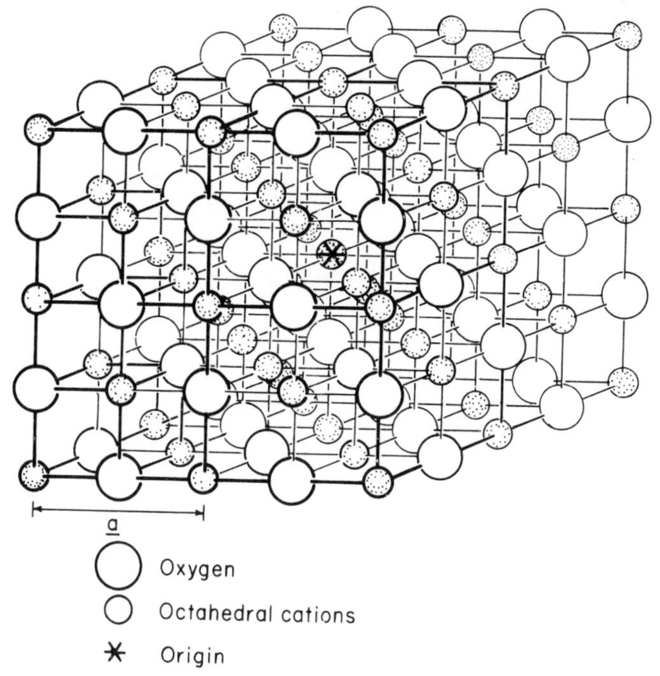

Figure L-1. A view of the monoxide (periclase, manganosite, ideal wüstite) structure, emphasizing the alternating (111) planes of oxygens and cations. Eight unit cells are shown to facilitate comparison with the spinel structure (Fig. L-3).

---

are occupied. Not all of both sets of sites can be filled, however, for this would require face-sharing between octahedra and tetrahedra--which is energetically unstable because of electrostatic repulsion of the cations. The filling of 8 tetrahedral sites precludes occupancy of more than 16 octahedral sites; conversely, the filling of 16 octahedral sites permits occupancy of only 8 of the 64 possible tetrahedral sites, if the symmetry of the space group is maintained.

Monoxides (Space group $Fm3m$)

Like periclase, manganosite (MnO) has the NaCl structure, with the divalent cations occupying the octahedral interstices. Viewed along the [111] axis, the structure consists of alternating (111) layers of oxygen and metal (Fig. L-1).

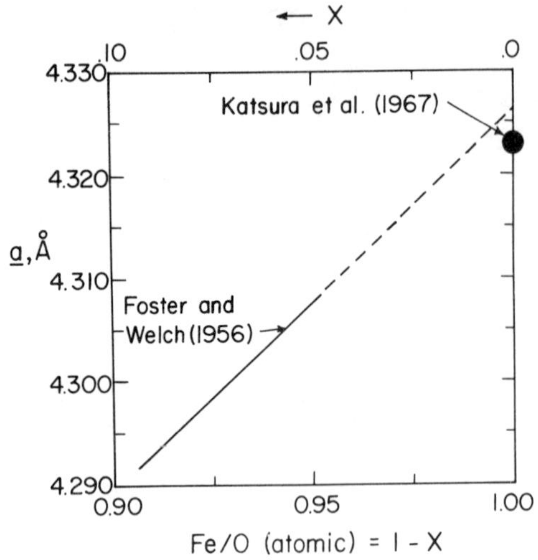

Figure L-2. Unit cell dimension of wüstite as a function of composition.

Although rare as a mineral (but see, for example, Walenta, 1960), the phase wüstite (nominally FeO) is of considerable interest. For years a controversy raged over whether stoichiometric FeO exists. There is now general agreement that stoichiometric FeO cannot exist as a stable phase at low pressures and that it is always cation-deficient. Charge balance is maintained by the replacement of 3x Fe^{2+} by 2x Fe^{3+}, so that the formula may be written $Fe^{2+}_{1-3x}Fe^{3+}_{2x}O$, or, more simply, $Fe_{1-x}O$. The most iron-rich wüstite stable at one atmosphere is about $Fe_{0.954}O$ (Darken and Gurry, 1945). Lattice constants *a* range from 4.309 Å for $Fe_{0.949}O$ to 4.292 Å for $Fe_{0.914}O$, the exact values depending on the thermal history of the samples (Foster and Welch, 1956); see Fig. L-2.

Katsura *et al.* (1967) reported synthesis of stoichiometric FeO at pressures above 36 kb. The value $a = 4.323 \pm 0.001$ Å determined by them is very close to that predicted for pure FeO by extrapolation (4.326 Å).

Roth (1960) has made a detailed study of the defect structure of wüstite by neutron diffraction. The diffraction patterns not only confirm the presence of vacancies in the octahedral iron positions, but indicate more such vacancies than are required by the chemical analysis. For example, the neutron diffraction data indicate 0.08 vacancies per octahedral iron site for the composition $Fe_{0.94}O$. Evidently 0.02 iron ions per formula unit must be accommodated in tetrahedral sites interstitial to the close-packed oxygen framework. Cubic symmetry is maintained if the unit cell is doubled; the resulting space group is *Fd3m*. The symmetry and oxygen content of such a cell are identical to that of magnetite, and, indeed, Roth visualized the local structure in the vicinity of octahedral vacancies as essentially that of magnetite.

Roth's model has been generally confirmed by single crystal x-ray diffraction studies (Koch *et al.*, 1966; Koch and Fine, 1967, Koch and Cohen, 1969) which suggested the presence of ordered complexes of octahedral vacancies and tetrahedral cations. Shirane *et al.* (1962) have found Mössbauer evidence that the *local* symmetry around the Fe^{2+} irons is less than cubic; they infer that the lower symmetry is due to the vacancies. Some studies have shown that quenched wüstites may have one of three slightly different structures (Vallet and Raccah, 1965; Kleman, 1965; Carel, 1967). Swaroof and Wagner (1967) could find no evidence of phase changes in the wüstite field at 950-1250°C; however, high-temperature x-ray diffraction studies by Manenc (1968) show that at least one of the ordered structures persists to 1000°C for Fe-poor compositions. Manenc suggests that at temperatures within the wüstite stability field, Fe-rich wüstites have the true NaCl structure (with the vacancies random) whereas Fe poor wüstites have an ordered structure.

Spinel group

A large number of oxide minerals, some sulfides, and many artificial substances crystallize with the spinel structure, which is extraordinarily flexible in terms of the cations it will accept. At least 30 different elements, with valences ranging from +1 to +6, can serve as cations in oxide spinels. Some geologically important spinel end members are listed in Table L-1. Of the iron-titanium oxides, the magnetite-ulvospinel solid solution series has the spinel structure, as do the cation-deficient oxides--at least as a first approximation--maghemite and titanomaghemite. The unit cell is face-centered, cubic, and in oxide spinels contains 32 oxygens, which form a nearly cubic-close-packed framework as viewed along the cube diagonals ([111]), the space group being *Fd*3*m*. The cations occupy interstices within the oxygen framework (Fig. L-3). In space group *Fd*3*m* there are two alternative sets of compatible tetrahedral and octahedral sites: 16*d* (octahedral) and 8*a* (tetrahedral) or 16*c* and 8*b* (International Tables, 1952, p. 340-341). These sets, although mutually exclusive, are identical in that a translation of the origin by $\frac{1}{2}$, $\frac{1}{2}$, $\frac{1}{2}$ will bring one set into coincidence with the other. By convention, the set 16*d* and 8*a* is chosen for the spinels. There is considerable variation in the literature regarding the notation used to describe these sites. Most workers have employed the Wyckoff (1922) notation, others that of the International Tables--which is followed here--and still others a notation based on the concept of magnetic substructures

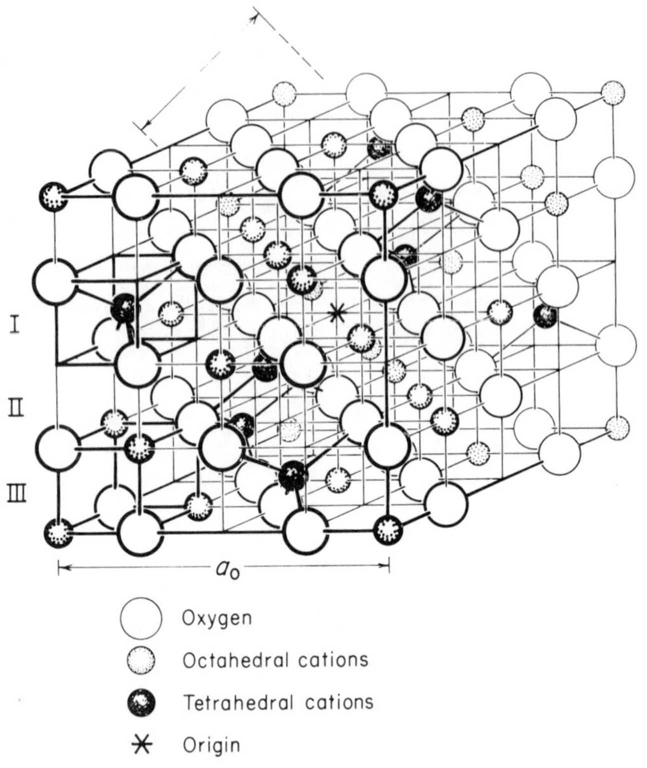

Figure L-3. The spinel unit cell, oriented so as to emphasize the (111) planes. Atoms are not drawn to scale; the circles simply represent the centers of atoms. The origin in this diagram lies at the center of symmetry, as recommended by the International Tables (1952); it differs by $(\frac{1}{8},\frac{1}{8},\frac{1}{8})$ from the origin used in much of the literature. The arrow at the top indicates the cation and oxygen layers shown in Figure L-5.

Table L-1. Some spinel end members.

Mineral Name	Formula	Cell Edge a, in Å	Oxygen Parameter u	Structure
Magnetite	$Fe^{3+}[Fe^{2+}Fe^{3+}]O_4$	8.396	0.2548	I
Magnesioferrite	$Fe^{3+}[Mg^{2+}Fe^{3+}]O_4$	8.383	0.257	I
Jacobsite	$Fe^{3+}[Mn^{2+}Fe^{3+}]O_4$	8.51		I
Chromite	$Fe^{2+}[Cr_2^{3+}]O_4$	8.378		N
Magnesiochromite	$Mg^{2+}[Cr^{3+}]O_4$	8.334	0.260	N
Spinel	$Mg^{2+}Al_2^{3+}O_4$	8.103	0.262	7/8 I
Hercynite	$Fe^{2+}[Al_2^{3+}]O_4$	8.135		N
Ulvöspinel	$Fe^{2+}[Fe^{2+}Ti^{4+}]O_4$	8.536	0.261	I

N, "normal" cation distribution, $X[Y_2]O_4$, where [ ] indicate octahedral cations. I, "inverse" distribution, $Y[XY]O_4$. Many spinels are probably intermediate between these extremes. Data from Burns (1970, p. 110).

A and B. For the convenience of those who wish to pursue the original literature, Table L-2 correlates these various notations.

Table L-2. Nomenclature of cation sites in spinels.

Tetrahedral	Octahedral	Example of Usage
f	*c*	Wyckoff (1922, 1965)
A	B	Néel (1948)
a	*d*	International Tables (1952)

The sites occupied by cations in the spinel unit cell are *special* sites; that is, they lie at the intersections of symmetry elements. These sites are therefore fixed, and their coordinates are given by rational fractions of the cell edge, a. Thus the coordinates x, y, z of one tetrahedral (a) site are $\frac{1}{8}a$, $\frac{1}{8}a$, $\frac{1}{8}a$ (abbreviated $\frac{1}{8}$, $\frac{1}{8}$, $\frac{1}{8}$). The coordinates of all the sites are given in Table L-3, based on an origin at the center of symmetry, rather than the more usual origin at $-\frac{1}{8}$, $-\frac{1}{8}$, $-\frac{1}{8}$. The 32 oxygens, on the other hand, occupy sites whose exact coordinates must be determined experimentally. Only one parameter, conventionally called u, needs to be determined; the oxygen

Table L-3. Coordinates of anion and cation sites in spinels. Space group *Fd3m*. Origin at center ($\bar{3}m$).

Type of Site	No. of Sites	Notation	Symmetry	Coordinates
Anion	32	*e*	3m	$u,u,u;\ u,\frac{1}{4}-u,\frac{1}{4}-u;\ \frac{1}{4}-u,u,\frac{1}{4}-u;$ $\frac{1}{4}-u,\frac{1}{4}-u,u;\ \bar{u},\bar{u},\bar{u};\ \bar{u},\frac{3}{4}+u,\frac{3}{4}+u;$ $\frac{3}{4}+u,\bar{u},\frac{3}{4}+u;\ \frac{3}{4}+u,\frac{3}{4}+u,\bar{u}$
Tetrahedral Cation	8	*a*	$\bar{4}3m$	$\frac{1}{8},\frac{1}{8},\frac{1}{8};\ \frac{7}{8},\frac{7}{8},\frac{7}{8}$
Octahedral Cation	16	*d*	$\bar{3}m$	$\frac{1}{2},\frac{1}{2},\frac{1}{2};\ \frac{1}{2},\frac{1}{4},\frac{1}{4};\ \frac{1}{4},\frac{1}{2},\frac{1}{4};\ \frac{1}{4},\frac{1}{4},\frac{1}{2}$

Note: Because of the face-centered lattice, there are four points equivalent to the origin:

$$0,0,0;\ 0,\frac{1}{2},\frac{1}{2};\ \frac{1}{2},0,\frac{1}{2};\ \frac{1}{2},\frac{1}{2},0$$

The full sets of sites are generated by applying the coordinates in the table to each of the equivalent points.

Reference: International Tables (1952, p. 341).

coordinates are then derived from symmetry considerations. The oxygens lie in 32 *e* sites whose coordinates in terms of u are given in Table L-3. If the oxygen parameter u happened to equal $\frac{1}{4}$, the oxygens would occupy special sites which correspond to rigorous cubic close-packing.* Inasmuch as u does not deviate greatly from $\frac{1}{4}$, it is a reasonable and useful approximation to view the oxygens as close-packed. Variations in u correspond to displacements of oxygens along cube diagonals, and reflect adjustments to the relative effective radii of cations in the tetrahedral and octahedral sites. An increase in u corresponds to an enlargement of the tetrahedral coordination polyhedra and a compensating diminution in the octahedra (see Fig. L-4). In addition, the entire oxygen framework (i.e., the unit cell) can expand or contract to accommodate cations of larger or smaller average effective radius. It is this flexibility of the

*Many workers choose an origin for the spinel structure at $\bar{4}3m$, which is $(-\frac{1}{8}, -\frac{1}{8}, -\frac{1}{8})$ removed from the origin at the center ($\bar{3}m$) adopted here. The values for u listed here should be increased by 0.125 to compare them with most values in the literature.

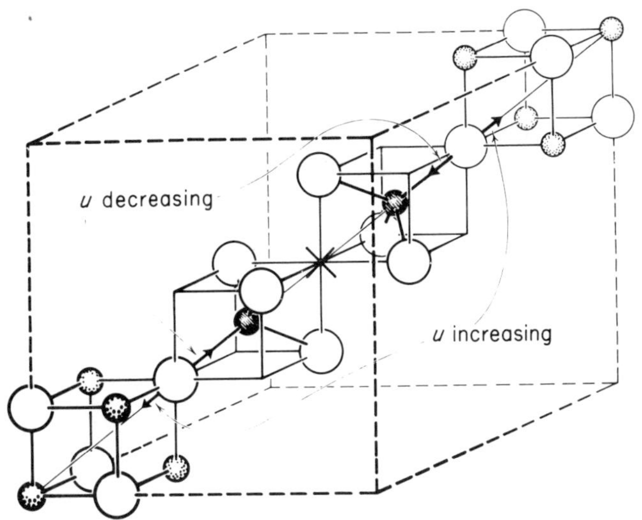

Figure L-4. Details along a body diagonal of the spinel unit cell in Fig. L-3, illustrating how changes in the oxygen parameter u change the relative sizes of the tetrahedral and octahedral sites.

---

oxygen framework that permits such a large number of elements to occur as important cations in oxide spinels.

The general chemical formula for ideal spinels is XY_2O_4 (or $X_8Y_{16}O_{32}$ per unit cell), where X and Y are cations of different valence. In magnetite X = Fe^{2+} and Y = Fe^{3+}; in ulvospinel, on the other hand, X = Ti^{4+} and Y = Fe^{2+}. The original determinations of the spinel structure (Bragg, 1915a,b; Nishikawa, 1915) were made on magnetite and on spinel (X=Mg^{2+}, Y=Al^{3+}); it was impossible to distinguish between the X and Y cations on the basis of x-ray intensities because of the similar scattering powers of Mg and Al and of Fe^{2+} and Fe^{3+}. For simplicity it was assumed that the eight X cations occupy the $8a$ sites and the sixteen Y cations occupy the $16d$ sites, giving a structural formula: $X[Y_2]O_4$, where the brackets denote cations in octahedral ($16d$) sites. Barth and Posnjak (1931,1932) discovered by careful x-ray intensity measurements that this "normal" structure was correct for several aluminate spinels, but not for many other spinels. The latter have 8 of the 16 Y cations in the $8a$ sites, yielding the structural formula $Y[YX]O_4$. Verwey and Heilmann (1947) termed these spinels "inversed," and at present they are known as "inverse spinels." It is now known that most spinels are intermediate between these two extremes. Table L-1 indicates which end member spinels are normal and which are inverse.

Verwey and Heilmann (1947) also gave empirical rules regarding the distribution of cations in the tetrahedral and octahedral sites of the spinel structure. Electrostatic and ionic radius considerations alone are insufficient to explain the observed distributions. Goodenough and Loeb (1955) showed that cations having a tendency to form hybrid sp^3 bonds--such as Fe^{3+}--are favored in tetrahedral sites. There have been suggestions that ions thus bound in tetrahedral sites are relatively immobile (Frölich and Stiller, 1963; O'Reilly and Banerjee, 1966), but self-diffusion experiments indicate otherwise (Lindner and Akerstrom, 1956).

Magnetite (Fe_3O_4). Both Bragg (1915a,b) and Nishikawa (1915) determined the magnetite structure using Laue x-ray photographs obtained from single crystals. They assumed that the structural formula was $Fe^{2+}[Fe_2^{3+}]O_4$, and that the oxygen parameter u was 0.25, although they recognized that it would vary with cation composition. Claassen (1926) refined the value of u to 0.254 ± 0.001. Hamilton (1958) further refined u to 0.2548 ± 0.0002 at 23°C from neutron diffraction data.

A wide range of values for the unit-cell parameter (a) of magnetite has been reported. Several reasons for this variation exist. Many of the magnetites were not adequately characterized chemically, and the discrepancies probably reflect the presence of cations other than Fe^{2+} and Fe^{3+}, or of cation vacancies. Precise x-ray work on natural and synthetic magnetites of known composition has yielded values of a ranging from 8.393 to 8.3963 Å.

The similarity in scattering power of Fe^{2+} and Fe^{3+} made it impossible for the early workers to determine the structural formula of magnetite by x-ray methods. Verwey and deBoer (1936) used measurements of electrical conductivity to demonstrate that magnetite has the inverse spinel structure, $Fe^{3+}[Fe^{2+}Fe^{3+}]O_4$. Néel (1948) predicted that the irons in the $8a$ and the $16d$ sites form two magnetic substructures with antiparallel moments along [111]. If magnetite had the normal structure, $Fe^{2+}[Fe_2^{3+}]O_4$, there would be 8x4 μ_B in the $8a$ sites and 8x(5+5) μ_B oppositely directed in the $16d$ sites, for a net moment of 48 μ_B per unit cell [=6 μ_B per formula unit]. In the case of inverse structure, $Fe^{3+}[Fe^{2+}Fe^{3+}]O_4$ the net moment would be 8x(4+5)-8x5 = 32 μ_B per unit cell [=4 μ_B per formula unit]. The empirical value (extrapolated to 0°K), of 4.07 μ_B (Néel, 1948, p. 179) per formula unit thus supports the Néel model regarding both the inverse structure and the ferrimagnetic structure. Néel assumed that the excess moment (.07 μ_B per formula unit) reflects an orbital contribution to the moment of Fe^{2+}; an alternative explanation is that the structure is not completely inverse. Final confirmation of the model was provided by the neutron diffraction experiments of Shull *et al.* (1951b).

For many purposes it is useful to view the magnetite structure along the [111] axis (Fig. L-5). It is along this axis that the (nearly) cubic close-packing of the oxygens is observed and that the ferrimagnetic moment tends to be directed. In the wüstite structure normal to [111], planes of oxygens alternate with planes of iron ions. In magnetite the sequence is somewhat more complex because tetrahedral and octahedral cations do not lie in the same planes: normal to [111] the sequence is $O\text{-}Fe^{VI}\text{-}O\text{-}Fe^{IV}\text{-}Fe^{VI}\text{-}Fe^{IV}\text{-}O\text{-}Fe^{VI}\ldots$, where the superscripts refer to coordination numbers. The distance along [111] between Fe^{IV} layers and the adjacent Fe^{VI} layers is small (0.675 Å, or 0.08 a),

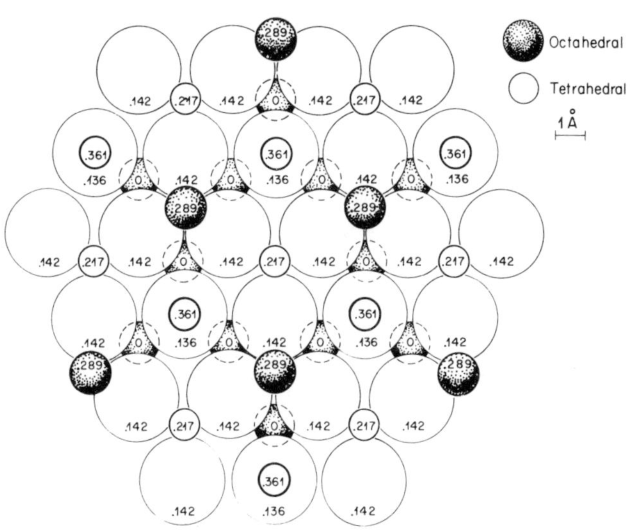

Figure L-5. Diagram, approximately to scale, showing an oxygen layer of a spinel and the cation layers on either side of it, projected onto a (111) plane. The large open circles are oxygens. The ionic radii (Shannon and Prewitt, 1969) are drawn for magnetite. Decimal fractions show height above the lower octahedral layer, expressed as a portion of the body diagonal of the unit cell ($a\sqrt{3}$). Note that the oxygen plane is puckered slightly; for ideal cubic-close-packing, the height of the oxygens would be 0.144 (= $\frac{1}{12}\sqrt{3}$).

and for some purposes it is a good approximation to consider the structure as successive planes of $O-Fe^{VI}-O-(Fe^{IV}Fe^{VI}Fe^{IV})-O-Fe^{VI}$.... The structure of magnetite as viewed in this way bears many similarities to that of hematite, and the (111) planes play an important role in the textural relations between these two minerals.

The Curie temperature of magnetite corresponds to a transition from long-range ferrimagnetic ordering at lower temperatures to disorder at higher temperatures. Values of the Curie temperature ranging from 570° to 581°C have been reported. The transition is accompanied by maxima in the coefficient of thermal expansion, the specific heat and the neutron scattering cross-section.

At low temperatures magnetite undergoes another transition which involves a decrease in crystallographic symmetry, electrical conductivity, and magnetic properties. Probably the best value of the transition temperature is 119.4°K, obtained for stoichiometric, synthetic single crystals (Calhoun, 1954).

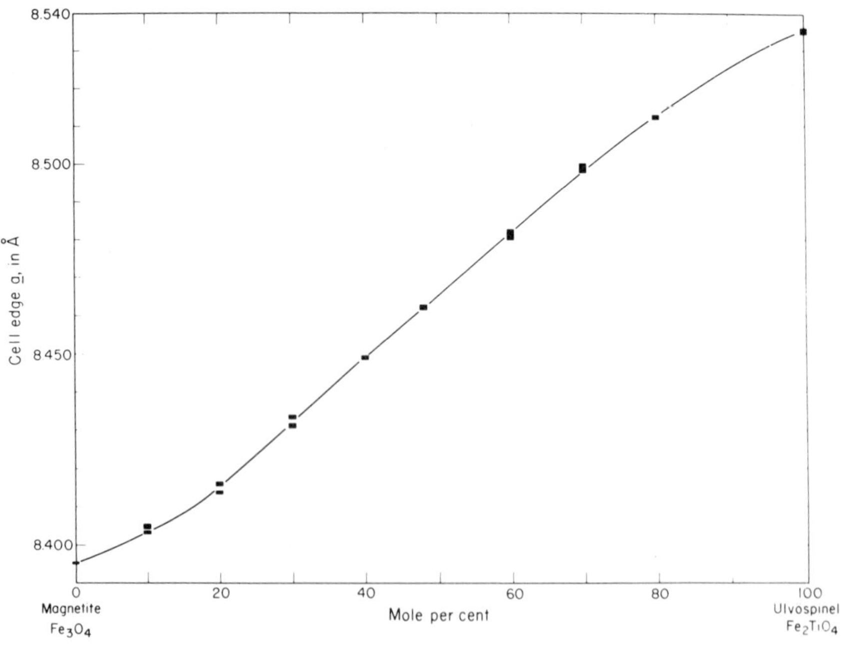

Figure L-6. Variation of cell dimension with composition in the $(1-x)Fe_3O_4$-xFe_2TiO_4 series. There appear to be changes in slope at or near x = 0.2 and x = 0.8. Height of symbols shows uncertainties (Lindsley, 1965).

Ulvöspinel (Fe_2TiO_4). Pure synthetic ulvöspinel was shown to have the inverse spinel structure ($Fe[FeTi]O_4$), with $u = 0.265 \pm 0.01$, by Barth and Posnjak (1932). The cell dimension a is 8.536 ± 0.001 (Lindsley, 1965; Akimoto and Syono, 1967). Forster and Hall (1965) determined u to be 0.261 $\pm$ 0.001 by neutron diffraction. The sequence of planes normal to [111] for an ideal $Fe[FeTi]O_4$ is $O-(Fe,Ti)^{VI}-O-Fe^{IV}-(Fe,Ti)^{VI}-Fe^{IV}-O$. The valence states of the iron and titanium are evidently $Fe^{2+}[Ti^{4+}Fe^{2+}]O_4$ (Gorter, 1957, p. 345; Verwey and Heilmann, 1947; Rossiter and Clark, 1965). The ideal inverse structure has 8 Fe^{2+} in $8a$ with spins antiparallel to 8 Fe^{2+} in $16d$ and is antiferromagnetic with a net moment of 0 μ_B.

Magnetite-ulvöspinel solid solutions. Solid solutions $(1-x)Fe_3O_4 \cdot x(Fe_2TiO_4)$ between magnetite and ulvöspinel have the spinel structure. Variations of the unit-cell parameter a with composition for synthetic solid solutions are presented in Figure L-6. Syono (1965, p. 136) reported that neutron diffraction measurements gave $u = 0.261 \pm 0.001$ for x = 0.99 and $u = 0.255 \pm 0.001$ for x = 0.56.

Several different structural formulae for the solid solution have been proposed. The Akimoto model assumes that Ti^{4+} always replaces the Fe^{3+} in $16d$ in magnetite whereas Fe^{2+} always replaces the Fe^{3+} in $8a$; the general formula would be $Fe^{3+}_{1-x}Fe^{2+}_{x}[Fe^{2+}Fe^{3+}_{1-x}Ti^{4+}_{x}]O_4$, yielding a predicted saturation moment of $4(1-x)$ μ_B (Akimoto, 1954); see Figs. L-7 and L-8. Néel (1955, p. 196) and Chevallier *et al.* (1955) assumed that all Ti^{4+} enters the d sites, and that Fe^{2+} first replaces Fe^{3+} in d sites and only then the Fe^{3+} in a sites. This model, which is in accord with the empirical rules of Verwey and Heilmann (1947) as well as the theoretical considerations of Goodenough and Loeb (1955), yields the structural formulae:

$$\begin{aligned} &Fe^{3+}[Fe^{2+}_{1+x}Fe^{3+}_{1-2x}Ti^{4+}_{x}]O_4 && \text{for } 0 \le x \le 1/2 \\ &Fe^{3+}_{2-2x}Fe^{2+}_{2x-1}[Fe^{2+}_{2-x}Ti^{4+}_{x}]O_4 && \text{for } 1/2 \le x \le 1. \end{aligned} \qquad (1i-1)$$

The corresponding saturation moments are $(4-6x)$ μ_B for $0 \le x \le 1/2$ and $(2-2x)$ μ_B for $1/2 \le x \le 1$. The observed moments for many synthetic solid solutions (Fig. L-7) fall about half-way between the values predicted for these two models (Akimoto *et al.*, 1957a, p. 173; Akimoto, 1962). A model that fits the magnetic data has been proposed by O'Reilly and Banerjee (1965; Banerjee and O'Reilly, 1966). Their model is identical with the Néel-Chevallier-Bolfa-Mathieu model over the composition ranges $0 \le x \le 0.2$ and $0.8 \le x \le 1.0$, but varies continuously between x = 0.2 and 0.8 with no break at x = 0.5. It may be expressed (see Fig. L-8):

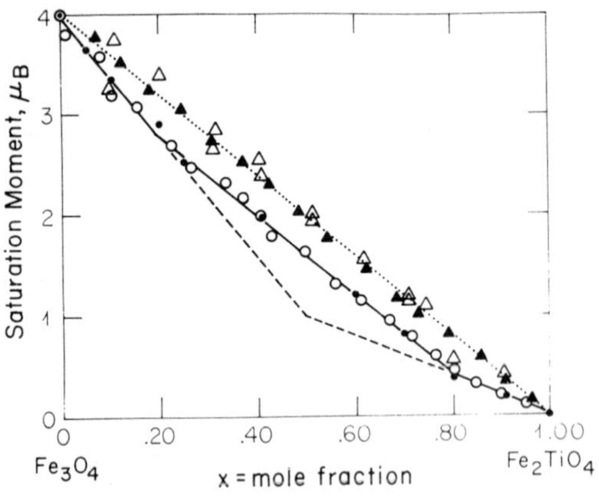

Figure L-7. Variation of saturation moment (in Bohr magnetons, μ_B) with composition in the $(1-x)Fe_3O_4$-xFe_2TiO_4 series. Dotted line, predicted values for Akimoto's (1954) model. Dashed line, predicted values for the Néel (1955) and Chevallier-Bolfa-Mathieu (1955) model. Solid line, predicted values for the O'Reilly-Banerjee (1965) model. Points labeled "O'Reilly *et al.* (1967)" should read "O'Reilly and Banerjee (1965)."

$$\left.\begin{array}{ll} Fe^{3+}[Fe^{2+}_{1\ x}Fe^{3+}_{1-2x}Ti^{4+}_{x}]O_4^{2-} & 0 \leq x \leq 0.2 \\ Fe^{3+}_{1.2-x}Fe^{2+}_{x-0.2}[Fe^{2+}_{1.2}Fe^{3+}_{0.8-x}Ti^{4+}_{x}]O_4^{2-} & 0.2 \leq x \leq 0.8 \\ Fe^{3+}_{2-2x}Fe^{2+}_{2x-1}[Fe^{2+}_{2-x}Ti^{4+}_{x}]O_4^{2-} & 0.8 \leq x \leq 1.0 \end{array}\right\} \quad (1i\text{-}2)$$

Although there is no theoretical reason why Fe^{2+} should begin to occupy the tetrahedral sites at $x \geq 0.2$, there is abundant permissive evidence for this model from the physical properties of the magnetite-ulvöspinel series: plots of electrical conductivity, activation energy of conduction, unit-cell dimensions, and Curie temperatures all show changes in slope at or near $x = 0.2$ and 0.8. Furthermore, at or near $x = 0.2$ there are breaks in the coercive force, remanent coercive force, remanent magnetization in uniform fields, and in magnetostriction and magnetocrystalline anisotropy constants.

However, there is some evidence that cation distribution in the magnetite-ulvöspinel series may be sensitive both to temperature of formation and to cooling rate. Stephenson (1969) showed that samples rapidly quenched from just below their melting temperatures have saturation moments close to $4(1-x)$ μ_B and thus appear to have the Akimoto distribution. Bleil (1971) found similar results for samples quenched from 1300°C. The Stephenson and Bleil results have remarkable implications: The Akimoto model (Fig. L-8a) requires that Fe^{3+} (octahedral) = Fe^{3+} (tetrahedral) for all values of x. A further requirement is that exactly eight of the 16 octahedral sites per unit cell be occupied by Fe^{2+}, again for all values of x. This strong suggestion of

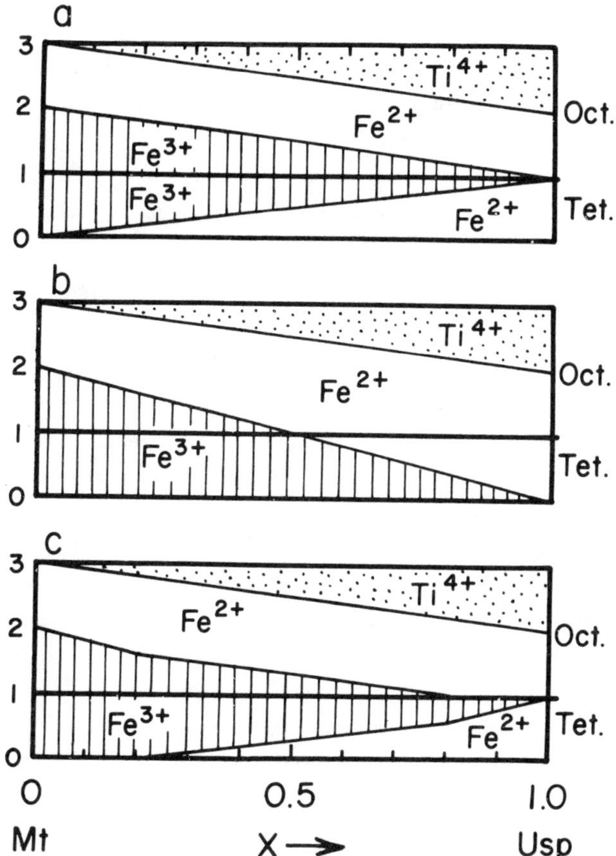

Figure L-8. Schematic representations of proposed cation distributions in the Fe_3O_4-Fe_2TiO_4 (Mt-Usp) series. a, Akimoto model; b, Néel model; c, O'Reilly-Banerjee model.

ordering is incompatible with space group *Fd3m* (because all 16*d* sites are symmetrically and thus also energetically equivalent). In addition, one would expect samples quenched from high temperatures to *disordered* rather than ordered. Jensen and Shive (1973) present quantum-mechanical arguments that it is impossible to quench tetrahedral Fe^{2+} in magnetite-ulvöspinel solid solutions; they claim that electrons will be lost to octahedral Fe^{3+} by tunneling.

Stephenson (1969) and Bleil (1971,1976) suggest that cation distributions in samples quenched more slowly or from lower temperatures will depart from the Akimoto model and will approach the Néel model (1i-1; Fig. L-8b) at very low temperatures.

Clearly, cation distribution in the magnetite-ulvöspinel series is not yet fully understood. Most probably it will be necessary to combine magnetic, Mössbauer, neutron-diffraction, and high-temperature x-ray studies on well characterized samples to resolve the ambiguities.

Maghemite (γ-Fe_2O_3). Maghemite is controversial. Its name, composition, stability, and very existence have been hotly debated, so it is hardly surprising that its structure has also been subject to disagreement. In an excellent summary of the early work on maghemite, Mason (1943) points out that much of the confusion undoubtedly arose from the failure of many workers to recognize maghemite as a monotropic polymorph of Fe_2O_3: it is always metastable with respect to hematite (α-Fe_2O_3). Its distinguishing characteristics are a composition close to Fe_2O_3 combined with high magnetic susceptibility and remanence. Much work on maghemite has been spurred by the discovery of its usefulness in making magnetic tapes.

Robbins (1859) showed that a "magnetic peroxide of iron," prepared by the oxidation of magnetite, had the composition Fe_2O_3 and thus was a polymorph of hematite. Hägg (1935a,b), Verwey (1935a,b), and Kordes (1935) independently proposed that maghemite has an iron-deficient spinel structure with a unit-cell formula $Fe^{3+}_{21.333}O_{32}$. The distribution of the vacancies--octahedral, tetrahedral, or both--has remained controversial. Hägg interpreted his x-ray intensity data (1935b) as indicating that the vacancies are randomly distributed in the octahedral 16*d* sites, yielding a structural formula $Fe^{3+}[Fe^{3+}_{1.67}\square_{0.33}]O_4$ and a unit-cell content $Fe^{3+}_8\ [Fe^{3+}_{13.33}\square_{2.67}]O_{32}$, where $\square$ indicates a cation vacancy. Verwey's interpretation of his intensity data was somewhat equivocal: In three papers (Verwey, 1935a,b; Verwey and vanBruggen, 1935) he states that the vacancies are distributed over all 24 cation positions (16*d* + 8*a* sites) whereas in one (1935a) he also points out that the absence of (111) reflections shows the vacancies are restricted to octahedral sites! As a result, the latter paper has been cited in subsequent literature as evidence for each interpretation. Hägg's structure is in accord with the preference of Fe^{+3} for tetrahedral sites because of its tendency toward sp^3 hybrid bonding (Goodenough and Loeb, 1955), and is further supported by magnetic saturation measurements (Henry and Boehm, 1956), neutron diffraction studies (Ferguson and Hass, 1958), and Mössbauer-effect studies (Armstrong *et al.*, 1966). The restriction of the vacancies to octahedral sites was generally accepted until 1967, when Kullerud and Donnay reported synthesis of a cation-deficient ferric oxide spinel that appeared to have an appreciable proportion of tetrahedral vacancies. Kullerud *et al.* (1968,1969) subsequently proposed that natural "maghemites"--cation-deficient iron oxide spinels--might be omission solid solutions between magnetite and two end members, one having octahedral vacancies and one, with tetrahedral vacancies, that was termed "kenotetrahedral magnetite." Mössbauer studies on "kenotetrahedral magnetite" show that the vacancies probably are actually distributed randomly over the tetrahedral and octahedral sites (Weber and Hafner, 1971).

Many recipes for making maghemite involve water--for example, dehydration of lepidocrocite or oxidation of magnetite under hydrous conditions--and some workers have suggested that hydrogen or hydroxyl is an integral part of the maghemite structure. There can be little doubt that the presence of water *promotes* the formation of maghemite, but it appears to serve only as a catalyst, not as an essential constituent. Colombo *et al.* (1964,1965) succeeded in oxidizing magnetite to maghemite in the absence of water. Probably the best conclusion is that maghemite can be anhydrous, but forms more readily in the presence of water and then may well contain some structural hydroxide.

As early as 1939, Haul and Schoon observed that some artificial maghemites gave x-ray reflections that are barred by space group $Fd3m$. Such reflections may be interpreted as due to ordering of vacancies within the octahedral sites or as due to a superstructure. In the 1950's many workers suggested that the superstructure was somehow due to structural water or hydroxide.

VanOosterhout and Rooijmans (1958) prepared maghemite that is tetragonal with $c/a = 3$; this cell has space group $P4_1$ or $P4_3$ and contains three "spinel" cells and 32 Fe_2O_3. There is no measurable distortion of the spinel structure, but the vacancies are so ordered that the minimum repeat along [001] is $3a$. Bernal *et al.* (1959) found that they could detect maghemites with four different structures--one tetragonal with $c/a = 3$; one with the true spinel structure; one primitive cubic with space group $P4_132$ (or $P4_332$); and a fourth one primitive cubic but showing reflections (110, 200, ...) not permitted for $P4_132$ or $P4_332$--depending on the process of preparation. The best conclusion at present seems to be that a variety of maghemite structures is possible, depending on the method of formation, and that these structures may tend to convert by ordering to the tetragonal structure.

Because it is metastable with respect to hematite, maghemite can form only by inheriting at least fragments of spinel-like structures. In the case of oxidized magnetite the source of the cubic-close-packed oxygen framework is obvious. Bernal *et al.* (1957) have found that [001], [110], and [$\bar{1}$10] of maghemite are parallel to [100], [010], and [001] of parent lepidocrocite. They show that this topotaxy results from the removal of one-half the hydroxyls from one (010) sheet and the H^+ from the adjacent (010) sheet in lepidocrocite (with a reduction in thickness normal to [010] from 12.57 Å to 8.85 Å, combined with a translation of 1/2 an O^{2-} width (1.94 Å) in [100]. The morphology of synthetic maghemite is important in industrial applications; acicular particles impart the best magnetic properties to recording tapes. The acicular form of the particles is inherited from the γ-FeOOH or α-FeOOH from which they are prepared.

In keeping with the general confusion regarding the structure of maghemite is the disagreement over the value of the unit-cell parameter a: values ranging from 8.30 to 8.35 Å have been reported. Undoubtedly part of the discrepancy results from the fact that a continuous (but metastable) solid solution series exists between magnetite (a = 8.396 Å) and maghemite, so that some of the higher values reported for maghemite probably are from maghemites containing some Fe_3O_4. Some values of a reported in the literature are listed in Table L-4; many of these values are for nonanalyzed material. Probably the best value lies between 8.330 and 8.340 Å.

Kullerud *et al.* (1968,1969) predicted that a (hypothetical) spinel having Fe_2O_3 bulk composition and with all vacancies in the *tetrahedral* sites would have a cell edge of approximately 8.45 Å. They suggested that "magnetites" with cell edges greater than 8.42 Å contain tetrahedral vacancies. It should be noted that *none* of the "magnetites" cited by them had been adequately characterized both chemically and structurally, so this suggestion must be viewed with reservation, particularly in view of the fact that Mössbauer data (Weber and Hafner, 1971) cast doubt on the existence of a "kenotetrahedral magnetite" end member.

Apparently no determinations of the oxygen parameter u for maghemite have been published. Gilleo (1958) lists u = 0.257 (0.382), but the source of this value is unclear. Ferguson and Hass (1958) assumed in interpreting their neutron diffraction data that u = 0.250 (0.375), but showed that a value of 0.255 (0.380) affected their intensity calculations less than one percent.

Table L-4. Cell dimensions (a), crystal systems, and space groups reported for maghemite (γFe_2O_3).

a,Å	Uncertainty	Crystal System	Space Group	Specimen	Reference
(8.339)		Cubic	$Fd3m$(?)	Synth.	Hägg (1935)
(8.32)		Cubic		Nat.	Newhouse and Glass (1936)
(8.343)		Cubic		Nat.	Newhouse and Glass (1936)
(8.333)	± 0.004	Cubic	(Superstructure)	Synth.	Haul and Schoon (1939)
(8.342)		Cubic		Synth.	Mason (1943)
(8.337)		Cubic		Nat.	Mason (1943)
8.33		Cubic	$P4_332$	Synth.	Braun (1952)
(8.348)		Cubic		Synth.	David and Welch (1956)
8.35	± 0.02	Cubic	(Prim)	Synth.	Davey and Scott (1957)
8.322		Tetrag.		Synth.	Behar and Collongues (1957)
8.33		Tetrag.	$P4_1$ ($P4_3$)	Synth.	Van Oosterhout & Rooijmans (1958)
8.34		Tetrag.	$P4_1$ ($P4_3$)	Synth.	Rooksby (1961)
8.33		Tetrag.	$P4_1$ ($P4_3$)	Synth.	Schrader and Büttner (1963)

Notes: *Å values in parenthesis are converted from presumed original kX units. Uncertainties are those given by original authors. Synth., synthetic; Nat., natural. Tetragonal $c = 3a$.*

The instability of maghemite at high temperatures makes it impossible to determine the Curie point directly. Two different values that differ by nearly 100°C are presently in use; each has been determined by extrapolation. Michel and Chaudron (1935) determined the value 675 ± 10°C by preparing a series of Na-stabilized "maghemites" and extrapolating their Curie temperatures to pure γ-Fe_2O_3. This value is accepted by most workers. However, in 1949, Maxwell *et al.* observed that "certain unreduced iron oxide catalysts" with Fe^{2+}/Fe^{3+} varying from 0.352 to 1.276 all had Curie temperatures equal, within a few degrees, to that of magnetite, 583°. From this they concluded that the Curie temperature of maghemite is likewise approximately 583°. This conclusion may be correct, but based on the information published (which is in an abstract only), it should be considered unproved. Aharoni *et al.* (1962) claim to have measured directly a Curie point of 590°C for maghemite; but as their major means of identifying the sample as maghemite is by the Curie temperature, their results must be viewed with caution. Frölich and Vollstaedt (1967) have calculated a Curie temperature for maghemite of 675°C, based on a measured value 650°C for $LiFe_5O_8$. The best conclusion at the present time is that none of the methods for determining the Curie temperature of maghemite is completely satisfactory, and as a result no published value can be considered definitive. However, the 675° value seems tentatively more acceptable.

Magnetite-maghemite solid solutions. Hägg (1935b) prepared three spinels intermediate between Fe_3O_4 and γ-Fe_2O_3 by partial oxidation of magnetite; the relation of their compositions to cell parameters is shown in Figure L-9. On this basis Hägg proposed that there can exist a continuous solid solution series between magnetite and maghemite--which corresponds to a continuous increase from 0 to 2.67 vacancies per unit cell.

Several workers (Feitknecht and Lehmann, 1959; Basta, 1959; Colombo *et al.*, 1965; and Davis *et al*, 1968) have proposed an oxidation mechanism that can produce such a continuous series. During oxidation, oxygen cannot be added *within* the existing structure which already consists of a (nearly) cubic-close-packed oxygen framework. However, free oxygen *at the surface* of a crystal can be ionized by the addition of two electrons which are derived from ferrous ions within the crystal by the reaction $2Fe^{2+} = 2Fe^{3+} + 2e^-$. The charge within the original crystal is no longer balanced; to restore the balance it is necessary to remove $2Fe^{3+}$ for every 3 oxygens added at the surface--corresponding to one of every three newly formed Fe^{3+} ions. The removal is accomplished by diffusion of the Fe^{3+} through the oxygen framework to the surface of the crystal, where one of several possible fates awaits the expelled Fe^{3+} and newly formed

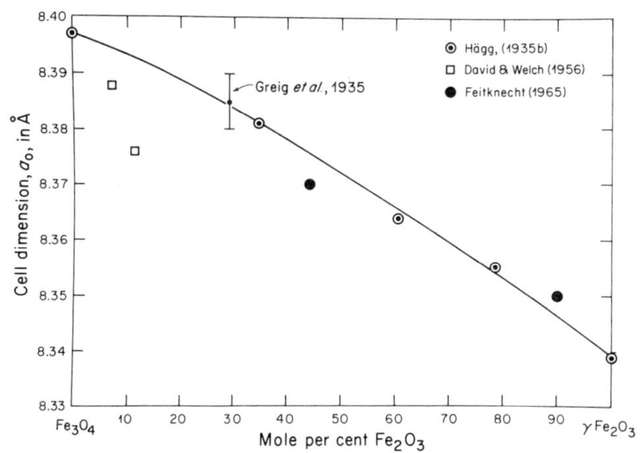

Figure L-9. Variation of cell dimension with composition for spinels in the Fe_2O_3-Fe_3O_4 series. The data of Hägg (1935b) and Feitknecht (1965) are for samples oxidized from magnetite at relatively low temperatures (<500°C). The data of Greig *et al.* (1935) and David and Welch (1956) are for specimens quenched from high temperatures (>1200°C). Hägg's data and those of David and Welch are converted from KX; Greig *et al.* (1935, pp. 296-297) reported that the cell edge of their specimen was approximately 0.01 Å smaller than that of pure magnetite, which is 8.394-8.396 Å.

O^{2-}: (1) They may be removed in solution if a liquid phase is present. (2) They may be deposited epitaxially as hematite on the surface of the crystal. (3) They may be deposited on the surface of the crystal as an extension of the original spinel structure, thus forming new maghemite. Ideally these three different mechanisms lead to three different products: Case (1) produces a crystal of maghemite containing the same number of oxygens, but smaller than, the parent magnetite, since the unit cell of maghemite is smaller than that of magnetite. Case (2) produces a similar maghemite crystal that is surrounded by hematite, and case (3) results in a crystal of maghemite containing the same number of iron atoms but 1/8 more oxygens than the parent magnetite; the crystal will have increased in size by $(\frac{9}{8})(a_{\text{maghemite}}/a_{\text{magnetite}})$, or by a factor of approximately 1.118. In practice, of course, it is unlikely that any one of these mechanisms would operate alone; more likely two or even all three would be involved. Colombo *et al.* (1965) have shown that case (3) is unlikely to obtain if there is any hematite in intimate contact with the magnetite; the Fe^{3+} and O^{2-} will crystallize as stable hematite if seed crystals are present.

It is worth noting that the Fe^{3+} ions expelled during oxidation are not necessarily those that have been converted from Fe^{2+} to Fe^{3+}. As Verwey *et al.* (1947) have pointed out, the eight Fe^{2+} and eight Fe^{3+} in octahedral sites per unit cell of maghemite are better viewed as a cloud of eight electrons distributed around 16 Fe^{3+}. The electrons that ionize the exterior oxygens are drawn from this cloud, not from any given Fe^{2+} ion. Probably some tetrahedral cations also migrate, since there may be some tetrahedral vacancies.

The cation-deficient Fe_2O_3 spinels produced by Kullerud and Donnay (1967) by reactions such as

$$3Fe_3O_4 + 2S = FeS_2 + 4Fe_2O_3 \text{ (spinel)}$$

must have formed in the absence of excess oxygen. In this case it is sulfur, rather than oxygen, that effectively removed electrons and iron from the magnetite. Kullerud and Donnay have argued that an important part of the vacancies in these spinels occur in tetrahedral sites. If this interpretation were correct, it would indicate a subtle difference between oxygen and sulfur in inducing diffusion of Fe^{3+} in magnetite; but it is difficult to envision a mechanism by which the difference would be effective, since the diffusing cations may be drawn from sites many unit cells away from the new anions.

Basta (1959) has tried to draw a distinction between the low-temperature phases produced by oxidation of magnetite and the limited solid solution of Fe_2O_3 in Fe_3O_4 at high temperatures (Greig *et al.*, 1935), calling the first the γ-Fe_2O_3-Fe_3O_4 series and the second the α-Fe_2O_3-Fe_3O_4 series. It is true that one of these series forms metastably at low temperatures and the other stably at high temperatures; but there seems to be no reason why they should be different structurally. In each the decrease in Fe/O must be accomplished by the formation of vacancies in the spinel structure. Basta argues (1959, p. 700) that the data of Greig *et al.* show there is less change in a with composition in the α-Fe_2O_3-Fe_3O_4 series than in the γ-Fe_2O_3-Fe_3O_4 series. A comparison of the Greig *et al.* data for magnetite containing 29 mole percent (22 weight percent) Fe_2O_3 with Hägg's data (Fig. L-9) shows no significant difference.

It is mainly a matter of personal preference whether spinels intermediate in composition between Fe_3O_4 and Fe_2O_3 are considered as omission solid solutions between magnetite and γ-(or α-)Fe_2O_3, or simply as cation-deficient magnetites. I find the latter concept more meaningful, especially insofar as it is extended to the titanium-bearing "titanomaghemites."

Titanomaghemites. Spinels intermediate in composition between the Fe_3O_4-Fe_2TiO_4 join and the Fe_2O_3-$FeTiO_3$ join are known to form in two ways: (1) by stable--but quite limited--solid solution of Fe_2O_3-$FeTiO_3$ in the Fe_3O_4-Fe_2TiO_4 series at temperatures above 1000°C, and (2) by metastable oxidation of essentially binary Fe_3O_4-Fe_2TiO_4 solid solutions at lower temperatures--probably well below 600°C. As is the case with spinels intermediate between magnetite and Fe_2O_3, there is no *a priori* reason to assume that the structures formed at high and at low temperatures are basically different. It is possible that those Fe-Ti spinels formed stably at high temperatures may have a more nearly random distribution of Fe^{2+}, Fe^{3+}, Ti^{4+}, and vacancies over the octahedral and tetrahedral sites; but apparently no x-ray work on quenched specimens (let alone at high temperatures) has been done to test this possibility. In the absence of any data on the structures of the high-temperature phases, this discussion will concentrate on the structures of those formed by oxidation at lower temperatures. These phases have been given a variety of names: "Generalized titanomagnetite," "abnormal" or "very abnormal titanomagnetite," "titanomaghemite," "γ-titanohematite," and "γ-phases." The term "titanomaghemite" will be used here to apply to any Fe-Ti spinels lying off the Fe_3O_4-Fe_2TiO_4 join--that is, spinels with vacancies in some cation positions--and will not be restricted, as has sometimes been done, to spinels lying outside the Fe_3O_4-Fe_2TiO_4-$FeTiO_3$ compositional field.

In the past, many workers have assumed that there was extensive, essentially binary, solid solution of $FeTiO_3$ in magnetite. To account for this supposed solid solution, Dunn and Dey (1937, p. 160) proposed a hypothetical end member--cubic $FeTiO_3$--by analogy with maghemite, cubic Fe_2O_3. This proposed end member was revived and named $\gamma FeTiO_3$ by Nichols (1955, p. 143) and by Chevallier *et al.* (1955, p. 322). As subsequent experimental work and analyses of volcanic Fe-Ti spinels have shown, however, there is no unique solid solution of $FeTiO_3$ in magnetite, but rather an extension of the field of Fe-Ti spinels from the Fe_3O_4-Fe_2TiO_4 join toward--and in some cases even past--the Fe_2O_3-$FeTiO_3$ join. Consequently, this field of spinel has been viewed by some workers as reflecting solid solution of γFe_2O_3-$\gamma FeTiO_3$ in Fe_3O_4-Fe_2TiO_4. There is nothing inherently incorrect in this viewpoint: The concept of omission solid solution is valid, and the use of a hypothetical end member is likewise acceptable. However, I question the utility of this approach, partly because it has led to pointless discussions as to the nature and structure of $\gamma FeTiO_3$, and partly because it tends to obscure the oxidation origin of titanomaghemites. Chemically, the concept of $\gamma FeTiO_3$ is of little use: It is sufficient, for example, to refer to *spinels* lying in the Fe_3O_4-Fe_2TiO_4-$FeTiO_3$-Fe_2O_3 field, without reference to a gamma end member. The concept of cation-deficiency, applied to spinels oxidized from essentially binary magnetite-ulvöspinel solid solutions, is a much more useful (although no more "correct") approach to the structure of the titanomaghemites.

In keeping with this viewpoint, compositions of titanomaghemites can conveniently be expressed in terms of two parameters, one that expresses in some way the Fe/Ti ratio, and one that indicates the degree of oxidation. The first parameter can be expressed either as an atomic ratio or as the mole fraction of Fe_2TiO_4 (or Fe_3O_4) in the parent binary solid solution, and is generally considered to remain constant during oxidation. The oxidation parameter can express (1) the number of Fe^{2+} ions oxidized (which equals twice the number of oxygens consumed), (2) the atomic fraction of oxidized Fe^{2+} relative to initial total Fe^{2+}, or (3) the number of cations per unit cell; this number plus the number of vacancies equals 24. Table L-5 correlates some of the parameters that have been used by various workers. The parameter $(Fe_2TiO_4)/(Fe_3O_4 + Fe_2TiO_4)$ (this is the x of the Japanese workers and of O'Reilly and Banerjee, and the q of Verhoogen) is particularly useful in that it is appropriate for the unoxidized magnetite-ulvöspinel series as well as the oxidized spinels. Perhaps the most useful oxidation parameter is z = (number of Fe^{2+} ions oxidized)/(initial number of Fe^{2+}).

A structurally significant aspect of the titanomaghemites is their cubic-close-packed oxygen framework, inherited from their titanomagnetite parents. Evidently the oxidizing oxygen is adsorbed to this framework and forms an extension of it, thereby creating new cation sites that are filled by diffusion of cations from the parent crystal (Sasajima, 1961, p. 201). Sites in the parent formerly occupied by the diffused cations become vacancies; and overall charge balance is maintained by the oxidation of two Fe^{2+} to Fe^{3+} for every oxygen added. Note that the analytical data of Akimoto and Katsura (1959)--which show that natural titanomagnetites retain their Fe/(Fe+Ti) ratios upon oxidation--demonstrate that the diffusing Fe^{3+} ions must be retained in the overall expanded structure; if they were not the Fe/(Fe+Ti) ratio would change with oxidation. In the absence of any detailed x-ray studies one assumes that titanomaghemites do retain the spinel structure; it is possible, however, that all or some of the structures reported for maghemite (ordered vacancies, primitive cubic, tetragonal) may ultimately be found. The most important aspect of the structure that remains to be determined, therefore, is the distribution of cations plus vacancies in the *8a* (tetrahedral) and *16d* (octahedral) sites of the spinel structure.

Several early models for cation distribution in titanomaghemites make use of assumed structural formulae for the hypothetical end member $\gamma FeTiO_3$, an approach that hindsight shows to be of little use. More appropriate are models

Table L-5. Compositional parameters used to describe Fe-Ti spinels.

Iron/Titanium Ratio		Oxidation Index		
Definition	Range	Definition	Range	Reference
$"x" = \frac{Fe_3O_4}{Fe_3O_4 + Fe_2TiO_4}$ (moles)	$0 \leqq "x" \leqq 1$	---		Chevallier et al. (1955)
$x = \frac{Fe_2TiO_4}{Fe_3O_4 + Fe_2TiO_4}$	$0 \leqq x \leqq 1$	---		Akimoto, Katsura, & Yoshida (1957)
$\frac{Fe}{Fe + Ti}$ (atomic)	1-- 2/3	$\frac{32(Fe + Ti)}{O}$	24-- 19.2	Akimoto and Katsura (1959) Katsura and Kushiro (1961) Sasajima (1961)
$q = \frac{Fe_2TiO_4}{Fe_3O_4 + Fe_2TiO_4}$ (moles)	$0 \leqq q \leqq 1$	n = # Fe^{2+} oxidized per initial 4 oxygens	$0 \leqq n \leqq 1 + q$	Verhoogen (1962)
$X = \frac{Fe}{Fe + Ti}$ (atomic)	$1 \geqq X \geqq 2/3$	$Y = \frac{32(Fe + Ti)}{O}$	$24 \geqq Y \geqq \frac{64}{4 - X}$	Zeller and Babkine (1965); Zeller et al. (1967)
$x = \frac{Fe_2TiO_4}{Fe_3O_4 + Fe_2TiO_4}$ (moles)	$0 \leqq x \leqq 1$	$z = \frac{\# Fe^{2+} \text{ oxidized}}{\text{original } \# Fe^{2+}}$	$0 \leqq z \leqq 1$	O'Reilly and Banerjee (1966, 1967) and many others
(same)	(same)	n = # Fe^{2+} oxidized per initial 4 oxygens	$0 \leqq n \leqq \frac{8}{7}x - \frac{1.6}{7}$ $(0.2 \leqq x \leqq 0.8)$	Schult (1968)
(same)	(same)	γ = "excess of oxygen" over 4 per formula unit		Hauptman (1974)
(same)	(same)	n = # Fe^{2+} oxidized per initial 4 oxygens	$0 \leqq n \leqq 1 + x$	Gidskehaug (1975)

Notes: Most workers have used x (= q of Verhoogen) to indicate the mole fraction Fe_2TiO_4. This x is related to the X of Zeller and co-workers by: $x = 3(1 - X)$; the oxidation parameter $Y = \frac{192}{8 + z + xz}$. z of O'Reilly and Banerjee = $\frac{n}{1 + q}$ of Verhoogen. γ of Hauptman = $\frac{z(1 + x)}{2}$. Hauptman's "excess of oxygen" index is emphatically not recommended.

that consider (1) the distribution of cations in the unoxidized parent spinel and (2) their redistribution by diffusion as a result of oxidation. For example, in the saturation magnetization values calculated for titanomaghemites by Verhoogen (1962, p. 178) there is the implicit assumption that all Fe^{3+} enters only tetrahedral sites until they have been filled. Although this assumption is supported by theoretical considerations (Goodenough and Loeb, 1955), it is probably not valid for some intermediate Fe_3O_4-Fe_2TiO_4 solid solutions ($0.2 \leq x \leq 0.8$; O'Reilly and Banerjee, 1965), and thus may not be generally applicable. Unfortunately, it is not possible to derive models for cation and vacancy distributions in titanomaghemites without similar assumptions, including the nature of the cation distribution in the parent titanomagnetites--and this itself is still a matter of some controversy. What follows are several such models to acquaint the reader with some of the problems involved; but none should be considered definitive.

O'Reilly and Banerjee (1966,1967) proposed a cation-distribution model based on the following assumptions: (1) The distribution in the parent Fe_3O_4-Fe_2TiO_4 binary series is that proposed by O'Reilly and Banerjee (1965); see equations 1i-2. Note that this model does not allow any Ti^{4+} in tetrahedral sites. (2) Only Fe^{2+} in octahedral sites is oxidized during the "first stage of oxidation" (during which the spinel structure is assumed to be retained). (3) Only oxidized Fe^{2+} ions diffuse, with the result that tetrahedral Fe^{2+} and Fe^{3+} as well as Ti^{4+} and primary octahedral Fe^{3+} remain immobile; and only octahedral vacancies are formed. According to O'Reilly and Banerjee (1967, p. 31) these assumptions lead to a unique cation distribution given by:

$$Fe^{3+}_{1-\delta}Fe^{2+}_{\delta}\left[Fe^{3+}_{7+\delta-\frac{16(3+x)}{8+z+xz}}\;Fe^{2+}_{\frac{8(9+x)}{8+z+xz}-8-\delta}\;Ti^{4+}_{\frac{8x}{8+z+xz}}\;\square_{\frac{3z(1+x)}{8+z+xz}}\right]O^{2-}_4$$

or, using the notation of Readman and O'Reilly (1971):

$$Fe^{3+}_{1-\delta}Fe^{2+}_{\delta}[Fe^{3+}_{(2-2x+z+zx)R-1+\delta}Fe^{2+}_{(1+x)(1-z)R-\delta}Ti^{4+}_{xR}\square_{3(1-R)}]O^{2-}_4$$

where $R = 8/[8+z(1+x)]$. This formula would yield a saturation magnetization at 0°K of $[2\delta-2+48(1-x)]/(8+z+xz)$ Bohr magnetons per four oxygens. The parameters x and z are as defined in Table L-5; δ and $1-\delta$ are, respectively, the numbers of Fe^{2+} and Fe^{3+} in tetrahedral sites as given by the O'Reilly-Banerjee (1965) formulae for the parent Fe_3O_4-Fe_2TiO_4 solid solution and, by hypothesis, in the resulting titanomaghemite as well. The number of octahedral Fe^{2+} is lowered by $z(1+x)$ as a result of oxidation, and the total number of Fe^{3+} increased accordingly. The number of oxygens added per initial 4 oxygens is $(z/2)(1+x)$, so the coefficients are multiplied by $4/[4+(z/2)(1+x)]$ (= R) to yield the values per 4 oxygens in the titanomaghemite.

Unfortunately, there are several objections to this potentially useful model. First, it is predicated on the implicit assumption that only octahedral Fe^{2+} ions can be oxidized because only octahedral ions can diffuse readily. However, oxidation of Fe^{2+} to Fe^{3+} need only involve the transfer of an electron to adsorbed oxygen at the crystal surface; local charge balance can then be restored by the subsequent diffusion of *any* sufficiently mobile cation to the surface. Second, for every formula unit (per 4 oxygens) in the parent crystal, $(z/2)(1+x)$ oxygens are added to the oxygen framework during oxidation, resulting in the creation of $(z/8)(1+x)$ new tetrahedral sites and $(z/4)(1+x)$ new octahedral sites. If only octahedral Fe^{2+} ions can be oxidized and diffuse into these new sites, the resultant spinel will consist of two phases, a titanomaghemite core, coextensive with the parent titanomagnetite and having the formula:

$$Fe^{3+}_{1-\delta}Fe^{2+}_{\delta}[Fe^{2+}_{(1+x)(1-z)-\delta}Fe^{3+}_{1+2x+(2/3)z(1+x)+\delta}Ti^{4+}_{x}\square_{(z/3)(1+x)}]O_4$$

surrounded by a crystallographically continuous mantle of γFe_2O_3 that contains $(z/8)(1+x)$ formulas of $Fe^{3+}[Fe^{3+}_{1.67}\square_{0.33}]O_4$ per four oxygens in the core. A minimum condition for the production of a *homogeneous* titanomaghemite therefore is that $x\,\frac{z(1+x)}{4+(z/2)(1+x)}$ Ti^{4+} ions diffuse out of the primary oxygen framework into the new framework and that additional diffusion of Fe ions and electron transfer take place. Third, there is no *a priori* reason why the ratio of Fe^{2+} to Fe^{3+} in the $(z/8)(1+x)$ new tetrahedral sites should equal $\delta/(1-\delta)$, the ratio for the unoxidized parent. The $z(1+x)R/32$ new tetrahedral sites per four total oxygens must be filled by diffusion, and since only an electron transfer is required to change the ratio in the original tetrahedral sites, there is no reason why a homogeneous titanomaghemite should be expected to inherit the Fe^{2+}/Fe^{3+} ratio in the tetrahedral sites of its parent. The results of Sakamoto *et al.* (1968) tend to confirm that at least some tetrahedral Fe^{2+} is, in fact, oxidized.

Readman and O'Reilly (1971) recognized these shortcomings and presented two new models that help overcome them. In their model 3 (described as "zero mobility tetrahedral Fe^{2+}-Fe^{3+} tetrahedral site preference tetrahedral electron hopping"--p. 334) they allow diffusion of Ti^{4+} into the new octahedral sites as well as transfer of electrons to establish an overall homogeneous charge balance. In their model 4 they also allow some diffusion of cations from the original tetrahedral sites. They conclude that the availability of tetrahedral Fe^{2+} for oxidation and diffusion is approximately 0.20 that of octahedral Fe^{2+}. If the availability term is α, then the final number of tetrahedral Fe^{2+} ions per formula unit is

$$\delta = \delta_o(1-z)R/(1-z+\alpha z)$$

Unfortunately, the saturation moment of such a titanomeghemite is a function of α and thus cannot uniquely characterize the values of x and of z.

One could apply similar reasoning for the oxidation of titanomaghemites that obey the models of Stephenson (1969) and of Bleil (1971), as was done, for example, by Stephenson (1972). Hauptman (1974) has made a detailed study of the stable high-temperature oxidation of $Fe_{2.4}Ti_{0.6}O_4$ (x = 0.6). Unfortunately, Hauptman expresses the oxidized spinel as $Fe_{2.4}Ti_{0.6}O_{4+\gamma}$ and speaks of the sample acquiring "an excess of oxygen." While such terms are not wrong in a strictly *chemical* sense, they tend to hide the fact that it is impossible to add excess oxygen into an already (nearly) close-packed structure; the concept of cation deficiency is much more useful in a *structural* sense than is oxygen excess.

A completely different approach to the determination of cation distribution has been used by Zeller *et al.* (1967a,b), who employed the bulk composition, Curie temperature, saturation magnetization, and "hyperbolic law of paramagnetic susceptibility" (Néel, 1948). The chemical analysis provides ΣFe^{2+}, ΣFe^{3+}, ΣTi^{4+} and the number of vacancies per unit cell. There are eight unknowns to be determined--the distribution of three cations plus vacancies over two different sites. Four relations are given by the fact that the sum of each cation species (or vacancy) in the octahedral and tetrahedral sites must equal the analyzed value; two more relations are given by the fact that the total number of cations plus vacancies in each type of site is fixed. The final two relations to yield eight simultaneous equations are those between cation distribution on one hand and the Curie temperature and saturation magnetization on the other. No *a priori* assumptions regarding site preferences were made; the only assumptions made regard homogeneity of the samples and the validity of Néel's laws. The results are remarkable: they appear to confirm the assumption of O'Reilly and Banerjee (1966, 1967) that essentially no Ti^{4+} or vacancies occur in tetrahedral sites. Furthermore, in the example given (x = 0.75 - 0.76), the relative proportions of Fe^{2+} and Fe^{3+} on tetrahedral sites are fixed over an oxidation range from essentially zero [32 (Fe+Ti)/O = 23.84; z = 0.03] to 32(Fe+Ti)/O = 22.22 (z = 0.37), as proposed by O'Reilly and Banerjee (1966, 1967). Only when the oxidation factor is quite high, 32(Fe+Ti)/O = 21.20 (z = 0.60), does the proportion of Fe^{2+} and Fe^{3+} assigned to tetrahedral sites change, and then in the "wrong" direction--the proportion of Fe^{2+} in tetrahedral sites is said to increase. As an interesting sidelight, the initial distribution of Fe^{2+} and Fe^{3+} in tetrahedral sites calculated by Zeller *et al.* ($\delta = 0.32$) is quite different from that predicted for unoxidized spinels from the formulae of O'Reilly and Banerjee (1965), $\delta = 0.55$. This independent approach of Zeller *et al.* needs to be applied to far more specimens to test its utility. It would be interesting to see a ninth relationship added: the unit-cell parameter. This might either be viewed as a degenerate ninth equation in eight unknowns to serve as a test for consistency; or alternatively, it might be used partly to compensate for the effect of minor constituents (Al,Mg,Mn, etc.) in the samples.

Data are available on the relationship between the unit-cell parameters of titanomaghemites and their composition. Akimoto *et al.* (1957a) first presented extensive data on this relationship, based on a number of synthetic titanomaghemites. Data from natural specimens were found to be in fair to good agreement (Akimoto and Katsura, 1959). However, it appears that at least some of their synthetic specimens were multi-phase, and the data of O'Reilly and Readman (1971) are probably to be preferred. Basta (1959,1960) presented data for additional natural titanomaghemites. Zeller and Babkine (1965) applied regression analysis to all available data--163 natural and synthetic specimens--to derive the relationship $a = 6.943 + 0.912X + 0.0776Y - 0.0551XY$, where X is x (see Table L-5) and 24-Y gives the number of vacancies. One may use the relationship $Y = 192/[8+z(x+1)]$ to convert this expression into terms of the O'Reilly-Banerjee compositional parameters. A minor shortcoming of the Zeller-Babkine approach is that it assumes a linear relationship between a and X for Y = 24 (that is, for unoxidized Fe_3O_4-Fe_2TiO_4 spinels); the data of Lindsley (1965) and Syono (1965) show a slightly nonlinear relationship. Nevertheless, the Zeller-Babkine expression should give a reasonable estimate of a for many Fe-Ti spinels.

Akimoto *et al.* (1957a) and O'Reilly and Readman (1971) also give data relating composition to Curie temperature for synthetic titanomagnetites and titanomaghemites.

Gidskehaug (1975) proposed using electron microprobe analyzes to determine x and n (= z) parameters in titanomaghemites. Unfortunately, the method is critically dependent on the oxide sums of the analyses to yield z.

THE RHOMBOHEDRAL OXIDES

Hematite and ilmenite form a solid solution series that is complete at high temperatures (certainly above 960°C and probably above 800°C) but interrupted by a miscibility gap at lower temperature. Unlike the spinel phases, there is no one name that conveniently refers to the entire series. Hematite and ilmenite have different space groups, so it is best not to extend the name of either end member to serve as a general term for the series. The same might also hold for the term "α-series" (Verhoogen, 1962), which has been extended from hematite, α-Fe_2O_3. A characteristic of the hematite-ilmenite series is its rhombohedral symmetry, and for convenience it will be referred to as the "rhombohedral series" and individual phases as "rhombohedral phases."

Two parameters are necessary to describe a rhombohedral unit cell. By convention the two used are the length of an edge of the unit rhombohedron, a_R, and the angle between two edges, α. Any rhombohedral unit cell can, of course, be converted to a (non-primitive) hexagonal cell with three times the volume. From the viewpoints of both crystal chemistry and magnetic properties, the hexagonal unit cell is often more convenient to use, and all indices will refer to the hexagonal cell unless explicitly stated otherwise.

Oxygen in the rhombohedral phases forms planes lying parallel to (0001) of the hexagonal cell [(111) of the rhombohedral cell] that are nearly--but not quite--hexagonal close packed. Within each plane are found "triplets" of touching oxygens; these "triplets" are slightly separated from each other, so that the average oxygen-oxygen distance between "triplets" is somewhat greater than that for close packing (Fig. L-10). Oxygen "triplets" in adjacent planes do not lie one above the other, but are offset so that above and below each "triplet" are three non-touching oxygens. In an ideally hexagonal-close-packed oxygen framework there are 18 octahedral interstitial sites per 18 oxygens. In the rhombohedral Fe-Ti oxides, two-thirds of these potential sites are occupied by cations. The six-fold oxygen-coordination polyhedra about the 12 occupied sites are distorted because of departures from hexagonal close packing. The polyhedra are usually referred to as octahedra, but it must be borne in mind they are not true octahedra because of the distortion. Two opposite faces of these octahedra lie in the (0001) planes, that is, in the planes of the oxygen layers. For each octahedron, one of these faces is shared and one is unshared. The three oxygens comprising the shared face are the "triplets" mentioned above, and are essentially close-packed--the oxygen-oxygen distance in hematite (Blake *et al.*, 1966) is 2.669 Å compared to the effective ionic radius of 1.36 Å (Shannon and Prewitt, 1968)--whereas those in the unshared face lie slightly farther apart (3.035 Å in hematite). The three oxygens in the unshared face are in fact corners of three adjacent *shared faces* lying in the same oxygen plane. From geometrical considerations it is clear that the cation will lie somewhat closer to the plane of the unshared face (it tends to fall down between

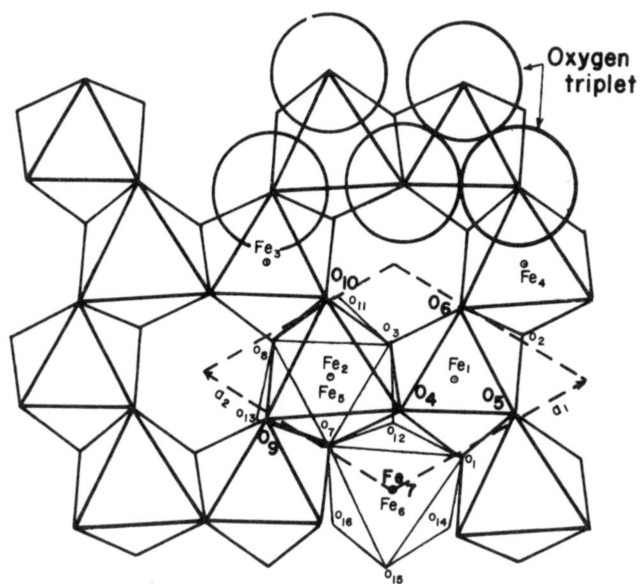

Figure L-10. The hematite structure viewed down the hexagonal *c* axis, emphasizing the coordination octahedra of oxygens around Fe^{3+} cations. Size of atomic symbol indicates its relative height along the *c* axis. The hexagonal unit cell is outlined. Modified from Blake *et al.* (1966). An oxygen "triplet" is shown.

Table L-6. Coordinates of O^{2-} and Fe^{3+} sites in the hematite structure. Space group $R\bar{3}c$. Origin at center ($\bar{3}$). Hexagonal axes.

Type of Site	No. of Sites	Notation	Symmetry	Coordinates
O^{2-}	18	*e*	2	$x,0,\frac{1}{4}$; $0,x,\frac{1}{4}$; $\bar{x},\bar{x},\frac{1}{4}$; $x,0,\frac{3}{4}$; $0,\bar{x},\frac{3}{4}$; $x,x,\frac{3}{4}$.
Fe^{3+} (octahedral)	12	*c*	3	$0,0,z$; $0,0,\bar{z}$; $0,0,\frac{1}{2}+z$; $0,0,\frac{1}{2}-z$

Note: There are three points - 0,0,0; $\frac{1}{3},\frac{2}{3},\frac{2}{3}$; $\frac{2}{3},\frac{1}{3},\frac{1}{3}$ *- equivalent to the origin. The full sets of sites are generated by applying the coordinates in the table to* *__each__* *of the equivalent points.*

the separated oxygens) than to that of the shared face. This is a result of electrostatic repulsion between the cations on either side of the shared face. For this reason the cations lie on either side of planes midway between the oxygen layers, rather than on those planes as they would for ideal hexagonal close packing.

Hematite

Crystal structure of hematite (Fe_2O_3). Although several earlier attempts were made, the first successful determination of the hematite structure was that of Pauling and Hendricks (1925). They reported $a_R = 5.43 \pm 0.01$ Å and $\alpha = 55°17'$, with space group $R\bar{3}c$ and $Z = 2Fe_2O_3$ per rhombohedral cell. Fe^{3+} ions occupy the $4c$ sites, and oxygens the $6e$ sites. These are special sites in space group $R\bar{3}c$--the c sites lie on 3-fold symmetry axes and the e sites on 2-fold axes--so that one parameter is necessary to locate the positions of each type of site in the unit cell. Coordinates of these sites are given in Table L-6. Pauling and Hendricks determined these parameters as 0.105 (or 0.355 relative to an origin at a center of symmetry in a hexagonal unit cell) for Fe^{3+} and 0.292 for oxygen. A refinement of x-ray diffraction data (Blake *et al.*, 1966) gives the Fe^{3+} parameter $z = 0.3553 \pm 0.0001$ and the oxygen parameter $x = 0.3059 \pm 0.0001$ (both relative to an origin at the center of a hexagonal unit cell). Thus, the distortion of the octahedra--which is also a measure of the deviation from hexagonal close packing--is greater than that reported by Pauling and Hendricks. A stereoscopic view of the hematite structure is given in Figure L-11.

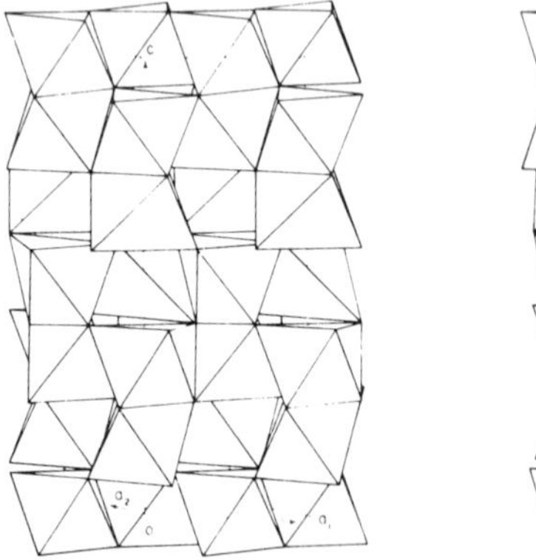
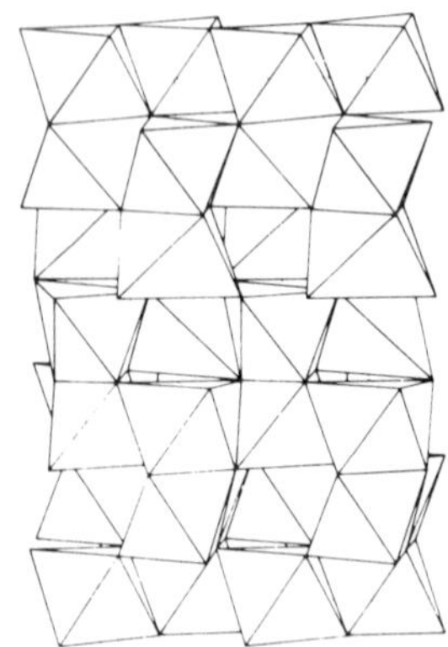

Figure L-11. A stereoscopic view of the hematite structure. From Blake *et al.* (1966).

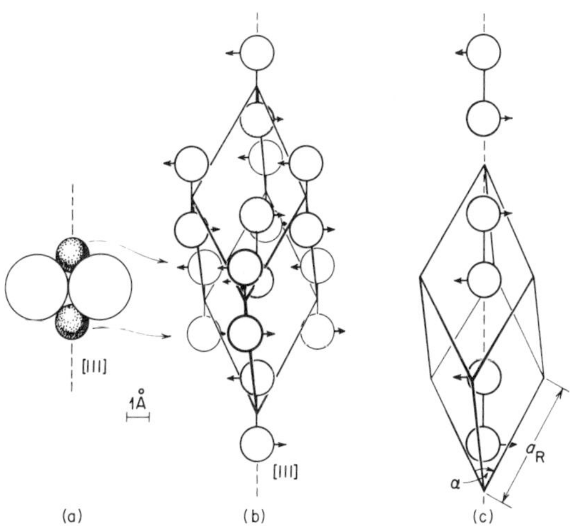

Figure L-12. The structure of hematite: rhombohedral aspects. a. Fe-O_3-Fe "pseudo-molecule." The "triplets" of three oxygens (large spheres) are essentially close-packed and lie in the (111) plane [= (0001) plane, hexagonal axes]. b. The rhombohedral cell of hematite according to the Pauling-Hendricks origin. Each corner and the center of the unit rhombohedron is occupied by the center of a Fe-O_3-Fe unit; the oxygen "triplets" are omitted for clarity. Only four Fe^{3+} ions lie within the cell. The magnetic structure is indicated schematically by the arrows: all Fe^{3+} ions in a given (111) layer have parallel spins but are antiparallel to those of the adjacent layers. c. The rhombohedral cell of hematite with the origin shifted by $(-\frac{1}{4}, -\frac{1}{4}, -\frac{1}{4})$ (Barth-Posnjak origin).

The hematite structure can be visualized as being made up of Fe-O_3-Fe units--"triplets" of close-packed oxygens with Fe^{3+} on either side (Figure L-12a). While it is not proposed that hematite be considered as a molecular structure (this view of the structure is outmoded in terms of modern crystallography), it is very useful in understanding the magnetic properties of hematite. The Fe^{3+} ions in each Fe-O_3-Fe unit have opposite spins; that is, they are antiferromagnetically coupled, a result of superexchange interaction through the intervening "triplet" of oxygens (Li, 1956). For the Pauling-Hendricks choice of the origin, each corner plus the center of the rhombohedral unit cell is occupied by the center of a Fe-O_3-Fe unit (Fig. L-12b). For the currently accepted origin at the center of symmetry (International Tables, 1952, pp. 274-275), two Fe-O_3-Fe units lie along [111] (Fig. L-12c). Unit cell dimensions reported for pure synthetic hematite are a_H = 5.0343 ± 0.0005 Å, c_H = 13.742 ± 0.005 Å (Lindsley, 1965). Corundum (Al_2O_3) has the hematite structure, with a_H = 4.758 Å, c_H = 12.991 Å.

Because the Fe^{3+} ions lie to either side of imaginary planes midway between the oxygen layers, the sequence of planes along [0001] rhombohedral [111] can be considered as ...O-Fe-Fe-O-Fe-Fe-O.... However, since all Fe^{3+} lying between two adjacent oxygen planes are ferromagnetically coupled, for magnetic purposes the sequence along [0001] is ...O-Fe-O-Fe-O....

Magnetic structure of hematite. It has long been known that some specimens of hematite are strongly magnetic (in the sense of being attracted to a permanent magnet). Plücker (1848) suggested that magnetic hematites might contain small amounts of FeO, an idea that was strongly endorsed by Sosman and Hostetter (1917), who had previously proposed the existence of complete solid solution between hematite and magnetite (1916). Many other workers merely assumed that the hematites exhibiting strong magnetism contained admixed grains of magnetite. Honda and Soné (1914), however, showed that the high susceptibility persisted up to 680°C, well above the Curie temperature of magnetite. Furthermore, Smith (1916) showed that the magnetic susceptibility of hematite varied with crystallographic direction. Both of these results seemed to indicate that the magnetic properties pertain to the hematite itself, rather than to admixed magnetite.

At this stage the discussion centered on whether hematite was inherently "ferromagnetic" or whether it was basically paramagnetic, with the observed properties being due to impurities. A major step in understanding resulted from Néel's suggestion that hematite is antiferromagnetic (1949), that is, alternate planes of Fe^{3+} along [0001] are magnetized in opposite directions.

Neutron diffraction studies confirmed Néel's antiferromagnetic model for hematite: Shull *et al.* (1951a) showed that at room temperature the moments of all Fe^{3+} lying between adjacent (0001) oxygen layers are parallel, lying in the (0001) plane. Alternate iron layers, however, are coupled antiferromagnetically, as illustrated diagrammatically in Figure L-12b, so that the net moment is essentially zero. The magnetic unit cell coincides with the crystallographic cell. It can be useful to consider hematite as having two magnetic substructures A and B that have equal and opposite moments. The sequence of iron layers along [0001], then, is ABABAB..., where the moments of the A layers can be labelled (+) and those of the B layers (-).

The elucidation of the antiferromagnetic structure of hematite in the early 1950's solved many problems regarding the magnetic structure of hematite, but the mystery of the "parasitic ferromagnetism" remained. Néel had suggested that it resulted from imperfect antiferromagnetic compensation because one magnetic sublattice contained more Fe^{3+} than the other (1953,1955).

A new hypothesis was offered by Dzyaloshinskii (1957; Dzyaloshinsky, 1958) that it is the antiferromagnetic *coupling*, rather than the *compensation*, that is imperfect. He showed that deviations of about 10^{-4} radians from perfect antiparallelism--commonly termed "spin-canting"--not only would explain the observed moment but would still meet the requirements of magnetic symmetry. Moriya (1960) subsequently provided a firm theoretical basis for this model, showing that the spin-canting is caused by anisotropic exchange interaction. Dunlop (1971) gave an exhaustive review of the magnetic properties of hematite.

Curie, Néel, and Morin temperatures of hematite. Honda and Soné (1914) observed that the susceptibility of hematite decreased to zero at 680°C. The value of 675°C (Forestier and Chaudron, 1925) was long accepted as the Curie temperature by those who considered the ferromagnetic moment as an inherent property of hematite. When Néel proposed the antiferromagnetic structure of hematite (1949), he suggested that 675°C is what we now call the Néel point (the temperature at which antiferromagnetic coupling is destroyed by thermal agitation) and that the Curie temperature happens to coincide with it. But although neutron diffraction studies (Shull *et al.*, 1951a) convincingly confirmed the Néel antiferromagnetic structure, they cast a shadow on this interpretation of the Néel and Curie points, for they seemed to show that a magnetic structure persists above 675°C.

In the 1960's there was much discussion as to whether the Curie and Néel temperatures do in fact coincide at approximately 675°C or whether the Néel point is some tens of degrees higher. Most workers have now accepted that the Curie and Néel temperatures of hematite coincide at a temperature somewhere in the range 675-690°C.

Another magnetic transition in hematite occurs below room temperature. It was first discovered by Honda and Soné (1914), found again by Charlesworth and Long (1939), and then simultaneously rediscovered in 1950 by Morin (1950) and Guillard (1951). (The fact that it is called the Morin transition offers mute but poignant testimony that immortality in science may depend less on being first than on publishing at an opportune time and in the "right" journal!) A rather wide range of values for the transition temperature have been determined; at least part of the range must reflect impurities or imperfections, or both. For example, Lin found that the transition takes place over a range of 120°C for Elba hematite (1960), but that it occurs sharply at 259°K for a synthetic single crystal (1961). Shull *et al.* (1951a) showed by neutron diffraction measurements that the moments of the Fe^{3+} ions lie in (0001) above the transition, but along [0001] below it. Dzyaloshinskii (1957,

1958) argued that upon cooling hematite from temperatures just above the transition, the spin-canting increases rapidly, so that the spin moments move continuously from (0001) to [0001]. Lin (1960,1961) has verified this model, which is now referred to as "spin flop." Blum *et al.* (1965) showed that a strong external magnetic field inhibits the Morin transition; at 80°K the spins flop from [0001] to (0001) when a field of 76.5 $\pm$ 3 kOe is applied.

A curious property of hematite related to the Morin transition is that of magnetic "memory." One would think that cooling a sample through the Morin transition with its accompanying spin flop would destroy an initial remanent magnetism. However, Haigh (1957) found that a portion of the initial moment returned upon heating the sample back to room temperature. Evidently the low-temperature magnetic structure, which according to Dzyaloshinskii's theory should have the antiferromagnetic moment rigorously parallel to [0001] and thus no net ferromagnetic component, somehow retains a "memory" of the room-temperature moment. A likely explanation is that the two magnetic substructures are imperfectly compensated, yielding a net moment similar to but much weaker than that originally proposed by Néel (1953) for the room-temperature moment. This component of the magnetization is known as "parasitic ferromagnetism" or the "defect moment."

Ilmenite

Crystal structure of ilmenite. On the basis of the close similarity of its powder pattern with that of hematite, Ewald and Hermann (1931) predicted that ilmenite would have essentially the same structure as hematite. However, the morphology of natural ilmenite crystals had shown that the space group must have a lower symmetry than that of hematite; $R\bar{3}$ (a subgroup of $R\bar{3}c$) was chosen for ilmenite on this basis. The structure and space group were confirmed by Barth and Posnjak (1934) through single-crystal x-ray studies. Compared to the hematite structure (Fig. L-12), one-half of the Fe ions are replaced by Ti in such a way that every $Fe-O_3-Fe$ unit becomes a $Fe-O_3-Ti$ unit, and the sequence of cations along any [0001] axis is ...Fe-Ti-□-Ti-Fe-□-Fe... where the hyphens represent intervening planes of oxygen layers. As in hematite, every third potential octahedral site is empty; the nearest cations to an empty site along [0001] are always two Fe or two Ti (Fig. L-13). The $Fe-O_3-Ti$ units are so arranged in (0001) that oxygen layers alternate with metal layers, which themselves alternate between layers of Fe and layers of Ti (Fig. L-13). As in hematite the metal layers are puckered, and are always separated by planes of oxygens. The sequence of metal layers in ilmenite is somewhat similar to the

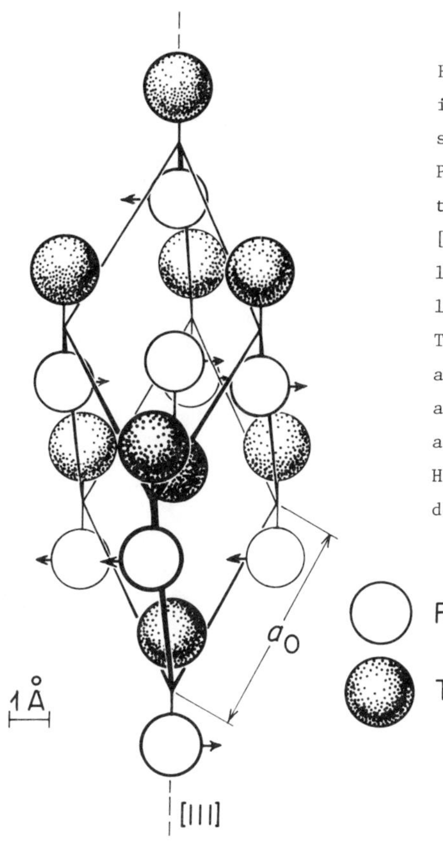

Figure L-13. The rhombohedral cell of ilmenite ($FeTiO_3$), with the origin shifted by $(\frac{1}{4},\frac{1}{4},\frac{1}{4})$ from that of Barth-Posnjak (see Fig. L-12c). The oxygen triplets are omitted for clarity. Along [111], layers of cations alternate with layers of oxygens; of the cation layers, layers of Fe^{2+} alternate with layers of Ti^{4+}. Below the Néel temperature (55°K) all Fe^{2+} spins lie parallel to [111] and are antiparallel in adjacent Fe layers, as indicated schematically by the arrows. However, the arrows should point up and down, *not* horizontally.

---

magnetic structure of hematite: the Fe planes correspond to the A substructure and the Ti planes to the B substructure. (It should be noted, however, that the magnetic structure of ilmenite does *not* correspond to that of hematite.)

Unit cell parameters determined for synthetic ilmenite are $a_H = 5.0881 \pm 0.0005$ Å, $c_H = 14.080 \pm 0.005$ Å, respectively (Lindsley, 1965). Although both a_H and c_H are larger than those of hematite, most of the difference is found in the basal spacing, c_H. Minerals of the ilmenite structure-type include pyrophanite ($MnTiO_3$), with $a_H = 5.137$ Å and $c_H = 14.29$ Å, and geikielite ($MgTiO_3$), with $a_H = 5.054$ Å, and $c_H = 13.898$ Å.

There was long a controversy regarding the valence of the metal ions in ilmenite: is the formula $Fe^{2+}Ti^{4+}O_3$ or $Fe^{3+}Ti^{3+}O_3$? Hámos and Stscherbina (1933) showed that the x-ray absorption edge of ilmenite was much closer to those of $Mg^{2+}Ti^{4+}O_3$ and $Ti^{4+}O_2$ than to that of $Ti_2^{3+}O_3$, and on this basis most workers have assumed the formula $Fe^{2+}Ti^{4+}O_3$. Confirmation of this formula was provided by Mössbauer-effect studies (Ruby and Shirane, 1961).

Table L-7. Site coordinates for ilmenite ($FeTiO_3$). Space group R$\bar{3}$. Origin at center ($\bar{3}$). Hexagonal axes.

Type of Site	No. of Sites	Notation	Symmetry	Coordinates
O^{2-}	18	*f*	1	$x,y,z;\ \bar{y},x-y,z;\ y-x,\bar{x},z$ $\bar{x},\bar{y},\bar{z};\ y,y-x,\bar{z};\ x-y,x,\bar{z}$
Fe^{2+}	6	*c*	3	$0,0,z;\ 0,0,\bar{z}$
Ti^{4+}	6	*c*	3	$0,0,z;\ 0,0,\bar{z}$

Note: The full sets of sites are generated by applying the coordinates in the table to each of three points equivalent to the origin: $(0,0,0);\ (\frac{1}{3},\frac{2}{3},\frac{2}{3});\ (\frac{2}{3},\frac{1}{3},\frac{1}{3})$.

The metal ions in ilmenite lie on special positions along 3-fold axes (*c* sites) but as the charges and radii differ, Fe^{2+} and T^{4+} will lie at different distances away from the oxygen planes, and thus separate parameters are needed to fix their positions. Furthermore, as a result of the lower symmetry of the space group R$\bar{3}$, the oxygens no longer lie on symmetry elements; they occupy general positions (*f* sites) that require three parameters to be specified completely. The coordinates of the sites are given in Table L-7. The five parameters necessary to describe the structure (Table L-8) were determined by Barth and Posnjak (1934); these values have been refined by neutron diffraction measurements (Shirane *et al.*, 1959; Shirane *et al.*, 1962).

Magnetic structure of ilmenite. At room temperature pure ilmenite is essentially paramagnetic, but at sufficiently low temperatures it becomes antiferromagnetic. Bizette and Tsai (1956) determined the Néel temperature of a natural ilmenite as 68°K; pure synthetic ilmenite, however, was found to have a Néel point at 55°K (Ishikawa and Akimoto, 1957), and this value has been

Table L-8. Atomic parameters reported for ilmenite.

OXYGEN: Rhombohedral Cell x	y	z	OXYGEN: Hexagonal Cell x	y	z	Fe	Ti	Reference
0.555	-0.055	0.250	0.305	0.000	0.250	0.358	0.142	Barth and Posnjak (1932)
0.555	-0.040	0.235	0.305	0.015	0.250	--	--	Shirane et al. (1959)
0.560	-0.050	0.230	0.313	0.017	0.247	0.356	--	Shirane et al. (1962)

confirmed by Stickler *et al.* (1967) who reported the value 56 ± 2°K. In antiferromagnetic ilmenite all Fe^{2+} in a given layer have parallel spins, but alternate iron layers have antiparallel spins. This structure requires a magnetic cell with c_H twice that of the crystallographic cell. Neutron diffraction studies (Shirane *et al.*, 1959) have shown that the moments are parallel to [0001], with alternating Fe layers being antiparallel. The magnetic structure of ilmenite is of little direct interest in rock magnetism, but it is of great importance as it affects the properties of hematite-ilmenite solid solutions.

Crystal and magnetic structure of hematite-ilmenite solid solutions

Hematite and ilmenite form a complete solid solution series at high temperatures. The composition may be expressed $Fe^{3+}_{2-2x}Fe^{2+}_{x}Ti_{x}O_3$, where x is the mole fraction of ilmenite. Mössbauer studies (Shirane *et al.*, 1959; Shirane *et al.*, 1962) and electrical conductivity measurements (Ishikawa, 1958a) indicate that the intermediate solid solutions contain Fe^{2+} and Fe^{3+} in amounts approximately proportional to x and $(2-2x)$, respectively, so it is reasonable to assume that titanium is present only as Ti^{4+}. The existence of continuous solid solution implies that the space group is the same for both end members at high temperatures. Although evidently no structural studies--x-ray or neutron diffraction--have been made at high temperature, there is fairly convincing evidence that the structure of ilmenite at high temperature is identical to that of hematite, $R\bar{3}c$. This space group implies complete disorder between Fe and Ti in ilmenite and intermediate compositions $Fe^{3+}_{2-2x}Fe^{2+}_{x}Ti_{x}O_3$ at high temperature, so that all metal layers are equivalent. This presumed high-temperature structure is apparently non-quenchable for compositions near $FeTiO_3$, but it can be preserved by quenching for intermediate compositions. Significant differences between quenched and annealed specimens with x near 0.5 have been found in magnetic properties, neutron diffraction patterns, and Mössbauer absorption spectra.

Ishikawa (1958b) found that the temperature of the transition from $R\bar{3}c$ to $R\bar{3}$ ranges from approximately 1100°C for x = 0.65 to 600°C at x = 0.45. For $x \geqq 0.6$ it was impossible to quench the $R\bar{3}c$ structure, whereas complete order ($R\bar{3}$ structure) was not obtained for $0.6 > x \geqq 0.5$, probably because diffusion rates are too slow at temperatures below the transition temperature. It might be noted that the $R\bar{3}c \rightarrow R\bar{3}$ transition for x = 0.5 is metastable with respect to exsolution into two phases, $hematite_{ss}$ and $ilmenite_{ss}$. Evidently diffusion rates for ordering Fe and Ti between the metal layers are considerably faster than those for segregation into discrete phases. Several workers have pointed out that ordering may take place by means of diffusion within each metal layer to form domains of Ti and of Fe. Exsolution, on the other hand, must involve diffusion of the cations across the intervening oxygen layers, since exsolved specimens consist of alternating ilmenite-rich and hematite-rich lamellae parallel to (001).

In the hematite structure all metal sites are equivalent, as they are related by symmetry elements. In ilmenite at room temperature there are two main types of sites--A = Fe and B = Ti--that occur in alternating layers. But as the layers are puckered, each contains two potentially different kinds of sites (Fig. L-13). Those ions at the top of $Fe-O_3-Ti$ units lie above the plane of the layer and occupy type 1 sites; those at the bottom of the units lie below the plane of the layer and occupy type 2 sites. Thus the cations of a given $Fe-O_3-Ti$ unit will either be an A_1-B_2 pair or a B_1-A_2 pair. This distinction is unnecessary for pure ilmenite, but it can be useful for the intermediate $Fe^{3+}_{2-2x}Fe^{2+}_{x}Ti_{x}O_3$ solid solutions. In the high-temperature, disordered, $R\bar{3}c$ structure all four sites are equivalent and thus all will statistically contain the same proportions of Fe^{2+}, Fe^{3+}, and Ti^{4+}. In the completely ordered $R\bar{3}$ structure for $x < 1$, on the other hand, only Fe^{2+} and Fe^{3+} occupy the A sites. The B sites, however, contain both Ti and Fe ions which need not be distributed equally between B_1 and B_2. The most favorable composition for ordered distribution of Fe and Ti between B_1 and B_2 is $x = 0.5$; complete order would mean that either B_1 or B_2 was completely occupied by Ti only, and the other would contain Fe only. However, neutron diffraction data for this composition show conclusively that such ordering between B_1 and B_2 does not take place (Shirane *et al.*, 1962).

Variation of cell dimensions with composition is plotted in Figure L-14.

The magnetic structures of the hematite-ilmenite series can only be summarized here. For more details the interested reader is referred to the papers of Ishikawa and of Shirane and their co-workers cited above and to Hoffman (1975). Evidently the magnetic structure of hematite, with the A and B substructures antiferromagnetically coupled, develops in any portion of the series that retains the high-temperature, disordered structure on cooling. Thus, the hematite magnetic structure is always found for $0.45 \leqq x \leqq 0$ (since in that range there is no significant ordering of Ti on the laboratory time scale), but can develop only in quenched specimens with $0.65 \leqq x \leqq 0.45$. For compositions close to ilmenite ($x \sim 1$), the nearly instantaneous ordering of Ti^{4+} on B sites upon cooling (even upon quenching) makes the B substructure magnetically inert, and a new magnetic structure--antiferromagnetic coupling of alternating A sites--is developed. Between room temperature and the Néel temperature (55°K or below) phases with $1 < x \leqq 0.8$ show magnetic coupling only in clusters, a property similar to superparamagnetism (Bean, 1955). For intermediate compositions with $0.8 \leq x \leq 0.45$, sufficient Fe remains in the B sites after ordering takes place that the antiparallel coupling between the A and B substructures is retained. However, the A substructure has a larger moment because it contains more iron, and thus the structure becomes ferrimagnetic. Bozorth *et al.* (1957), and Ishikawa (1962) have shown that intermediate compositions, upon annealing, develop saturation magnetizations that approach the theoretical value of 4x Bohr magnetons (per three oxygens) predicted for this structure.

The picture shown here is oversimplified for the intermediate compositions. Ishikawa and Syono (1963) have shown that a metastable iron-enriched phase forms during ordering of phases with x near 0.5, and that it is this iron-enriched phase that is responsible for the celebrated self-reversal properties of the intermediate solid solutions. Hoffmann (1975) proposed a mechanism by which this "x-phase" might form.

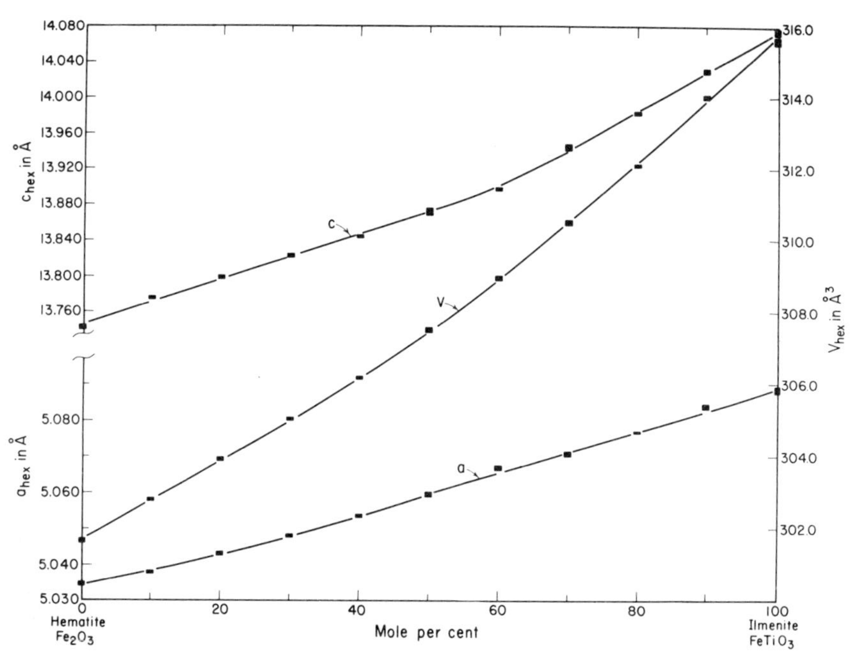

Figure L-14. Cell dimensions of the hematite-ilmenite (Fe_2O_3-$FeTiO_3$) series as functions of composition for synthetic specimens (Lindsley, 1965).

The behavior of Curie temperature with composition appears simple, but is highly complex in detail. In general the Curie temperature decreases rather uniformly with increasing x from 675°C of hematite towards -218°C, the Néel temperature of ilmenite. But in the region $0.8 \leqq x \leqq 0.45$ the Curie temperature depends upon the previous heat treatment (that is, on the degree of ordering of the sample, whereas in the superparamagnetic region $1.0 < x \leqq 0.8$ there is, of course, no measurable Curie temperature.

It should be emphasized that the foregoing discussion of magnetic structures refers to synthetic or volcanic minerals that have cooled relatively rapidly. More slowly cooled plutonic minerals exsolve into two phases and may be expected to have very different magnetic properties (Carmichael, 1961).

ORTHORHOMBIC OXIDES - THE PSEUDOBROOKITE GROUP

The structure of pseudobrookite (Fe_2TiO_5)

The crystal structure of pseudobrookite was determined by Pauling (1930) from x-ray diffraction patterns of natural single crystals. The x-ray data permit three possible space groups: *Bbmm*, $Bb2_1m$, and *Bbm2*; however, the holohedral morphology exhibited by pseudobrookite crystals (Doss, 1892) permits only the first space group. The unit cell contains $4Fe_2TiO_5$ at the sites given in Table L-9.

Table L-9. Structural parameters for pseudobrookite (Fe_2TiO_5). Origin at center ($\frac{2}{m}$). (Data from Pauling, 1930.) Space group *Bbmm*.

No. of Sites	Notation	Symmetry	Coordinates
8	*f*	m	(0.045,0.110,0);(-0.045,-0.110,0); (0.045,0.390,0);(-0.045,0.610,0)
8	*f*	m	(0.310,0.095,0);(-0.310,-0.095,0); (0.310,0.405,0);(-0.310,0.595,0)
4	*c*	mm	(0.730,0.250,0);(-0.730;0.750,0)
8	*f*	m	(0.135,0.560,0);(-0.135,-0.560,0); (0.135,-0.060,0);(-0.135,0.060;0)
4	*c*	mm	(0.190,0.250,0);(-0.190;0.750,0)

Note: Because the unit cell is b-centered, the points (000) and ($\frac{1}{2}$ 0 $\frac{1}{2}$) are equivalent to the origin. The full sets of sites are obtained by taking the coordinates given above with respect to both points.

Pauling interpreted his data as indicating that the *8f* sites are occupied by Fe^{3+} and the *4c* sites by Ti^{4+}. This assignment was retained by Hamelin (1958). Lind and Housley (1972) argued that Hamelin's data are more compatible with four Fe^{3+} in *4c* and four Fe^{3+} and four Ti^{4+} in *8f*. The *c* sites are sometimes called M1 and the *f* sites M2. A view of the armalcolite structure is given in Figure L-15. Each M1 octahedron shares six edges with M2 octahedra; it also shares two opposite corners with M1 octahedra to form chains along the *c* axis. Each M2 octahedron shares one edge with another M2 octahedron and three edges with M1 octahedra. The oxygens in pseudobrookite do not approach close packing.

The unit cell dimensions of pseudobrookite are a = 9.767 Å, b = 9.947 Å, c = 3.717 Å (Akimoto *et al.*, 1957). The structure determination by Pauling confirms that the formula of pseudobrookites is Fe_2TiO_5, not $Fe_4Ti_3O_{12}$ as had previously been assumed. Pauling showed that the excess TiO_2 indicated by chemical analyses was in the form of minute inclusions of rutile.

The Fe_2TiO_5-$FeTi_2O_5$ Series

Akimoto *et al.* (1957b) prepared synthetic $FeTi_2O_5$ ("ferropseudobrookite") and four phases intermediate between Fe_2TiO_5 and $FeTi_2O_5$. They were able to index the powder x-ray pattern of each in terms of an orthorhombic unit cell, and the continuous variation of the cell parameters suggests that the basic pseudobrookite structure is retained throughout the solid solution series (Fig. L-16). Shirane *et al.* (1962) studied the Mössbauer spectra of $FeTi_2O_5$ and have confirmed that the iron is present as Fe^{2+}. Not knowing the detailed structure, these workers assumed the space group *Cmcm* (= *Bbmm* for the conventional axes for pseudobrookite) and considered two possibilities: (1) $8Ti^{4+}$ in *8f* and $4Fe^{2+}$ in *4c*, or (2) 4 Ti and 4 Fe in *8f*, with 4 Ti in *4c*. They prefer the second arrangement (which is somewhat analogous to an inverse spinel), as it is more compatible with the existence of complete solid solution with pseudobrookite. However, they note that the existence of slight asymmetries in the Mössbauer absorption spectra may indicate a small amount of Fe^{2+} in the *4c* sites.

Members of the Fe_2TiO_5-$FeTi_2O_5$ series are paramagnetic at room temperature (Akimoto *et al.*, 1957b; Chevallier and Mathieu, 1958) and thus they do not play an important role in rock magnetism.

The lunar mineral armalcolite ($Fe^{2+}_{0.5}Mg_{0.5}Ti_2O_5$) has been shown to have the pseudobrookite structure (Lind and Housley, 1972; Wechsler *et al.*, 1976). Lind and Housley argued strongly that armalcolite has Ti^{4+} in *8f* (M2) and Fe^{2+}

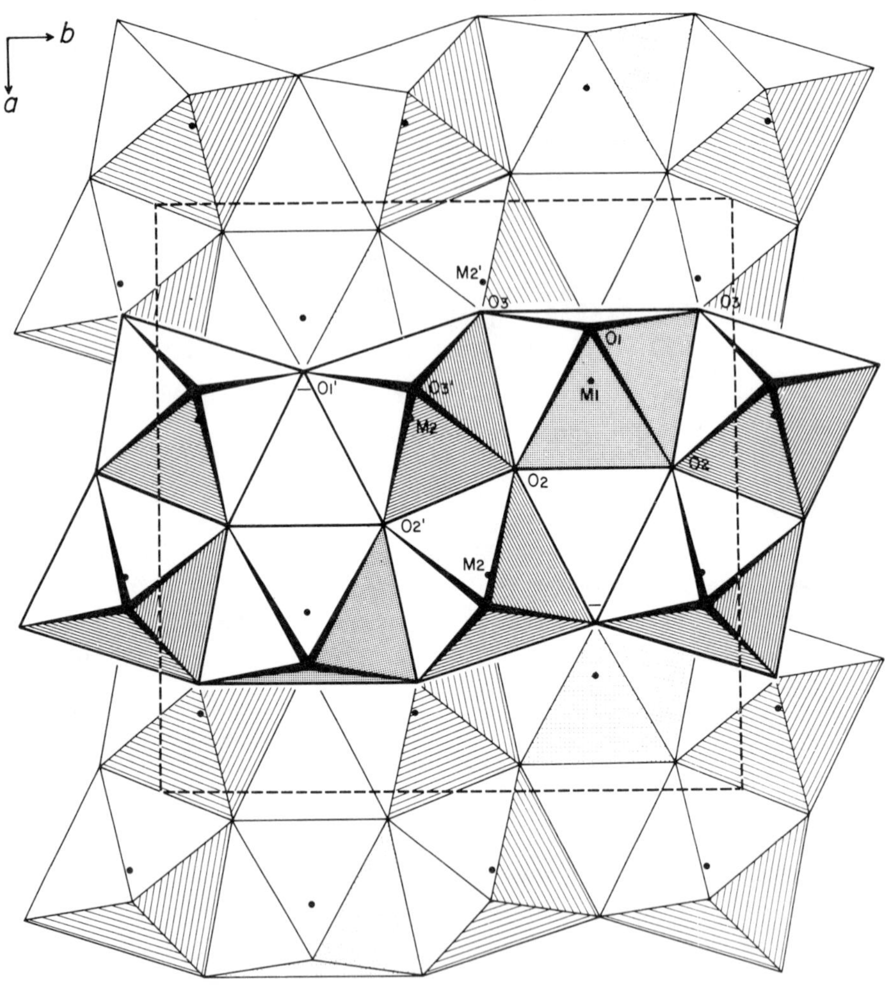

Figure L-15. The pseudobrookite structure projected down the *c* axis for armalcolite, $(Fe,Mg)Ti_2O_5$. *a* and *b* of the unit cell are shown in dashed outline. Wechsler *et al.* (1976).

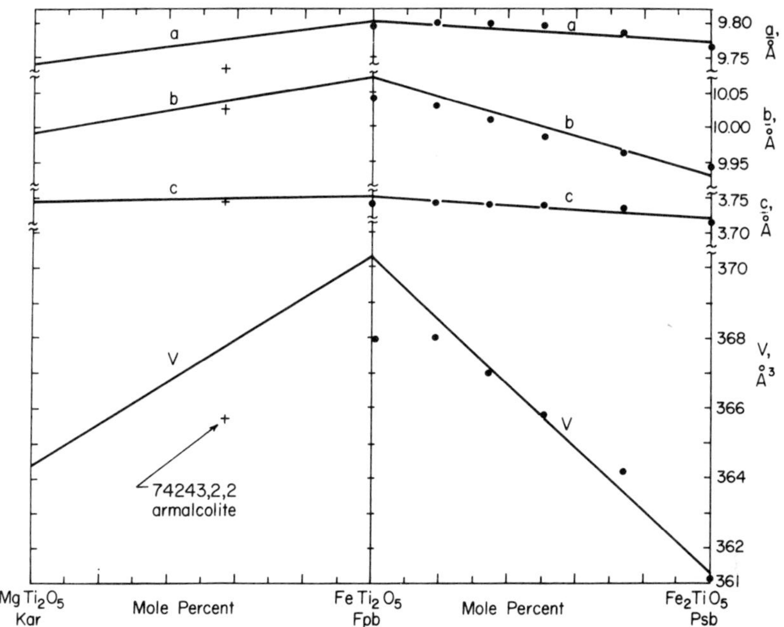

Figure L-16. Unit-cell dimensions for the Fe_2TiO_5-$FeTi_2O_5$ (Psb-Fpb) series and the $FeTi_2O_5$-$MgTi_2O_5$ (Fpb-Kar) series. Lines for the Fpb-Kar series are linear least-square fits to the data of Lindsley *et al.* (1974). The crosses are data for a natural armalcolite from lunar sample 74243,2,2; it contains Al, Cr, Mn, and probably Ti^{3+} (Wechsler *et al.*, 1976). Lines for the Psb-Fpb series are "eyeball" fits to the data (dots) of Akimoto *et al.* (1957), but are constrained to the Fpb values of Lindsley *et al.* (1974).

---

and Mg^{2+} randomly distributed over the *4c* (M1) sites. Wechsler *et al.* agree with this arrangement for natural and annealed specimens, but claim that substantial disorder is present in armalcolites quenched from high temperatures. Unit-cell parameters for the $FeTi_2O_5$-$MgTi_2O_5$ series are shown in Figure L-16.

STRUCTURES OF RUTILE, ANATASE AND BROOKITE

The structures of the three TiO_2 polymorphs--rutile, anatase, and brookite--are discussed together to emphasize the similarities and differences between them. Brookite is orthorhombic, whereas the others are both tetragonal, with $c/a < 1$ for rutile and $c/a > 1$ for anatase.

The first determination of the structures of rutile and anatase was made by Vegard (1916a,b), who observed that each could be considered as made up of O-Ti-O "molecules." In rutile the O-Ti-O units lie in (001), but in anatase they are parallel to [001]; Vegard considered this the explanation for the differences in *c*/*a*. The now more familiar view of these structures as sequences of coordination polyhedra was set forth by Pauling (1928; Pauling and Sturdivant, 1928). In both structures Ti is coordinated by six oxygens, and each oxygen by three titaniums; the difference is in the sequence of the coordination octahedra. In rutile only two edges of the octahedra--those lying in (001)--are shared (Fig. L-17), whereas in anatase four edges are shared (Fig. L-18). As would be expected, the shared edges are shorter than the unshared edges. The relationships between the "molecular" structures and the coordination structures are also illustrated in the figures. In rutile the octahedra form chains parallel to [001]; the chains are linked by the sharing of corners. In anatase, on the other hand, there are no distinct chains (Fig. L-18).

On the basis of Vegard's data, Niggli (1921) assigned rutile to space group $P4_2/mnm$ and anatase to $I4_1/amd$; Parker (1923) provided an exhaustive confirmation of the latter assignment. In each structure the Ti ions occupy unique special sites, but the positions of the oxygen ions must be fixed by one experimentally determined parameter (Table L-10). Values of the unit cell

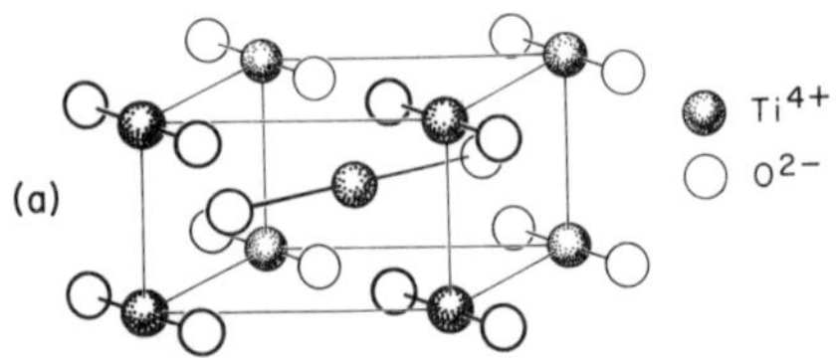

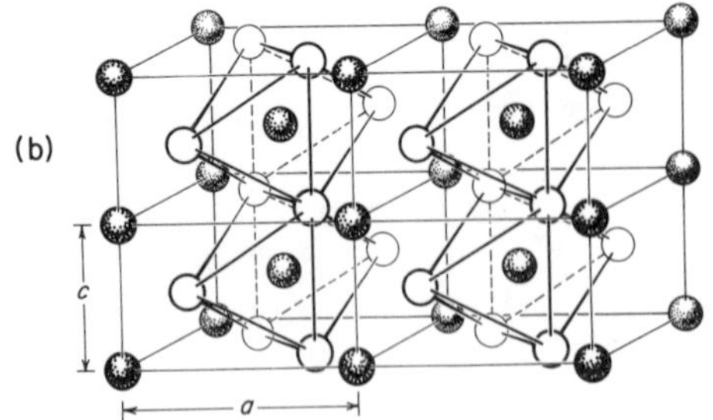

Figure L-17. Two aspects of the structure of rutile (TiO_2). The circles show only the centers of ions and are not to scale.
a. The tetragonal unit cell as determined by Vegard (1916a;1926), showing the O-Ti-O "molecules" lying in (001).
b. Four unit cells showing the coordination octahedra of oxygens about the Ti at the center of each cell. Only those two edges of the octahedra that lie in (001) are shared with adjacent octahedra. The shared edges are shorter than the unshared edges.

Table L-10. Atomic coordinates for rutile, anatase, and brookite.

Mineral	Space Group	Origin	Type of Ion	No. of Sites	Notation	Symmetry	Coordinates
Rutile	$P\frac{4_2}{m}nm$	mmm	Ti^{4+}	2	*a*	mmm	$(0,0,0),(\frac{1}{2},\frac{1}{2},\frac{1}{2})$
			O^{2-}	4	*f*	mm	$(x,x,0);(\bar{x},\bar{x},0);(\frac{1}{2}+x,\frac{1}{2}-x,\frac{1}{2});$ $(\frac{1}{2}-x,\frac{1}{2}+x,\frac{1}{2})$.
Anatase	$I\frac{4_1}{a}md$	$\frac{2}{m}$	Ti^{4+}	4	*a*	$\bar{4}2m$	$(0,\frac{3}{4},\frac{1}{8});(0,\frac{1}{4},\frac{7}{8})^*$
			O^{2-}	8	*e*	mm	$(0,\frac{1}{4},z);(0,\frac{3}{4},\frac{1}{4}+z);(0,\frac{3}{4},\bar{z});(0,\frac{1}{4},\frac{3}{4}-z)^*$
Brookite	$Pbca$	$\bar{1}$	Ti^{4+}	8	*c*	1	$(x,y,z);(\frac{1}{2}+x,\frac{1}{2}-y,\bar{z});$ $(\bar{x},\frac{1}{2}+y,\frac{1}{2}-z);(\frac{1}{2}-x,\bar{y},\frac{1}{2}+z);$
			O^{2-}	8	*c*	1	
			O^{2-}	8	*c*	1	$(\bar{x},\bar{y},\bar{z},);(\frac{1}{2}-x,\frac{1}{2}+y,z);$ $(x,\frac{1}{2}-y,\frac{1}{2}+z);(\frac{1}{2}+x,y,\frac{1}{2}-z)$

* *To get full sets of sites take these coordinates from each of the two equivalent points (0,0,0);* $(\frac{1}{2},\frac{1}{2},\frac{1}{2})$.

Reference: International Tables (1952). Oxygen parameter for rutile: u = 0.3056 ± 0.0006; for anatase, u = 0.2066 ± 0.0009 (Cromer and Herrington, 1955). See Table L11 for brookite.

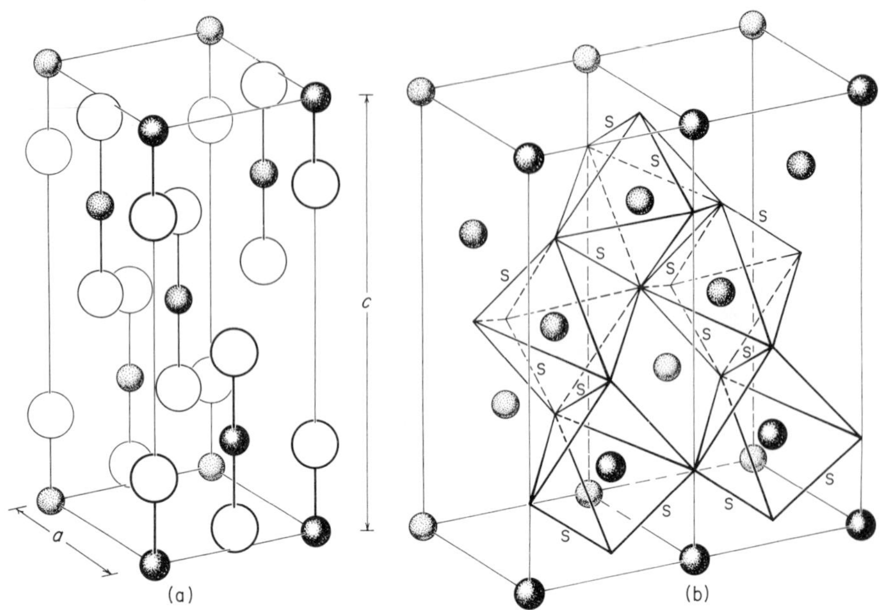

Figure L-18. Two aspects of the structure of anatase (TiO_2). The circles show only the centers of ions and are not to scale. a. The tetragonal unit cell as determined by Vegard (1916b;1926), showing the O-Ti-O "molecules" parallel to [001]. b. Two unit cells of anatase showing distorted octahedra of oxygens about Ti. The oxygens are omitted for clarity; their centers lie at the apices of the octahedra. The front halves of the two lower octahedra are omitted; they lie in front of the two cells illustrated. The shared edges (*s*) are shorter than the unshared edges.

---

dimensions for rutile are a = 4.59373 ± 0.00005 Å, c = 2.95812 ± 0.00005 Å (Straumanis *et al.*, 1961) and the oxygen parameter u = 0.3056 ± 0.0006 (Cromer and Herrington, 1955). For anatase, a = 3.785 ± 0.001 Å, c = 9.514 ± 0.0006 Å, u = 0.2066 ± 0.0009 (Cromer and Herrington, 1955).

The structure of brookite (Pauling and Sturdivant, 1928) is considerably more complex than those of rutile and anatase. Pseudomolecules of O-Ti-O are not found. The 8 Ti and 16 oxygen ions per unit cell occupy general (8*c*) sites of space group *Pbca*, and as a result nine different parameters must be specified to describe the structure (Table L-11). The prediction and verification of these parameters must have been an early triumph for Pauling's coordination theory. The deformed coordination octahedra form staggered chains

Table L-11. Atomic parameters for brookite.

Atom	x	y	z	Reference
Ti	0.1290 ±0.0004	0.0972 ±0.0008	0.8629 ±0.0008	Baur (1961)
O(1)	0.0101 ±0.0008	0.1486 ±0.0015	0.1824 ±0.0015	
O(2)	0.2304 ±0.0008	0.1130 ±0.0015	0.5371 ±0.0015	

that are cross-linked by shared edges (Fig. L-19). The Ti ions are slightly displaced from the centers of the octahedra, a feature that has been emphasized by refinements of the structure (Weyl, 1959; Baur, 1961). Unit cell dimensions are a = 9.174 ± 0.001, b = 5.456 ± 0.002, c = 5.138 ± 0.001 (Swanson *et al.*, 1964).

The framework of oxygens in brookite approaches double-hexagonal close packing, and that in anatase approximates cubic close packing. In rutile, on the other hand, the oxygens are not even approximately close packed. Nevertheless, rutile is much the densest of the three.

Rutile can take extensive amounts of niobium and tantalum into solid solution.

Upon heating under reducing conditions, rutile first loses oxygen (TiO_{2-x}) and then converts to a host of other phases that are termed Magnéli phases. The structures of these phases will not be considered here; the reader is referred to a number of papers on the subject (Asbrink and Magnéli, 1957; Andersson *et al.*, 1957a,b; Straumanis *et al.*, 1961).

Simons and Dachille (1967) and Jamieson and Olinger (1968) have described yet another TiO_2 polymorph (TiO_2-II) as the product of high-pressure experiments. TiO_2-II has space group *Pbcn*, with a = 4.515 Å, b = 5.497 Å, and c = 4.939 Å.

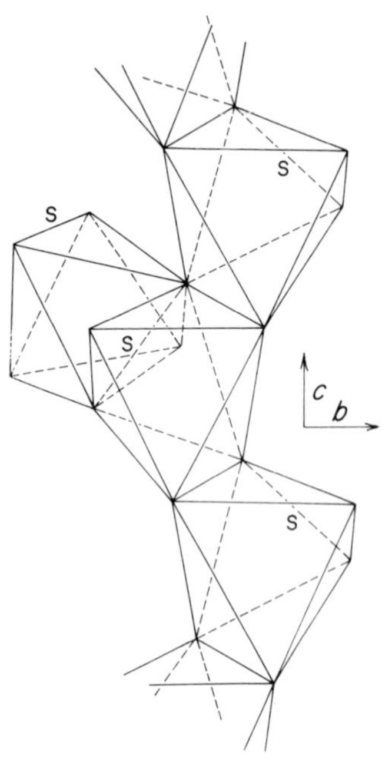

Figure L-19. Part of the brookite structure, showing a chain of distorted coordination octahedra of oxygens about Ti's. Each octahedron shares two edges with two octahedra in the chain and a third edge with an octahedron--one of which is shown--of an adjacent chain. Other shared edges are shown by *s*. After Pauling and Sturdivant (1928).

REFERENCES

Aharoni, A., E. H. Frei, and M. Schieber (1962) Curie point and origin of weak ferromagnetism in hematite. *Phys. Rev. 127*, 439-441.

Akimoto, S. (1954) Thermo-magnetic study of ferromagnetic minerals contained in igneous rocks. *J. Geomag. Geoelectricity 6*, 1-14.

________ (1962) Magnetic properties of FeO-Fe_2O_3-TiO_2 system as a basis of rock magnetism. *J. Phys. Soc. Jpn. 17*, Suppl. B1, 706-710.

________, and T. Katsura (1959) Magneto-chemical study of the generalized titanomagnetite in volcanic rocks. *J. Gemag. Geoelectricity 3*, 69-90.

________, T. Katsura, and M. Yoshida (1957a) Magnetic properties of $TiFe_2O_4$-Fe_3O_4 system and their change with oxidation. *J. Geomag. Geoelectricity 9*, 165-178.

________, T. Nagata, and T. Katsura (1957b) The $TiFe_2O_5$-Ti_2FeO_5 solid solution series. *Nature 179*, 37-38.

________, and Y. Syono (1967) High-pressure decomposition of some titanite spinels. *J. Chem. Phys. 47*, 1813-1817.

Anderson, P. W. (1950) Antiferromagnetism. Theory of superexchange interaction. *Phys. Rev. 79*, 350-356.

Andersson, S., B. Collén, G. Kruuse, U. Kuylenstierna, A. Magńeli, H. Pestmalis, and S. Asbrink (1957a) Identification of titanium oxides by x-ray powder patterns. *Acta Chem. Scand. 11*, 1653-1657.

Andersson, S., B. Collén, U. Kuylenstierna, and A. Magnéli (1957b) Phase analysis studies on the titanium-oxygen system. *Acta Chem. Scand. 11*, 1641-1652.

Armstrong, R. J., A. H. Morrish, and G. A. Sawatzky (1966) Mössbauer study of ferric ions in a spinel. *Phys. Lett. 23*, 414-416.

Asbrink, S. and A. Magnéli (1957) Note on the crystal structure of trititanium pentoxide. *Acta Chem. Scand. 11*, 1606-1607.

Banerjee, S. K. and W. O'Reilly (1966) Coercivity of Fe^{++} in octahedral sites of Fe Ti spinels. *IEEE, Trans. Magnetics Mag-2*, 463-467.

Barth, T. F. W. and E. Posnjak (1931) The spinel structure, an example of variate atom equipoints. *J. Washington Acad. Sci. 21*, 255-258.

________, ________ (1932) Spinel structures: with and without variate atom equipoints. *Z. Kristallogr. A 82*, 325-341.

________, ________ (1934) The crystal structure of ilmenite. *Z. Kristallogr. A 88*, 265-270.

Basta, E. Z. (1959) Some mineralogical relationships in the system Fe_2O_3-Fe_3O_4 and the composition of titanomaghemite. *Econ. Geol. 54*, 698-719.

________ (1960) Natural and synthetic titanomagnetites (the system Fe_3O_4-Fe_2TiO_4-$FeTiO_3$). *Neues Jahrb. Mineral. Abh. 94, Festband Famdohr*, 1017-1048.

Baur, W. H. (1961), Atomabstände und Bindungswinkel im Brookit, TiO_2. *Acta Crystallogr. 14*, 214-216.

Bean, C. P. (1955) Hysteresis loops of mixtures of ferromagnetic micropowders. *J. Appl. Phys. 26*, 1381-1383.

Behar, I. and R. Collongues (1957) Sur la transformation ordre-désordre dans la phase sesquioxyde de fer cubique. *C. R. Acad. Sci. (Paris) 244*, 617-619.

Bernal, J. D., D. R. Dasgupta, and A. L. Mackay (1957) Oriented transformations in iron oxides and hydroxides. *Nature (London) 180*, 645-647.

________, ________, ________ (1959) The oxides and hydroxides of iron and their structural interrelationships. *Clay Mineral. Bull. 4*, 15-30.

Bizette, H. and B. Tsai (1956) Susceptibilitiés magnétiques principales d'un cristal natural d'ilmenite ($TiFeO_3$). *C. R. Acad. Sci. (Paris) 242*, 2124-2127.

Blake, R. L., R. E. Hessevick, T. Zoltai, and L. W. Finger (1966) Refinement of the hematite structure. *Am. Mineral. 51*, 123-129.

Bleil, U. (1971) Cation distribution in titanomagnetites. *Z. Geophys. 37*, 305-319.

________ (1976) An experimental study of the titanomagnetite solid solution series. *Pure and Appl. Geophys. 114*, 165-175.

Blum, N., A. J. Freeman, J. W. Shaner, and L. Grodzins (1965) Mössbauer studies of spin flop in antiferromagnetic hematite. *J. Appl. Phys. 36*, 1169-1170.

Bozorth, R. M. (1951) *Ferromagnetism*. Van Nostrand, New York.

________, D. E. Walsh, and A. J. Williams (1957) Magnetization of ilmenite-hematite system at low temperatures. *Phys. Rev. 108*, 157-158.

Bragg, W. H. (1915a) The structure of magnetite and the spinels. *Nature 95*, 561.

________ (1915b) The structure of the spinel group of crystals. *Phil. Mag. 30*, 305-315.

Braun, P. (1952) A superstructure in spinels. *Nature (London) 170*, 1123.

Burns, R. G. (1970) *Mineralogical Applications of Crystal Field Theory.* Cambridge University Press, Cambridge, xiii + 224p.

Carel, C. (1967) Recherches expérimentales et théoriques sur le diagramme d'etat de la wüstite solide au-dessus de 910°C. *Mem. Sci. Rev. Metal. 64*, 734-744, 821-836.

Carmichael, C. M. (1961) The magnetic properties of ilmenite-haematite crystals. *Proc. Roy. Soc., Ser. A, 263*, 508-530.

Charlesworth, G. and F. A. Long (1939) Note on the magnetic properties of ferric oxide at low temperatures. *Proc. Leeds Phil. Soc. 3*, 515-519.

Chevallier, R., J. Bolfa, and S. Mathieu (1955) Titanomagnétites et ilménites ferromagnétiques. (1) Etude optique, radiocristallographique, chimique. *Bull. Soc. Franc. Mineral. Cristallogr. 78*, 307-346.

________, and S. Mathieu (1958) Susceptibilité magnétique specifique de pyroxènes monocliniques. *Bull. Soc. Chim. France*, 726-729.

Claassen, A. A. (1926) The scattering power of oxygen and iron for x-rays. *Proc. Phys. Soc. 38*, 482-487.

Colombo, U., G. Fagherazzi, F. Gazzarrini, G. Lanzavecchia, and G. Sironi (1964) Mechanisms in the first stage of oxidation of magnetites. *Nature 202*, 175-176.

________, F. Gazzarrine, G. Lanzavecchia, and G. Sironi (1965) Magnetite oxidation: a proposed mechanism. *Science 147*, 1033.

Cromer, D. T. and K. Herrington (1955) The structures of anatase and rutile. *J. Am. Chem. Soc. 77*, 4708-4709.

Darken, L. S. and R. W. Gurry (1945) The system iron-oxygen. I. The wüstite field and related equilibria. *J. Am. Chem. Soc. 67*, 1398-1412.

Davey, P. T. and T. R. Scott (1957) Preparation of maghaemite by electrolysis. *Nature 179*, 1363.

David, I. and A. J. E. Welch (1956) The oxidation of magnetite and related spinels. Constitution of gamma ferric oxide. *Trans. Faraday Soc. 52*, 1642-1650.

Davis, B. L., G. Rapp, Jr., and M. J. Walawender (1968) Fabric and structural characteristics of the martitization process. *Am. J. Sci. 266*, 482-496.

Doss, B. (1892) Ueber eine zufällige Bildung von Pseudobrookit, Hämatit und Anhydrit als Sublimationsproducte, und über systematische stellung des ersteren. *Z. Kristallogr. 20*, 566-587.

Dunlop, D. J. (1971) Magnetic properties of fine-particle hematite. *Ann. Geophys. 27*, 269-293.

Dunn, J. A. and A. K. Dey (1937) Vanadium-bearing titaniferrous iron ores in Singhbhum and Mayurbhanji, India. *Trans. Mining and Geol. Inst., India, 31*, 117-184.

Dzyaloshinskii, I. E. (1957) Thermodynamic theory of "weak" ferromagnetism in antiferromagnetic substances. *Soviet Physics JETP 5*, 1259-1272 [trans. from *Zhu. Eksper. Teor. Fiziki 32*, 1547-1562].

Dzaloshinsky, I. E. (1958) A thermodynamic theory of "weak" ferromagnetism of antiferromagnetics. *J. Phys. Chem. Solids 4*, 241-255.

Ewald, P. P. and C. Hermann (1931) Strukturbericht, 1913-1928 [*Z. Kristallogr.*]. Akademische Verlagsgesellschaft, Leibzig.

Feitknecht, W. (1965) Einfluss der teilchen grösse auf den Mechanismus von Festkörperreactionen. *Pure and Appl. Chem. 9*, 423-440.

Feitknecht, W. and H. W. Lehmann (1959) Über die oxidation von Magnetit zu γ-Fe_2O_3. *Helv. chim. Acta 42*, 2035-2039.

Ferguson, G. A. and M. Hass (1958) Magnetic structure and vacancy distribution in γ-Fe_2O_3 by neutron diffraction. *Phys. Rev. 112*, 1130-1131.

Forestier, H. and G. Chaudron (1925) Points de transformation magnétique dans le system sesquioxyde de fer-magnésie. *C. R. Acad. Sci. (Paris) 181*, 509-511.

Forster, R. H. and E. O. Hall (1965) A neutron and x-ray diffraction study of ulvöspinel, Fe_2TiO_4. *Acta Crystallogr. 18*, 857-862.

Foster, P. K. and A. J. E. Welch (1956) Metal-oxide solid solutions. Part 1. Lattice constant and phase relationships in ferrous oxide (wüstite) and in solid solutions of ferrous oxide and manganous oxide. *Trans. Faraday Soc. 52*, 1626-1635.

Frölich, F. and H. Stiller (1963) The nature of chemical bond of magnetite and consequences. *Geof. Pura e Appl. 55*, 91-100.

________, and H. Vollstaedt (1967) Untersuchungen zur Bestimmung der Curie-temperatur von maghemit (γ-Fe_2O_3). *Monatsber. Deut. Akad. Wiss. (Berlin) 9*, 180-186.

Gidskehaug, Arne (1975) Method to determine the degree of non-stoichiometry of iron-titanium oxides. *R. Astron. Soc., Geophys. J. 41*, 255-269.

Gilleo, M. A. (1958) Superexchange interaction energy for Fe^{3+}-O^{2-}-Fe^{3+} linkages. *Phys. Rev. 109*, 777-781.

Goodenough, J. B. and A. L. Loeb (1955) Theory of ionic ordering, crystal distortion, and magnetic exchange due to covalent forces in spinels. *Phys. Rev. 98*, 391-408.

Gorter, E. W. (1957) Chemistry and magnetic properties of some ferrimagnetic oxides like those occurring in nature. *Adv. Phys. 6*, 336-361.

Greig, J. W., E. Posnjak, H. E. Merwin, and R. B. Sosman (1935) Equilibrium relationships of Fe_3O_4, Fe_2O_3, and oxygen. *Am. J. Sci. 30*, 239-316.

Guillaud, C. (1951) Magnetic properties of αFe_2O_3. *J. Phys. Rad. 12*, 489-491.

Hägg, G. (1935a) The spinels and the cubic sodium-tungsten bronzes as new examples of structures with vacant lattice points. *Nature 135*, 874.

________ (1935b) Die Kristallstruktur des magnetischen Ferrioxyds, γ-Fe_2O_3. *Z. Physik. Chem. 29 B*, 95-103.

Haigh, G. (1957) Observations on the magnetic transition of hematite at -15°C. *Phil. Mag., Ser. 8, 2*, 877-890.

Hamelin, M. (1958) Structure du composé $TiO_2 \cdot Al_2O_3$. Comparison avec la pseudo-brookite. *Bull. Soc. Chim. France, 1958*, 1559-1566.

Hamilton, W. C. (1958) Neutron diffraction investigation of the 119°K transition in magnetite. *Phys. Rev. 110*, 1050-1057.

Hämos, L. V., and W. Stscherbina (1933) Über die Röntgenabsorptionskante von Ti in Ti-verbindungen und über die Konstitution des Ilmenites. *Nachr. Ges. Wiss. Gottingen. Math. physik Kl., Fachgruppen*, p. 232.

Haul, R. and Th. Schoon (1939) Zur struktur des ferromagnetischen Eisen (III)-Oxyds γ-Fe_2O_3. *Z. Physik. Chem. 44, B*, 216-226.

Hauptman, Z. (1974) High-temperature oxidation, range of non-stoichiometry and Curie point variation of cation deficient titanomagnetite $Fe_{2.4}Ti_{0.6}O_{4+\gamma}$. *R. Astron. Soc., Geophys. J. 38*, 29-47.

Henry, W. E. and M. J. Boehm (1956) Intradomain magnetic saturation and magnetic structure of γ-Fe_2O_3. *Phys. Rev. 101*, 1253-1255.

Hoffman, K. A. (1975) Cation diffusion processes and self-reversal of thermoremanent magnetization in the ilmenite-hematite solid solution series. *R. Astron. Soc., Geophys. J. 41*, 65-80.

Honda, K. and T. Soné (1914) Uber die magnetische untersuchung der Strukturänderungen in Eisen- und Chromverbindungen bei höheren Temperaturen. *Sci. Rept., Tôhuku Imperial University 3*, 223-234.

Hulthén, L. (1936) Uber das antiferromagnetische Austauschproblem bei tiefen temperaturen. *Proc. Akad. Weten. Amsterdam 39*, 190-200.

International Tables for X-Ray Crystallography (1952) International Union of Crystallography. The Kynoch Press, Birmingham.

Ishikawa, Y. (1958a) Electrical properties of $FeTiO_3$-Fe_2O_3 solid solution series. *J. Phys. Soc. Jpn. 13*, 37-42.

________ (1958b) An order-disorder transformation phenomenon in the $FeTi_3$-Fe_2O_3 solid solution series. *J. Phys. Soc. Jpn. 13*, 828-837.

________ (1962) Magnetic properties of ilmenite-hematite system at low temperature. *J. Phys. Soc. Jpn. 17*, 1835-1844.

________, and S. Akimoto (1957) Magnetic properties of the $FeTiO_3$-Fe_2O_3 solid solution series. *J. Phys. Soc. Jpn. 12*, 1083-1098.

________, and Y. Syono (1963) Order-disorder transformation and reverse thermoremanent magnetism in the $FeTiO_3$-Fe_2O_3 system. *J. Phys. Chem. Solids 24*, 517-528.

Jamieson, J. C. and B. Olinger (1969) Pressure-temperature studies of anatase, brookite, rutile, and TiO_2(II): a discussion. *Am. Mineral. 54*, 1477-1481.

Jensen, S. D. and P. N. Shive (1973) Cation distribution in sintered titanomagnetites. *J. Geophys. Res. 78*, 8474-8480.

Katsura, T. and I. Kushiro (1961) Titanomaghemite in igneous rocks. *Am. Mineral. 46*, 134-145.

________, B. Iwasaki, S. Kimura, and S. Akimoto (1967) High-pressure synthesis of the stoichiometric compound FeO. *J. Chem. Phys. 47*, 4559-4560.

Kelman, M. (1965) Propriétés thermodynamiques du protoxyde de fer sous forme solide. Application de résultats experimentaux au tracé du diagramme d'équilibre. *Mem. Sci. Rev. Met. 62*, 457-469.

Koch, F. and J. B. Cohen (1969) The defect structure of $Fe_{1-x}O$. *Acta Crystallogr. B25*, 275-287.

________, M. E. Fine, and J. B. Cohen (1966) O[r]dering in $Fe_{1-x}O$ (abstr). *Bull. Am. Phys. Soc. 11*, 473.

________, and M. E. Fine (1967) Magnetic properties of Fe_xO as related to the defect structure. *J. Appl. Phys. 38*, 1470-1471.

Kordes, E. (1935) Kristallchemische Untersuchungen über Aluminium verbindungen mit spinellartigem Gitterbau und über γ-Fe_2O_3. *Z. Kristallogr. 91*, 193-228.

Kramers. H. A. (1934) L'interaction entre les atomes magnétogenes dans un cristal paramagnetique. *Physica 1*, 182-192.

Kullerud, G. and G. Donnay (1967) Sulfide-oxide relations. *Carnegie Inst. Washington Year Book 65*, 356-357.

Kullerud, G., G. Donnay, and J. D. H. Donnay (1968) Omission solid solution in magnetite. *Carnegie Inst. Washington Year Book 66*, 497-498.

________, ________, and ________ (1969) Omission solid solution in magnetite: kenotetrahedral magnetite. *Z. Kristallogr. 128*, 1-17.

Li, Y.-K. (1956) Superexchange interactions and magnetic lattices of the rhombohedral sesquioxides of the transition element and their solid solutions. *Phys. Rev. 102*, 1015-1020.

Lin, S. T. (1960) Magnetic behavior in the transition region of a hematite solid crystal. *J. Appl. Phys., Suppl. 31*, 273S-274S.

________ (1961) Remanent magnetization of a synthetic hematite single crystal. *J. Appl. Phys., Suppl. 32*, 394S-395S.

Lind, M. D. and R. M. Housley (1972) Crystallization studies of lunar igneous rocks: Crystal structures of synthetic armalcolite. *Science 175*, 521-523.

Lindner, R. and A. Akerstrom (1956) Selbstdiffusion und Reaktion in Oxyd- und Spinellsysteme. *Z. Physikal. Chem. 6*, 162-177.

Lindsley, D. H. (1965) Iron-titanium oxides. *Carnegie Inst. Washington Year Book 64*, 144-148.

________, S. E. Kesson, M. J. Hartzman, and M. K. Cushman (1974) The stability of armalcolite: Experimental studies in the system MgO-Fe-Ti-O. *Proc. Lunar Sci. Conf. 5th, Suppl. 5, Geochim. Cosmochim. Acta 1*, 521-534.

Manenc, J. (1968) Structure du protoxyde de fer, résultats récents. *Bull. Soc. franc. Mineral. Cristallogr. 91*, 594-599.

Mason, Brian (1943) Mineralogical aspects of the system $FeO-Fe_2O_3-MnO-Mn_2O_3$. *Geol. Fören. Förh. (Stockholm) 65*, 97-180.

Maxwell, L. R., J. S. Smart, and S. Brunauer (1949) Dependence of the intensity of magnetization and the Curie point of certain iron oxides upon the ratio of Fe^{++}/Fe^{+++} (abstr.). *Phys. Rev. 76*, 459-460.

Michel, A. and G. Chaudron (1935) Étude du sesquioxyde de fer cubique stabilisé. *C. R. Acad. Sci. (Paris) 201*, 1191-1193.

Morin, F. J. (1950) Magnetic susceptibility of αFe_2O_3 and αFe_2O_3 with added titanium. *Phys. Rev. 78*, 819-820.

Moriya, Tôru (1960) Anisotropic superexchange interaction and weak ferromagnetism. *Phys. Rev. 120*, 91-98.

Nagata, T. (1961) *Rock Magnetism*, revised ed. Maruzen Co., Ltd., Tokyo, 350p.

Néel, L. (1948) Propriétés magnétiques des ferrites; ferrimagnétisme et antiferromagnétisme. *Ann. de Phys. 3*, 137-198.

________ (1949) Essai d'interpretation des propriétés magnétiques du sesquioxyde de fer rhomboédrique. *Ann. de Phys. 4*, 249-268.

________ (1953) Some new results on antiferromagnetism and ferromagnetism. *Rev. Mod. Phys. 25*, 58-63.

________ (1955) Some theoretical aspects of rock magnetism. *Adv. Phys. 4*, 191-243.

Newhouse, W. H. and J. P. Glass (1936) Some physical properties of certain iron oxides. *Econ. Geol. 31*, 699-711.

Nichols, G. D. (1955) The mineralogy of rock magnetism. *Adv. Phys. 4*, 113-190.

Niggli, P. (1921) Aüszuge. *Z. Kristallogr. 56*: Rutil, p. 119, Anatas, p. 120.

Nishikawa, S. (1915) The structure of some crystals of the spinel group. *Proc. Math. Phys. Soc. Tokyo 8*, 199-209.

O'Reilly, W. and S. K. Banerjee (1965) Cation distribution in titanomagnetites $(1-x)Fe_3O_4-xFe_2TiO_4$. *Phys. Lett. 17*, 237-238.

________, and ________ (1966) Oxidation of titanomagnetites and self-reversal. *Nature (London) 211*, 26-28.

________, and ________ (1967) The mechanism of oxidation in titanomagnetites: A magnetic study. *Mineral. Mag. 36*, 29-37.

________, and P. W. Readman (1971) The preparation and unmixing of cation deficient titanomagnetites. *Z. Geophys. 37*, 321-327.

Parker, R. L. (1923) Zur Kristallographie von Anatas und Rutile. II. Die Anatasstruktur. *Z. Kristallogr. 59*, 1-54.

Pauling, L. (1928) The sizes of ions and their influence on the properties of salt-like compounds. *Z. Kristallogr. 67*, 377-404.

________ (1930) The crystal structure of pseudobrookite. *Z. Kristallogr. 73*, 97-112.

________, and S. B. Hendricks (1925) The crystal structures of hematite and corundum. *J. Am. Chem. Soc. 47*, 781-790.

________, and J. H. Sturdivant (1928) The crystal structure of brookite. *Z. Kristallogr. 68*, 239-256.

Plücker, J. (1848) Uber Intensitäts bestimmung der magnetischen und diamagnetischen Kräfte. *Poggendorff's Annalen 74*, 321-379.

Readman, P. W. and W. O'Reilly (1971) Oxidation processes in titanomagnetites. *Z. Geophys. 37*, 329-338.

Robbins, J. (1859) Magnetic peroxide of iron. *Chem. News 1*, 11-12.

Rooksby, H. P. (1961) Oxides and hydorixdes of aluminum and iron. In, G. Brown, ed., Chap. 10, *The X-Ray Identification and Crystal Structures of Clay Minerals*, Jarrold and Sons, Norwich, England.

Rossiter, M. J. and P. T. Clark (1965) Cation distribution in ulvöspinel Fe_2TiO_4. *Nature (London) 207*, 402.

Roth, W. L. (1960) Defects in the crystal and magnetic structures of ferrous oxide. *Acta Crystallogr. 13*, 140-149.

Ruby, S. L. and G. Shirane (1961) Magnetic anomaly in $FeTiO_3$-αFe_2O_3 system by Mössbauer effect. *Phys. Rev. 123*, 1239-1240.

Sasajima, S. (1961) γ-titanohematites in nature and the role they play in rock magnetism. *J. Geomag. Geoelectricity 12*, 190-215.

Schrader, R. and G. Büttner (1963) Untersuchungen über γ-Eisen (III)-oxid. *Z. anorg. allgem. Chemie 320*, 205-219.

Schult, A. (1968) Self-reversal of magnetization and chemical composition of titanomagnetites in basalts. *Earth Planet. Sci. Lett. 4*, 57-63.

Shannon, R. D. and C. T. Prewitt (1968) Effective ionic radii in oxides and fluorides. *Acta Crystallogr. B25*, 925-946.

Shirane, G., S. J. Pickart, R. Nathans, and Y. Ishikawa (1959) Neutron-diffraction study of antiferromagnetic $FeTiO_3$ and its solid solutions with α-Fe_2O_3. *J. Phys. Chem. Solids 10*, 35-43.

Shirane, G., D. E. Cox, and S. L. Ruby (1962) Mössbauer study of isomer shift, quadruple interaction, and hyperfine field in several oxides containing Fe^{57}. *Phys. Rev. 125*, 1158-1165.

Shull, C. G., W. A. Stauser, and E. O. Wollan (1951a) Neutron diffraction by paramagnetic and antiferromagnetic substances. *Phys. Rev. 83*, 333-345.

________, E. O. Wollan, and W. C. Koehler (1951b) Neutron scattering and polarization by ferromagnetic materials. *Phys. Rev. 84*, 912-921.

Simons, P. Y., and F. Dachille (1967) The structure of TiO_2-II, a high-pressure phase of TiO_2. *Acta Crystallogr. 23*, 334-335.

Smith, T. T. (1916) The magnetic properties of hematite. *Phys. Rev. 8*, 721-737.

Sosman, R. B. and J. C. Hostetter (1916) The oxides of iron. I. Solid solution in the system Fe_2O_3-Fe_3O_4. *J. Am. Chem. Soc. 38*, 807-833.

________, and J. C. Hostetter (1917) The ferrous iron content and magnetic susceptibility of some artificial and natural oxides of iron. *Am. Inst. Mining Engrs. Bull. 126*, 907-943.

Stacey, F. D. and S. K. Banerjee (1974) *The Physical Properties of Rock Magnetism.* Elsevier, Amsterdam, vi + 195p.

Stephenson, A. (1969) The temperature dependent cation distribution in titanomagnetites. *Roy. Astron. Soc., Geophys. J. 18*, 199-210.

________ (1972) Spontaneous magnetization curves and Curie points of spinels containing two types of magnetic ion. *Phil. Mag. 25*, 1213-1232.

Stickler, J. J., S. Kern, A. Wold, and G. S. Heller (1967) Magnetic resonance and susceptibility of several ilmenite powders. *Phys. Rev. 164*, 765-767.

Straumanis, M. E., T. Ejima, and W. J. James (1961) The TiO_2 phase explored by the lattice constant and density method. *Acta Crystallogr. 14*, 493-497.

Swanson, H. E., M. C. Morris, E. H. Evans, and L. Ulmer (1964) Standard x-ray diffraction powder patterns. *U. S. Natl. Bur. Standards Monograph 25*, Sec. 3, 64p.

Swaroof, B. and J. B. Wagner (1967) On the vacancy concentrations of wüstite (FeOx) near the p to n transition. *Trans. Metal. Soc. AIME 239*, 1215-1218.

Syono, Y. (1965) Magnetocrystalline anisotropy and magnetostriction of Fe_3O_4-Fe_2TiO_4 series--with special application to rock magnetism. *Jpn. J. Geophys. 4*, 71-143.

Vallet, P. and P. Raccah (1965) Contribution à l'étude des propriétés thermodynamiques du protoxyde du fer solide. *Mem. Sci. Rev. Metal. 62*, 1-29.

Van Oosterhout, G. W. and C. J. M. Rooijmans (1958) A new superstructure in gamma-ferric oxide. *Nature 181*, 44.

Vegard, L. (1916a) Results of crystal analysis, II. *Phil. Mag. 32*, 65-96.

________ (1916b) Results of crystal analysis, III. *Phil. Mag. 32*, 505-518.

________ (1926) Results of crystal analysis. *Phil. Mag. 1*, 1151-1193.

Verhoogen, J. (1962) Oxidation of iron-titanium oxides in igneous rocks. *J. Geol. 70*, 168-181.

Verwey, E. J. W. (1935a) The crystal structure of γ-Fe_2O_3 and γ-Al_2O_3. *Z. Kristallogr. 91*, 65-69.

________ (1935b) Incomplete atomic arrangement in crystals. *J. Chem. Phys. 3*, 592-593.

Verwey, E. J. W. and J. H. deBoer (1936) Cation arrangement in a few oxides with crystal structures of the spinel type. *Recueil des Travaux Chemiques des Pays-Bas, 4th Ser., 55*, 531-540.

________, P. W. Haayman, and F. C. Romeijn (1947) Physical properties and cation arrangement of oxides with spinel structures. II. Electric conductivity. *J. Chem. Phys. 15*, 181-187.

________, and E. L. Heilmann (1947) Physical properties and cation arrangement of oxides with spinel structures. I. Cation arrangement in spinels. *J. Chem. Phys. 15*, 174-180.

________, and M. G. VanBruggen (1935) Structure of solid solutions of Fe_2O_3 in Mn_3O_4. *Z. Kristallogr. 92*, 136-138.

Walenta, K. (1960) Natürliches Eisen (11)-Oxyd (Wüstit) aus der vulkanischen tuffbreccie von Scharnhausen bei Stuttgart. *Neues Jahrb. Mineral. Mh.*, 151-159.

Weber, H. P. and S. S. Hafner (1971) Vacancy distribution in non-stoichiometric magnetites. *Z. Kristallogr. 133*, 327-340.

Wechsler, B. A., C. T. Prewitt, and J. J. Papike (1976) Chemistry and structure of lunar and synthetic armalcolite. *Earth Planet. Sci. Lett. 29*, 91-103.

Weyl, R. (1959) Präzisionsbestimmung der Kristalstruktur des Brookites, TiO_2. *Z. Kristallogr. 111*, 401-420.

Wyckoff, R. W. G. (1922) The analytical presentation of the results of the theory of space groups. *Carnegie Inst. Washington Publ. 318.*

________ (1965) *Crystal Structures*, 2nd ed. Interscience, New York, Vol. 3.

Zeller, C. and J. Babkine (1965) Mise en évidence d'une loi reliant le paramètre cristalline au chimisme des titanomagnetites. *C. R. Acad. Sci. (Paris) 250*, 1375-1378.

________, J. Hubsch, and J. Bolfa (1967a) Relations entre les propriétés magnétiques et la structure d'une titanomagnétite au cours son oxydation. *C. R. Acad. Sci. B265*, 1034-1036.

________, J. Hubsch, J. C. Reithler, and J. Bolfa (1967b) Determination de la structure spinelle à partir de l'aimantation absolue, des solutions solides $xMe_2TiO_4(1-x)MeFe_2O_4$. *C. R. Acad. Sci. (Paris) B265*, 1335-1338.

EXPERIMENTAL STUDIES of OXIDE MINERALS

Donald H. Lindsley

Chapter 2

INTRODUCTION

Experimental studies of oxide minerals are made for a variety of reasons, including: ore processing, effects on magnetic properties, and determining the conditions under which the oxide minerals formed. This review will concentrate mainly on the last topic. The oxide minerals in a rock can, in many cases, be considered as a "subsystem" of the rock, more or less independent of the silicates or other minerals present. The advantage of this approach is that the oxide "subsystem" tends to be chemically simpler than the whole rock and thus is easier to handle in both experiments and theoretical analysis. Most experiments to date have followed this approach. But of course the oxides cannot be completely independent of the rock containing them, and many recent studies have been made on either whole rocks or chemically complex synthetic systems. The electron microprobe, which permits chemical analysis of very small grains, has made this latter approach feasible.

It is important to bear in mind that many experimental studies that appear to report equilibrium phase relations are in fact *synthesis* studies--that is, the reported phase or phases were grown from oxide mixes or other materials of relatively high free energy. Sometimes synthesis experiments produce equilibrium assemblages, but altogether too often they do not. The reader is urged to treat such results with caution and to rely on phase diagrams that have been verified by *reversed reactions*, that is, where phase or assemblage *A* is converted to *B* and then *B* is converted back to *A*.

Control of experimental conditions

Since a major purpose of the experiments is to reproduce or otherwise identify the conditions under which certain minerals have formed (or have been modified), the experimenter must be able to control or monitor a variety of parameters. Temperature and pressure come immediately to mind; and while there are various technical difficulties in controlling them (see, for example, various chapters in Ulmer, 1971, for details), they are relatively simple conceptually and do not need further discussion here.

Oxygen fugacity. Because most oxide minerals contain one or more transition metals that can exist in more than one oxidation state (for example, Fe, Mn, V, Ti, Cr), it is critical that redox conditions be controlled in experiments on them. These conditions can be expressed in various ways such as Eh, but by far the most common parameter used in high-temperature studies is *oxygen fugacity* (f_{O_2}), which to all intents and purposes is numerically equal to the *partial pressure* of oxygen (p_{O_2}) for geologically realistic values. Objections have been raised that the numerical values of f_{O_2} are so low (sometimes 10^{-25} atm or even lower) that the equivalent partial pressures of oxygen are physically unrealistic. Nevertheless, f_{O_2} remains a convenient expression even though the experimenter has to devise various techniques to impose the desired value on his experiments. The reader should always bear in mind that a particular value of f_{O_2} is of little meaning unless a temperature is also specified. Thus an f_{O_2} value of 10^{-15} atm would be strongly oxidizing for most minerals at 500°C, but would be very reducing at 1000°C (Fig. L-20). Values of f_{O_2} are often expressed in logarithms (base 10).

Rarely in geological environments are oxygen fugacities greater than that of pure air ($f_{O_2} \approx p_{O_2} = 0.21$ atm) (Fig. L-20) encountered. Indeed, for most processes other than weathering and surficial oxidation of volcanic rocks, the appropriate f_{O_2} values are much lower (see, for example, Sato and Wright, 1966; Sato, 1972; Sato *et al.*, 1973). To a limited extent one can lower the f_{O_2} simply by decreasing the air pressure and assuming that the proportion of oxygen remains at 0.21. But this method is restricted by the limitations of the vacuum system: $p_{O_2} \approx 10^{-6}$ to 10^{-8} atm for good mechanical pumps; and while diffusion pumps can achieve lower pressures, it is unlikely that the proportion of oxygen remains that of air. Accordingly, other methods are necessary.

Most experiments on oxide minerals have used one of two techniques for controlling f_{O_2}: mixtures of reactive gases (usually but not necessarily at a total pressure (p_{Tot}) of one atm) or solid-solid oxygen "buffers"--assemblages of phases that impose a fixed f_{O_2} on the system. In the gas-mixture technique (see for example, Nafziger *et al.*, 1971), a relatively oxidizing gas such as CO_2 or H_2O (see Fig. L-20) is mixed with a reducing gas (CO or H_2) and passed through the furnace in the vicinity of the oxide sample. The gases react, producing various species, including oxygen. The amount of oxygen (and hence its fugacity) is controlled by the mixing ratio of the gases and the temperature of the furnace. Deines *et al.* (1974) tabulate oxygen fugacity values for a wide range of C-H-O gas mixtures and temperatures. The oxygen buffer

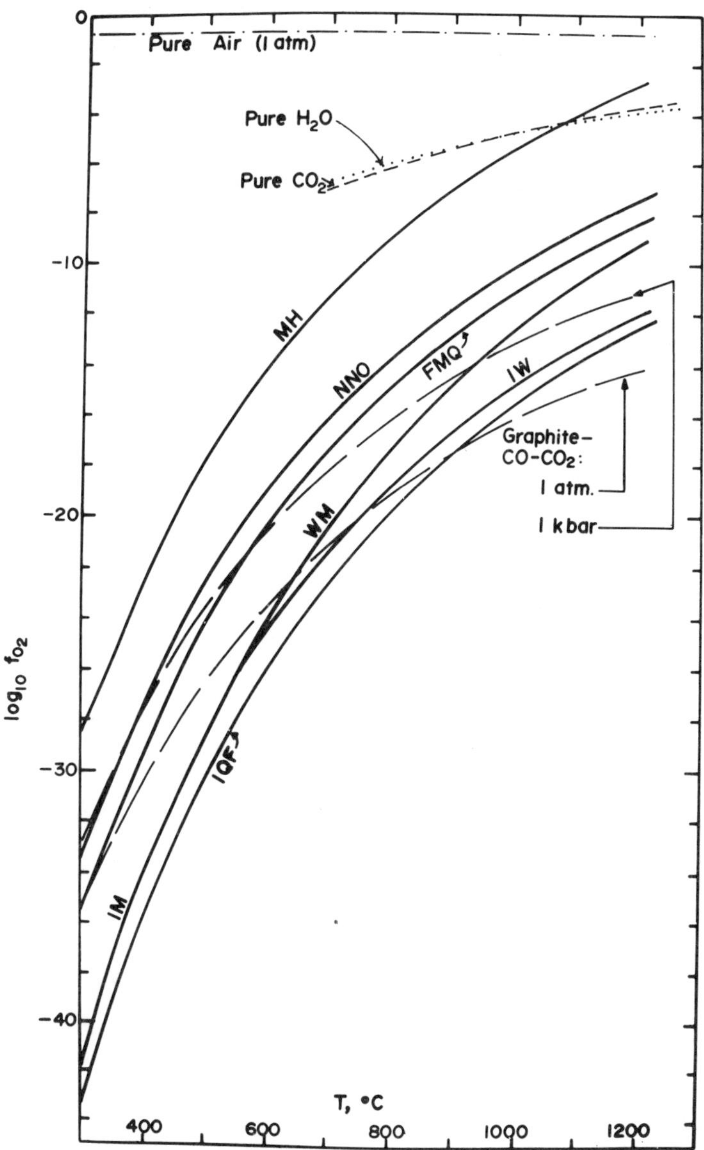

Figure L-20. Oxygen fugacities (f_{O_2}, in atm) as functions of temperature for selected solid-solid buffers, pure air, pure CO_2, pure H_2O, and the assemblage graphite-CO-CO_2. Pressures are 1 atm except where noted. Pressure effects are very small for solid-solid buffers, but are appreciable for the gas mixtures; increasing total pressure increases f_{O_2}. Abbreviations for solid-solid buffers: MH, magnetite-hematite; NNO, nickel-nickel oxide; FMQ, fayalite-magnetite-quartz; WM, wüstite-magnetite; IM, iron-magnetite; IW, iron-wüstite, IQF, iron-quartz-fayalite.

technique uses assemblages of solid phases--such as magnetite + hematite (MH) or fayalite + magnetite + quartz (FMQ)--that contain a transition element in two different oxidation states. A further requirement of such a buffer assemblage is that the number of solid phases equals the number of components--that is, the minimum number of chemical entities required to describe the assemblage. Thus the MH buffer can be described in terms of two components (for example, Fe-O, or $FeO-Fe_2O_3$), whereas the FMQ buffer requires three ($Fe-O-SiO_2$). Application of the Gibbs Phase Rule shows that the variance of such assemblages is two; and since two degrees of freedom are lost when the experimenter fixes temperature and pressure, all other intensive variables--including oxygen fugacity--are fixed so long as all phases of the buffer assemblage are present. The use of such buffers is reviewed by Huebner (1971); typical buffer curves appear in Figure L-20.

Container problems. Perhaps surprisingly to the non-experimentalist, relatively few materials can serve as satisfactory containers for oxide minerals during experiments. Many commonly used containers react with the sample, thereby changing its composition. For example, at all but very high oxygen fugacities, platinum reacts with iron-bearing oxides and silicates (see, for example, Merrill and Wyllie, 1973):

$$FeO \text{ (in oxide or silicate)} + Pt = \text{Fe-Pt alloy} + \tfrac{1}{2}O_2$$

This reaction not only depletes the remaining sample in iron, it also releases oxygen which must be removed or buffered to prevent an unwanted increase in f_{O_2}. There is not room here to discuss the advantages and drawbacks of various container materials that are used; but the reader is cautioned that some experimental studies that appear useful are in fact of dubious validity because inappropriate containers were used. Among the containers that have been found most useful are silver and silver-rich Ag-Pd alloys (Muan, 1963) at relatively low temperatures; pure iron at very low oxygen fugacities; loops of very fine platinum wire (Ribaud and Muan, 1962); and Fe-Pt alloys of composition chosen to be inert with respect to the sample (Huebner, 1973).

Experiments at very high pressures or high temperatures

I have arbitrarily chosen to limit this summary mainly to experiments made at pressures below 10 kbar; that is, corresponding approximately to the earth's crust. For examples of the special problems and properties of oxides at very high pressures, see Drickamer *et al.* (1969) and Mao (1974). I have also placed the emphasis on studies below 1200°C. This choice, which eliminates many studies of high quality made at higher temperatures, reflects a prejudice that the data obtained at lower temperatures are more directly applicable to most geologic problems.

Minerals and phases considered

Names, ideal formulas, and abbreviations used in the text and figures are given in Table L-12 for the minerals and phases considered in this chapter.

Table L-12. Names, ideal formulas, and abbreviations of oxide minerals and phases discussed in this chapter.

Name	Ideal Formula	Abbreviation
Spinels		
Magnetite	Fe_3O_4	Mt
Maghemite	γFe_2O_3	Mgh
Ulvospinel	Fe_2TiO_4	Usp
Titanomaghemite	$(Fe^{2+},Fe^{3+},Ti^{4+},\square)_3O_4$	Ti-Mgh
Hercynite	$FeAl_2O_4$	Hc
Chromite	$FeCr_2O_4$	Chr
Picrochromite	$MgCr_2O_4$	--
Rhombohedral series		
Corundum	Al_2O_3	Cor
Hematite	αFe_2O_3	Hem
Ilmenite	$FeTiO_3$	Ilm
Geikielite	$MgTiO_3$	Geik
Orthorhombic series		
Armalcolite	$Fe_{0.5}Mg_{0.5}Ti_2O_5$	Arm
Karrooite	$MgTi_2O_5$	Kar
Ferropseudobrookite	$FeTi_2O_5$	Fpb
Pseudobrookite	Fe_2TiO_5	Psb
Others		
Metallic iron	Fe	Fe,I
Wüstite	$Fe_{1-x}O$	Wüs
Rutile	TiO_2	Rut
Anatase	TiO_2	--
Brookite	TiO_2	--
Liquid	--	Liq,L

Mt_{ss}, solid solutions close to Fe_3O_4; $(Usp\text{-}Mt)_{ss}$, solid solutions on or near the Fe_2TiO_4-Fe_3O_4 join; and similarly for other phases.

Many oxide minerals fall in or close to the compositional triangle Fe-Ti-O; these include magnetite, hematite, ilmenite, and rutile as well as a number of others.

Fe-O join

The Fe-O join (Fig. L-21) includes magnetite and hematite; metallic Fe and wüstite (both known as minerals but rare); and maghemite, which does not appear on the equilibrium diagram because it is always metastable with respect to hematite. Figure L-21 has been constructed from various sources, the most important being Darken and Gurry (1945,1946), Greig *et al.* (1935), Phillips and Muan (1960), and Crouch *et al.* (1971). It should be noted that f_{O_2} (and p_{O_2}) vary greatly over Figure L-21, ranging from less than 10^{-30} atm in the lower

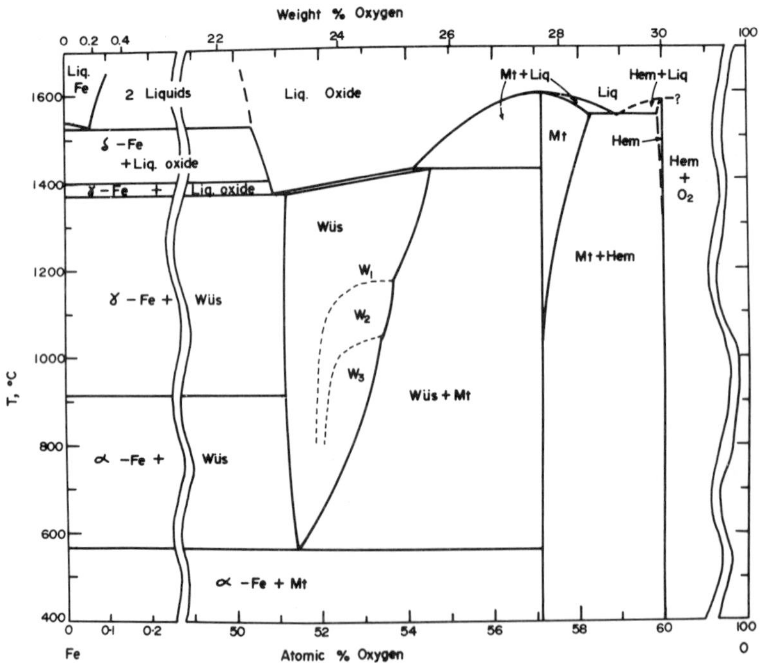

Figure L-21. Phase diagram for the join Fe-O at low pressures. Note breaks in the compositional scale and also the expanded scale near Fe. The three wüstites (W_1,W_2,W_3) are controversial, and it is possible that not all exist at high temperature. See Table L-12 for abbreviations.

left to more than 10 atm in the upper right. Muan and Osborn (1965, Fig. 13) show oxygen isobars for part of this diagram.

Wüstite. Wüstite is never stoichiometric at low pressures. It invariably contains some Fe^{3+} plus vacancies, the amount depending on temperature and f_{O_2}. Different ordering schemes of the Fe^{3+} and the vacancies produce three different wüstites in quenched samples; there is fair evidence that at least two and possibly three of these are also present at high temperatures, and thus should appear on the phase diagram (Kleman, 1965; Vallet and Raccah, 1964,1965; Koch, 1967; Koch and Cohen, 1961; Manec, 1968). Swaroof and Wagner (1967) have argued, however, that if such wüstites do exist at high temperatures, they are not separated by first-order transitions.

The rarity of wüstite in nature stems from several causes: It requires low oxygen fugacity and temperatures above 570°C, but most importantly it reacts with SiO_2 to form silicates in all but the most silica-undersaturated environments.

Fe_2O_3 in magnetite. Only at very high temperatures is there appreciable *stable* solubility of Fe_2O_3 in magnetite. Sosman and Hostetter (1916) had reported complete solubility between hematite and magnetite, but their results were disproved by Greig *et al.* (1935). At temperatures below approximately 600°C, non-equilibrium oxidation of magnetite yields metastable, cation-deficient magnetite-maghemite solid solutions whereas equilibrium oxidation produces hematite. For examples of a vast and contentious literature on the subject, see Columbo *et al.* (1964,1965), Feitknecht and Mannweiler (1967), Feitknecht and Gallagher (1970), Heizmann and Baro (1967), Huguenin (1973a,b), Johnson and Merrill (1972), Johnson and Jensen (1974), and Kushiro (1960).

TiO_2

Three polymorphs of TiO_2--anatase, brookite, and rutile--occur in nature, and a fourth, named TiO_2-II, has been synthesized at high pressures (Dachille and Roy, 1962; Bendeliani *et al.*, 1966; McQueen *et al.*, 1967; Jamieson and Olinger, 1968). Of the natural polymorphs probably only rutile is stable, with both anatase and brookite being metastable with respect to it. Anatase and brookite tend to form in low-temperature, hydrothermal environments. Although many studies of the synthesis and inversion of these polymorphs have been made (for example, Beard and Foster, 1967; Czanderna *et al.*, 1958; Dachille *et al.*, 1968,1969; Glemser and Schwartzmann, 1956; Izumi and Fujiki, 1975; Jamieson and Olinger, 1969; Keesman, 1966; Knoll, 1961,1963,1964; Matthews, 1976; Osborn,

1953; Rao, 1961; Rao *et al.*, 1961; and Shannon and Pask, 1969), no phase diagram is given here because most (probably all) published "phase boundaries" between rutile and anatase or brookite reflect kinetics rather than equilibrium. TiO_2-II, on the other hand, may well have a true stability field.

Relations on the join Ti-TiO_2 are highly complex and do not require detailed discussion here. At high temperatures and very low oxygen fugacities, Ti_2O_3 and Ti_3O_5 can form limited solid solutions in ilmenite and ferropseudobrookite, respectively.

FeO-TiO_2 join

Minerals within the FeO-TiO_2 join (Fig. L-22) include ilmenite, ulvöspinel, and ferrospeudobrookite, the last two being rare except as components in solid solution series. Figure L-22 has been constructed from the data of MacChesney

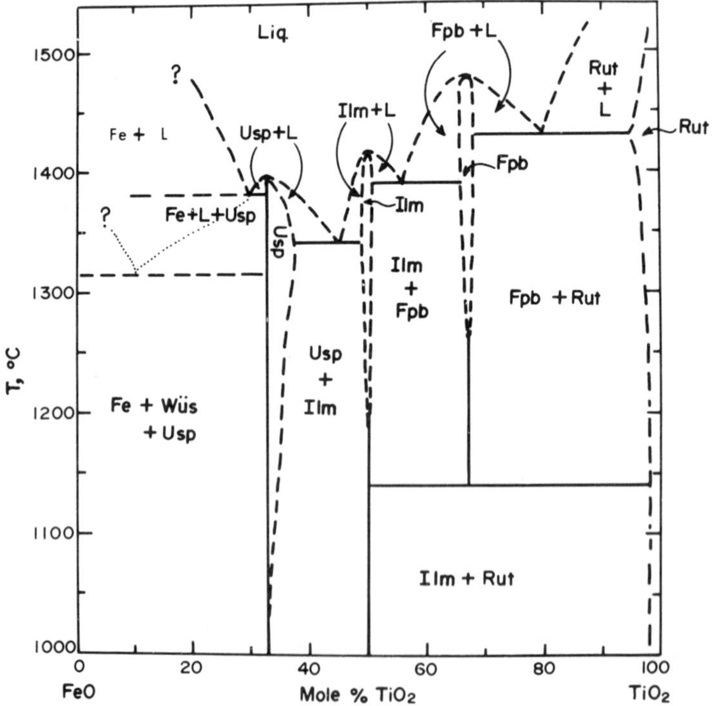

Figure L-22. Approximate phase diagram for the join FeO-TiO_2 at low pressures. Relations near the FeO end are ternary because of the instability of pure FeO. Abbreviations as in Table L-12.

and Muan (1960) and of Taylor (1963,1964), with the low-temperature instability of ferropseudobrookite from Lindsley (1965). In the presence of metallic Fe, Fe_2TiO_5 takes Ti_3O_5 into solid solution and is thereby stabilizied down to 1075°C (Hartzman and Lindsley, 1973; Saha and Biggar, 1974; Lindsley *et al.*, 1974; Lipin and Muan, 1974; Simons, 1974; see also Fig. L-26b). Its instability at lower temperatures, combined with its high TiO_2 content, make ferropseudobrookite very rare as a mineral.

Fe_2O_3-TiO_2 join

In addition to the end members hematite and rutile, the Fe_2O_3-TiO_2 join (Fig. L-23) includes pseudobrookite. Figure L-23 is based mainly on MacChesney and Muan (1959) and Taylor (1963,1964). Pure Fe_2TiO_5 is unstable below 585°C (Lindsley, 1965).

FeO-Fe_2O_3-TiO_2(-Ti_2O_3) join

Equilibrium phase relations in the join FeO-Fe_2O_3-TiO_2 have been studied at liquidus temperatures (Taylor, 1963), 1300°C (Taylor, 1964), 1200°C (Webster and Bright, 1961) and 1000°C (Schmahl *et al.*, 1960) at 1 atm; and under hydrothermal conditions by Buddington and Lindsley (1964), Lindh (1972), Lindsley (1962,1963,1965,1973), Lindsley and Lindh (1974) and Matsuoka (1971). Other studies include those of Haggerty and Lindsley (1970), Saha and Biggar (1974), Simons (1974) and Taylor *et al.* (1972).

1300°C isotherm. Although the temperature is a bit high relative to most crustal environments, the 1300°C isotherm (Fig. L-24) illustrates several important aspects of mineral relations in the FeO-Fe_2O_3-TiO_2 system. Dominating the diagram are three solid solution series, each important in the mineralogy of the Fe-Ti oxides: the orthorhombic series (Fpb-Psb), the rhombohedral series (Ilm-Hem) and the spinel series (Usp-Mt). At 1300°C, the solid solutions can depart from stoichiometry; that is, the single-phase fields have a measurable width. It is noteworthy that for all three series departures from stoichiometry are in the direction of *cation deficiency* (above the binary joins in Fig. L-24). However, even at 1300°C, the spinel field is not sufficiently wide to account for many natural ilmenite-magnetite intergrowths by a simple process of exsolution. [Magnetite grains containing up to 50% ilmenite lamellae are widespread in igneous and metamorphic rocks and were long interpreted as due to exsolution from $FeTiO_3$-Fe_3O_4 solid solutions. Schmahl *et al.* (1960) reported finding 50% $FeTiO_3$ solid solution in magnetite at 1000°C. Their

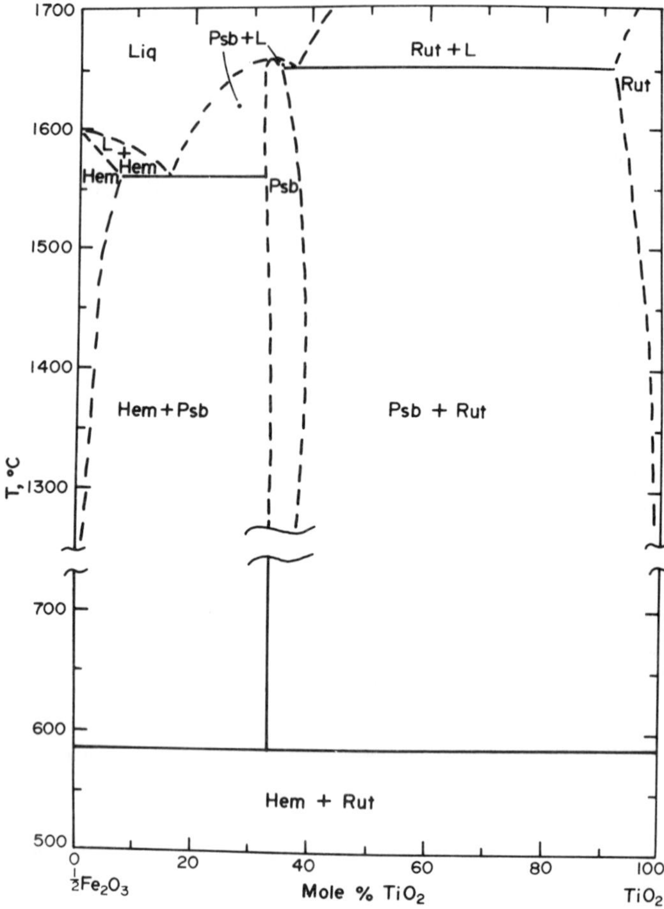

Figure L-23. Approximate phase diagram for the join Fe_2O_3-TiO_2 at low pressures. Oxygen pressures greater than 1 atm are required to prevent reduction of Hem and Psb at high temperatures. Abbreviations are as in Table L-12.

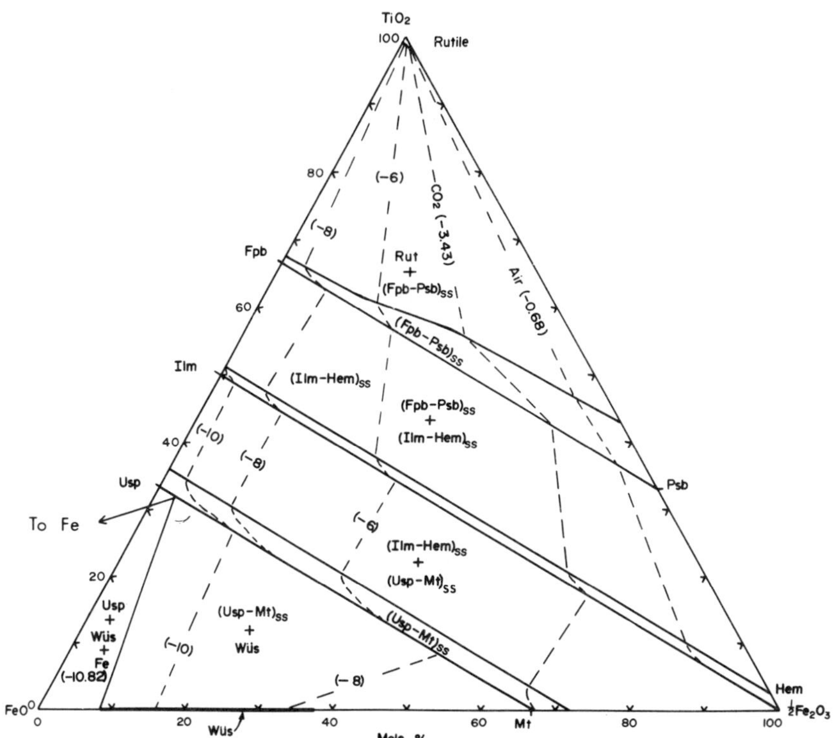

Figure L-24. Phase relations for the system $FeO-Fe_2O_3-TiO_2$ at 1300°C and 1 atm total pressure (Taylor, 1964). Light dashed lines are representative oxygen isobars and are labeled with values of log f_{O_2} in parentheses. Abbreviations as in Table L-12.

findings are incompatible with more recent studies, and it appears that their interpretation of their experimental data was strongly influenced by the mineralogical gospel of the time.]

Between the solid solution series are the two-phase fields $(Usp-Mt)_{ss}$ + $(Ilm-Hem)_{ss}$ and $(Ilm-Hem)_{ss}$ + $(Fpb-Psb)_{ss}$. Tie lines connecting the compositions of the coexisting phases are oxygen isobars. As one would suspect, the oxygen fugacity decreases from right to left in the diagram, that is, as the phases lose Fe_2O_3 and gain FeO. The isobars can and generally do curve within the one-phase fields. At 1300°C, the tie lines between $(Usp-Mt)_{ss}$ and $(Ilm-Hem)_{ss}$ tend to be parallel (except for very Fe_2O_3-rich compositions). At lower temperatures, however, this parallelism is destroyed; in particular, Ilm_{ss}

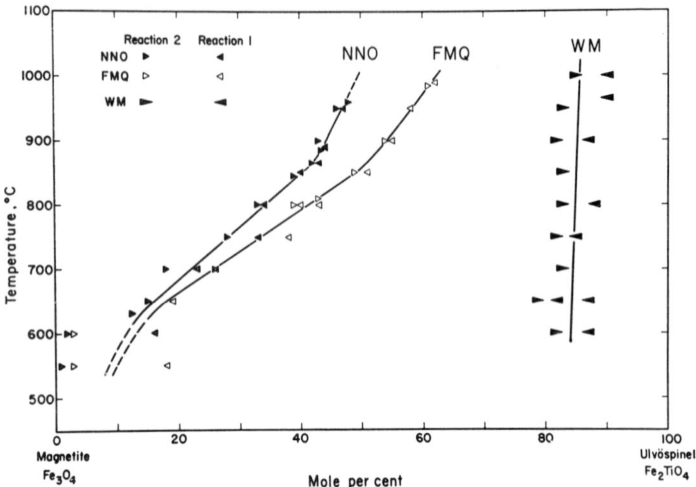

Figure L-25A-C. Compositions of coexisting $(Usp\text{-}Mt)_{ss}$ and $(Ilm\text{-}Hem)_{ss}$. (A) Compositions of $(Usp\text{-}Mt)_{ss}$ coexisting with the $(Ilm\text{-}Hem)_{ss}$ shown in (B) for several buffers. To the left of each curve, single-phase spinels are stable. Spinel compositions to the right of each curve will oxidize to $(Ilm\text{-}Hem)_{ss}$ plus that $(Usp\text{-}Mt)_{ss}$ whose composition is given by the appropriate buffer curve at the temperature of reaction.

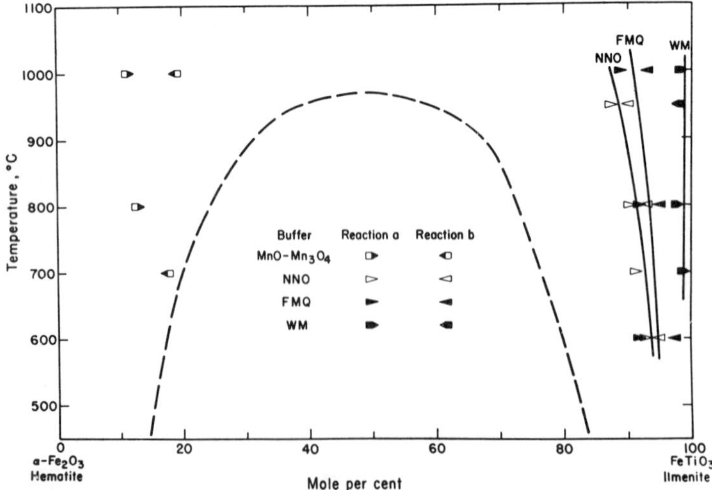

(B) Compositions of Ilm_{ss} coexisting with the $(Usp\text{-}Mt)_{ss}$ shown in (A). To the right of each solid curve, single-phase spinels are stable at the f_{O_2} of the corresponding buffer. Ilmenite compositions to the left of each curve will oxidize to $(Mt\text{-}Usp)_{ss}$ plus that Ilm_{ss} whose composition is given by the appropriate buffer curve at the temperature of reaction. The dashed, hoop-shaped curve is the miscibility gap of Carmichael (1961) now considered to be incorrect.

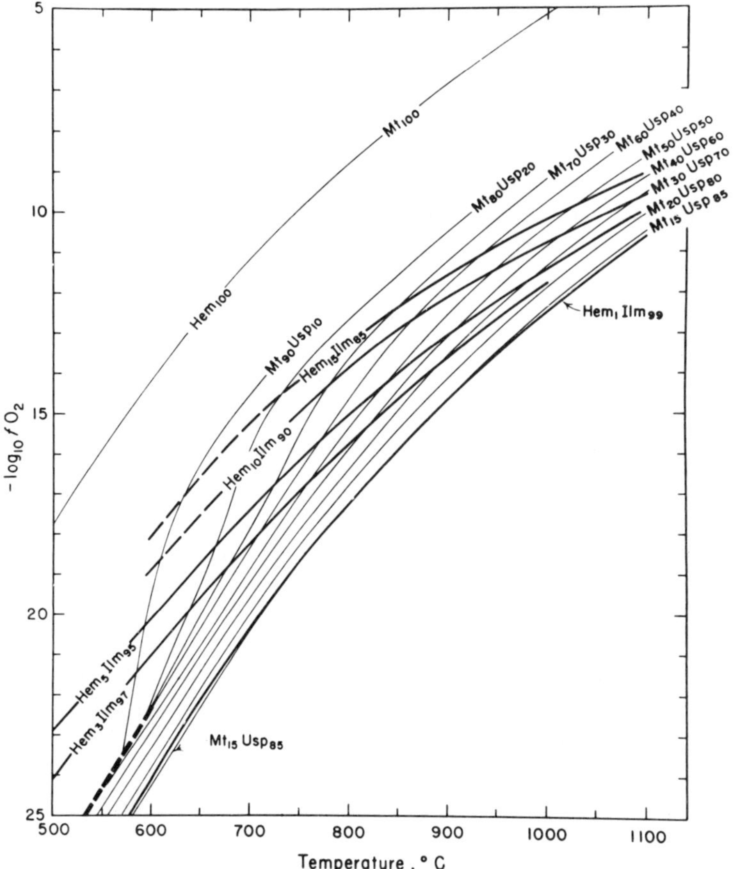

Figure L-25C. Compositions (mole percent) of coexisting $(Usp\text{-}Mt)_{ss}$ [light contours] and $(Ilm\text{-}Hem)_{ss}$ [heavy contours] plotted as functions of T and f_{O_2}.

close to $FeTiO_3$ can coexist with a wide range of $(Usp\text{-}Mt)_{ss}$. The compositions of the coexisting phasea at 600-1000°C and for several buffers are shown in Figures L-25A-B. Both sets of compositions are plotted as functions of temperature and f_{O_2} in Figure L-25C. The intersection of two contours--say $Mt_{60}Usp_{40}$ and $Hem_{10}Ilm_{90}$--defines the unique set of temperature and f_{O_2} at which those two phases can coexist: 825 ± 30°C and $10^{-12.7 \pm 0.5}$ atm. This is the basis of the magnetite-ilmenite geothermometer-oxygen barometer (Buddington and Lindsley, 1964).

The slopes of most of the contours in Figure L-25C are steeper than the buffer curves (Fig. L-20). This means that upon cooling, either along a buffer curve or in the presence of a fluid of constant composition, a given $(Usp\text{-}Mt)_{ss}$ will be oxidized. The resultant texture is that of ilmenomagnetite--lamellae of ilmenite lying in the (111) planes of magnetite--which for years had been considered convincing evidence of exsolution from $FeTiO_3$-Fe_3O_4 solid solutions. Buddington and Lindsley (1964) named this process *oxidation-"exsolution."* Conversely, a Fe_2O_3-rich ilmenite cooled under similar conditions will be *reduced*, yielding lmaellae of Ti-magnetite in the basal (0001) planes of the ilmenite. The process could be termed *reduction-"exsolution."* Taken together, these two processes are expressed by the counterclockwise rotation of the tie lines in Figures L-26C-E. This may also be considered a special type of exchange reaction:

$$\underset{(Usp\text{-}Mt)_{ss}}{(Fe^{2+}+Ti^{4+})} + \underset{Ilm_{ss}}{2(Fe^{3+})} = \underset{Ilm_{ss}}{(Fe^{2+}+Ti^{4+})} + \underset{(Usp\text{-}Mt)_{ss}}{2(Fe^{3+})}$$

provided one keeps in mind that while the bulk Fe/Ti ratio will remain constant during the exchange, the oxygen content and Fe^{2+}/Fe^{3+} probably will not.

Phase relations in the FeO-Fe_2O_3-TiO_2 system at various temperatures are summarized in Figures L-26A-F. At the higher temperatures the component Ti_2O_3 is included in order to show relations under very reducing conditions, data of particular interest for some lunar rocks. At 1300°C (Fig. L-26A) all three solid solution series are complete. At 1130°C (L-26B), pure Fe_2TiO_5 is no longer stable, being replaced by the assemblage Ilm + Rut. The Fpb phase can, however, be stabilized by the addition of small amounts of either Ti_3O_5 (reduction) or Fe_2TiO_5 (oxidation). The instability of $FeTi_2O_5$ heralds the demise of the $(Fpb\text{-}Psb)_{ss}$ series. With decreasing temperature the series retreats towards Fe_2TiO_5 leaving in its stead coexisting Rut + $(Ilm\text{-}Hem)_{ss}$ (Fig. L-26B-E). Simultaneously, the tie lines from $(Usp\text{-}Mt)_{ss}$ to $(Ilm\text{-}Hem)_{ss}$ fan out, presaging the Ilm-Hem miscibility gap. The consolute temperature of this gap is not known, but is probably below 800°C (Fig. L-27). At 600°C (Fig. L-26E) the three-phase assemblage Rut + Ilm_{ss} + Hem_{ss} and Ilm_{ss} + Hem_{ss} + Mt_{ss} are shown. At some lower temperature--not yet determined experimentally but probably below 400°C--Hem_{ss} and Ilm_{ss} can no longer exist: They are replaced by the assemblage Mt + Rut. Below 600°C the $(Usp\text{-}Mt)_{ss}$ series is also interrupted by a miscibility gap.

Ti-Mgh. Titanomaghemites are believed to form at relatively low temperatures (generally below 600°C) by non-equilibrium oxidation of $(Usp\text{-}Mt)_{ss}$. Figure L-26G shows the compositional range possible for these cation-deficient phases. Because of their great importance in rock magnetism, they have been studied intensively. Some references in addition to those given in Chapter 1 of this volume are: Day *et al.* (1970), Johnson and Merrill (1973), Ozima and Ozima (1972), and Sakamoto *et al.* (1968). Most workers argue that water vapor must be present to permit oxidation to Ti-Mgh; in its absence the stable phase(s) $(Ilm\text{-}Hem)_{ss}$ and/or Psb_{ss} tend to form. Other workers claim that water vapor, while helpful, is not essential to the formation of Ti-Mgh.

Reduction of Fe-Ti oxides. Interest in the reduction of Fe-Ti oxides stems from industrial applications as well as from lunar samples. Typical studies include Grey and Ward (1973), Grey *et al.* (1974), Lipin and Muan (1975), McCallister and Taylor (1973), and Saha and Biggar (1974).

Figure L-26A-G. Changes of phase assemblages in the system $FeO-Fe_2O_3-TiO_2(-Ti_2O_3)$ with decreasing temperature. Most of the phase relations are relatively independent of total pressure; the pressures given are those at which the experiments were performed. Abbreviations as in Table L-12 and Figure L-20.

The heavy black line in A-C is the metallic Fe saturation surface, taken mainly from Simons (1974). Dashed tie lines in C-E connect the $(Usp-Mt)_{ss}$ and $(Ilm-Hem)_{ss}$ that coexist at the f_{O_2} of the buffer indicated. Note the retreat of the $(Fpb-Psb)_{ss}$ series toward Psb. Relations in F are mainly inferred from nature. G shows the possible range of titanomaghemites. Ti-Mgh to the right of the $FeTiO_3-Fe_2O_3$ join is relatively rare, but does exist. Oxidation takes place along lines of constant Fe/Ti, which are horizontal in this diagram; two examples (with arrows) are shown.

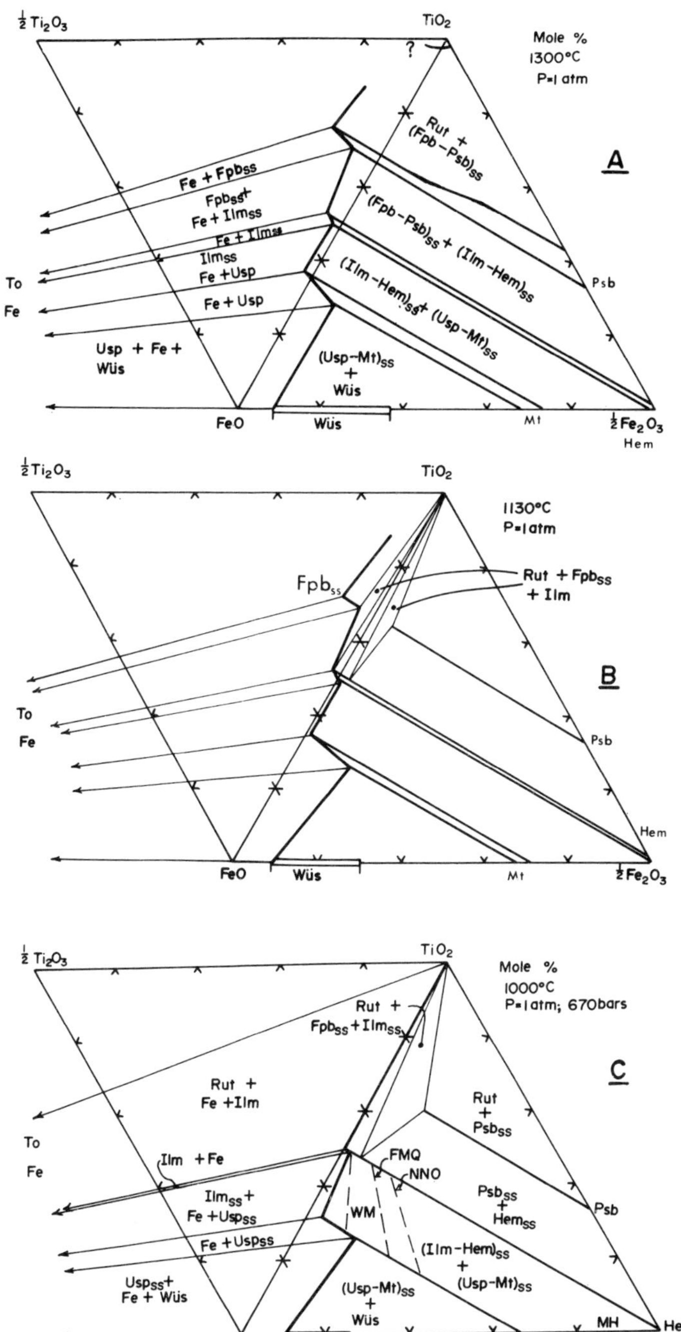

$\frac{1}{2}Ti_2O_3$
TiO_2
Mole %
1300°C
P=1 atm
?
A
Rut + (Fpb-Psb)ss
Fe + Fpbss
Fpbss + Fe + Ilmss
Fe + Ilmss
Ilmss Fe + Usp
(Fpb-Psb)ss + (Ilm-Hem)ss
(Ilm-Hem)ss + (Usp-Mt)ss
To Fe
Fe + Usp
Usp + Fe + Wüs
(Usp-Mt)ss + Wüs
Psb
FeO
Wüs
Mt
$\frac{1}{2}Fe_2O_3$
Hem
$\frac{1}{2}Ti_2O_3$
TiO_2
1130°C
P=1 atm
Fpbss
Rut + Fpbss + Ilm
B
To Fe
Psb
Hem
FeO
Wüs
Mt
$\frac{1}{2}Fe_2O_3$
$\frac{1}{2}Ti_2O_3$
TiO_2
Mole %
1000°C
P=1 atm; 670bars
Rut + Fpbss + Ilmss
C
Rut + Fe + Ilm
Rut + Psbss
To Fe
Ilm + Fe
FMQ
NNO
Ilmss + Fe + Uspss
WM
Psbss + Hemss
Psb
Fe + Uspss
(Ilm-Hem)ss + (Usp-Mt)ss
Uspss + Fe + Wüs
(Usp-Mt)ss + Wüs
MH
Hem
FeO
Wüs
Mt
$\frac{1}{2}Fe_2O_3$

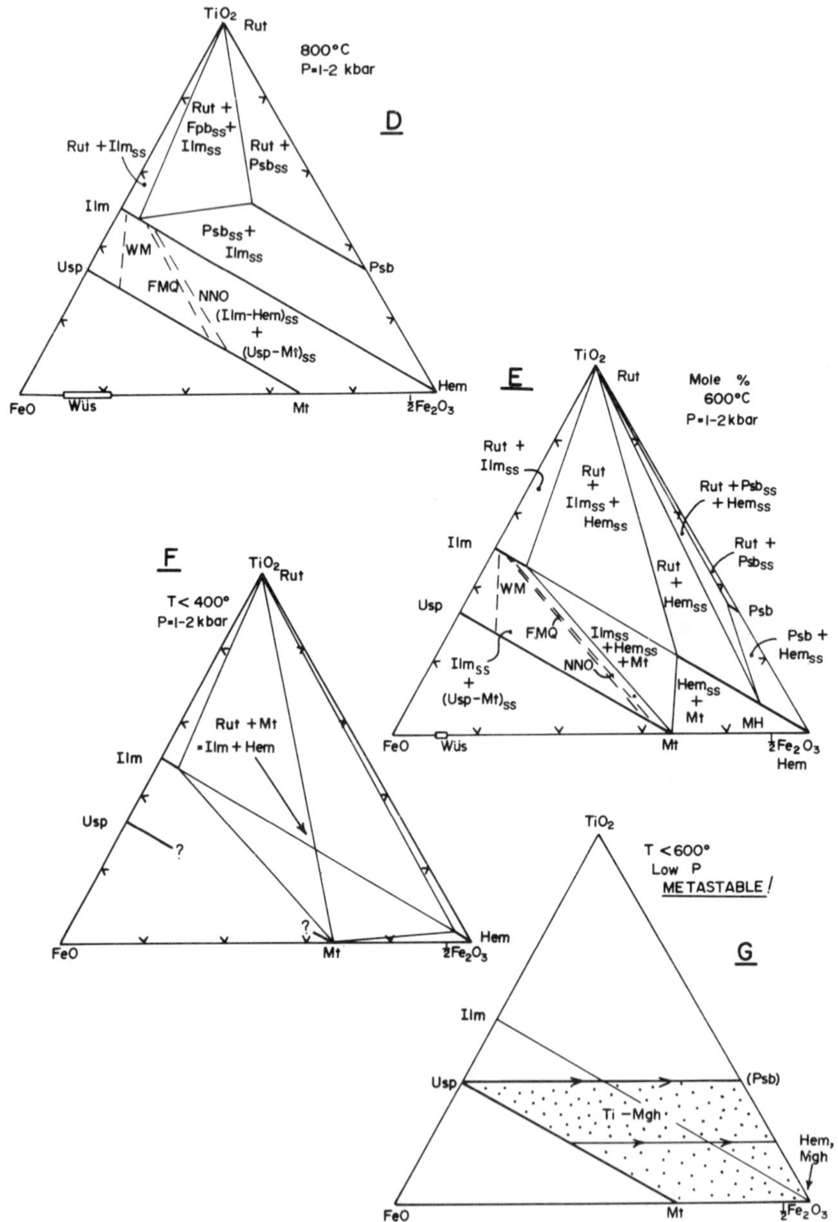
D
TiO2
Rut
800°C
P=1-2 kbar
Rut +
Fpbss+
Ilmss
Rut +
Psbss
Rut + Ilmss
Ilm
Psbss+
Ilmss
WM
Usp
FMQ
NNO
Psb
(Ilm-Hem)ss
+
(Usp-Mt)ss
Hem
FeO
Wüs
Mt
½Fe2O3
E
TiO2
Rut
Mole %
600°C
P=1-2kbar
Rut +
Ilmss
Rut
+
Ilmss +
Hemss
Rut + Psbss
+ Hemss
Ilm
Rut +
Psbss
Rut
+
Hemss
Usp
WM
Psb
FMQ
Ilmss
+Hemss
+Mt
Psb +
Hemss
Ilmss
+
(Usp-Mt)ss
NNO
Hemss
+
Mt
MH
FeO
Wüs
Mt
½Fe2O3
Hem
F
TiO2
Rut
T< 400°
P=1-2 kbar
Rut + Mt
= Ilm + Hem
Ilm
Usp
?
?
FeO
Mt
½Fe2O3
Hem
TiO2
T <600°
Low P
METASTABLE!
G
Ilm
Usp
(Psb)
Ti-Mgh
Hem,
Mgh
FeO
Mt
½Fe2O3

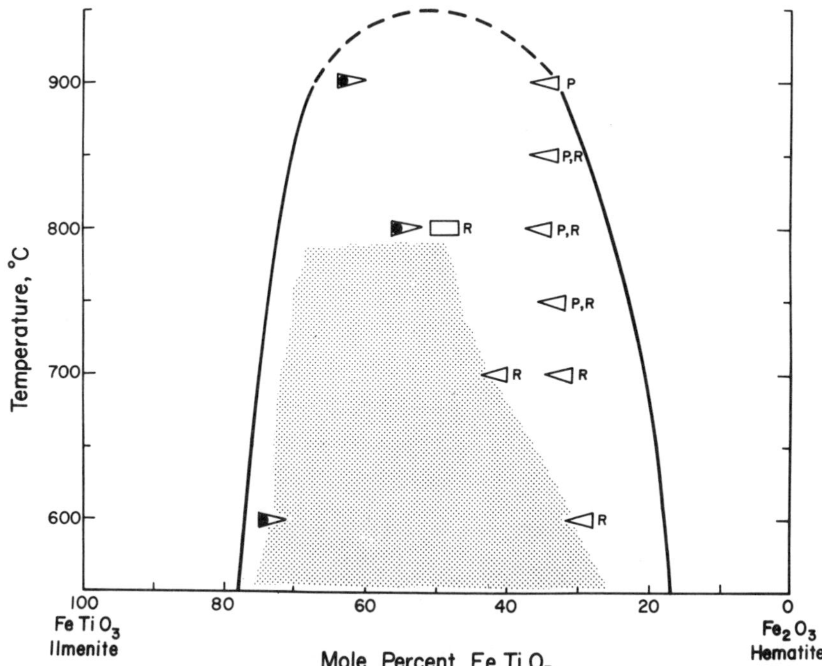

Figure L-27. The join Fe_2O_3-$FeTiO_3$ showing the Hem-Ilm miscibility gap of Carmichael (1961) (hoop-shaped curve) and Lindsley's estimate of the region (stippled area) within which the miscibility gap must lie, based on the homogenization experiments shown (1973). The actual location of the miscibility gap within the stippled area has not yet been determined. P means that Psb_{ss}, R that rutile, was present in the run products.

FE-O-MGO-TIO_2 SYSTEM

The addition of MgO to the Fe-Ti oxides is of interest for two main reasons. Many terrestrial magnetite-ilmenite pairs--especially those from basaltic rocks--contain appreciable amounts of MgO, and successful application of the geothermometer-oxygen barometer technique depends on how this component is considered. In addition, oxide minerals from lunar basalts tend to be rich in MgO: The mineral armalcolite is ideally $Fe_{0.5}Mg_{0.5}Ti_2O_5$, although in fact there is considerable variation in Fe/Mg.

FeO-Fe_2O_3-MgO join

The FeO-Fe_2O_3-MgO join includes the spinel phase magnesioferrite, which is the earliest oxide mineral in some basalts. The join has been studied at high temperatures by Speidel (1967) and by Willshee and White (1967).

FeO-MgO-TiO_2(-Ti_2O_3) join

Many lunar ilmenites and ulvöspinels as well as armalcolites fall near the FeO-MgO-TiO_2 join. In some cases, particularly for armalcolites, the presence of Ti^{3+} requires the component Ti_2O_3 as well. Experimental studies include those of Lind and Housley (1972), Hartzman and Lindsley (1973), and Lindsley *et al.* (1974). Ideal armalcolite ($Fe_{0.5}Mg_{0.5}Ti_2O_5$) is stable down to 1010 ± 20°C; below that temperature it breaks down to Mg-ilmenite + rutile (Fig. L-28). Kesson and Lindsley (1975) showed that armalcolite is stabilized to still lower temperatures by Al^{3+}, Cr^{3+}, and Ti^{3+}.

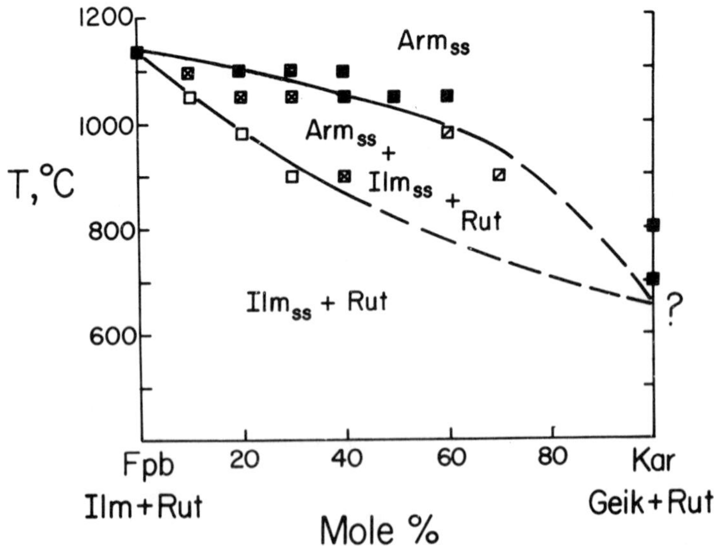

Figure L-28. Phase relations along the join $FeTi_2O_5$-$MgTi_2O_5$ at low pressures (Lindsley *et al.*, 1974), showing the stability field for armalcolite. Only the upper curve is binary; the lower curve marks the intersection of the join with the Ilm_{ss}-Rut edge of Arm_{ss}-Ilm_{ss}-Rut three-phase triangles in the system FeO-MgO-TiO_2. Abbreviations as in Table L-12.

$FeO-Fe_2O_3-MgO-TiO_2$ join

The $FeO-Fe_2O_3-MgO-TiO_2$ join was studied in air by Woerman *et al.* (1969) and at 1160-1300°C and a variety of oxygen fugacities by Speidel (1970). Microprobe analysis of coexisting $(Usp-Mt)_{ss}$ and Ilm in Speidel's experiments indicated that Mg preferentially enters the magnetite, in contrast to most natural pairs in which Mg is enriched in the ilmenite. Pinckney and Lindsley (1976) studied the join hydrothermally at 700-1000°C using reversed exchange equilibria and found that Mg preferentially enters the ilmenite, the preference becoming more pronounced with decreasing temperature. It is not clear whether the discrepancy with Speidel's results is simply a temperature effect or whether his experiments may have failed to reach equilibrium.

$FEO-FE_2O_3-AL_2O_3$ SYSTEM

Portions of the $FeO-Fe_2O_3-Al_2O_3$ system were studied hydrothermally by Turnock and Eugster (1962). They determined the extent of the miscibility gap between magnetite and hercynite and also the composition of magnetite in equilibrium with hematite and corundum (Fig. L-29). These results are particularly applicable to the genesis of emery deposits.

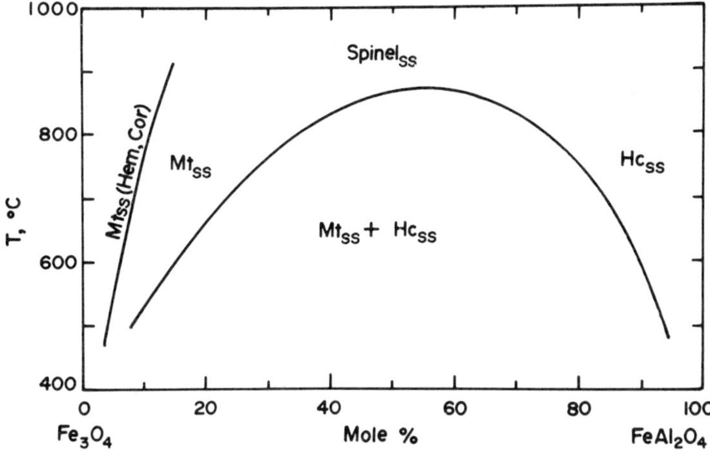

Figure L-29. The join $Fe_3O_4-FeAl_2O_4$ (Turnock and Eugster, 1962) showing the $Mt_{ss}-Hc_{ss}$ miscibility gap. The curve labeled $(Mt_{ss}$(Hem,Cor) shows the composition of Mt_{ss} in equilibrium with hematite and corundum. Pressures ranged from 1-4 kbar.

Chromites are ideally solid solutions between $FeCr_2O_4$ and $MgCr_2O_4$, but most terrestrial chromites also contain appreciable amounts of $FeAl_2O_4$, $MgAl_2O_4$, Fe_3O_4 and $MgFe_2O_4$. Lunar chromites contain little or no Fe^{3+} but instead contain Fe_2TiO_4 and Mg_2TiO_4. Thus a large number of components is necessary to treat chromites experimentally. To date, many of the limiting subsystems have been studied; these have been recently reviewed by Muan (1975) and need not be covered in detail here. Some examples are presented below.

$FeO-Fe_2O_3-Cr_2O_3$ system

The $FeO-Fe_2O_3-Cr_2O_3$ system (Fig. L-30) was studied at 1300°C by Katsura and Muan (1964). Of interest is the complete solid solution between chromite and magnetite. The extensive cation deficiency shown is possible only at

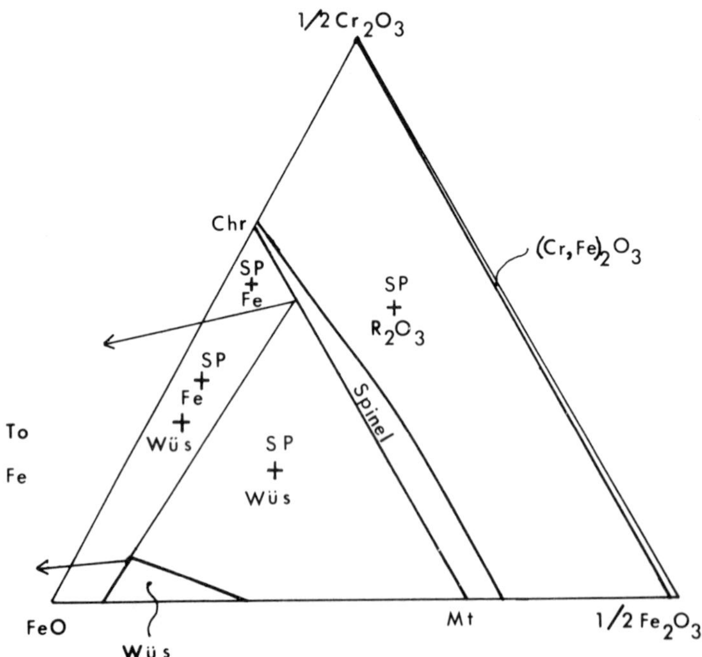

Figure L-30. The join $FeO-Fe_2O_3-Cr_2O_3$ at 1300°C and 1 atm total pressure (Katsura and Muan, 1964) with f_{O_2} ranging from 0.21 to $10^{-10.82}$ atm. SP means $(Chr-Mt)_{ss}$; R_2O_3 means $(Cr,Fe)_2O_3$ solid solutions. Other abbreviations are as in Table L-12.

rather high values of f_{O_2} (approximately 10^{-2}-10^{-4} atm) and is unlikely to be important in natural chromites.

<u>$MgAl_2O_4$-Mg_2TiO_4-$MgCr_2O_4$ system</u>

There is complete solid solution between $MgAl_2O_4$ and $MgCr_2O_4$ and between $MgCr_2O_4$ and Mg_2TiO_4, but a miscibility gap between $MgAl_2O_4$ and Mg_2TiO_4 (Muan *et al.*, 1972). The miscibility gap extends further into the ternary system with decreasing temperature (Fig. L-31). This system serves as a model for lunar spinels: In some samples Al,Cr-rich spinels coexist with ulvöspinels, although of course the effects of FeO-bearing components must be considered for detailed interpretations.

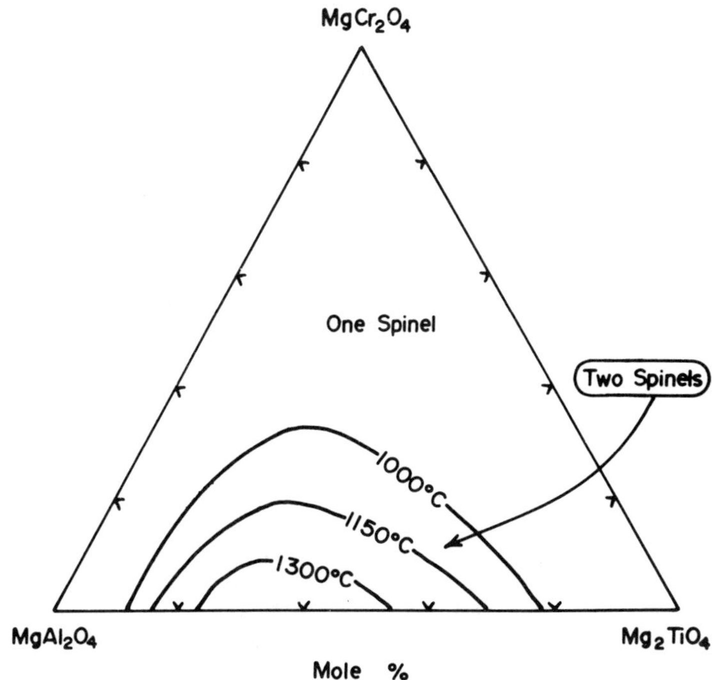

Figure L-31. The join $MgAl_2O_4$-Mg_2TiO_4-$MgCr_2O_4$, showing the ternary miscibility gap at three temperatures (after Muan *et al.*, 1972).

$FeCr_2O_4$-Fe_3O_4-$FeAl_2O_4$ system

Studies at low temperatures are needed to aid in interpreting the metamorphism of chromites. Cremer (1969) studied the $FeCr_2O_4$-Fe_3O_4-$FeAl_2O_4$ system at 500-1000° and reported an extensive ternary miscibility gap. Unfortunately, his diagram appears to be based on synthesis experiments only, and it is unclear whether equilibrium was attained. For example, his miscibility gap for Fe_3O_4-$FeAl_2O_4$ is less extensive than that determined by Turnock and Eugster (1962) using reversed experiments. It is recommended that Cremer's results be used with caution, if at all.

REFERENCES

Beard, W. C. and W. R. Foster (1967) High-temperature formation of anatase. *J. Am. Ceram. Soc. 50*, 493.

Bendeliani, N. A., S. V. Popova, and L. F. Vereshchagin (1966) New modification of titanium dioxide obtained at high pressures. *Geochem. Internat. 3*, 387-390.

Buddington, A. F. and D. H. Lindsley (1964) Iron-titanium oxide minerals and synthetic equivalents. *J. Petrol. 5*, 310-357.

Carmichael, C. M. (1961) The magnetic properties of ilmenite-haematite crystals. *Proc. Roy. Soc. A, 263*, 508-530.

Colombo, U., G. Fagherazzi, F. Gazzarrini, G. Lanzavecchia, and G. Sironi (1964) Mechanisms in the first stage of oxidation of magnetites. *Nature 202*, 175-176.

________, F. Gazzarrini, G. Lanzavecchia, and G. Sironi (1965) Magnetite oxidation: a proposed mechanism. *Science 147*, 1033.

Cremer, V. (1969) Die Mischkristallbildung im System Chromit-Magnetit-Hercynit zwischen 1000°C und 500°C. *Neues Jahrb. Mineral. Abh. 111*, 184-205.

Crouch, A. G., K. A. Hay, and R. T. Pascoe (1971) Magnetite-hematite-liquid equilibrium conditions at oxygen pressures up to 53 bars. *Nature 234* 132-133.

Czanderna, A. W., C. N. R. Rao, and J. M. Honig (1958) The anatase-rutile transition. Part 1, Kinetics of the transformation of pure anatase. *Trans. Farad. Soc. 54*, 1069-1073.

Dachille, F. and R. Roy (1962) A new high-pressure form of titanium dioxide (abstr.). *Bull. Ceram. Soc. Am. 41*, 225.

________, P. Y. Simons, and R. Roy (1968) Pressure-temperature studies of anatase, brookite, rutile, and TiO_2-II. *Am. Mineral. 53*, 1929-1939.

________, ________, and ________ (1969) Pressure-temperature relations of of anatase, brookite, rutile and TiO_2(II): a reply. *Am. Mineral. 54*, 1481-1482.

Darken, L. S. and R. W. Gurry (1945) The system iron-oxygen. I. The wüstite field and related equilibria. *J. Am. Chem. Soc. 67*, 1398-1412.

________, and ________ (1946) The system iron-oxygen. II. Equilibrium and thermodynamics of liquid oxide and other phases. *J. Am. Chem. Soc. 68*, 798-816.

Day, R., W. O'Reilly, and S. K. Banerjee (1970) Rotational hysteresis study of oxidized basalts. *J. Geophys. Res. 75*, 375-386.

Deines, P., R. H. Nafziger, G. C. Ulmer, and E. Woermann (1974) Temperature-oxygen fugacity tables for selected gas mixtures in the system C-H-O at one atmosphere total pressure. *Bulletin of the Earth and Mineral Sciences Experiment Station No. 88*, The Pennsylvania State University, University Park, Pa., 129p.

Drickamer, H. G., G. K. Lewis, Jr., and S. C. Fung (1969) The oxidation state of iron at high pressure. *Science 163*, 885-890.

Feitknecht, W. and K. J. Gallagher (1970) Mechanisms for the oxidation of Fe_3O_4. *Nature 228*, 548-549.

________, and U. Mannweiter (1967) Der Mechanismus der Umwandlung von γ- zu α-Eisensesquioxyd. *Helv. Chim. Acta 50*, 570-581.

Glemser, O. and E. Schwartzmann (1956) Zur Polymorphie des Titandioxyds. *Angew. Chem. 68*, 791.

Greig, J. W., E. Posnjak, H. E. Merwin, and R. B. Sosman (1935) Equilibrium relationships of Fe_3O_4, Fe_2O_3, and oxygen. *Am. J. Sci. 30*, 239-316.

Grey, I. E., A. F. Reid and D. G. Jones (1974) Reaction sequences in the reduction of ilmenite: 4-interpretation in terms of the Fe-Ti-O and Fe-Mn-Ti-O phase diagrams. *Trans. Inst. Mining Mineral. 83*, C105-111.

________, and J. Ward (1973) An X-ray and Mössbauer study of the $FeTi_2O_5$-Ti_3O_5 system. *J. Solid State Chem. 7*, 300-307.

Haggerty, S. E. and D. H. Lindsley (1970) Stability of the pseudobrookite (Fe_2TiO_5)-ferropseudobrookite ($FeTi_2O_5$) series. *Carnegie Inst. Washington, Year Book 68*, 247-249.

Hartzman, M. J. and D. H. Lindsley (1973) The armalcolite join ($FeTi_2O_5$-$MgTi_2O_5$) with and without excess Fe°: Indirect evidence for Ti^{+3} on the moon (abstr.). *Geol. Soc. Am. Abstr. with Progr., 1973 Ann. Mtgs.*, 653-654.

Heizmann, J. J. and R. Baro (1967) Relations topotaxiques entre des cristaux naturels de magnétite Fe_3O_4 et l'hématite Fe_2O_3 qui en est issue par oxydation chimique. *C. R. Acad. Sci., Paris, 265, sér. D*, 777.

Huebner, J. S. (1971) Buffering techniques for hydrostatic systems at elevated pressures. In, G. C. Ulmer, Ed., *Research Techniques for High Pressure and High Temperature*, p. 123-177. New York: Springer-Verlag, 367p.

________ (1973) Experimental control of wüstite activity and mole fraction. *Geol. Soc. Am. Abstr. with Progr., 5, 1973 Ann. Mtgs.*, 676-677.

Huguenin, R. L. (1973a) Photostimulated oxidation of magnetite, 1, kinetics and alteration phase identification. *J. Geophys. Res. 78*, 8481-8493.

________ (1973b) Photostimulated oxidation of magnetite, 2, mechanism. *J. Geophys. Res. 78*, 8495-8506.

Izumi, F. and Y. Fujiki (1975) Hydrothermal growth of anatase (TiO_2) crystals. *Chem. Lett. 1975*, 77-78.

Jamieson, J. C. and B. Olinger (1968) High-pressure polymorphism of titanium dioxide. *Science 161*, 893-895.

________, and ________ (1969) Pressure-temperature studies of anatase, brookite, rutile, and TiO_2(II): A discussion. *Am. Mineral. 54*, 1477-1481.

Johnson, H. P. and S. D. Jensen (1974) High temperature oxidation of magnetite to maghemite (abstr.). *EOS (Am. Geophys. Union Trans.) 55*, 233.

________, and R. T. Merrill (1972) Magnetic and mineralogical changes associated with low-temperature oxidation of magnetite. *J. Geophys. Res. 77*. 334-341.

Katsura, T. and A. Muan (1964) Experimental study of equilibria in the system $FeO-Fe_2O_3-Cr_2O_3$ at 1300°C. *Trans. Am. Inst. Mining Metal. Engr.* *230*, 77-84.

Keesman, I. (1966) Zur hydrothermalen Synthese von Brookit. *Z. anorg. u. allgem. Chem.* *346*, 30-43.

Kesson, S. E. and D. H. Lindsley (1975) The effects of Al^{3+}, Cr^{3+}, and Ti^{3+} on the stability of armalcolite. *Proc. Lunar Sci. Conf. 6th*, 911-920.

Kleman, M. (1965) Propriétés thermodynamiques do protoxyde de fer sous forme solide. Application de résultats experimentaux au tracé du diagramme d' équilibre. *Mém. Sci. Rev. Metal.* *62*, 457-469.

Knoll. H. (1961) Zur Bildung von Brookit. *Naturwiss.* *48*, 601.

________ (1963) Umwandlung von Anatas in Brookit. *Naturwiss.* *50*, 546.

________ (1964) Darstellung von Brookit. *Angew. Chem.* *76*, 592.

Koch, F. B. (1967) Ordering in non-stoichiometric Fe_xO. Ph.D. Dissertation, Northwestern University (Univ. Microfilms 67-15, 266).

________, and J. B. Cohen (1969) The defect structure of $Fe_{1-x}O$. *Acta Crystallogr.* *B25*, 275-287.

Kushiro, Ikuo (1960) γ-α transition in Fe_2O_3 with pressure. *J. Geomag. & Geoelectr.* *11*, 148-151.

Lind, M. D. and R. M. Housley (1972) Crystallization studies of lunar igneous rocks: crystal structures of synthetic armalcolite. *Science* *175*, 521-523.

Lindh, A. (1972) A hydrothermal investigation of the system FeO, Fe_2O_3, TiO_2. *Lithos* *5*, 325-343.

Lindsley, D. H. (1962) Investigations in the system $FeO-Fe_2O_3-TiO_2$. *Carnegie Inst. of Washington, Year Book* *61*, 100-106.

________ (1963) Fe-Ti oxides in rocks as thermometers and oxygen barometers. *Carnegie Inst. of Washington, Year Book* *62*, 60-66.

________ (1965) Iron-titanium oxides. *Carnegie Inst. of Washington, Year Book* *64*, 144-148.

________ (1973) Delimitation of the hematite-ilmenite miscibility gap. *Geol. Soc. Am. Bull.* *84*, 657-661.

________, S. E. Kesson, M. J. Hartzman, and M. K. Cushman (1974) The stability of armalcolite: experimental studies in the system MgO-Fe-Ti-O. *Proc. Lunar Sci. Conf. 5th*, 521-534.

________, and A. Lindh (1974) A hydrothermal investigation of the system FeO, Fe_2O_3, TiO_2: a discussion with new data. *Lithos* *7*, 65-68.

Lipin, B. R. and A. Muan (1974) Equilibria bearing on the behavior of titanate phases during crystallization of iron silicate melts under strongly reducing conditions. *Proc. Lunar Sci. Conf. 5th*, 535-548.

________, and ________ (1975) Equilibrium relations among iron-titanium oxides in silicate melts: the system $CaMgSi_2O_6$-"FeO"-TiO_2 in equilibrium with metallic iron. *Proc. Lunar Sci. Conf. 6th*, 945-958.

MacChesney, J. B. and A. Muan (1959) Studies in the system iron oxide-titanium oxide. *Am. Mineral.* *44*, 926-945.

________, and ________ (1960) The system iron oxide-TiO_2-SiO_2 in air. *J. Am. Ceram. Soc.* *43*, 586-591.

Manenc, J. (1968) Structure du protoxyde de fer, résultats récents. *Bull. Soc. Franc. Mineral. Cristallogr.* *91*, 594-599.

Mao, H. K. (1974) A discussion of the iron oxides at high pressure with implications for the chemical and thermal evolution of the earth. *Carnegie Inst. of Washington, Year Book 73*, 510-518.

Matsuoka, K. (1971) Syntheses of iron-titanium oxides under hydrothermal conditions. *Bull. Chem. Soc. Jpn. 44*, 719-722.

Matthews, A. (1976) The crystallization of anatase and rutile from amorphous titanium dioxide under hydrothermal conditions. *Am. Mineral. 61*, 419-424.

McCallister, R. H. and L. A. Taylor (1973) The kinetics of ulvöspinel reduction: synthetic study and applications to lunar rocks. *Earth Planet. Sci. Lett. 17*, 357-364.

McQueen, R. G., J. C. Jamieson, and S. P. Marsh (1967) Shock-wave compression and x-ray studies of titanium dioxide. *Science 155*, 1401-1404.

Merrill, R. B. and P. J. Wyllie (1973) Absorption of iron by platinum capsules in high pressure rock melting experiments. *Am. Mineral. 58*, 16-20.

Muan, A. (1963) Silver-palladium alloys as crucible material in studies of low-melting iron silicates. *Am. Ceram. Soc. Bull. 42*, 344-347.

________ (1975) Phase relations in chromium oxide-containing systems at elevated temperatures. *Geochim. Cosmochim. Acta 39*, 781-802.

________, J. Hauck, and T. Löfall (1972) Equilibrium studies with a bearing on lunar rocks. *Proc. Lunar Sci. Conf. 3rd*, 185-196.

________, and E. F. Osborn (1965) *Phase Equilibria Among Oxides in Steelmaking.* Addison-Wesley Publishing Co., xx + 236p.

Nafziger, R. H., G. C. Ulmer, and E. Woermann (1971) Gaseous buffering for the control of oxygen fugacity at one atmosphere. In, G. C. Ulmer, Ed., *Research Techniques for High Pressure and High Temperature*, p. 9-41. New York: Springer-Verlag, 367p.

Osborn, E. F. (1953) Subsolidus relationships in oxide systems in presence of water at high pressures. *J. Am. Ceram. Soc. 36*, 147-151.

Ozima, M. and M. Ozima (1972) Activation energy of unmixing of titanomaghemite. *Phys. Earth Planet. Inter. 5*, 87-89.

Phillips, B. and A. Muan (1960) Stability relations of iron oxides: phase equilibria in the system Fe_3O_4-Fe_2O_3 at oxygen pressures up to 45 atmospheres. *J. Phys. Chem. 64*, 1451-1453.

Pinckney, L. R. and D. H. Lindsley (1976) Effects of magnesium on iron-titanium oxides. *Geol. Soc. Am. Abstr. with Progr., 1976 Ann. Mtgs.* (in press).

Rao, C. N. R. (1961) Kinetics and thermodynamics of the crystal structure transformation of spectroscopically pure anatase to rutile. *Can. J. Chem. 39*, 498-500.

________, S. R. Yoganarasimhan, and P. A. Faeth (1961) Studies on the brookite-rutile transformation. *Trans. Farad. Soc. 57*, 504-510.

Ribaud, P. V. and A. Muan (1962) Phase equilibria in a part of the system "FeO"-MnO-SiO_2. *Trans. Am. Inst. Mining Metal. Engr. 224*, 27-33.

Saha, P. and G. M. Biggar (1974) Subsolidus reduction equilibria in the system Fe-Ti-O. *Indian J. Earth Sci. 1*, 43-59.

Sakamoto, N., P. I. Ince, and W. O'Reilly (1968) Effect of wet grinding on oxidation of titanomagnetites. *Geophys. J., Roy. Astron. Soc. 15*, 509-515.

Sato, M. (1972) Intrinsic oxygen fugacities of iron-bearing oxide and silicate minerals under low total pressure. In, *Studies in Mineralogy and Precambrian Geology, Geol. Soc. Am. Mem. 135*, 289-307.

________, N. L. Hickling, and J. E. McLane (1973) Oxygen fugacity values of Apollo 12, 14, and 15 lunar samples and reduced state of lunar magmas. *Proc. Lunar Sci. Conf. 4th*, 1061-1079.

________, and T. L. Wright (1966) Oxygen fugacities directly measured in magmatic gases. *Science 153*, 1103-1105.

Schmahl, N. G., B. Frisch, and E. Gargartner (1960) Zur Kenntnis der Phasenverhaltnisse im System Fe-Ti-O bei 1000°C. *Z. anorg. u. allgem. Chem. 305*, 40-54.

Shannon, R. D. and J. A. Pask (1965) Kinetics of the anatase-rutile transformation. *J. Am. Ceram. Soc. 48*, 391-398.

Simons, B. (1974) *Zusammensetzung und Phasenbreiten der Fe-Ti-Oxide im Gleichgewicht mit metallischem Eisen.* Diplomarbeit, Institut für Kristallographie, Technische Hochschule Aachem, 104p.

Sosman, R. B. and J. C. Hostetter (1916) The oxides of iron. I. Solid solution in the system Fe_2O_3-Fe_3O_4. *J. Am. Chem. Soc. 38*, 807-833.

Speidel, D. H. (1967) Phase equilibria in the system MgO-FeO-Fe_2O_3: the 1300°C isothermal section and extrapolations to other temperatures. *J. Am. Ceram. Soc. 50*, 243-248.

________ (1970) Effect of magnesium on the iron-titanium oxides. *Am. J. Sci. 268*, 341-353.

Swaroof, B. and J. B. Wagner (1967) On the vacancy concentrations of wüstite (FeOx) near the p to n transition. *Trans. Metallurg. Soc. AIME 239*, 1215-1218.

Taylor, L. A., J. Williams, and R. H. McCallister (1972) Stability relations of ilmenite and ulvöspinel in the Fe-Ti-O system and application of these data to lunar mineral assemblages. *Earth Planet. Sci. Lett. 16*, 282-288.

Taylor, R. W. (1963) Liquidus temperatures in the system FeO-Fe_2O_3-TiO_2. *J. Am. Ceram. Soc. 46*, 276-279.

________ (1964) Phase equilibria in the system FeO-Fe_2O_3-TiO_2 at 1300°C. *Am. Mineral. 49*, 1016-1030.

Turnock, A. C. and H. P. Eugster (1962) Fe-Al oxides: phase relationships below 1000°C. *J. Petrol. 3*, 533-565.

Ulmer, G. C., Ed. (1971) *Research Techniques for High Pressure and High Temperature.* New York: Springer-Verlag, 367p.

Vallet, P. and P. Raccah (1964) Sur les limites du domaine de la wüstite solide et le diagramme général qui en résulte. *C. R. Acad. Sci., Paris, 258*, 3679-3682.

________, and ________ (1965) Contribution à l'étude des propriétés thermodynamiques du protoxyde du fer solide. *Mem. Sci. Rev. Metal. 62*, 1-29.

Webster, A. H. and N. F. H. Bright (1961) The system iron-titanium-oxygen at 1200°C and oxygen partial pressures between 1 atm and 2 x 10^{-14} atm. *J. Am. Ceram. Soc. 44*, 110-116.

Willshee, J. C. and J. White (1967) An investigation of equilibrium relationships in the system MgO-FeO-Fe_2O_3 up to 1750° in air. *Trans. Brit. Ceram. Soc. 66*, 541-555.

Woermann, E., B. Brežný, and A. Muan (1969) Phase equilibria in the system MgO-iron oxide-TiO in air. *Am. J. Sci. 267A (Schairer Vol.)*, 463-479.

OXIDE MINERALS in METAMORPHIC ROCKS

Douglas Rumble, III

Chapter 3

INTRODUCTION

The oxide minerals occupy an important place in the history of metamorphic petrology for it was in the phase equilibria of these minerals that the capacity of metamorphic rocks to buffer volatile components was first recognized (James and Howland, 1955). Furthermore, the oxide minerals, especially assemblages of magnetite and ilmenite, have played an important role in geothermometric studies of metamorphic rocks (Buddington and Lindsley, 1964). Nowadays, the interest in oxide mineral parageneses remains high because of their utility in helping to deduce the properties of the fluid phase of metamorphism. The advent of the electron microprobe analyzer has contributed to the importance of oxide minerals to present-day metamorphic petrology in a curiously indirect way. It is virtually impossible to estimate the ferric iron content of chemically complex, hydrous silicate minerals from an electron microprobe analysis; thus, in most petrologic studies, information on an important degree of chemical variability is lost. The lost information may be partially recovered, however, by electron microprobe analysis of the oxide minerals because their simple stoichiometry makes possible accurate estimation of ferric iron content.

MINERALOGY

Spinel solid solutions

The term "spinel" is used loosely to refer to the oxide minerals having the formula AB_2O_4. End member components of spinel solid solutions include magnetite (Fe_3O_4), ulvöspinel (Fe_2TiO_4), chromite ($FeCr_2O_4$), spinel *sensu stricto* ($MgAl_2O_4$), gahnite ($ZnAl_2O_4$), franklinite ($ZnFe_2O_4$), and jacobsite ($MnFe_2O_4$) (Table R-1). Because of their wide range in composition, spinel solid solutions accurately reflect the bulk composition of the rocks in which they occur. Pure magnetite is found in banded iron formations (Annersten, 1968), pelitic schists, and quartzites (Rumble, 1973). Solid

Table R-1. Chemical analyses of spinels

	1	2A	2B	3	4	5
SiO_2	0.57	...	...	...	...	0.10
TiO_2	0.63	...	...	0.04	...	0.25
Al_2O_3	0.23	0.6	58.2	60.20	55.4	0.29
Cr_2O_3	0.15	6.6	8.6	0.01	...	...
V_2O_3	...	...	...	...	...	2.83
Fe_2O_3	65.43*	64.2*	5.4*	...	2.42*	65.08*
FeO	31.28*	20.7*	4.2*	32.60**	12.10*	31.34*
MnO	0.58	0.6	0.2	1.2	0.22	0.02
ZnO	...	...	...	0.14	26.9	...
MgO	0.12	6.3	24.3	6.60	2.25	0.04
Totals	98.99	99.00	100.9	100.79	99.29	99.95

*Recalculated electron microprobe analysis.
**Fe as FeO.

1. Magnetite coexisting with Mn-rich ilmenite from olivine-grunerite-garnet-magnetite-graphite-ilmenite rock (Huntington, 1975, p. 63, Table 9).
2. A,B. Coexisting spinels from olivine-orthopyroxene-hornblende-aluminous spinel-iron and chromium spinel-chlorite-talc-serpentine rock (Springer, 1974, p. 177, 178, Tables 8 and 9).
3. Pleonaste from garnet-spinel-hornblende-chlorite rock (Knauer *et al.*, 1974).
4. Zn-rich spinel from quartz-biotite-garnet-spinel-plagioclase-ilmenite rock (Frost, 1973, p. 832, Table 1).
5. V-rich magnetite from hornblende-orthopyroxene-clinopyroxene-plagioclase-ilmenite-magnetite rock (E. F. Stoddard, personal communication, 1976).

Note: See Evans and Frost (1975, p. 966, Table 2) for analyses of chromian spinel.

solutions of the components Fe_3O_4-$FeCr_2O_4$-$MgCr_2O_4$-$MgAl_2O_4$-$FeAl_2O_4$ are present in metamorphosed ultramafic rocks (Evans and Frost, 1975). Pleonaste ($MgAl_2O_4$-$FeAl_2O_4$ solid solution) forms in desilicated hornfels at the contacts of high-temperature intrusive rocks (Stewart, 1942). Gahnite-rich spinels are known from metamorphosed ore bodies (Nemec, 1972).

The extent of solid solution is limited between many of the end members as evidenced by the variety of lamellar exsolution intergrowths of spinel minerals. In metamorphic rocks lamellar intergrowths and coexisting pairs of the following types have been noted: magnetite-pleonaste (Read, 1931; Stewart, 1942; Barker, 1964; Smith, 1965; Meng and Moore, 1972; Nixon *et al.*, 1973); magnetite-ulvöspinel (Simmons *et al.*, 1974); and, in rocks with an appreciable content of Cr_2O_3, Fe-Cr spinels coexisting with Al-rich ones (Springer, 1974). Lamellar intergrowths of gahnite in franklinite host and hetaerolite ($ZnMn_2O_4$) in franklinite have also been reported (Frondel and Klein, 1965). The exsolution lamellae preferentially lie in the cube faces (100) of the host (*cf.* Ramdohr, 1969, p. 902,903, Figs. 535,536).

Complete miscibility exists along the joins Fe_3O_4-$Fe_{0.9}Mg_{0.1}Cr_2O_4$ and $Fe_{0.9}Mg_{0.1}Cr_2O_4$-$Mg_{0.8}Fe_{0.2}Al_2O_4$ at metamorphic temperatures (Evans and Frost, 1975). Complete miscibility probably is present along the join $MgAl_2O_4$-$FeAl_2O_4$-$ZnAl_2O_4$ in metamorphic rocks (*cf.* Winchell, 1941).

The lamellar intergrowths of ferrianilmenite in magnetite (termed ilmeno-magnetite by Buddington *et al.*, 1963) characteristic of igneous rocks are only rarely present in metamorphic rocks. Reported occurrences are either from rocks metamorphosed at sillimanite or higher grade (Buddington and Lindsley, 1964; Annersten, 1968; Dougan, 1974; Simmons *et al.*, 1974) or from partially recrystallized metamorphosed igneous rocks (Uytenbogaardt, 1953; Reed and Morgan, 1971).

Magnetite forms euhedral or subhedral porphyroblasts, notably free of inclusions, in pelites, quartzites, and basic schists (Banno and Kanehira, 1961; Rumble, 1973). Magnetites from weathered outcrops usually show traces of supergene alteration to specular hematite parallel to their octahedral faces and in extreme cases are completely converted to martite pseudomorphs (*cf.* Ramdohr, 1969, p. 907,909, Figs. 539,541a,b).

Hematite-ilmenite solid solutions

Hematite-ilmenite solid solutions are found in all metamorphic zones in a wide variety of rock types. The minerals are chiefly a ternary solid solution of the end members Fe_2O_3, $FeTiO_3$, and $MnTiO_3$ (pyrophanite end member)

(Table R-2) (Evans and Guidotti, 1966; Guidotti, 1970,1974; Ghent and DeVries, 1972). In magnesian rocks, however, such as magnesian hornfels or dolomitic calc-silicate rocks an appreciable content of $MgTiO_3$ (geikielite end member) is noted (Springer, 1974; Cook, 1974). There is limited miscibility at metamorphic temperatures along the join Fe_2O_3-$FeTiO_3$. In consequence, exsolution lamellae of ilmenite in hematite host or lamellae of hematite in ilmenite host are observed (*cf.* Ramdohr, 1969, p. 960,962, Figs. 564,567). The lamellae occur as severely flattened ellipsoids lying parallel to the basal plane of the host. The frequency distribution of lamellae thickness is usually bimodal, the thicker lamellae measuring 5 μm and the thinner ones, 1 μm or less. The lamellae and host are intergrown in nearly coherent fashion with

Table R-2. Chemical analyses of ilmenite-hematite

	1	2	3A	3B	4	5	6
SiO_2	0.19	...	0.15	0.32	0.23	0.09	...
TiO_2	51.03	2.83	45.39	14.23	52.09	62.53	1.66
Al_2O_3	0.08	0.20	0.18	0.43	0.18	0.04	0.36
Fe_2O_3	...	97.14*	11.92**	68.48**	0.16**	...	92.59*
Cr_2O_3	...	0.08	...	0.16	...	...	...
FeO	43.14†	...	40.03**	13.08**	25.66**	16.01†	...
MnO	2.90	0.30	0.87	0.10	21.00	0.69	4.13††
MgO	0.08	...	...	...	0.06	20.75	0.21
Totals	97.42	100.55	98.54	96.80	99.38	100.11	98.95

*Total Fe as Fe_2O_3.
**Recalculated electron microprobe analyses.
†Total Fe as FeO.
††As Mn_2O_3.

1. Biotite-zone ilmenite from quartz-plagioclase-chlorite-muscovite-ilmenite-rutile-potash feldspar biotite rock (Ramsay, 1973, p. 45, Table 1).
2. Low-grade hematite from phengite-paragonite-chlorite-chloritoid-garnet-quartz-hematite-rutile rock (Kramm, 1973, p. 188, Table 7).
3. A,B. Coexisting hemo-ilmenite and ilmeno-hematite from quartz-muscovite-staurolite-kyanite-chlorite-chloritoid-ilmeno-hematite-hemo-ilmenite-magnetite quartzite (Rumble, unpublished).
4. Mn-rich ilmenite from olivine-grunerite-garnet-pyroxmangite-kutnahorite-ilmenite rock (Huntington, 1975, p. 64, Table 10).
5. Mg-rich ilmenite from dolomite-spinel-clinochlore-ilmenite-brookite rock (Cook, 1974).
6. Mn-rich hematite from viridine-spessartine-manganphyllite-alurgite-hematite-quartz-braunite-hornfels (Abraham and Schreyer, 1975, p. 12, Table 6).

crystallographic axes of both host and lamellae mutually parallel according to x-ray diffraction studies of single crystals (D. Rumble and L. W. Finger, unpublished). Transmission electron micrographs of the intergrowth show hexagonal arrays of dislocations at the lamellae-host interface (Lally *et al.*, 1974).

The terminology of Buddington *et al.* (1963) is convenient to use in discussing the minerals:

Ferrianilmenite is a one-phase grain of ilmenite containing varying amounts of Fe_2O_3 in solid solution.

Titanhematite is a one-phase grain of hematite containing varying quantities of $FeTiO_3$ in solid solution.

Hemo-ilmenite is a two-phase grain consisting of titanhematite lamellae enclosed in a ferrian-ilmenite host.

Ilmeno-hematite is a two-phase grain consisting of ferrianilmenite lamellae enclosed in a titanhematite host.

Petrologic interpretation of exsolution textures provides information on ambient conditions at depth during metamorphism as well as during cooling and uplift to the earth's surface. The bulk composition of hemo-ilmenite or ilmeno-hematite grains is believed to represent the composition of the primary mineral before exsolution (Rumble, 1973). Thus, the bulk compositions of coexisting hemo-ilmenite and ilmeno-hematite grains are the compositions of formerly homogeneous ilmenite and hematite on the solvus under the conditions of metamorphism. The compositions of individual lamellae, however, record later episodes of exsolution that occurred during cooling.

The ilmenite-hematite solid solutions occur in pelitic schists as subhedral to euhedral plates, terminated by the basal pinacoid, that are lath-shaped in cross section. The plates lie parallel to the schistosity of the rock as defined by the preferred orientation of mica grains. In amphibolites, however, the minerals may lose their lath-like shape and are usually charged with inclusions of silicates. Polysynthetic twinning on $(10\bar{1}1)$ is occasionally observed (*cf.* Ramdohr, 1969, p. 947, Fig. 557).

Supergene alteration of the minerals is common in weathered outcrops. The alteration strongly affects ferrianilmenite and the ferrianilmenite lamellae and host of ilmeno-hematite and hemo-ilmenite, producing fine-grained intergrowths of rutile and hematite or rutile alone. Hematite (essentially pure Fe_2O_3) is often seen as a supergene replacement of magnetite with its basal pinacoid parallel to the octahedral face of magnetite (*cf.* Ramdohr, 1969, p. 907, Fig. 539).

Pseudobrookite solid solutions

Pseudobrookite solid solutions are found in rocks formed by high-temperature contact metamorphism and partial melting of pelites caused by mafic intrusions (Smith, 1965) or by high-temperature contact metamorphism of weathered basalt (Agrell and Langley, 1958). The minerals are a quaternary solid solution containing the end members Fe_2TiO_5 (pseudobrookite), $FeTi_2O_5$ (ferropseudobrookite), $MgTi_2O_5$ (karrooite), and Al_2TiO_5 (Table R-3). Pseudobrookite occurs as euhedral to subhedral grains with relict cores of ilmenite.

Table R-3. Chemical analyses of pseudobrookites

	1	2	3
SiO_2	...	0.1	0.21
TiO_2	64.9	44.2	41.18
Al_2O_3	5.3	7.2	7.22
Fe_2O_3	11.7*	41.8*	46.21
FeO	9.7*	4.9*	4.70
MgO	8.4	1.8	0.95
Totals	100.0	100.0	100.47

*Recalculated electron microprobe analysis.

1. Magnesian pseudobrookite from bytownite-spinel-ilmenite-pseudobrookite rock, Tievebulliagh, N. Ireland (Smith, 1965, p. 2009, Table 10).
2. Aluminous pseudobrookite, microprobe analysis, from cordierite-spinel-magnetite-pseudobrookite rock, Stihean Sluaigh, Scotland (Smith, 1965, p. 2009, Table 10).
3. Aluminous pseudobrookite, gravimetric analysis, from cordierite-mullite-magnetite-spinel-pseudobrookite-iron corundum rock (Smith, 1965, p. 2007, Table 9).

Rutile and polymorphs

Rutile is found in rocks from all metamorphic zones and in a wide variety of rock types. The mineral is usually almost pure TiO_2 with Fe_2O_3 as the most common impurity; however, Ghent (1975) has reported metamorphic rutile with up to 4.2 weight % Nb_2O_5 (Table R-4).

Primary rutile occurs in metamorphic rocks as euhedral columnar or needle-like crystals or as subhedral mineral grains. In pelites from the chlorite zone it is charged with quartz inclusions. Rutile also occurs as lamellae oriented parallel to the host's $(22\bar{4}3)$ in ilmeno-hematite (*cf.* Ramdohr, 1969,

Table R-4. Chemical analyses of rutiles

	1	2	3	4
SiO_2	0.23	...	0.66	0.3
TiO_2	96.07	96.3	98.87	91.6
Al_2O_3	0.15	...	0.33	0.3
Cr_2O_3	...	...	...	0.4
V_2O_3	...	...	...	0.6
FeO	0.27*	1.2*	0.38*	...
Fe_2O_3	...	...	...	1.4**
Nb_2O_5	...	...	...	4.2
Totals	96.72	97.5	100.25	98.80

*Total Fe as FeO.
**Total Fe as Fe_2O_3.

1. Low-grade rutile from quartz-plagioclase-chlorite-muscovite-ilmenite-rutile-potash feldspar-biotite rock (Ramsay, 1973, p. 45, Table 1).
2. Rutile from biotite-plagioclase-iron dolomite-calcite-rutile rock (P. H. Thompson, 1973, p. 73, Table 3).
3. Rutile from quartz-muscovite-biotite-staurolite-chlorite-garnet-rutile-ilmenite-sulfides schist (Rumble, unpublished).
4. Nb-bearing rutile from quartz-muscovite-biotite-garnet-staurolite-kyanite-plagioclase-ilmenite-rutile rock (Ghent, personal communication, 1976).

p. 964, Fig. 569) or as lamellae in titanhematite (*cf.* Ramdohr, 1969, p. 964, Fig. 570). Rutile is present in rocks collected from weathered outcrops as a supergene alteration product of ferrianilmenite. Such alteration may make it difficult to distinguish primary from secondary rutile; however, study of unweathered samples from artificial exposures serves to make a clear distinction between the two modes of occurrence.

Anatase is known from diagenetic, anchimetamorphic, and chlorite-zone metamorphic rocks (McNamara, 1965; Mielke and Schreyer, 1972; Frey, 1974).

OXIDE MINERALS IN RELATION TO METAMORPHIC MINERAL ZONES

Oxide mineral assemblages of metamorphic rocks are listed in Table R-5. The following notes provide information on critical assemblages and on the variation of the chemical composition of the minerals from zone to zone.

Table R-5. Oxide mineral assemblages in metamorphic rocks

Oxide mineral assemblage	Low-intermediate pressure metamorphic zones					High-pressure metamorphism
	Chlorite and biotite	Garnet and staurolite	Sillimanite	Sillimanite-potash feldspar	High-temperature contact metamorphism	Sanbagawa terrane, glaucophane and epidote-amphibolite facies
Anatase	7,8,9*	...	...	...	...	...
Rutile-ilmenite	5,42	14,42,48,50	17,44,47	23,26	...	12
Rutile-hematite	10	11,14,50	17,45	26	...	12
Ilmenite	8	12,15,17,41	17,22	23	...	12
Hematite	10,40	17,43	17	...	...	12,38
Rutile-magnetite	5,6,7	...	19	...	...	...
Hematite-ilmenite	...	7,50	19	26	...	12
Magnetite-ilmenite	5	11,12,13,14,15,17	17,18,19,20,21,46	26,49	39	12
Magnetite-hematite	1,2,3,4,48	11,12,15,16,17,48	17,18,19,48	...	...	12
Ilmenite-hematite-magnetite	...	11,14,17,50	17,18,19	26	...	12
Gahnite-ilmenite	...	...	36,37	...	...	...
Pleonaste-hematite-magnetite	...	...	...	24,25	35	...
Pleonaste-ilmenite	...	...	...	...	28,29,30,31	...
Pleonaste-ilmenite-magnetite	...	...	...	...	34	...
Pleonaste-magnetite	...	...	...	...	32,33	...
Pseudobrookite-magnetite-pleonaste	...	...	...	...	27	...
Pseudobrookite-pleonaste	...	...	...	...	27	...
Pseudobrookite-hematite-magnetite	...	...	...	...	39	...
Pseudobrookite-ilmenite-magnetite	...	...	...	...	39	...

*References:

1. French, 1973
2. Condie, 1967
3. Zen, 1960
4. Fisher, 1970
5. Rumble, unpublished
6. Southwick, 1968
7. Mielke and Schreyer, 1972
8. Frey, 1974
9. McNamara, 1965
10. Kramm, 1973
11. Harte, 1970
12. Kanehira *et al.*, 1964
13. Hollander, 1970
14. Uytenbogaardt, 1953
15. Hounslow and Moore, 1967
16. Klein, 1973
17. Rumble, 1973
18. Chinner, 1960
19. Annersten, 1968
20. Carmichael, 1970
21. Huntington, 1975
22. Guidotti, 1970, 1974
23. Evans and Guidotti, 1966
24. Nixon *et al.*, 1973
25. Meng and Moore, 1972
26. Dougan, 1974
27. Smith, 1965
28. Dickey and Obata, 1974
29. Springer, 1974
30. Loomis, 1972
31. Propach, 1971
32. Stewart, 1942
33. Read, 1931
34. Abraham and Schreyer, 1973
35. Barker, 1964
36. Beach, 1973
37. Frost, 1973
38. Banno and Kanehira, 1961
39. Agrell and Langley, 1958
40. Zen, 1974
41. Ghent, 1975
42. Ramsay, 1973
43. P. H. Thompson, 1973
44. Robinson and Jaffe, 1969
45. Sethuraman and Moore, 1973
46. Reinhardt, 1968
47. Schwarcz, 1966
48. James, 1955
49. Buddington and Lindsley, 1964
50. Westra, 1970

Metamorphic zones of low to intermediate pressure

The metamorphic zones listed below are defined by the first appearance of the indicated mineral in pelitic rocks. Low-pressure metamorphic zones include those of contact metamorphism as well as regional terranes characterized by the succession andalusite to sillimanite with increasing grade. Intermediate metamorphic zones show the succession kyanite to sillimanite with increase in grade.

Chlorite and biotite zones. The assemblage anatase-titanhematite, formed during diagenesis of red beds, gives way to rutile-titanhematite in the Alpine metamorphic belt. Reduction with progressive metamorphism eradicates titanhematite according to the reaction

$$\text{Al-rich chlorite} + \text{titanhematite} = \text{chloritoid} + \text{ripodolite} + \text{rutile} + H_2O + O_2$$

and in the biotite zone only rutile + ilmenite assemblages are found. The titanhematite of these low-grade rocks contains up to 7% $FeTiO_3$ (g.f.w. basis) according to x-ray diffraction studies (Frey, 1974). Electron microprobe analyses show that titanhematite contains up to 9% $FeTiO_3$ (Kramm, 1973).

The assemblage rutile-magnetite occurs in chlorite-zone pelites of the Hercynian metamorphic belt. With the incoming of biotite and almandine garnet, however, magnetite disappears and assemblages of titanhematite and ferrianilmenite occur instead (Mielke and Schreyer, 1972). Magnetite-rutile assemblages have been reported from rocks of the chlorite zone by Southwick (1968). The assemblage is also present in chlorite-zone pelites of eastern Vermont (Rumble, unpublished).

Garnet and staurolite zones. Garnet-zone oxide mineral assemblages are given in Table R-5. In the garnet and staurolite zones are found coexisting ilmeno-hematite and hemo-ilmenite grains, standing in reaction relationship to the rutile-magnetite assemblages of the chlorite zone. Electron microprobe analyses of some 50 examples of ferrianilmenite and titanhematite coexisting in ilmeno-hematite and hemo-ilmenite grains give the compositions $Ilm_{90}Hem_{10}$ and $Ilm_{30}Hem_{70}$ (Westra, 1970). The bulk compositions of ilmeno-hematite and hemo-ilmenite grains in contact with each other measured by the electron microprobe are $Ilm_{30}Hem_{70}$ and $Ilm_{90}Hem_{10}$, respectively (Rumble, 1973, and unpublished). The titanhematite host of ilmeno-hematite grains has a composition of $Ilm_{19}Hem_{81}$ as determined by electron microprobe and wet chemical analysis (Harte, 1970). Magnetite is essentially pure Fe_3O_4 (Westra, 1970; Rumble, 1973).

Sillimanite zone. The bulk compositions of coexisting ilmeno-hematite and hemo-ilmenite grains in the lowermost sillimanite zone are not significantly different from those measured from the staurolite zone (Rumble, 1973). Determination of host composition in ilmeno-hematite grains by x-ray diffraction gives $Ilm_{15}Hem_{85}$ (Annersten, 1968). Analyzed magnetites are nearly pure Fe_3O_4 (Annersten, 1968; Rumble, 1973; Huntington, 1975).

Sillimanite-potash feldspar zone. In the sillimanite-potash feldspar zone the bulk compositions of ilmeno-hematite grains as determined by electron microprobe range from $Ilm_{47}Hem_{53}$ to $Ilm_{33}Hem_{67}$ (Asokan and Chinner, 1973). The compositions of titanhematite host and ferrianilmenite in an ilmeno-hematite grain are $Ilm_{34}Hem_{66}$ and $Ilm_{87}Hem_{13}$, respectively (Nixon *et al.*, 1973). Electron microprobe analysis of hemo-ilmenite gives the values $Ilm_{96}Hem_4$ for the host and $Ilm_{40}Hem_{60}$ for the lamellae (Bolfa *et al.*, 1961). Determination of the chemical composition of titanhematite host in ilmeno-hematite grains by x-ray diffraction gives a value of $Ilm_{20}Hem_{80}$ (Meng and Moore, 1972). In contrast to these results are those of Dougan (1974), who measured the compositions of coexisting titanhematite and ferrianilmenite to be $Ilm_{21-26}Hem_{79-74}$ and $Ilm_{67-72}Hem_{33-28}$, respectively, using x-ray fluorescence analysis of mineral separates. Magnetite of the sillimanite-potash feldspar zone may contain appreciable Fe_2TiO_4 (Dougan, 1974).

Chromian spinel composition in relation to metamorphic zones. In a recent review of the subject, Evans and Frost (1975) describe the changes in chemical composition of spinels in metamorphosed ultramafic rocks in a progressive metamorphic succession. The correlation of ultramafic metamorphic mineral assemblages with the mineral zones of pelites used below is given by Trommsdorff and Evans (1974). In the biotite and garnet zones, the spinel of brucite and diopside-olivine-antigorite serpentinites ranges from nearly pure Fe_3O_4 to 40% $FeCr_2O_4$-60% Fe_3O_4 (g.f.w. %). Tremolite-olivine-antigorite rocks in the staurolite zone contain Fe_3O_4-$FeCr_2O_4$ solid solutions with up to 75% $FeCr_2O_4$. In the upper part of the staurolite zone, the spinels of talc-olivine and magnesiocummingtonite/anthophyllite-olivine rocks show a complete range of compositions along the join $FeCr_2O_4$-Fe_3O_4; in addition, the more chromian ones contain up to 25% $MgAl_2O_4$. Enstatite-bearing, metamorphosed ultramafic rocks of the sillimanite zone contain spinels that range continuously in composition along the join Fe_3O_4-$FeCr_2O_4$, with minor amounts of the components $MgCr_2O_4$ and $MgFe_2O_4$, and along the join $FeCr_2O_4$-$MgAl_2O_4$. Spinels of the granulite facies (sillimanite-potash feldspar zone) are a ternary solid solution of the components $Fe_2Cr_2O_4$, $MgAl_2O_4$, and $FeAl_2O_4$ (Medaris, 1975). Spinels of metamorphosed ultramafic rocks are distinct from those of layered intrusions and alpine peridotites by virtue of their relatively higher Fe^{2+}/Mg ratios (Evans and Frost, 1975).

High-temperature contact metamorphism. The oxide minerals of rocks formed by high-temperature contact metamorphism comprise a distinctive paragenesis (Table R-5). The composition of pleonaste in the presence of magnetite is given as 7.3% Fe_3O_4-1.4% Fe_2TiO_4-10.3% $MgAl_2O_4$-80.9% $FeAl_2O_4$ (g.f.w. %) by Stewart (1942) and as 7.4% Fe_3O_4-1.5% $ZnAl_2O_4$-10.1% $MgAl_2O_4$-81.1% $FeAl_2O_4$ by Abraham and Schreyer (1973). Determination of coexisting compositions of pleonaste and magnetite by combining the results of chemical analysis of exsolution intergrowths with x-ray diffraction measurements gives the values 10-15% Fe_3O_4 (g.f.w. %) in pleonaste and 5-10% ($FeAl_2O_4$ + $MgAl_2O_4$) in magnetite (Smith, 1965).

Both of the nonopaque oxide minerals corundum and mullite contain Fe_2O_3 in solid solution in high-temperature contact metamorphic environments. In assemblages with pseudobrookite, magnetite, and pleonaste, corundum contains up to 6.98 weight % Fe_2O_3 and mullite as much as 3.18% (Smith, 1965). In assemblages with pseudobrookite, magnetite, and hematite or ilmenite, corundum contains up to 9.5 weight % Fe_2O_3 and mullite 5.93% (Agrell and Langley, 1958).

High-pressure metamorphism

The oxide mineral assemblages of the glaucophane schist and epidote-amphibole facies rocks of the high-pressure Sanbagawa terrane show the same variety as is found in rocks from lower pressure metamorphic belts (Table R-5). The more oxidized assemblages such as rutile-titanhematite or titanhematite alone are much more common, however, in the Sanbagawa belt than they are in the low-pressure metamorphic rocks of the Abukama district (Kanehira *et al.*, 1964). Determination of host compositions in coexisting ilmeno-hematite and hemo-ilmenite grains by x-ray diffraction gives values of $Ilm_{94}Hem_6$ for ferrianilmenite and $Ilm_{13}Hem_{87}$ for titanhematite (Banno and Kanehira, 1961).

PHASE PETROLOGY

Element partitioning

Data on the partitioning of elements between phases provide insight into the attainment of chemical equilibrium by mineral assemblages. In the examples cited below, the partitioning of elements between oxide minerals is emphasized. Data on the partitioning of elements between oxide and silicate minerals are given by Schwarcz (1966), Annersten (1968), Hollander (1970), Guidotti (1974), Evans and Frost (1975), Frost (1975), and Ghent (1975).

The partitioning of Mn and Mg between coexisting ferrianilmenite or hemo-ilmenite and titanhematite or ilmeno-hematite is shown in Figure R-1 where it may be seen that Mn is favored by ferrianilmenite but Mg is partitioned equally between the two minerals. Ilmenite is strongly enriched in Mn relative to coexisting magnetite (Fig. R-2). In assemblages of pleonaste and ilmenite, Mg is partitioned into pleonaste but Mn is somewhat favored by ilmenite (Fig. R-2). Aluminous spinel coexisting with Fe-Cr spinel is strongly enriched in Mg (Springer, 1974). The work of Annersten and Ekström (1971) on assemblages of hematite and magnetite shows systematic enrichment of magnetite in Ni and Cu and of hematite in Ti at the parts per million level; their results for Mn and V are more scattered but tend to show enrichment of magnetite in Mn and of hematite in V. In assemblages of magnetite and ilmenite from granulite facies rocks, magnetite is enriched in both Al and V relative to coexisting ilmenite (E. F. Stoddard, personal communication, 1976).

The systematic partitioning of elements between coexisting oxide minerals implies that they are able to achieve mutual chemical equilibrium during metamorphism. Moreover, the systematic partitioning between coexisting oxide and silicate minerals as described in the references listed above implies that the two groups of minerals also attain equilibrium.

Oxide mineral equilibria

The Fe-Ti oxide mineral assemblages listed in Table R-5, according to their metamorphic zones, may be summarized in phase diagrams (Figs. R-3 and R-4). Chlorite- and biotite-zone assemblages are related to those of the garnet and staurolite zones by the reaction

rutile + magnetite = ferrianilmenite + titanhematite.

The report of rutile-magnetite assemblages in sillimanite-zone rocks by Annersten (1968) stands in contrast to the majority of cases (see Table R-5). Kanehira *et al.* (1964), in a comprehensive study of Japanese metamorphic terranes, concluded that ferrianilmenite-titanhematite was the stable assemblage throughout the middle grades of metamorphism in both low- and high-pressure environments. The reader should be warned that it is merely a convenient idealization to think of these minerals strictly in terms of the

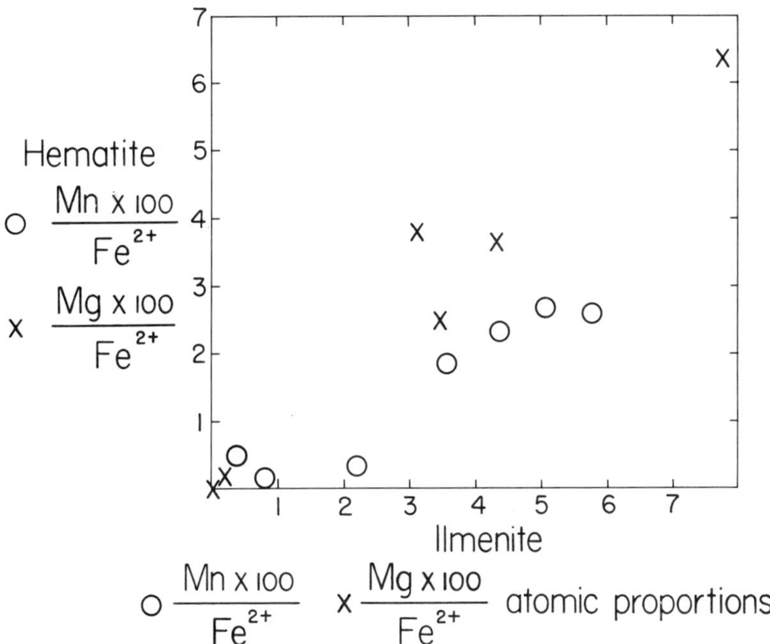

Figure R-1. Partitioning of Mn and Mg between coexisting ilmenite and hematite. Data from Dougan (1974), Nixon *et al.* (1973), and Rumble (1973 and unpublished).

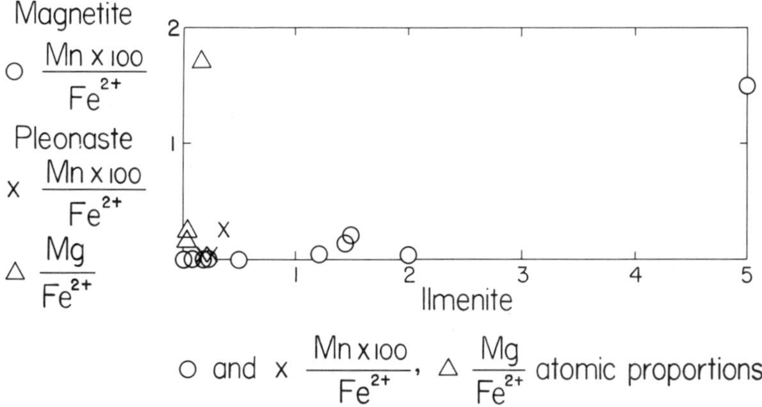

Figure R-2. Partitioning of Mn and Mg between coexisting ilmenite and magnetite or pleonaste. Data from Huntington (1975), Dickey and Obata (1974), Rumble (1973), Abraham and Schreyer (1973), and Propach (1971).

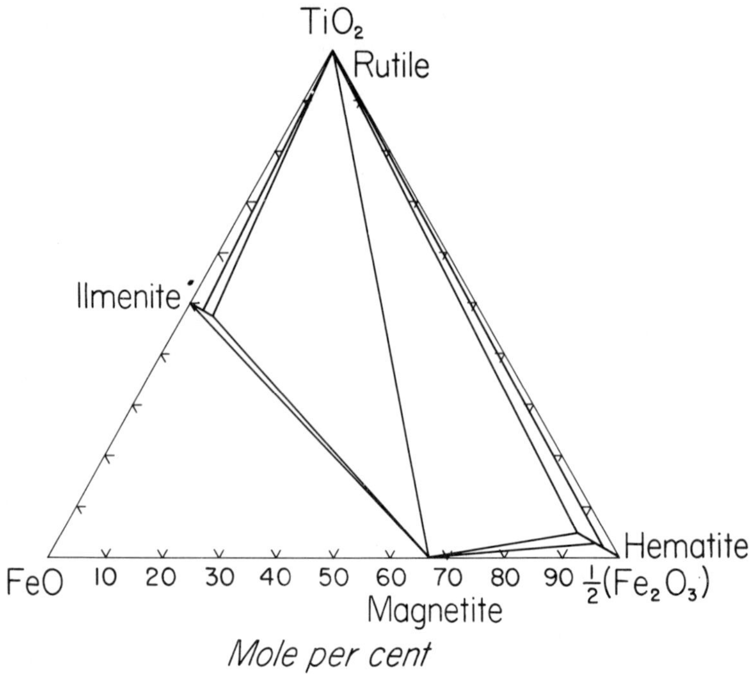

Figure R-3. Oxide mineral assemblages of the chlorite and biotite metamorphic mineral zones.

---

chemical system $FeO-Fe_2O_3-TiO_2$. In natural occurrences, ilmenite is almost always preferentially enriched in Mn or Mg or both. Thus, the four-phase assemblage rutile-magnetite-ferrianilmenite-titanhematite would be at least divariant and could conceivably exist stably over a range of P and T in rocks of appropriate bulk composition. In such a case, contrasting assemblages of rutile-magnetite and ferrianilmenite-titanhematite might be found that were related simply by a shift in rock bulk composition rather than a univariant chemical reaction.

It is also to be noted that ferrianilmenite and titanhematite from garnet- and staurolite-zone rocks have a wider range of mutual solubility than do those from the chlorite and biotite zones.

Ferrianilmenite and titanhematite from the sillimanite and sillimanite-potash feldspar zones show still wider ranges of mutual solubility of Fe_2O_3 and $FeTiO_3$ than do those from the garnet and staurolite zones. Magnetite

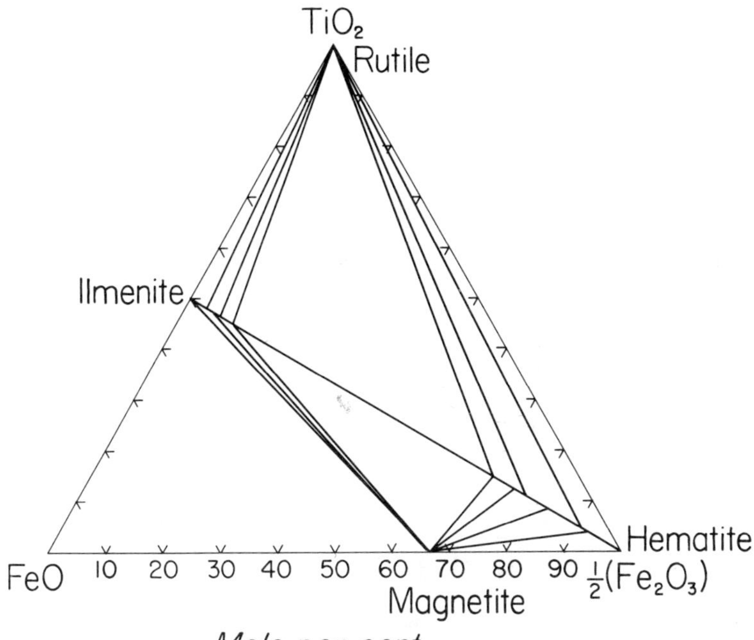

Figure R-4. Oxide mineral assemblages of the garnet, staurolite, and higher metamorphic mineral zones.

— — — — — — — — — — — — — — — — — —

from the sillimanite-potash feldspar zones has more Fe_2TiO_4 and $(Fe,Mg)Al_2O_4$ in solution that it does at lower grade.

Oxide-silicate mineral equilibria

The phase equilibria of oxide-silicate mineral assemblages are governed chiefly by the fact that the ferrous end members of the common Fe^{2+}-Mg silicate solid solutions are unstable in an oxidizing environment. Therefore, a positive correlation is observed between the Mg/(Mg + Fe) ratio of silicates and the μ_{O_2} defined by coexisting oxide minerals (Chinner, 1960; Mueller, 1960; Kanehira *et al.*, 1964; Hounslow and Moore, 1967; Annersten, 1968; Rumble, 1974). A similar relationship is found in Mn-rich rocks (Smith and Albee, 1967; Kramm, 1973). An example of such a correlation is shown in Figure R-5, where it may be seen that the Mg/(Fe + Mg) ratio of chlorite increases with increasing $Fe_2O_3/(Fe_2O_3 + FeTiO_3)$ in ilmenite in the assemblage quartz-muscovite-staurolite-chloritoid-chlorite-magnetite-ilmenite (Rumble,

1974). It is important to note that large changes in silicate mineral composition can be caused by variations in μ_{O_2} that are independent of *P* and *T*.

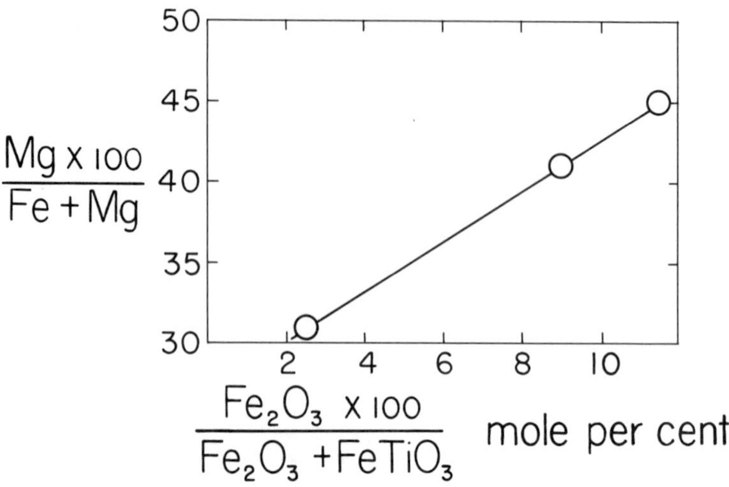

Figure R-5. Correlation between Mg/(Mg + Fe) ratio of chlorite and Fe_2O_3/(Fe_2O_3 + $FeTiO_3$) ratio of ilmenite from the assemblage quartz-muscovite-staurolite-chloritoid-chlorite-ilmenite-magnetite.

Deduction of conditions of metamorphism

The oxide minerals are of great value in deducing the conditions of metamorphism; indeed, their value is out of all proportion to their modal abundance in typical rocks, for they simultaneously record information on both the ambient temperature and the chemical potential of oxygen (μ_{O_2}) during metamorphism. This feature comes about because oxygen itself, not as an oxide, is a component of the chemical system containing the minerals. Not only are the usual temperature-sensitive features of phase diagrams met with, but also, for most pairs of minerals, a stoichiometric relation such as

$$O_2 = 6Fe_2O_3 - 4Fe_3O_4$$

may be written between the components of coexisting minerals that defines μ_{O_2} at given *P* and *T*.

The oxide minerals have great potential as geothermometers because of their ubiquity and simple chemistry, yet their practical utility is limited by the absence of complete experimental calibration. Information on temperatures of metamorphism is recorded by the many exsolution intergrowths of oxide minerals. The widening of mutual solubility ranges, recorded in the previous section, qualitatively records increasing temperature in a progressive metamorphic succession. Quantitative temperature estimates may be made from solvi calibrated in the laboratory. Unfortunately, the experimentally determined solvus along the join Fe_3O_4-$FeAl_2O_4$ does not adequately represent natural assemblages, as noted by Turnock and Eugster (1962), because the effect of substitution of $MgAl_2O_4$ is unknown. The use of coexisting Fe-Cr spinels with Al spinels (Springer, 1974) in geothermometry requires additional experimental work. Experimental determination of the solvus on the join Fe_2O_3-$FeTiO_3$ does not agree with the shape of the solvus observed in natural specimens. The experimental solvus (Carmichael, 1961; Lindh, 1972) is nearly symmetrical (see, however, Lindsley, 1973), whereas it is found to be asymmetric toward $FeTiO_3$ in nature. The rapidly developing field of study of exsolution microstructures using transmission electron microscopy offers the promise of quantitatively estimating cooling rates of rocks (Lally *et al.*, 1974).

Geothermometry is not limited to the foregoing exsolution-type mineral equilibria. The work of Buddington and Lindsley (1964) on coexisting magnetite and ilmenite has found wide application in this regard. Unfortunately, the magnetite-ilmenite geothermometer is severely limited in its usefulness in low- and intermediate-grade metamorphic rocks because the composition of natural magnetite remains fixed at Fe_3O_4. Only at the highest grade of metamorphism is the geothermometer applicable as magnetite is able to accommodate appreciable Fe_2TiO_4 in solid solution.

Although the geothermometric potential of the oxide minerals in metamorphic rocks has yet to be fully realized, the minerals are without peer in their capacity to record the chemical potential of oxygen during metamorphism. This feature of the mineral equilibria is emphasized in Figures R-6 and R-7, whose phase compatibilities correspond to those given in Figures R-3 and R-4, respectively. Comparison of Figures R-6 and R-7 with the distribution of mineral assemblages presented in Table R-5 reveals that the chemical potential of oxygen at any given grade of metamorphism ranges from values appropriate to the wüstite-magnetite buffer to those of the hematite-magnetite buffer.

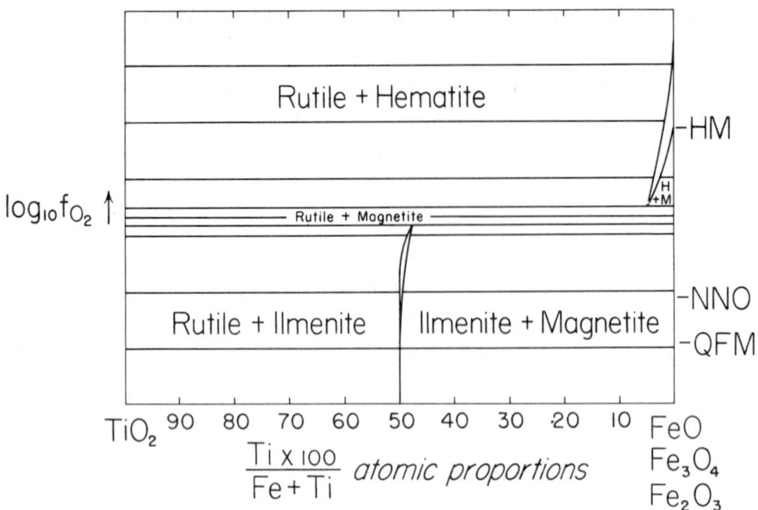

Figure R-6. Isothermal, isobaric oxygen fugacity vs. composition diagram for oxide mineral assemblages of the chlorite and biotite metamorphic mineral zones.

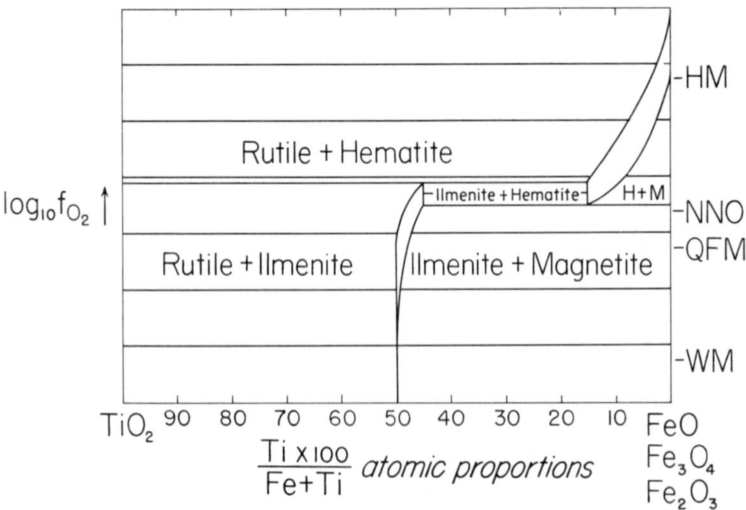

Figure R-7. Isothermal, isobaric oxygen fugacity vs. composition diagram for oxide mineral assemblages of the garnet, staurolite, and higher metamorphic mineral zones.

The oxide minerals play an important role in the ongoing discussion of the importance of large-scale oxidation-reduction processes in metamorphic rocks for they directly record the ambient μ_{O_2} of metamorphism. Consideration of rock bulk chemistry and the overall geochemical cycle has led Eugster (1972) to conclude that large-scale reduction of metamorphic rocks must occur either through the agency of organic matter buried with the sediments or by diffusion of H_2 from the mantle into the crust. J. B. Thompson (1972), however, has argued that the widespread occurrence of mineral assemblages defining widely varying values of μ_{O_2} during metamorphism in both low- and high-grade metamorphic terranes is inconsistent with large-scale reduction (*cf.* Table R-5). Indeed, many of the mineral reactions cited as evidence of reduction with metamorphic grade can be rewritten as dehydration reactions without doing great violence to the reported petrography (J. B. Thompson, 1972). In at least one example, the Littleton Formation of western New Hampshire, measurement of bulk rock composition indicates no significant change in the proportions of Fe_2O_3 and FeO with increasing metamorphic grade (Shaw, 1956).

The student of metamorphism is thus faced with an apparent conflict between two well-argued positions: Do metamorphic rocks experience progressive reduction with metamorphism, or do they retain the oxidation state inherited from their sedimentary or igneous parents? The concept of buffer capacity provides a logical route to harmonization of the conflicting viewpoints. The assemblage hematite-magnetite has a large buffer capacity with respect to reduction by a coexisting gas phase (Eugster, 1972). Consider such an assemblage subjected to progressive metamorphism and concommitant reduction by H_2-rich aqueous fluid. So long as the quantity of hematite is not exhausted by reduction, the chemical potential of oxygen is defined at a relatively high value by virtue of the relation

$$\mu_{O_2} = 6\mu_{Fe_2O_3} - 4\mu_{Fe_3O_4}$$

As reduction proceeds, however, the proportion of magnetite increases at the expense of hematite. Thus, a bulk chemical study of the rock would reveal that the ratio Fe_2O_3/FeO decreased with progressive metamorphism, whereas consideration of the phase equilibria would show that μ_{O_2} remained buffered at a relatively high value throughout metamorphism. Finally, it should be noted that the relatively oxidized parageneses listed in Table

R-5 have attracted considerable attention because of their unusual mineralogical and petrological nature, not because of their volumetric abundance. In western New Hampshire, for example, oxidized assemblages such as rutile-hematite and hematite-magnetite are restricted to a single, thin stratigraphic unit consisting of beach sand deposits. The overwhelmingly preponderant mica schists and amphibolites contain the relatively reduced assemblages rutile-ilmenite-graphite, rutile-ilmenite, or ilmenite alone (Rumble, 1973).

REFERENCES

Abraham, K., and W. Schreyer (1973) Petrology of a ferruginous hornfels from Riekensglück, Harz Mountains, Germany, *Contrib. Mineral. Petrol. 40*, 275-292.

________, and ________ (1975) Minerals of viridine hornfels from Darumstadt, Germany, *Contrib. Mineral. Petrol. 49*, 1-20.

Agrell, S. O., and J. M. Langley (1958) The dolerite plug at Tievebulliagh near Cushendall, Co. Antrim. *Proc. R. Ir. Acad., Sect. B, 59*, 93-127.

Annersten, H. (1968) A mineral chemical study of a metamorphosed iron formation in northern Sweden, *Lithos 1*, 374-397.

________, and T. Ekström (1971) Distribution of major and minor elements in coexisting minerals from a metamorphosed iron formation, *Lithos 4*, 185-204.

Asokan, S., and G. A. Chinner (1973) Local modification of gas composition by graphite-iron oxide buffering, *Nature (London) Phys. Sci. 246*, 75-76.

Banno, S., and K. Kanehira (1961) Sulfide and oxide minerals in schists of the Sanbagawa and central Abukuma metamorphic terranes, *Jpn. J. Geol. Geogr. 32*, 331-348.

Barker, F. (1964) Reaction between mafic magmas and pelitic schist, Cortlandt, New York, *Am. J. Sci. 262*, 614-634.

Beach, A. (1973) The mineralogy of high temperature shear zones at Scourie, N. W. Scotland, *J. Petrol. 14*, 231-248.

Bolfa, J., H. de la Roche, R. Kern, M. Capitant, and K. D. Phan (1961) Sur la nature minéralogique exacte d'exsolutions dans les "ilmenites" de Vohibarika (Madagascar) déterminée à la microsonde électronique, *Bull. Soc. franc. Mineral. Cristallogr. 84*, 400-401.

Buddington, A. F., J. Fahey, and A. Vlisidis (1963) Degree of oxidation of Adirondack iron oxide and iron-titanium oxide minerals in relation to petrogeny, *J. Petrol. 4*, 138-169.

________, and D. H. Lindsley (1964) Iron-titanium oxide minerals and synthetic equivalents, *J. Petrol. 5*, 310-357.

Carmichael, C. M. (1961) The magnetic properties of ilmenite-hematite crystals, *Proc. R. Soc. London, Ser. A, 263*, 508-530.

Carmichael, D. M. (1970) Intersecting isograds in the Whetstone Lake area, Ontario, *J. Petrol. 11*, 147-181.

Chinner, G. A. (1960) Pelitic gneisses with varying ferrous/ferric ratios from Glen Clova, Angus, Scotland, *J. Petrol. 1*, 178-217.

Condie, K. C. (1967) Oxygen, carbon dioxide, and sulfur fugacities during diagenesis and low-grade metamorphism of late Precambrian subgraywackes from northern Utah, *Am. Mineral. 52*, 1153-1160.

Cook, L. P. (1974) Metamorphosed carbonates of the Nashoba Formation, eastern Massachusetts, Ph.D. thesis, Harvard University.

Dickey, J. S., Jr., and M. Obata (1974) Graphitic hornfels dikes in the Ronda high-temperature peridotite massif, *Am. Mineral. 59*, 1183-1189.

Dougan, T. W. (1974) Cordierite gneisses and associated lithologies of the Guri area, Northwest Guayana Shield, Venezuela, *Contrib. Mineral. Petrol. 46*, 169-188.

Evans, B. W., and B. R. Frost (1975) Chrome-spinel in progressive metamorphism--a preliminary analysis, *Geochim. Cosmochim. Acta 39*, 959-972.

________, and C. V. Guidotti (1966) The sillimanite-potash feldspar isograd in western Maine, U.S.A., *Contrib. Mineral. Petrol. 12*, 25-62.

Eugster, H. P. (1972) Reduction and oxidation in metamorphism (II), *Int. Geol. Congr., 24th, Montreal, Sect. 10*, 3-11.

Fisher, G. W. (1970) The metamorphosed sedimentary rocks along the Potomac River near Washington, D. C. In, G. W. Fisher *et al.*, eds., *Studies of Appalachian Geology: Central and Southern*, John Wiley and Sons, Inc., New York, pp. 299-315.

French, B. M. (1973) Mineral assemblages in diagenetic and low-grade metamorphic iron formation, *Econ. Geol. 68*, 1063-1074.

Frey, M. (1974) Alpine metamorphism of pelitic and marly rocks of the central Alps, *Schweiz. Mineral. Petrogr. Mitt. 54*, 489-506.

Frondel, C., and C. Klein, Jr. (1965) Exsolution in franklinite, *Am. Mineral. 50*, 1670-1680.

Frost, B. R. (1973) Ferroan gahnite from quartz-biotite-almandine schists, Wind River Mountains, Wyoming, *Am. Mineral. 58*, 831-834.

________ (1975) Contact metamorphism of serpentinite, chloritic blackwall, and rodingite at Paddy-Go-Easy Pass, Central Cascades, Washington, *J. Petrol. 16*, 272-313.

Ghent, E. (1975) Temperature, pressure, and mixed-volatile equilibria attending metamorphism of staurolite-kyanite bearing assemblages, Esplanade Range, British Columbia, *Geol. Soc. Am. Bull. 86*, 1654-1660.

________, and C. D. S. DeVries (1972) Plagioclase-garnet-epidote equilibrium in hornblende-plagioclase bearing rocks from the Esplanade Range, British Columbia, *Can. J. Earth Sci. 9*, 618-635.

Guidotti, C. V. (1970) The mineralogy and petrology of the transition from the lower to the upper sillimanite zone in the Oquossoc area, Maine, *J. Petrol. 11*, 277-336.

________ (1974) Transition from staurolite to sillimanite zone, Rangeley quadrangle, Maine, *Geol. Soc. Am. Bull. 85*, 475-490.

Harte, B. (1970) Iron ore assemblages in pelites from Barrow's zones, Scotland, and their use in the evaluation of varying oxygen fugacity, *Collected Abstracts of the 7th General Meeting of the International Mineralogical Association*, p. 222.

Hollander, N. B. (1970) Distribution of chemical elements among mineral phases in amphibolites and gneiss, *Lithos 3*, 93-111.

Hounslow, A. W., and J. M. Moore, Jr. (1967) Chemical petrology of Grenville schists near Fernleigh, Ontario, *J. Petrol. 8*, 1-28.

Huntington, J. C. (1975) Mineralogy and petrology of metamorphosed iron-rich beds in the Lower Devonian Littleton Formation, Orange area, Massachusetts, M.S. thesis, Department of Geology, University of Massachusetts.

James, H. L. (1955) Zones of regional metamorphism in the Precambrian of northern Michigan, *Geol. Soc. Am. Bull. 66*, 1455-1487.

________, and A. L. Howland (1955) Mineral facies in iron- and silica-rich rocks, *Geol. Soc. Am. Bull. 66*, 1580-1581.

Kanehira, K., S. Banno, and K. Nishida (1964) Sulfide and oxide minerals in some metamorphic terranes in Japan, *Jpn. J. Geol. Geogr. 35*, 175-191.

Klein, C., Jr. (1973) Changes in mineral assemblages with metamorphism of some banded Precambrian iron formations, *Econ. Geol. 68*, 1075-1088.

Knauer, E., M. Okrusch, P. Richter, and K. Schmidt (1974) Die metamorphe Basit-Ultrabasit Assoziation in der Ballsteiner Gneisskuppel, Odenwald, *Neues Jahrb. Mineral. Abh. 122*, 186-228.

Kramm, U. (1973) Chloritoid stability in Mn-rich low-grade metamorphic rocks, Venn-Stavelot Massif, Ardennes, *Contrib. Mineral. Petrol. 41*, 179-196.

Lally, J. S., A. H. Heuer, and G. L. Nord, Jr. (1974) Transmission electron microscopy study of precipitation in hematite-ilmenite (abstr.), *Geol. Soc. Am. Abstr. Progr. 6*, 835.

Lindh, A. (1972) A hydrothermal investigation of the system FeO, Fe_2O_3, TiO_2, *Lithos 5*, 325-343.

Lindsley, D. H. (1973) Delimitation of the hematite-ilmenite miscibility gap, *Geol. Soc. Am. Bull. 84*, 657-662.

Loomis, T. P. (1972) Contact metamorphism of pelitic rock by the Ronda ultramafic intrusion, southern Spain, *Geol. Soc. Am. Bull. 83*, 2449-2474.

McNamara, M. (1965) The lower greenschist facies in the Scottish Highlands, *Geol. Foeren. Stockholm Foerh. 87*, 347-389.

Medaris, L. G., Jr. (1975) Coexisting spinel and silicates in alpine peridotites of the granulite facies, *Geochim. Cosmochim. Acta 39*, 947-958.

Meng, L. K., and J. M. Moore, Jr. (1972) Sapphirine-bearing rocks from Wilson Lake, Labrador, *Can. Mineral. 11*, 777-790.

Mielke, H., and W. Schreyer (1972) Magnetite-rutile assemblages in metapelites of the Fichtelgebirge, Germany, *Earth Planet. Sci. Lett. 16*, 423-428.

Mueller, R. F. (1960) Compositional characteristics and equilibrium relations in mineral assemblages of a metamorphosed iron formation, *Am. J. Sci. 258*, 449-497.

Nemec, D. (1972) Das Vorkommen der Zn-Spinelle in der Böhmischen Masse, *Tschermaks Mineral. Petrogr. Mitt. 19*

Nixon, P. H., A. J. Reedman, and L. K. Burns (1973) Sapphirine-bearing granulites from Labwor, Uganda, *Mineral. Mag. 39*, 420-428.

Propach, G. (1971) Hercynit und Ilmenit aus dem Korund-Spinell-Fels von Plössberg (Opf.), *Neues Jahrb. Mineral. Abh. 115*, 120-122.

Ramdohr, P. (1969) *The Ore Minerals and Their Intergrowths*, Pergamon: Oxford, 1174 pp.

Ramsay, C. R. (1973) The origin of biotite in Archean meta-sediments near Yellowknife, N.W.T., Canada, *Contrib. Mineral. Petrol. 42*, 43-54.

Read, H. H. (1931) On corundum-spinel xenoliths in the gabbro of Huddo House, Aberdeenshire, *Geol. Mag. 68*, 446-453.

Reed, J. C., Jr., and B. A. Morgan (1971) Chemical alteration and spilitization of the Catoctin greenstones, Shenandoah National Park, Virginia, *J. Geol. 79*, 526-548.

Reinhardt, E. W. (1968) Phase relations in cordierite-bearing gneisses from Gananoque area, Ontario, *Can. J. Earth Sci. 5*, 455-482.

Robinson, P., and H. W. Jaffe (1969) Aluminous enclaves in gedrite-cordierite gneiss from southwestern New Hampshire, *Am. J. Sci. 267*, 389-421.

Rumble, D. (1973) Fe-Ti oxide minerals from regionally metamorphised quartzites, *Contrib. Mineral. Petrol. 42*, 181-195.

________ (1974) Gradients in the chemical potentials of volatile components between sedimentary beds of the Clough Formation, Black Mountain, New Hampshire, *Carnegie Inst. Washington Year Book 73*, 371-380.

Schwarcz, H. P. (1966) Chemical and mineralogic variations in an arkosic quartzite during progression regional metamorphism, *Geol. Soc. Am. Bull. 77*, 509-531.

Sethuraman, K., and J. M. Moore, Jr. (1973) Petrology of metavolcanic rocks in the Bishop Corners-Donaldson area, Grenville Province, Ontario, *Can. J. Earth Sci. 10*, 589-614.

Shaw, D. M. (1956) Geochemistry of pelitic rocks. Part III. Major elements and general geochemistry, *Geol. Soc. Am. Bull. 67*, 919-934

Simmons, E. C., D. H. Lindsley, and J. J. Papike (1974) Phase relations and crystallization sequence in a contact-metamorphosed rock from the Gunflint Iron Formation, Minnesota, *J. Petrol. 15*, 539-565.

Smith, D. G. W. (1965) The chemistry and mineralogy of some emery-like rocks from Sithean Sluaigh, Strachur, Argyllshire, *Am. Mineral. 50*, 1982-2022.

Smith, D., and A. L. Albee (1967) Petrology of piemontite-bearing gneiss, San Gorgonio Pass, California, *Contrib. Mineral. Petrol. 16*, 189-203.

Southwick, D. L. (1968) Mineralogy of a rutile- and apatite-bearing ultramafic chlorite rock, *U. S. Geol. Surv. Prof. Pap. 600-C*, C38-C44.

Springer, R. K. (1974) Contact metamorphism of ultramafic rocks in the western Sierra Nevada foothills, California, *J. Petrol. 15*, 160-195.

Stewart, F. H. (1942) Chemical data on a silica-poor argillaceous hornfels and its constituent minerals, *Mineral. Mag. 36*, 260-266.

Thompson, J. B., Jr. (1972) Oxides and sulfides in regional metamorphism of pelitic schists, *Int. Geol. Congr., 24th, Montreal, Sect. 10*, 27-35.

Thompson, P. H. (1973) Mineral zones and isograds in impure calcareous rocks, an alternative means of evaluating metamorphic grade, *Contrib. Mineral. Petrol. 42*, 63-80.

Trommsdorff, V., and B. W. Evans (1974) Alpine metamorphism of peridotitic rocks, *Schweiz. Mineral. Petrogr. Mitt. 54*, 333-352.

Turnock, A. C., and H. P. Eugster (1962) Fe-Al oxides: phase relationships below 1000°C, *J. Petrol. 3*, 533-565.

Uytenbogaardt, W. (1953) On the opaque mineral constituents in a series of amphibolitic rocks from Norva Storfjüllet, Västerbotten, Sweden, *Ark. Mineral. Geol. 1*, 527-543.

Westra, L. (1970) *The Role of Fe-Ti Oxides in Plurifacial Metamorphism of Alpine Age in the South-eastern Sierra de los Filabres, SE Spain*, Academisch Proefschrift, Vrije Universiteit de Amsterdam, 82 pp.

Winchell, A. N. (1941) The spinel group, *Am. Mineral. 26*, 422-428.

Zen, E-An (1960) Metamorphism of Lower Paleozoic rocks in the vicinity of the Taconic Range in west-central Vermont, *Am. Mineral. 45*, 129-175.

________ (1974) Prehnite- and pumpellyite-bearing mineral assemblages, west side of the Appalachian metamorphic belt, Pennsylvania to Newfoundland, *J. Petrol. 15*, 197-242.

OXIDATION of OPAQUE MINERAL OXIDES in BASALTS

Stephen E. Haggerty

Chapter 4

INTRODUCTION

It is now well established that the earth's magnetic field passes through periodic and episodic cycles of polarity reversals. However, the possibility still exists, and research was heightened during the earth part of the last decade which tended to suggest that some rocks would show, or perhaps undergo, a self-reversal magnetic process that would result in a direction that was contrary to the direction of the ambient geomagnetic field. The problem of self-reversal has still not been resolved but an interesting consequence of these studies has resulted in a more articulate and intensive mineralogical approach to an important and quantitative magnetic parameter. The inescapable fact is that rocks carry a magnetic signature because of the Fe-Ti oxides which are present and it is implicit in magnetic measurements that these oxides reflect the state of the magnetic field at the time the rocks were formed. This is basically the major supposition of paleomagnetic studies and for the most part it is well founded and confirmed by measurements of modern extrusives. Somewhat less well established, however, are the times at which magnetic acquisition takes place, for example in red beds, or the retentive capacities of a single mineral, a complex intergrowth, or an entire rock unit.

If the mineral recorder is modified to any extent, partially destroyed, or totally reconstituted, it follows that any subsequent magnetic measurement must of necessity also be modified. The time dependencies of these mineral modifications are basic and essential components of a much broader and complex framework that is required for complete interpretation of magnetic property data. The major and undoubtedly the most important modifying mineralogical parameters are oxidation and exsolution. At valency levels the detection limits are pronounced in magnetic property measurements, but beyond the resolution of the ore-microscopist. Notwithstanding this limitation the heterogeneity and complexity of a major proportion of Fe-Ti oxides in all eruptive, hypabyssal and plutonic rocks is decidedly on the side of the ore-petrologist. Mineral chemistry plays a key role and in spite of the advances made to date, an integrated approach to magnetomineral chemistry remains an open and attractive field for future study.

The obvious starting point is to establish in detail the progressive sequences of Fe-Ti oxide mineral modification during initial cooling. Such time scales are many orders of magnitude shorter than those influencing the earth's magnetic field so that any magnetic property variation ought to be consistent with the observed mineralogy. If inconsistencies exist, post-deuteric processes are justifiably and commonly invoked, or wild excursions of the earth's magnetic field proposed. The rapid extrusion of lava flows and slow cooling of large plutons have provided important estimates of magnetic secular variations and evidence for polarity reversals. However, compositional differences, size or thickness and hence cooling rate, viscosity and trapped volatiles are known to induce major oxidation variations and hence magnetic mineral modification. The starting point therefore is not to distinguish among units or rock types as discussed in the previous chapter but to trace the evolution of oxide phases within single units. Few systematic studies of this type have been attempted and the most detailed studies are on basalts. Although the impetus for such studies was largely from the magnetic standpoint it should be evident also that the sensitivity of the Fe-Ti oxides to redox potential yields a measure of the oxidizing conditions which prevailed during initial cooling, thus providing an additional and important petrological parameter.

In the sections which follow an attempt has been made to establish and review a classification for Fe-Ti oxide mineral modification based primarily on oxidation. This review has not previously been published in detail. It is intended to provide the basic observations, uncertainties in interpretation, and a guide to future experimental studies. The review is by no means complete, nor is it intended to be so. Many basic and perhaps simplistic explanations have been overlooked or disregarded: The problems of magnetic self-reversals are less prominent now, but continue, and will remain a significant unanswered mineral-geophysical question unless a renewed and imaginative frontal attack is made. These sections include detailed reflection microscopic descriptions of oxide assemblages accompanied by photomicrographic coverage and electron microprobe analyses of major and minor components, and in all it is hoped that the data will enable the student, with a minimum of ore microscopic training, to recognize and to apply the oxidation classification as outlined. Clearly, chemical analyses can neither be practically obtained on each and every mineral grain in a basalt nor on each and every sample in a collection; a well-characterized optical data base is an essential and prime pre-requisite: The microscope continues to serve an important function even in this age of black box technology.

The data presented here are based on extensive suites of basalts from Iceland (Dagley *et al.*, 1967; Watkins and Haggerty, 1967, 1968; Wilson *et al.*, 1968; Wilson *et al.*, 1972), from Oregon (Lindsley *et al.*, 1971; Lindsley and Haggerty, 1971), and from Hawaii (Sato and Wright, 1966; Grommé *et al.*, 1969; Haggerty, 1971). The oxidation classification is an integrated, modified and refined version of previously published classifications by Wilson and Ade-Hall (1963), Buddington and Lindsley (1964), Wilson and Haggerty (1966), Wilson and Watkins (1967), Watkins and Haggerty (1968), Ade-Hall *et al.* (1969), and Haggerty (1968,1971). The review will draw freely on these publications, and the serious student is strongly encouraged to refer back to these earlier investigations for the details of the relationships of the oxides to magnetic parameters, and of the oxides to petrogenesis.

OXIDATION PARAGENESIS

Primary oxide mineralogy

The more detailed discussion of the primary opaque mineralogy of basalts covered in the previous chapter necessitates that only a brief summary statement be made here. The oxides consist of ilmenite and spinels. Spinels are of two major composition groups: chrome-spinels in the multicomponent system $FeCr_2O_4$-$FeAl_2O_4$-Fe_3O_4-Fe_2TiO_4-$MgCr_2O_4$-$MgAl_2O_4$-$MgFe_2O_4$; and Fe-Ti spinels in the system magnetite (Fe_3O_4)-ulvöspinel (Fe_2TiO_4). Chromian spinels are paragentically early, complex in composition, and rarely exceed 0.5% by volume of the rock; Fe-Ti spinels are later than chromian spinels, follow or co-crystallize with ilmenite and constitute between 1 and 5% by volume of basalts. Primary Fe-Ti spinels are most commonly in the range $Usp_{50}Mt_{50}$ to $Usp_{80}Mt_{20}$ and generally contain minor concentrations of MgO (<1-3 wt %), Al_2O_3 (<1-4 wt %), and MnO (<1-2 wt %). Titanian-spinel mantles on chromites, however, differ markedly from these ranges and in general have a limited distribution. Primary ilmenites, by contrast, adhere fairly rigorously to the join $FeTiO_3$-Fe_2O_3 although ilmenites in many basalts do exhibit minor solid-solubility towards $MgTiO_3$ and $MnTiO_3$. Ilmenite compositions range between $Ilm_{80}Hem_{20}$ to $Ilm_{95}Hem_5$ and volumes vary between <1 and 5% of the rock. The ratio of primary ilmenite:primary Fe-Ti spinel is typically close to unity, but both f_{O_2} and a_{SiO_2} are internally controlling factors which may bias this ratio in favor of one or the other oxide.

Oxidation of ulvöspinel-magnetite solid solutions (Usp-Mt_{ss})

Cubic titanomagnetite (Usp-Mt_{ss}) may be oxidized by two alternative mechanisms to produce rhombohedral Ilm-Hem_{ss}:

(a) Oxidation at low pressures and between 400-600°C to yield cationic-deficient spinels of the metastable titanomaghemite series (Usp-Mt-γFe_2O_3), which subsequently inverts to members of the Hem-Ilm_{ss} series.

(b) Oxidation at low to moderate pressures and above 600°C with the *direct* formation of Ilm-Hem_{ss}.

The latter has been conclusively demonstrated experimentally (Lindsley, 1962) and the former remains hypothetical (Nicholls, 1955; Verhoogen, 1962). The mechanism of oxidation discussed in this chapter will therefore be restricted to the oxidation of Usp-Mt_{ss} at or above 600°C.

$Ilmenite_{ss}$ intergrowths in titanomagnetite may be divided into the following textural forms according to the Buddington and Lindsley (1964) classification:

(a) Trellis types (Fig. Hg-1).

(b) Composite (granular) types (Fig. Hg-2).

(c) Sandwich types (Fig. Hg-3).

The ilmenite contents of basaltic titanomagnetites are extremely variable; in some grains ilmenite may reach an estimated concentration of 70%, whereas in other crystals only a few lamellae or a single inclusion of ilmenite are observed. The composite types are the least abundant and in some instances it can be demonstrated that these ilmenites are primary precipitates from the melt; the trellis types are the most abundant and are clearly oxidation products of Mt-Usp_{ss}; the sandwich types are relatively common and their origin, as is the case with composite types, may be either the result of oxidation or the result of a primary ilmenite nucleus around which titanomagnetite has crystallized.

Trellis type

The orientation of ilmenite lamellae in titanomagnetite has been determined by single crystal analysis on material from the Skaergaard intrusion (Bernal *et al.*, 1957); oxygen layers parallel to {111} in the spinel structure, and lattice oxygen parallel to {0001} in the ilmenite structure are almost equivalent, which accounts for the similarity in texture that these intergrowths bear to true exsolution fabrics. Because of these similarities, terms such as "oxidation-exsolution" or "oxyexsolution" (Buddington and Lindsley, 1964) may conveniently be used to describe the textural habit and origin of ilmenite intergrowths in titanomagnetite. The use of these terms is restricted in this study to trellis ilmenite along {111} planes in titanomagnetite because of the uncertainty which remains for composite and sandwich intergrowths.

A complete transition from fine (<1-10 μm) spindles of ilmenite along one set of {111} planes to crowded lamellae along all sets of the octahedral planes are commonly observed (Fig. Hg-1) and have received wide exposure in the literature. This intergrowth is identical in appearance to the well-known Widmanstätten texture in iron meteorites. Ilmenite lamellae which are parallel to any one set of octahedral planes are in optical continuity, show identical degrees of anisotropy and similar pleochroic schemes. Oxyexsolved lamallae show sharp, well defined contacts with their titanomagnetite hosts and although the lamellae are smooth in outline they are rarely parallel. Tapered terminations develop at the intersection of two or more lamellae which is indicative and typical of diffusion, but it is important to note that tapering also occurs in ilmenite lamellae which extend to the limits of titanomagnetite grain boundaries, a feature which is in marked contrast to the sandwich laths described below. The width of ilmenite lamellae in any one grain tends to be fairly uniform (Fig. Hg-1b-c) but a super trellis work of larger laths (10-20 μm) infilled by several generations of finer lamellae (1-10 μm) are also common (Fig. Hg-1e-f).

Ilmenite lamellae are often concentrated along cracks, around silicate inclusions and, most significantly of all, along titanomagnetite grain boundaries (Fig. Hg-1a). These textural features strongly support the experimental evidence that oxidation and not exsolution is responsible for the formation of ilmenite lamellae from primary $Usp\text{-}Mt_{ss}$. Ilmenite in these oxidized lamellar zones increases in size and abundance towards the grain boundaries (Fig. Hg-1a). Lamellae that project into the titanomagnetite are sharply tapered and gradually disappear, on a microscopic scale, as the unoxidized areas are approached. There are significant changes in color, reflectivity and in the composition of the oxidized titanomagnetite: the mineral becomes whiter with an increase in reflectivity as titanium diffuses from the host and the iron is oxidized; the ratios of Fe:Ti and $Fe^{3+}:Fe^{2+}$ increase. If a basic lamellar framework is established at the edge of a grain, microcapillary access to the center becomes possible and continued oxidation of the crystal interior may result largely be propagation along these lattice discontinuities.

From the foregoing discussion the basic premise being made is that oxidation of $Mt\text{-}Usp_{ss}$ in basalts results in "exsolved" lamellae of ilmenite, and this is confirmed experimentally (Buddington and Lindsley, 1964). Composite

Figure Hg-1. Trellis type. Stages C2 and C3. (a) Subhedral grain of titanomagnetite showing a preferred concentration of ilmenite trellis lamellae along the grain boundary. Note the differences in color between the Usp_{ss}-poor area with ilmenite and the Usp_{ss}-rich central core which is ilmenite free. (b) Fine trellis lamellae of ilmenite radiating from cracks in a titanomagnetite crystal. The largest concentration of ilmenite is adjacent to the cracks and lamellae gradually thin out towards the grain boundary. (c) Two dominant sets of ilmenite lamellae along {111} planes and a third minor set which has developed in areas of maximum ilmenite concentration. (d) Short discontinuous lamellae of ilmenite along two sets of {111} spinel planes and a third set (along a direction NW-SE) which appears to be earlier. The adjacent titanomagnetite crystal (lower left) shows ilmenite in a rectangular pattern; these are not {100} planes but result from the direction of sectioning. (e) Well developed early sets of coarse ilmenite lamellae infilled by several generations of finer lamellae. Note the lensoidal form of the coarse lamellae and their constricted nature at contacts with other lamellae. (f) Very coarse ilmenite lamellae illustrating offsets of other lamellae, lamellae-interference, and the development of a fine lamellae set along one {111} plane. Grains (a) and (b) are a clear demonstration of oxidation exsolution and would be classified as C2; the remaining grains would be classified as C3. Scale bar = 25 μm.

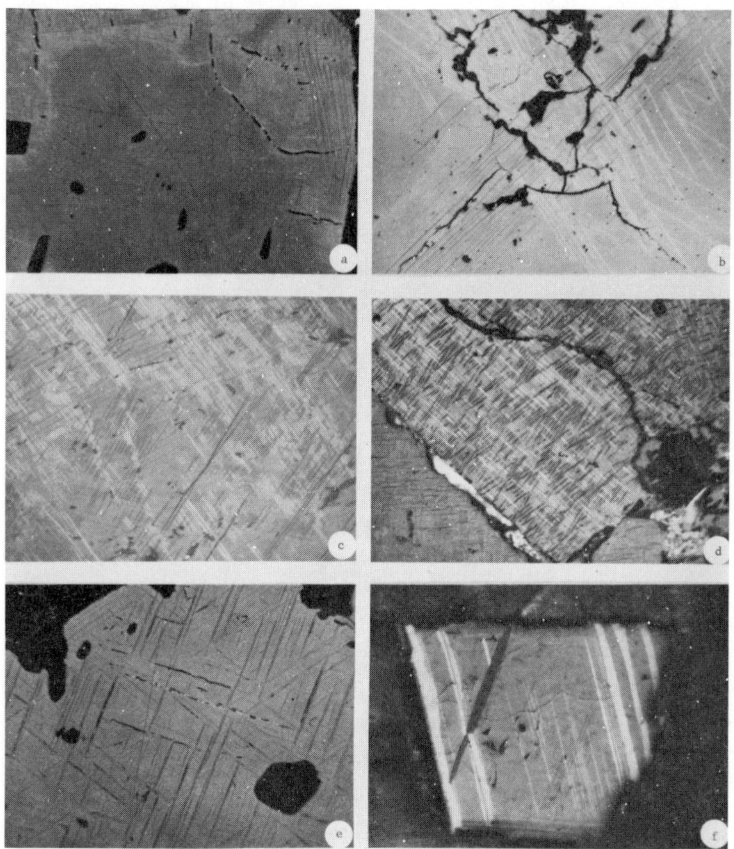
a
b
c
d
e
f

and sandwich inclusions of ilmenite on the other hand, as discussed in the following section, cannot be regarded as indicative of advanced oxidation, whereas the variations in density and abundance of finer lamellae in {111} networks, in agreement with Buddington and Lindsley (1964) do represent advanced stages of diffusion and are therefore a qualitative function of the intensity of oxidation.

The stages of oxidation defined by the oxidation "exsolution" of $Usp\text{-}Mt_{ss}$ are therefore as follows; each stage is prefixed by the letter C (cubic) to distinguish this classification from the discrete ilmenite R (rhombohedral) classification described in a later section.

C1 stage. Optically homogeneous ulvöspinel-rich magnetite solid solutions.

C2 stage. Magnetite-enriched ulvöspinel solid solutions with a small number of "exsolved" ilmenite lamellae parallel to {111} spinel parting planes.

C3 stage. Ulvöspinel-poor magnetite solid solutions with densely crowded "exsolved" ilmenite along {111} parting planes of the cubic host.

Typical reactions that apply to the C2 and C3 assemblages, with the partial and the complete oxidation of ulvöspinel are as follows:

$$\text{C2:}\quad 6Fe_2TiO_4 + O_2 = 6FeTiO_3 + 2Fe_3O_4$$
$$\text{C3:}\quad 4Fe_2TiO_4 + O_2 = 4FeTiO_3 + 2Fe_2O_3$$

These reactions represent progressively higher states of oxidation, and thermodynamic calculations by Verhoogen (1962) show that the equilibrium f_{O_2} for these reactions at 600 and 800°C are as follows:

	600°C	800°C
Reaction C2	$10^{-20.8}$ atm.	$10^{-15.4}$ atm.
Reaction C3	$10^{-18.3}$ atm.	$10^{-13.0}$ atm.

Continued and more intense stages of oxidation of C2 and C3 assemblages are defined as *pseudomorphic oxidation assemblages* and are discussed in a later section.

Composite types

Euhedral to anhedral inclusions of ilmenite are frequently present in titanomagnetite. These inclusions show sharp contacts with their titanomagnetite hosts and when compared with trellis lamellae are rarely oriented along either of the major {100} or {111} planes. The inclusions are termed internal or external, depending on whether the ilmenite is partially or totally included in the titanomagnetite (Buddington and Lindsley, 1964). The term composite, rather than "granule-exsolution" as used by Buddington and Lindsley, is preferred because the ilmenite inclusions may be either primary precipitates or oxidation-"exsolution" products of Mt-Usp_{ss}.

The question as to whether composite intergrowths are the result of diffusion at high temperatures (Vincent *et al.*, 1954; Wright, 1961), the result of increasing degrees of oxidation and diffusion (Buddington and Lindsley, 1964), or the result of sequential or contemporaneous crystallization of titanomagnetite and ilmenite is an important factor in applications of either the Fe-Ti oxide geothermometer, or the magnetomineral oxidation classification as discussed in this chapter. Vincent (1960, p. 1606) for example, considers that composite intergrowths of magnetite and ilmenite are "clearly fundamental to a full understanding of the petrological role of the opaques," and some observations and the evidence in support of their primary origin in basalts now follows. It should be emphasized, however, that many basalts contain composite inclusions for which there is neither definitive evidence for oxidation or for primary precipitation. The criteria for unequivocal distinction are undefined in the broadest sense but can be evaluated in those cases where detailed microprobe data are available.

The terms *internal* and *external* granule "exsolution," proposed by Buddington and Lindsley (1964), were specifically intended to emphasize the oxidation "exsolution" nature of these intergrowths with higher rates of diffusion being operative at higher temperatures, and with increasing degrees of oxidation. By analogy with sulphide exsolution textures (Brett, 1964), Buddington and Lindsley (p. 323) argued that increasing degrees of oxidation and diffusion should result in a systematic series of fabrics, and their proposed scheme is as follows:

1. Homogeneous titanomagnetite.
2. Trellis intergrowths of thin ilmenite lamellae along all sets of {111} planes in the host.
3. Sandwich intergrowths of thick lamellae, predominantly in one set of {111} planes.
4. Granule intergrowths within the titanomagnetite (internal composite).
5. Granules or occasional lamellae of ilmenite on the external borders of the magnetite (external composite).

While accepting that composite intergrowths of ilmenite may develop by slow diffusion in deep seated intrusives (Wright, 1961; Buddington and Lindsley, 1964), their formation in volcanic rocks can be demonstrated to result by oxidation or by primary precipitation, and the evidence for both mechanisms follows:

1. The sequence of crystallization for the Fe-Ti oxides, from a basaltic magma, is ilmenite followed by titanomagnetite (Wright and Weiblen, 1967), which

Figure Hg-2. Composite type. (a) Coarse subhedral, internal composite inclusions of ilmenite which have crystal terminations that closely parallel those of the octahedral planes of the host titanomagnetite. Finer sets of ilmenite trellis lamellae are also present. (b) An external composite inclusion of ilmenite which has at least two crystal terminations which are precisely parallel to {111} as shown from the directions of associated trellis lamellae. Note that the composite extends into the silicate matrix and that the cuspate contact with the matrix is continuous with the contact which is formed by the titanomagnetite. (c) Internal and external composite inclusions of ilmenite which have both irregular and sharp contacts with the titanomagnetite host. (d) An external composite inclusion with a well-developed euhedral arm extending into the silicate matrix. The dark blebs within the ilmenite are inclusions of glass. (e) External composite ilmenite occupying approximately 30% of the titanomagnetite crystal with abundant trellis lamellae along {111} planes. (f) An external composite crystal of ilmenite extending into the silicate matrix in a coarse graphic texture. A second external composite grain (upper right) is contained within the titanomagnetite but shows an irregular contact with the silicates. Fine trellis lamellae are also present.

The conformity of ilmenite crystal faces and titanomagnetite crystallographic planes in grains (a) and (b) could argue for either "exsolution" oxidation or for topotactic growth of titanomagnetite on ilmenite. The case for grain (c) is ambiguous, whereas (d), (e), and (f) are almost certainly primary. Scale bar = 25 μm.

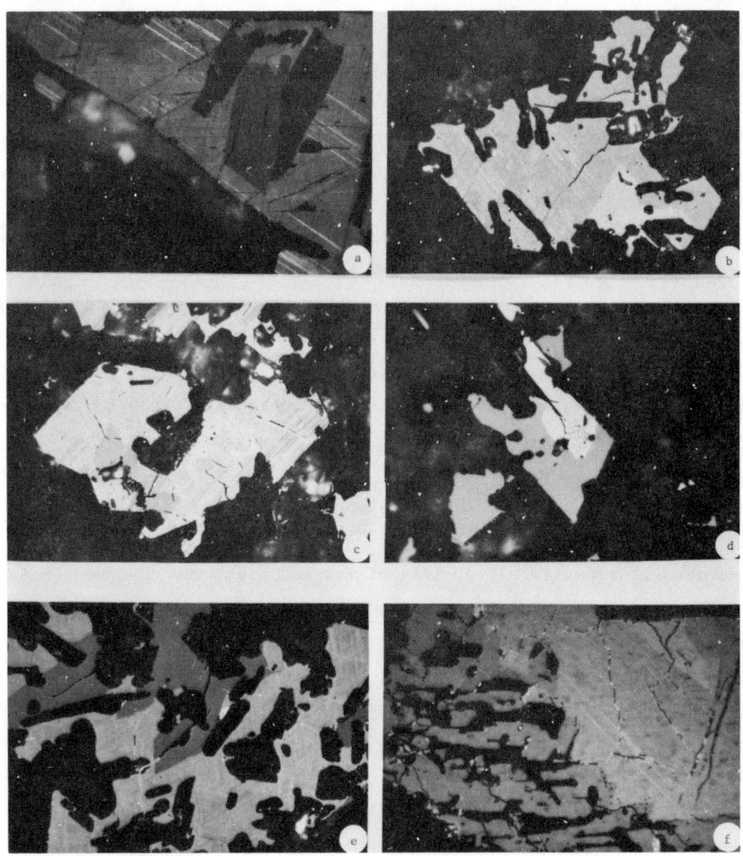
a
b
c
d
e
f

means that titanomagnetite may nucleate on earlier formed ilmenite (Fig. Hg-2a); the orientation of some composite ilmenites suggests that the overgrowth may be epitactic (Fig. Hg-2a-c).

2. Protruding arms, from external composite grains, form either eudhedral plates of ilmenite extending into the groundmass, or graphic intergrowths within the silicates. Extensions of this type can obviously not result by subsolidus diffusion but are more probably the product of primary crystallization (Fig. Hg-2d-f).

3. It has now been clearly demonstrated that ilmenite does not exsolve from titanomagnetite in the classic sense but is the result of a subsolidus oxidation process. Concentric diffusion-rim textures similar to those produced by the solute in exsolved sulfide solid solutions are not observed, and as far as is known, have not previously been recognized. The external composites are often subhedral or equant (Fig. Hg-2b) and as such do not resemble the textures that result from the unmixing of sulfide solid solutions.

4. It will be shown subsequently that advanced stages of oxidation are required for the formation of exsolved alumino-spinels in titanomagnetite. These spinels are only present in titanomagnetite grains that contain large concentrations of ilmenite along {111} planes and are *never* observed in composite ilmenite-titanomagnetite intergrowths, unless lamellar ilmenite is also present; the implication being that composite intergrowths are not the result of advanced oxidation.

5. The presence of all three textural types (trellis, sandwich, composite) of ilmenite in titanomagnetite precludes the possibility that composite inclusions are the result of advanced oxidation. Composite inclusions are often in the center of a framework of thick ilmenite laths and/or within a network of finer lamellae. The lamellae terminate sharply at the composite, titanomagnetite-ilmenite interface, and there is no evidence to suggest that inclusions are enlarged by the diffusion of material from oriented lamellae. In all cases it is judged that lamellar structures post-date composite forms of ilmenite. If large composite inclusions develop under conditions of high diffusion rates, and presumably also at high temperatures, most if not all of the ulvöspinel component would be oxidized at a relatively early stage, and further diffusion at lower temperatures to form a second generation of ilmenite would be inhibited.

There are no differences either compositionally or in the density of ilmenite lamellae which develop in either composite or non-composite types, indicating that the magnetite of composite types is as titaniferous as the magnetite of non-composite types.

6. Basalts in which all or nearly all of the titanomagnetite grains are densely packed with ilmenite lamellae along {111} planes will always contain a few grains in which the "exsolved" ilmenite has been further oxidized to ferrian rutile + titanohematite. If composite inclusions of ilmenite are the result of advanced oxidation, one might expect to see some evidence of oxidation in the composite, but this is not the case; titanohematite and ferrian-rutile only develop in composite ilmenite when discrete trellis lamellae are also present.

Composite inclusions of ilmenite therefore cannot be oxidized independently of their otherwise homogeneous titanomagnetite hosts, *i.e.*, without the cubic phase being affected. Once the titanomagnetite contains a well-established framework of ilmenite lamellae, both textural forms of ilmenite (trellis plus composite) may then follow a sequence of more advanced oxidation side by side. Furthermore, this indicates (supporting 5 above) that although large inclusions of ilmenite may be present in the titanomagnetite, the titanomagnetite is effectively and outwardly unoxidized, until {111} ilmenite lamellae appear.

7. If stoichiometric ulvöspinel, which is more titaniferous than the compositions which are typical of basaltic titanomagnetites, is completely oxidized to ilmenite, ilmenite and magnetite form in 3:1 molecular proportions according to the following equation:

$$3Fe_2TiO_4 + \frac{1}{2}O_2 = 3FeTiO_3 + Fe_3O_4$$

This 3:1 molecular ratio corresponds to a 2.2:1 volume ratio (Jensen, 1966), and many composite grains in basalts have a volume ratio of ilmenite:titanomagnetite which is substantially larger than this value.

8. Overlapping step scans for Fe, Ti, Mg, Al and Cr at 1 μm intervals by electron microprobe across composite and lamellar ilmenite in titanomagnetite show that distinct diffusion gradients for these elements are apparent for ilmenite lamellae and for some composite ilmenites, but are absent in other composites. Quantitative electron probe analyses also show that the compositions of composite and non-composite titanomagnetites within the same sample are indistinguishable, although in some basalts this similarity may also be extended to include trellis lamellae.

The evidence therefore strongly suggests that composite ilmenite-titanomagnetite intergrowths in many basalts are probably not the result of advanced oxidation and diffusion, but rather the result of primary inclusions. Such inclusions should perhaps therefore not be used when applying the Fe-Ti oxide geothermometer unless independent evidence is available substantiating that oxidation exsolution has taken place. In addition, these intergrowths should

Figure Hg-3. Sandwich types. (a) This is an extreme example of a sandwich texture but is a clear demonstration of an early crystal of ilmenite on which titanomagnetite has nucleated. (b) Two thick laths of ilmenite which appear to have a crude {111} orientation. Note that the laths are approximately parallel-sided and that the laths terminate in edges parallel to the titanomagnetite grain boundaries. (c) An internally truncated lath and a second lath which extends to the edges of the titanomagnetite crystal and shows increased width at the spinel grain boundary. (d) A thick ilmenite sandwich lath with irregular sides and with second and tertiary trellis sets along {111} planes. The sandwich lath conforms to one of the {111} planes. (e) Well-developed sandwich laths of ilmenite extending into the silicate matrix as euhedral crystals. (f) A lensoidal sandwich lath of ilmenite which is terminated at one edge of the titanomagnetite crystal but continues into the silicate matrix at the other edge in a graphic texture. The graphic ilmenite contains olivine + plagioclase and the apparent crystal terminations result from a coarse ophitic texture of plagioclase + pyroxene.

Grains (a) and (e) contain primary ilmenite whereas the ilmenite in grains (b), (c), and (d) may be either primary or may have resulted by oxidation "exsolution." The lath in grain (f) may have "exsolved" but the graphic extension to this grain is clearly primary. Scale bar = 25 μm.

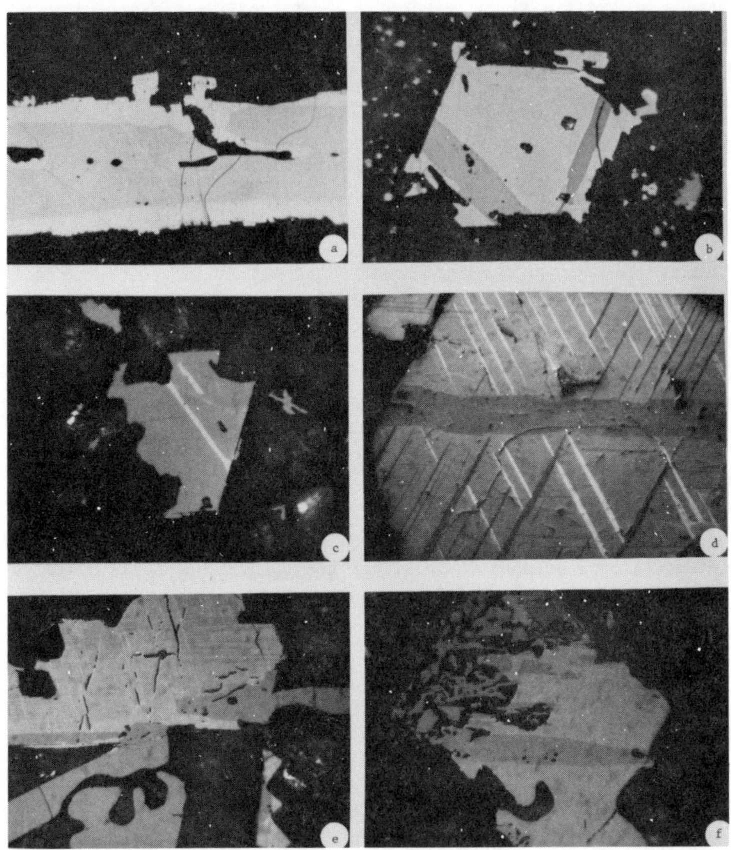
a
b
c
d
e
f

not be included within the class of titanomagnetite grains which contain large concentrations of trellis ilmenite but should, rather, be classified as a subset of homogeneous titanomagnetite when the magnetomineral oxidation classification is applied.

Sandwich type

Thick (25-50 μm) sandwich laths of ilmenite along one set of the octahedral planes are commonly observed (Fig. Hg-3). These laths generally occur in small numbers and a single lath is the most common form. These laths have sharply defined contacts with the titanomagnetite, in common with composite ilmenite, and rarely have parallel sides or tapered terminations typical of trellis ilmenite. Sandwich laths that coexist with trellis lamellae predate these lamellae, but sandwich laths that coexist with composite inclusions are rarely in contact, and their relative paragenesis is therefore indeterminate.

The results of electron microprobe scans to determine major element diffusion gradients across sandwich ilmenite-titanomagnetite contacts are inconclusive; diffusion gradients are observed in some cases but are absent in others. These sandwich laths may therefore be primary inclusions or may result by oxidation of a primary cubic spinel host; there are certain similarities with both the composite and the trellis types but their origin is not as clearly defined.

In summary, trellis lamellae clearly result by oxidation exsolution whereas sandwich and composite ilmenite are products either of oxidation or of primary crystallization from the melt.

Oxidation of titanomagnetite-ilmenite intergrowths

Two distinct textural assemblages of Fe-Ti oxides develop from the oxidation of magnetite-ulvöspinel$_{ss}$. These are (a) oxidation-"exsolution" lamellae of ilmenite along {111} planes in titanomagnetite, and (b) the pseudomorphic-oxidation products rutile, titanohematite and pseudobrookite. Both assemblages develop above 600°C but at radically different oxygen fugacities. The latter group develop paragenetically later than the former and at higher values of f_{O_2}. For example, at 800°C the equilibrium f_{O_2} for the reaction $6Fe_2TiO_4 + O_2 = 2Fe_3O_4 + 6FeTiO_3$ is $10^{-15.4}$ atms. and for the reaction $4FeTiO_3 + O_2 = 4TiO_2 + 2Fe_2O_3$ the corresponding value is $10^{-8.1}$ atms. (Verhoogen, 1962); slightly lower values ($10^{-12.0}$) have been reported for the latter reaction (Carmichael and Nicholls, 1967) but the differences in f_{O_2} are still sufficiently far removed to categorize the first group as oxidation-exsolution and the second as pseudomorphic oxidation.

Ilmenite which is produced by oxidation-exsolution is structurally controlled within the titanomagnetite host. Phases which develop subsequently from this ilmenite, with more intense oxidation, are also controlled but the control is largely compositional rather than structural. Hence rutile develops preferentially along planes of pre-exsolved ilmenite, and hematite develops preferentially in regions of pre-existing titanomagnetite. In spite of extensive cationic reorganization even the most advanced stages of oxidation reflect relic {111} pseudomorphic planes. These relic features, coupled with the higher concentrations of titanohematite that result from oxidized $Mt\text{-}Usp_{ss}$, permit such pseudomorphs to be distinguished from those originating from primary discrete ilmenite.

The oxidation trends for "exsolved" ilmenite are texturally and mineralogically similar to those of discrete primary ilmenite. These paragenetically distinct ilmenites oxidize in parallel, except at the incipient stages when the "exsolved" ilmenite is generally more susceptible to oxidation that its discrete counterpart. This difference is due to the initial variations in Fe^{2+} and Fe^{3+} and in the stabilizing effects of Mg and Al. With more intense oxidation the titanomagnetite host is gradually depleted in the ulvöspinel component as larger concentrations of ilmenite develop. The redistribution of cations also results in strong partitioning coefficients of minor stabilizing elements such as Mg and Al which are both preferentially accommodated in the spinel structure. The "exsolved" ilmenite continues to follow its path of oxidation without any dramatic effects on the titanomagnetite host. A saturation point is finally reached, however, which results in the exsolution of pleonaste ($FeAl_2O_4$-$MgAl_2O_4$) magnesioferrite ($MgFe_2O_4$) solid solutions and with the oxidation of the residual titanomagnetite to titanohematite.

In detail the following oxide assemblages develop with more intense oxidation from either the C2 or C3 stages (titanomagnetite + oxidation exsolved ilmenite) described in the previous section.

C4 stage

The first sign of post "exsolution" oxidation that is observed optically is an indistinct mottling of the ilmenite (lamellar)-titanomagnetite (intralamellar) intergrowth (Fig. Hg-4a). This mottling is due (a) to the fine serrations that develop at the "exsolution" interfaces, (b) to the formation of minute *exsolved* transparent spinels in the titanomagnetite, and (c) to the development of ferri-rutile in the ilmenite (metailmenite).

With increasing oxidation the "exsolved" metailmenite becomes lighter in color (Hem_{ss}), and the titanomagnetite changes from tan to dark brown (Mt_{ss}).

Figure Hg-4. Oxidation of titanomagnetite-ilmenite intergrowths. Stages C4, C5, and C6. (a-d) This series illustrates the progressive decomposition of fine- and coarse-grained trellis ilmenite lamellae to ferrian rutile + ferrian ilmenite and to rutile + Hem_{ss}. The lamellae are mottled in (a) and (b) but well-developed rutile lenses can be discerned in (c) and (d). Grain (d) also contained an original composite inclusion, the large central area which is now R + Hem_{ss}. The host magnetite (dark gray in the photomicrographs but brown in reflected light oil immersion) contains abundant spinel lamellae (pleonaste-magnesio-$ferrite_{ss}$) which are black and are oriented along {100} spinel planes. This series is typical of stage C4. (e) Coarse lenses of rutile + Hem_{ss} replacing original ilmenite lamellae but extending now to replace the magnetite also. Spinel lamellae in magnetite are black, and the white lamellae are hematite. (f) A completely pseudomorphed grain of titanomagnetite + ilmenite. The assemblage is R + Hem_{ss} with associated disseminated blebs of black $pleonaste_{ss}$. The original {111} fabric is still maintained with larger concentrations of rutile in areas originally occupied by ilmenite and higher concentrations of Hem_{ss} in areas which were originally titanomagnetite. This is a perfect example of stage C5. (g) This grain would be classified as C6 because of the first appearance of Pb_{ss} (irregular and medium gray, lower left). Residual magnetite is present and the remainder of the grain consists of R + Hem_{ss}. (h) The dark gray irregular lamellae are Pb_{ss}, the fine medium-gray lenses are rutile and the gray to white host is Hem_{ss}. An original {111} fabric is still evident with Pb_{ss} and with R + Hem_{ss} tending to concentrate along relic ilmenite. The fine mottled black specks are $pleonaste_{ss}$ and the larger black inclusions are silicates. This is a C6 grain. (i) The entire sequence from C4 to C7 may be followed in this original titanomagnetite + ilmenite grain starting in the upper right (with magnetite + metailmenite) and proceeding anticlockwise to R + Hem_{ss} (C5) to Pb_{ss} + Hem_{ss} + R (C6), and to Pb_{ss} + Hem_{ss} (C7).

Scale bar = 25 μm.

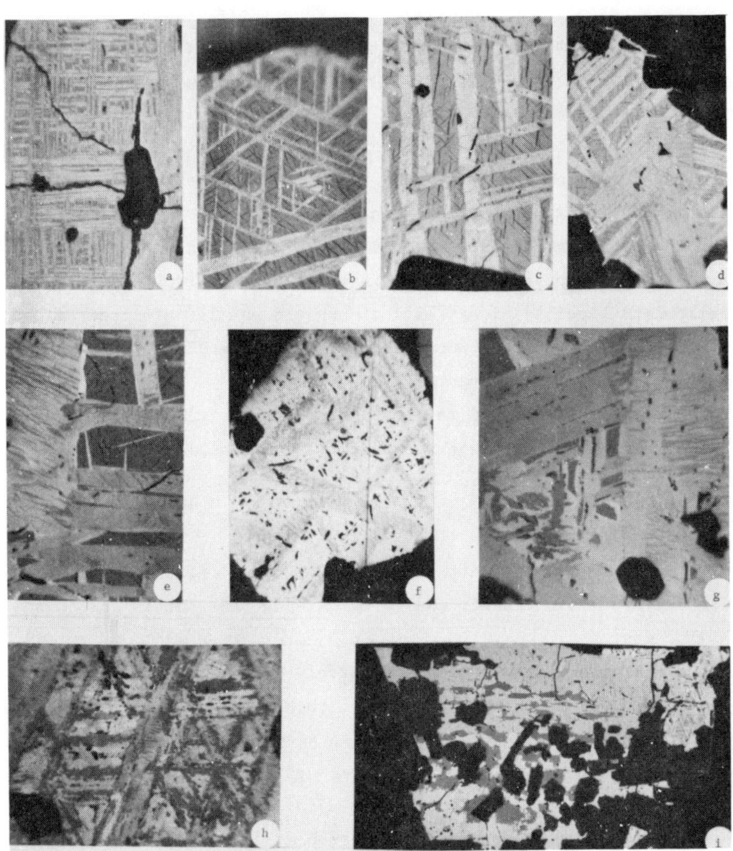
a
b
c
d
e
f
g
h
i

There is a considerable increase in reflectivity in metailmenite, internal reflections are apparent, and the lamellae show variable degrees of anisotropy caused by finely disseminated ferri-rutile. Lenses of ferri-rutile are oriented parallel to the length of metailmenite lamellae {0001} and are also present at an acute angle to the lamellae (probably $\{01\bar{1}1\}$). The lamellar assemblage corresponds to the R2 and R3 stages of discrete ilmenite which are defined in a later section. Exsolved rods of spinel in the titanomagnetite are black, and their orientation is along {100} (Fig. Hg-4b-d).

Although the entire grain is subjected to the oxidation process, the most rapid changes observed optically are within the metailmenite. Coarse lamellae and fine reticulate lamellae along {111} planes are equally affected, although lamellae towards the edges of titanomagnetite grains are always more intensely oxidized than those towards the center.

The spinel rods, the internal reflections and the mottled appearance of the metailmenite lamellae are the characteristic microscopic properties of this oxidation stage.

C5 stage

This stage of oxidation is characterized by rutile + titanohematite. Ferri-rutile may appear in transitionally oxidized grains but is absent in the more advanced stages.

Rutile and titanohematite develop extensively within the "exsolved" metailmenite lamellae and complete replacement of these lamellae are observed. This stage of oxidation is identical to the R5 stage of discrete ilmenite; the optical, textural form and orientation properties are similar, but the ratio of titanohematite:rutile is slightly greater. The serrated contacts between the "exsolved" lamellar planes and the host titanomagnetite are more pronounced than those observed at the C4 stage. With more intense oxidation the contacts become irregular and the lamellar rutile-titanohematite assemblage extends into and begins to develop within the titanomagnetite. The contacts have the appearance of replacement-fronts, although at least part of the rutile probably originates from the decomposition of titanomagnetite. The progressive enlargement of the lamellar assemblage and the complete breakdown of the titanomagnetite are illustrated in Figure Hg-4e. Although replacement at this stage is complete, the pre-"exsolved" {111} texture may still persist as a pseudomorphic relic.

The relic {111} fabric is optically prominent, particularly under crossed-nicols, and is dominantly controlled by the ratio of rutile:titanohematite which may be several orders of magnitude greater in the lamellar zones (after ilmenite) than it is in the intralamellar sectors (after titanomagnetite). Other contributory factors to its prominence are the darker color of the lamellar titanohematite and the optical continuity displayed by relic {111} lamellar sets (Fig. Hg-4f). It appears that there is little or no diffusion across these {111} "exsolved" planes and that the primary establishment of the Ti-rich zones is maintained with more intense oxidation. In other cases the relic {111} fabric is less apparent because there are significantly larger concentrations of rutile in the intralamellar areas; these types probably reflect initially lower degrees of oxidation "exsolution" followed by rapid intense oxidation of the "unexsolved" ulvöspinel component.

C6 stage

This stage, in common with the R6 stage of discrete ilmenite, is defined by the incipient formation of pseudobrookite$_{ss}$ (Pb_{ss}) from rutile + titanohematite.

Pb_{ss} develop almost exclusively along {111} relic planes (Fig. Hg-4g-h); its preferential development along these planes is a reflection of the initially higher Ti content and the correspondingly higher rutile:titanohematite ratio that results as a consequence of intense oxidation of pre-existing "exsolved" ilmenite lamellae. Pseudobrookite$_{ss}$ are characteristically heterogeneous in color, reflectivity and degree of optical anisotropy. The rutile lenses are much finer than those observed at the same stage of oxidation in discrete ilmenite, and although selective replacement of rutile by Pb_{ss} (refer to R5) may occur, it is much more difficult to observe or follow. Electron microprobe analyses of these color-contrasted pseudobrookites are inconclusive although there is the suggestion of Fe^{2+}-Fe^{3+} variability. The lamellar Pb_{ss} are typically jagged with finely textured cuspate contacts against the intralamellar areas.

Total replacement of the original titanomagnetite by rutile + titanohematite is not a prerequisite for the formation of Pb_{ss}: Relic areas of titanomagnetite tend to be highly resistant to oxidation--particularly towards the central-most regions of grains. These areas are either triangular or rectangular in shape (Fig. Hg-4g-i), contain large concentrations of exsolved aluminous-magnesio-ferrite and are isotropic and dark brown in color.

The development of members of the Pb_{ss} series are indicative of more intense oxidation, and the three-phase assemblage Pb_{ss} + rutile + titanohematite with or without unoxidized titanomagnetite (plus exsolved spinel) defines this stage of oxidation.

Figure Hg-5. Oxidation of titanomagnetite-ilmenite intergrowths. Stage C7. (a) The trellis portion of this grain consists of Pb_{ss} + Hem_{ss} and the upper left-hand portion consists largely of R + Hem_{ss} + associated Pb_{ss}. (b) A completely pseudomorphed grain consisting of Pb_{ss} along {111} relic planes in a host of Hem_{ss}. (c) Thick original trellis lamellae and composite inclusions of ilmenite now oxidized to Pb_{ss}. The white areas are Hem_{ss} and the medium-gray inclusions in Pb_{ss} are largely R although Hem_{ss} is also present. (d) An original sandwich ilmenite lath pseudomorphed by Pb_{ss} + Hem_{ss}. The Pb_{ss} area to the lower right was probably a composite inclusion; the inclusions are R + associated Hem_{ss}. $Pleonaste_{ss}$ blebs are present in the upper left of the photomicrograph and the remaining dark patches are polishing artifacts and silicate inclusions. Some degree of recrystallization of the Pb_{ss} has taken place (compare with grain (b) for example). (e) Although a crude linear fabric of the Pb_{ss} is present in this grain the original titanomagnetite probably underwent very rapid oxidation which resulted in the breakdown directly to Pb_{ss} + Hem_{ss} + associated R, rather than through the intermediate stages of C3 to C6. (f) This is another example of very rapid oxidation of original titanomagentite. There are no residual spinels present and rutile is sparse. MgO and Al_2O_3 are now largely in the Pb_{ss} and this grain as well as grain (e) should be contrasted with the examples shown in Figure Hg-6 where the effects of the partitioning of Mg + Al by oxidation "exsolution" are well illustrated (*i.e.*, by the formation of $pleonaste_{ss}$).

Scale bar = 25 μm.

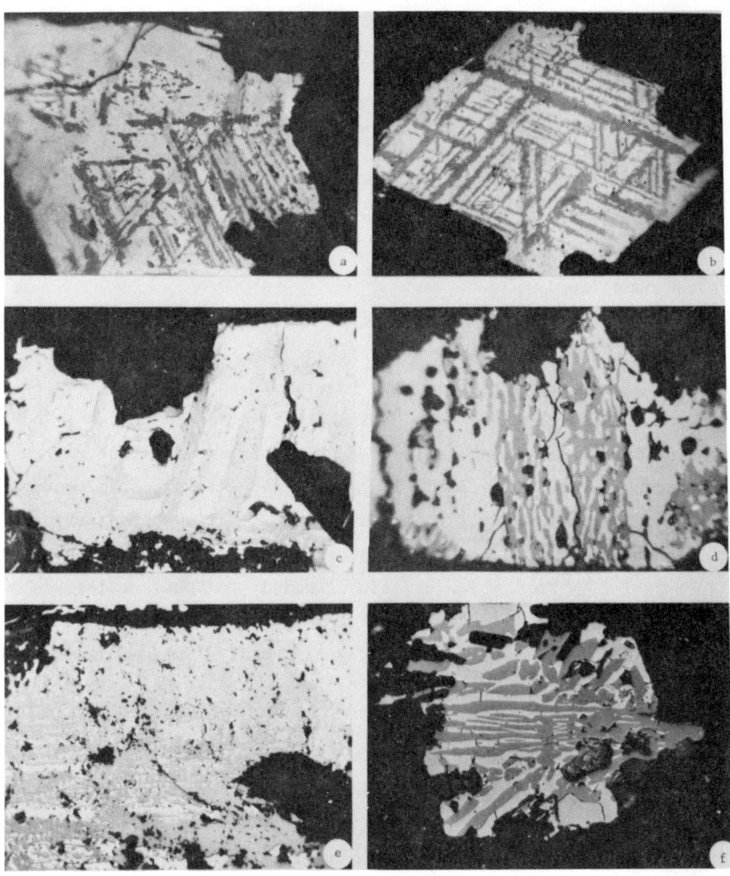

C7 stage

This stage is characterized by the assemblage pseudobrookite$_{ss}$ + hematite$_{ss}$ and is the most advanced stage of oxidation in the titanomagnetite classification.

Pb$_{ss}$ are typically present in two distinct textural forms: either as pseudomorphic lamellae along relic {111} planes (Fig. Hg-5a-c) or as graphic intergrowths with titanohematite (Fig. Hg-5d-f). These textural forms have slightly different origins, although the transition of lamellar to graphic fabrics can be traced and is the result of a progressive redistribution of Pb$_{ss}$ within titanohematite.

Lamellar pseudobrookite appears to develop most commonly in grains which have previously undergone extensive oxidation "exsolution"; such grains have high Ti lamellae and high Fe hosts with the result that Pb$_{ss}$ develop selectively in those areas that are compositionally favorable. In cases where initial oxidation "exsolution" is limited, or in titanomagnetite grains containing sandwich or composite inclusions, the spinel host continues to maintain large concentrations of Ti which may result in titanohematite + rutile (C6) or in pseudobrookite. Such Pb$_{ss}$ intergrowths are graphic, but these are always accompanied by Pb$_{ss}$ which are also lamellar in form, reflecting an original {111} "exsolved" ilmenite fabric.

Inclusions in lamellar Pb$_{ss}$ are dominantly rutile, whereas inclusions in intralamellar Pb$_{ss}$ are dominantly titanohematite. These inclusions may result, in part, from decomposition of Pb$_{ss}$ with slow cooling, but are more likely the result of incomplete reaction of ilmenite and of titanomagnetite, respectively. Both forms of Pb$_{ss}$ are optically similar, both show variations in color, and both types are polycrystalline; the variations in color and degree of anisotropy are not due to random crystal orientation, because the contacts between light bluish-gray and dark gray areas are typically gradational. Lamellar Pb$_{ss}$ contacts generally retain their jagged cuspate form (C6) although with extensive redistribution relatively coarse polyhedral grains with well defined contacts against the titanohematite may result (Fig. Hg-5f).

Titanohematite is considerably whiter than that observed in either C5 or C6; it is more intensely anisotropic, and red internal reflections are apparent in thin plates. Titanohematite associated with graphic Pb$_{ss}$ is typically polycrystalline, whereas in association with lamellar Pb$_{ss}$ it is either polycrystalline or in laminated optical continuity.

The C7 oxidation stage of titanomagnetite is analogous to the R7 stage of discrete ilmenite, and both are present in close association. High concentrations of titanohematite in C7 contrast with trace amounts of titanohematite in R7. These variations are distinctive and allow completely pseudomorphed crystals to be distinguished as either titanomagnetite or as ilmenite. In addition to these modal variations, compositional differences, the preservation of {111} relic planes and crystal morphology are also usefully distinguishing features.

An alternative sequence to the high-temperature oxidation stages described above is observed in some basalts, but is more typical of picritic suites and ankaramites in which the titanomagnetites contain high initial values of Mg and Al. Oxidation "exsolution" of ilmenite from these titanomagnetites enriches the spinel to an even greater extent in Mg and Al, and pleonaste-magnesioferrite$_{ss}$ members exsolve. This preferred partitioning into the spinel is similar to that of basaltic titanomagnetite but because the concentrations are substantially higher in the picritic suites the magnetite is stabilized to the point that it resists oxidation even to the limits at which Pb_{ss} members develop. The sequence of oxidation for these suites is illustrated in Figure Hg-6, and the salient features are as follows: (1) Ilm_{ss} which form by oxidation "exsolution" decompose to metailmenite (R + Hem_{ss} in most cases) and large concentrations of pleonaste$_{ss}$ are visible in the Mt_{ss} host (Fig. Hg-6a); (2) Pb_{ss} develop directly from R + Hem_{ss} and either partially (Fig. Hg-6a) or completely pseudomorph pre-existing Ilm_{ss} lamellae; (3) with progressive oxidation and with increasingly larger proportions of exsolved pleonaste$_{ss}$ the host Mt_{ss} is gradually depleted in Mg and Al, becomes less stable and is oxidized to Hem_{ss} (Fig. Hg-6c-e); the exsolved pleonaste$_{ss}$ cannot be accommodated in the Hem_{ss} and these spinels appear as disaggregated undigested blebs (Fig. Hg-6c); (4) if oxidation continues and cooling is prolonged the original Mt_{ss} is replaced by Hem_{ss} and subhedral cyrstals of magnesioferrite$_{ss}$ result (Fig. Hg-6f-g). The end product of this oxidation is similar to that of basaltic titanomagnetite (Pb_{ss} + Hem_{ss}) with the addition of magnesioferrite$_{ss}$ and with substantially larger concentrations of MgO in Hem_{ss} and with MgO and Al_2O_3 in Pb_{ss} (see section on phase chemistry).

Figure Hg-6. Oxidation of Mg-Al titanomagentites. (a) A subhedral grain of magnetite containing {111} metailmenite lamellae and the incipient formation of Pb_{ss} (medium gray) along the largest of the thick lamallae. The magnetite contains coarse lamellae of pleonaste-magnesioferrite$_{ss}$ oriented along {100} of the spinel host planes. (b) A more advanced oxidation stage from the assemblage in grain (a) with both trellis and composite ilmenite oxidized to Pb_{ss} + Hem_{ss}. (c) Oxidation is now extended to include the magnetite which is replaced by hematite, and the exsolved spinel lamellae are scattered throughout as finely disaggregated blebs. (d-e) These photomicrographs are at a lower and higher magnification, respectively, illustrating the oxidation of original ilmenite to Pb_{ss}, the exsolution of pleonaste$_{ss}$ in magnetite, the replacement of Mt_{ss} by Hem_{ss}, and the resistance of unresorbed spinels which are dispersed within the Hem_{ss}. (f) A subhedral grain of original titanomagnetite with satellite skeletal extensions consisting of Pb_{ss} + Hem_{ss} and two large rectangular areas of magnesioferrite$_{ss}$ (black, lower left). None of the original magnetite is any longer present. (g) A higher magnification photomicrograph of a grain similar to that of grain (f) but with the magnesioferrite$_{ss}$ now being partially replaced along {111} by hematite.

Scale bar = 25 μm.

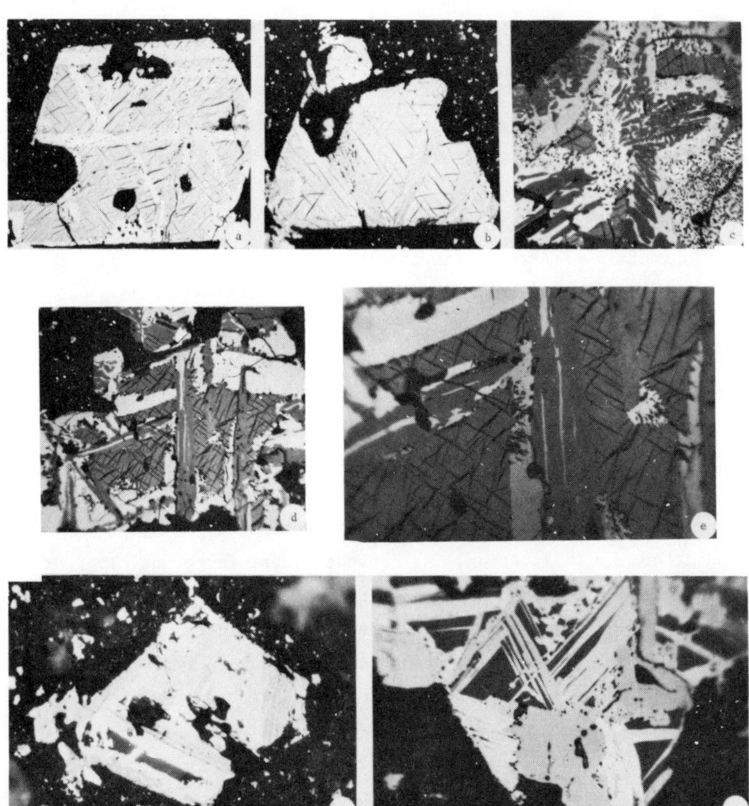
a
b
c
d
e
f
g

Oxidation of discrete primary ilmenite

Seven clearly defined oxidation assemblages emerge for ilmenites in the same suites of basalts and in the same specimens described for titanomagnetite above. Examples of these assemblages are illustrated in Figures Hg-7-9, and the classification of assemblages with progressive high temperature oxidation is as follows:

1) homogeneous ilmenite
2) ferrian ilmenite + ferrian rutile
3) ferrian rutile + (ferrian ilmenite)
4) rutile + titanohematite + ferrian rutile + ferrian ilmenite
5) rutile + titanohematite
6) rutile + titanohematite + (pseudobrookite)
7) pseudobrookite + (rutile + titanohematite)

Although this is the most commonly observed sequence, an alternative sequence of ilmenite oxidation results in the formation of pseudobrookite directly, thus bypassing other intermediate assemblages. Factors controlling the development of either of these progressive sequences are largely temperature dependent and by implication, f_{O_2} dependent; the former develops at lower temperatures and the latter at correspondingly higher temperatures ($\pm$ 800°C). Shielded ilmenite, or ilmenite which has remained unreacted at elevated temperatures, will follow the sequence of assemblages outlined above provided that levels of f_{O_2} are still sufficiently high and with the provision that slow cooling ensues.

Ilmenite oxidation classification

Details of the seven stages of ilmenite oxidation observed optically are as follows; each stage is prefixed by R (rhombohedral) to distinguish this classification from the cubic (C) ulvöspinel-magnetite classification.

R1 stage. Homogeneous, unoxidized primary ilmenite is the lowest state of oxidation in this classification (Fig. Hg-7a).

R2 stage. The first signs of ilmenite oxidation observed optically are an increase in reflectivity and a change in the color of ilmenite from reddish-brown to light-tan. Fine wisp-like sigmoidal lenses (1-5 μm) of ferrian rutile develop along {0001} and $\{01\bar{1}1\}$ parting planes of the ilmenite. These lenses generally develop in equal concentrations throughout the entire ilmenite crystal although aggregations towards the grain boundaries and along fractures are also present. There is a tendency for lamellae parallel to {0001} to be coarser than those developing along the rhombohedral planes, and by analogy with exsolution in ilmenite-hematite$_{ss}$ these finer lenses may develop at lower temperatures and under conditions of lower diffusion activity (Fig. Hg-7b). The distribution of larger and smaller lenses, however, is not an ordered or seriate arrangement as is commonly the case in Ilm_{ss} or Hem_{ss} which have exsolved from Ilm-Hem_{ss}.

R3 stage. With more advanced diffusion the lenses become thicker and more abundant. There are slight changes in color from pale gray-white to white as the lenses thicken, with marked increases in reflectivity, the appearance of white internal reflections, and a pronounced increase in anisotropy. Lenses within any one crystallographic plane are in optical continuity, are doubly terminated, and are relatively well defined at each of the progressive growth stages. Ferrian rutile lenses are mantled by haloes of bleached ferrian ilmenite that are gradationally lighter in color as the ferrian ilmenite host is approached. This variation in color is due in part to differential concentrations of Fe^{+2} and Fe^{+3} in the host, and in part to the concomitant diffusion of Ti from the ferrian ilmenite to the ferrian rutile lamellae. The {0001} and $\{01\bar{1}1\}$ lenses rarely intersect or continue along precisely the same planes for any distance; tapering occurs on contact and the sigmoidal form of the lenses gives rise to classic syneusis textures (Fig. Hg-7c-d).

R4 stage. This stage of ilmenite oxidation is considerably more complex than the preceding stage or the stage that follows. The assemblage is a four-phase metastable assemblage comprising ferrian ilmenite, titanohematite, ferrian rutile and rutile. The rhombohedral ferrian ilmenite and titanohematite are present in approximately equal concentrations and constitute the host; ferrian rutile and rutile are present as sigmoidal lenses or as finely disseminated lamellae and these phases are oriented along presumed {0001} and $\{01\bar{1}1\}$ planes of the original ilmenite. Optical continuity of phases along

Figure Hg-7. Oxidation of primary ilmenite. Stages R1 to R4. (a) Homogeneous discrete crystal of primary ilmenite. Stage R1. (b) Subhedral crystal of ilmenite showing one coarse and many fine lamellae of ferrian rutile. Stage R2. (c) Fine sigmoidal lenses of ferrian rutile in ferrian ilmenite. Stage R3. (d) Similar to grain (c) but with one particularly well-developed set of ferrian rutile lamellae along {0001} parting planes of ferrian ilmenite. The subsidiary planes are $\{01\bar{1}1\}$ or $\{01\bar{1}2\}$. Stage R3. (e–g) Three metailmenite crystals illustrating the progressive transformation of ferrian rutile to rutile, and of ferrian ilmenite to titanohematite. All grains contain the four-phase metastable assemblage typical of stage R4. The light-gray sigmoidal lamellae are rutile, the lighter-gray lamellae (*cf.* grains (b), (c), (d)) are ferrian rutile and the host phases are ferrian ilmenite (medium gray) and titanohematite (gray to white). The oxide complex to the upper right of grain (g) is an oxidized olivine crystal (see Fig. Hg-17).

Scale bar = 25 μm.

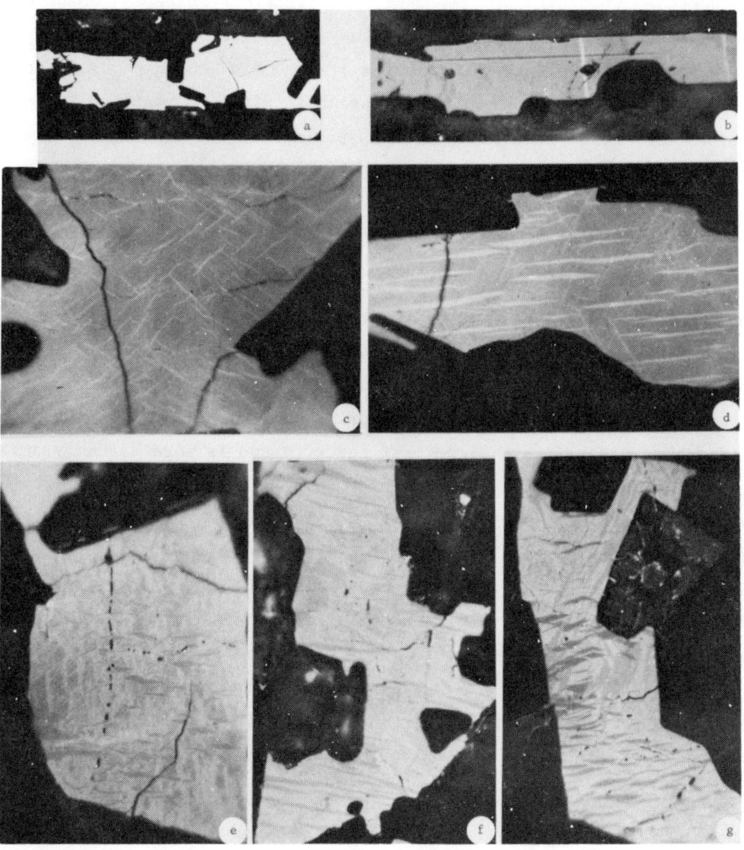
a
b
c
d
e
f
g

these planes is maintained, but there are marked reflectivity increases in both the host and the lamellar components when this assemblage is compared with the R3 assemblage. Diffusion rates and the redistribution of Fe^{2+}, Fe^{3+} and Ti are highly variable even within the limits of a single crystal, and as a consequence the compositions of phases are also highly variable. The process reflected in this four-phase metastable assemblage is the transition of ferrian ilmenite to titanohematite and of ferrian rutile to rutile (Fig. Hg-7e-3).

R5 stage. Rutile and titanohematite develop extensively at this stage of oxidation (Fig. Hg-8a). Ferrian rutile and ferrian ilmenite may appear as finely disseminated unreacted phases but these gradually disappear with increased diffusion and oxidation. This stage is represented by a simple two-phase assemblage: The titanohematite host is, in general, optically homogeneous and each grain has the properties of a single crystal. Titanohematite is considerably whiter, has a higher reflectivity and is more strongly anisotropic than its incipient counterpart in R4. These properties result from an increase in Fe^{+3} and a loss of TiO_2. The ferrian rutile lenses are sharply defined and crystallographically controlled along $\{0001\}$ and $\{01\bar{1}1\}$ parting planes. These lenses are in optical continuity along their respective orientations, are strongly anisotropic, and show distinct yellow or yellow-red internal reflections. From a textural standpoint stages R3, R4 and R5 are identical, insofar as each assemblage contains a member of the ilmenite-hematite series and each assemblage is characterized by lenses of a Ti-rich oxide which are crystallographically controlled within the host.

R6 stage. This stage is characterized by the development of Pb_{ss} from the R5 rutile-titanohematite assemblage. In the incipient stages, pseudobrookite is concentrated along cracks and grain boundaries and gradually replaces the central-most regions of the grain by preferential migration along rutile lenses. Both titanohematite and rutile participate in the reactions, but the development of pseudobrookite in frond-like protuberances is most clearly controlled by the distribution and orientation of rutile (Fig. Hg-8 b-e). Pseudobrookite develops as a result of solid-state diffusion and counter diffusion of cations, with the addition of oxygen, between the rutile and the titanohematite. Because the development of pseudobrookite is controlled by pre-existing rutile and because of the compositional differences between the two primary phases, the resulting pseudobrookite is optically and compositionally heterogeneous. The pseudobrookite is polycrystalline and varies from pale to dark gray in color, being slightly darker in close proximity to rutile and lighter in regions adjacent to or in contact with titanohematite. These optical differences reflect higher TiO_2 contents and correspondingly higher $FeTi_2O_5$ contents in those pseudobrookites developing from nuclei of rutile. Similarly, pseudobrookites which develop predominantly by excess TiO_2 diffusion into titanohematite result in an enriched Fe_2TiO_5 component. Variations in color are the dominant differences in these pseudobrookites; reflectivity and the degree of anisotropy are almost identical, as are the reddish internal reflections.

The well-defined lensoidal texture of rutile is totally disrupted by the replacement of pseudobrookite. Replacement is rarely complete and even in the most advanced stages unreacted subgraphic inclusions of rutile and titanohematite tend to persist in the pseudobrookite host. These inclusions are neither ordered in distribution nor crystallographically controlled, which is a useful distinguishing textural feature between generative pseudobrookite (*i.e.*, Pb_{ss} from pre-existing assemblages) and pseudobrookite which is undergoing decomposition (to rutile + $hematite_{ss}$).

R7 stage. This is the most advanced stage of oxidation in the ilmenite classification and is represented by the predominance of pseudobrookite (Fig. Hg-8f-h). Intermediate stages between complete pseudomorphs and the R6 assemblage (ferrian rutile + titanohematite + pseudobrookite) are frequently observed in single polished sections; in fact, the entire spectrum from totally unoxidized ilmenite to pseudomorphs of pseudobrookite after ilmenite although common in many basalts are also occasionally present in single hand specimens.

Somewhat similar optical and textural variations as those noted in R6 apply here. Pseudobrookites are optically heterogeneous, and faint ghost-like traces of original ferrian rutile are sometimes apparent; these relic lenses are particularly evident in crossed-nicols, and two distinct sets of extinction occur on rotation of the microscope stage. Pseudobrookite generally constitutes 80-90% of the pseudomorph, ferrian rutile is next in abundance, and titanohematite is a minor component or totally absent. Ferrian rutile and titanohematite occur as drop-like inclusions or as subgraphic intergrowths and show no preferred orientation in the pseudobrookite. As noted in R6, these textures do not result from the breakdown of pseudobrookite but simply reflect the incomplete formation of pseudobrookite from a pre-existing assemblage (viz. ferrian rutile + titanohematite). Our isothermal decomposition experiments on $pseudobrookite_{ss}$ (Haggerty and Lindsley, 1970) show that

Figure Hg-8. Oxidation of primary ilmenite. Stages R5-R7. (a) This grain is a classic example of stage R5 with coarse rutile lenses in a host of titanohematite. (b-c) Pb_{ss} (gray) mantles on the assemblage R + Hem_{ss} after original primary ilmenite. Grain (b) contains abundant rutile in the mantle assemblage whereas grain (c) is mostly pseudobrookite. These mantles develop progressively from the two-phase assemblage illustrated by grain (a), and the assemblage is characteristic of the R6 stage of oxidation. (d) In contrast to grains (b) and (c) Pb_{ss} has developed along cracks in an ilmenite grain which is otherwise at the R5 stage (R + Hem_{ss}). Note that the protruding crystals of Pb, which are normal to the cracks, are parallel to the directions of the rutile lamellae. Stage R6. (e) The assemblage in this grain is similar to those in grains (b), (c) and (d) and illustrates the disruption of rutile lenses by the encroaching Pb_{ss} which replaces the assemblage R + Hem_{ss}. Cuniform inclusions in Pb_{ss} are rutile (light gray) and Hem_{ss} (white). Stage R6. (f) The core of this oxidized crystal consists of R + Pb_{ss} with associated Hem_{ss}, whereas the outermost part of the crystal consists of a Hem_{ss} matrix with Pb_{ss} lamellae along a crude {111} relic fabric. The relative ratios of R:Pb_{ss} and Hem_{ss}:Pb_{ss} suggest that the core region was originally Ilm_{ss} and that the mantle was originally Usp-Mt_{ss}, an example comparable to the relationship illustrated in Figure Hg-3a. Stages C7 and R7. (g-h) Completely pseudomorphed crystals after original primary ilmenite. Dark gray is Pb_{ss}, medium gray is rutile, and the whitish areas are Hem_{ss}. Stage R7.

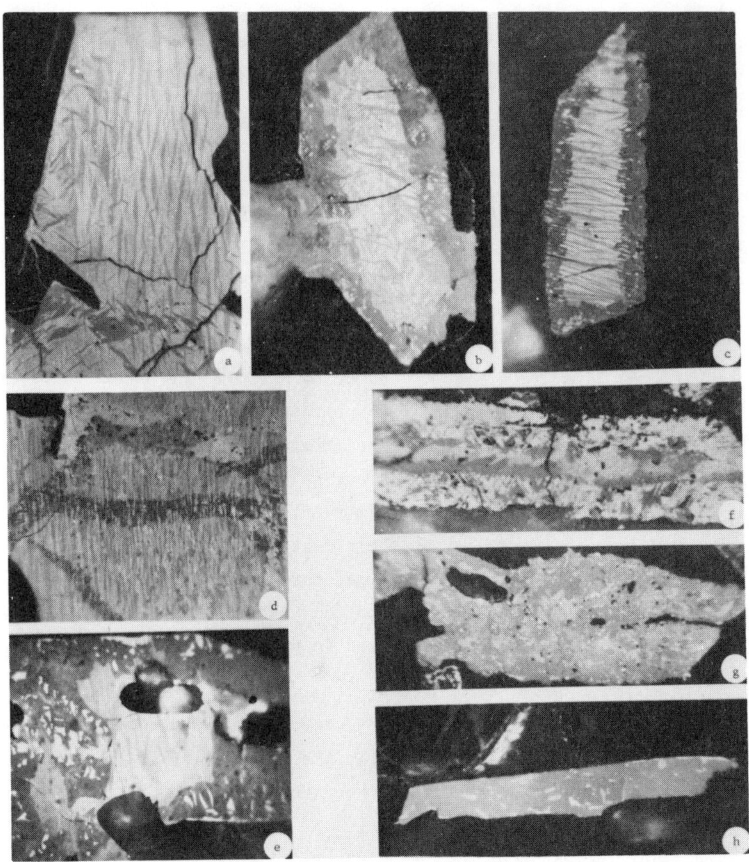
a
b
c
d
e
f
g
h

rutile + hematite$_{ss}$ or rutile + ilmenite$_{ss}$ develop along {100} or {110} planes as fine lamellae or as lenses in the pseudobrookite host; textures of this form do exist in basaltic ilmenite and must therefore result by decomposition at lower temperatures.

An alternative oxidation trend to the seven stages described above are present in a small percentage of basalts and examples and are illustrated in Figure Hg-9. In this trend Pb_{ss} develop directly from ilmenite, and other intermediate assemblages are not observed except in these ilmenite crystals which are partially oxidized, or in ilmenites which have escaped initial oxidation. The distinction between these trends appears to be temperature dependent, with the direct formation of Pb_{ss} developing, at higher temperatures, and the rutile-titanohematite to Pb_{ss} trend developing at somewhat lower temperatures (*i.e.*, above 800°C and below 800°C). The assemblage rutile + titanohematite which may be present in partially reacted ilmenite is paragenetically later than Pb_{ss} developing directly from ilmenite, suggesting that oxidation is not isothermal but is continuous with cooling. The fact that second generation Pb_{ss} (from rutile + titanohematite) are rarely observed in basalts displaying the direct formation of Pb_{ss} from ilmenite suggests furthermore that high temperature oxidation is kinetically rapid and that lower-temperature oxidation is expectedly sluggish.

There are several distinctive features in pseudobrookites which develop at elevated temperatures:

a) Crystals are a darker gray, rarely show the typical reddish internal reflections, have slightly higher reflectivities and are distinctly less anisotropic than those pseudobrookites associated with the indirect formation of Pb_{ss} from rutile + titanohematite.

b) The major included phase in pseudobrookite is rutile. A crude lineation of bleb-like inclusions is sometimes present but generally no well-defined crystallographic ordering is apparent.

c) The pseudobrookite pseudomorphs are either single crystals or are coarsely polycrystalline.

d) Pseudobrookite is considerably more homogeneous and shows no preferred orientation in the ilmenite that it replaces.

e) The directly formed Pb_{ss} are enriched in $FeTi_2O_5$, whereas the indirectly formed Pb_{ss} contain higher concentrations of Fe^{3+}, lower concentrations of Fe^{2+} and TiO_2 and contain more of the Fe_2TiO_5 component.

The distinctions are in accord with experimental data for the join $FeTi_2O_5$-Fe_2TiO_5 (Lindsley, 1967; Haggerty and Lindsley, 1970) and on heating experiments of natural ilmenites which confirm that the direct formation of Pb_{ss} is rate controlled and temperature dependent (Haggerty, unpublished).

PHASE CHEMISTRY OF OXIDATION ASSEMBLAGES

Introduction

There are wide variations in the compositions of the oxidation assemblages which develop from primary titanomagnetite and from ilmenite. These variations are dependent upon and result from: (a) the composition of the initially crystallizing magma; (b) the composition of the primary oxides; (c) the degree to which decomposition has proceeded; and (d) the absolute values of T°C and f_{O_2} of initial crystallization and of deuteric cooling. In most instances a single variation is difficult to attribute to a single chemical or kinetic parameter but in order to outline in general the basic principles of oxide decomposition, a small but judicious choice of examples will be considered. These examples do not cover the entire, or the expected, compositional ranges of the assemblages discussed in the previous section, nor indeed does the choice address many of the outstanding problems in textural interpretation.

Discussion is limited to the mineral chemistry of two basalts from Iceland, an olivine basalt from western Iceland (WA14-1), and a tholeiitic basalt from eastern Iceland (JS-10); two coarse-grained diabasic basalts from Oregon (DHL 296/69 and DHL 296/69); and a picritic basalt from Hawaii (H-5). The Iceland basalts are chosen because of their initial compositional variations, because both basalt samples contain a wide variation in oxidation assemblages and because JS-10 is one sample from a continuous sampling of a single lava profile displaying systematic oxidation variations from base to top. The Snake River plateau basalts illustrate an equally wide variation in oxidation assemblages, but the distinction for these samples is that oxidation can be demonstrated to be related to joints and contrasts with the Iceland examples where oxidation is particularly pervasive towards the central-most regions of lava flows. The Hawaii sample is highly oxidized and although neither the primary mineralogy nor other intermediate assemblages are present, the end products of oxidation are particularly well developed, coarse grained, closely approximate equilibrium, and are more magnesian and aluminous than the other examples. Representative analytical data of a selection of assemblages for these samples are given in Tables Hg-1 through Hg-4, and these data are plotted in Figures Hg-10 through Hg-12.

(Text resumes on p. Hg-46.)

Figure Hg-9. Rapid oxidation of primary ilmenite. (a-f) This series of photomicrographs illustrates the formation of Pb_{ss} members directly from Ilm_{ss} members bypassing oxidation stages R2 to R6. Grains (a) and (b) show Pb_{ss} in association with large areas of partially oxidized ferrian ilmenite which contain oriented lamellae of ferrian rutile. In these grains, and in grains (d) and (e), the Pb_{ss} is the product of rapid oxidation at high temperatures, the unoxidized portions of the ilmenite follow the sequence of progressive oxidation stages illustrated in Figures Hg-7 and Hg-8, so that grains (a) and (b) contain Pb_{ss} typical of stage R7 as well as areas classified as R3; grain (c) contains R7 as well as R4 (uppermost portion: R + Hem_{ss} + ferrian rutile + ferrian ilmenite); grain (d) has gone one stage further and the unoxidized portion is now at R6 (R + Hem_{ss}) with the remainder of the grain at R7. For grains (e) and (f) high temperature oxidization proceeded very nearly to completion, with the entire grain advancing uniformly to the R7 stage.

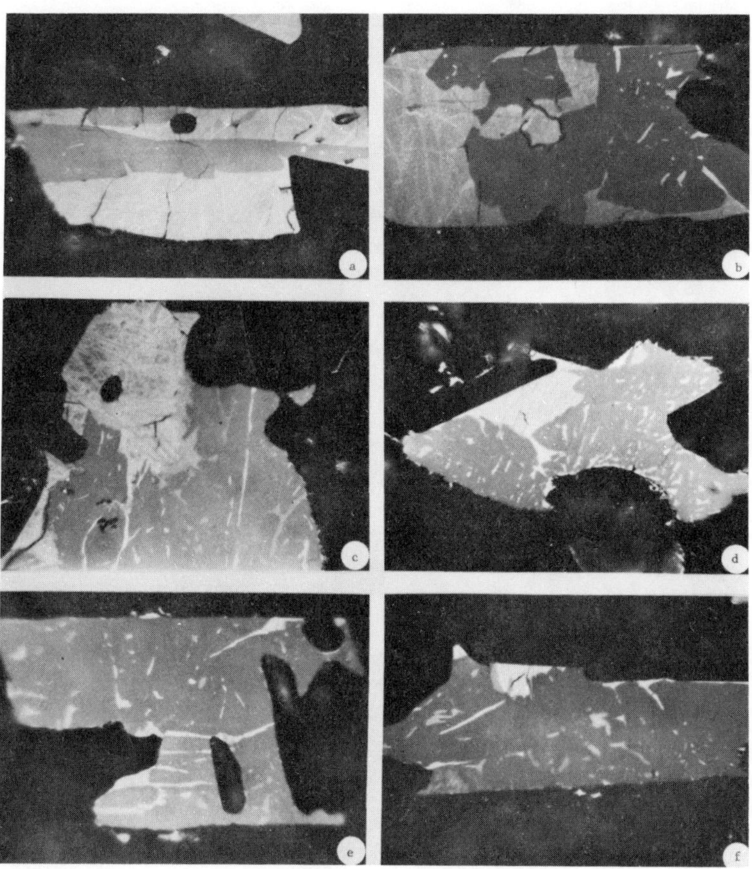
a
b
c
d
e
f

Table Hg-1. Representative electron microprobe data for coexisting oxides in sample WA14-1. HI = homogeneous primary ilmenite; C = composite; T = trellis; S = sandwich; I (+R) = ferrian ilmenite + ferrian rutile; Mt (+C), (+S), (+T) = Mt_{ss} with composite, sandwich and trellis ilmenite, respectively; Sp = pleonaste-magnesioferrite$_{ss}$ + H + R = Hem_{ss} + rutile. Fe_2O_3 has been calculated assuming mineral stoichiometry and a perfect electron microprobe analysis. R + H analyses are broad beam analyses and F_2O_3 is calculated on the Pb structural basis with 3 cations and 5 oxygens.

	Ilmenites							Titanomagnetites					
	HI 1	C 2	T 3	S 4	I(+R) 5	H+R 6	Pb 7	Mt(+C) 8	Mt(+T) 9	Mt(+S) 10	Mt(+Sp) 11	Mt(+R+H) 12	H+R 13
MgO	2.32	2.17	2.71	2.59	2.50	2.13	2.04	1.34	0.93	0.95	1.46	3.73	1.56
Al_2O_3	0.18	0.10	0.08	0.94	0.12	0.03	0.15	1.92	2.54	1.56	2.83	3.39	0.61
TiO_2	49.18	49.84	49.59	48.94	45.95	47.96	36.94	20.01	12.96	19.70	7.88	0.97	21.76
Cr_2O_3	0.06	0.01	0.01	0.00	0.02	0.00	0.00	0.06	0.07	0.05	0.13	0.09	0.05
MnO	0.54	0.63	0.65	0.50	0.68	0.64	0.65	0.52	0.52	0.47	0.37	1.51	0.34
FeO	39.59	40.41	39.19	38.88	36.23	7.96	0.25	47.55	41.34	47.80	36.35	25.59	0.00
Fe_2O_3	7.89	6.06	6.95	8.27	13.54	40.30	59.04	28.49	40.55	29.83	51.33	65.23	75.37
Total	99.76	99.22	99.18	100.12	99.04	99.02	99.07	99.89	98.91	100.36	100.35	100.51	99.69
Oxygens	3	3	3	3	3	5	5	32	32	32	32	32	5
Mg	0.086	0.081	0.101	0.096	0.094	0.125	0.120	0.591	0.416	0.419	0.644	1.624	0.091
Al	0.005	0.003	0.002	0.002	0.003	0.001	0.007	0.670	0.833	0.544	0.990	1.166	0.028
Ti	0.921	0.939	0.931	0.920	0.868	1.398	1.093	4.472	2.937	4.393	1.758	0.214	0.641
Cr	0.001	0.000	0.000	0.000	0.000	0.000	0.000	0.013	0.018	0.012	0.031	0.021	0.001
Mn	0.011	0.013	0.014	0.010	0.014	0.022	0.022	0.131	0.134	0.119	0.093	0.373	0.011
Fe^{2+}	0.825	0.847	0.819	0.813	0.762	0.253	0.001	11.765	10.425	11.856	9.021	6.247	0.000
Fe^{3+}	0.148	0.114	0.131	0.155	0.256	1.198	1.749	6.342	9.199	6.658	11.464	14.327	2.222
Cations	1.997	1.997	1.998	1.996	1.997	2.997	2.992	23.984	23.962	24.001	24.001	23.972	2.994

Table Hg-2. Representative electron microprobe data for coexisting oxides in sample JS-10. See caption to Table Hg-1 for explanation of abbreviations. Pb (Gr) = gray Pb_{ss}; Pb (Br) = brownish Pb_{ss} under oil-immersion optics.

	Ilmenites					Titanomagnetites							
	HI 14	C 15	T 16	I(+R) 17	H(+R) 18	Mt(+C) 19	Mt(+T) 20	Mt(+Sp) 21	R+H 22	R+H 23	Pb(Gr) 24	Pb(Br) 25	H(+Pb+R) 26
MgO	2.22	1.12	1.59	3.07	3.89	0.44	0.95	1.51	1.90	1.14	1.56	0.99	1.47
Al_2O_3	0.20	0.09	0.16	0.14	0.13	1.93	2.56	3.71	0.25	0.06	0.69	0.61	0.66
TiO_2	48.61	49.21	50.62	44.98	20.17	17.34	15.01	1.84	40.29	32.50	38.56	38.94	10.15
Cr_2O_3	0.10	0.09	0.08	0.00	0.08	0.15	0.13	0.13	0.09	0.11	0.06	0.05	0.05
MnO	0.62	0.64	0.86	0.63	1.15	1.05	0.49	0.56	0.65	0.22	0.42	0.38	0.46
FeO	39.24	41.69	41.99	34.37	10.11	45.68	43.23	30.31	1.47	0.00	1.08	2.68	6.20
Fe_2O_3	8.89	6.88	4.59	17.57	64.16	33.13	36.84	61.04	55.97	65.03	57.04	55.60	80.26
Total	99.88	99.72	99.89	100.76	99.69	99.72	99.21	99.10	100.62	99.06	99.41	99.25	99.25
Oxygens	3	3	3	3	3	32	32	32	5	5	5	5	3
Mg	0.082	0.042	0.059	0.113	0.147	0.196	0.428	0.677	0.109	0.067	0.091	0.059	0.057
Al	0.006	0.003	0.005	0.004	0.004	0.682	0.904	1.313	0.001	0.028	0.032	0.028	0.020
Ti	0.910	0.929	0.949	0.833	0.384	3.907	3.381	0.370	1.172	0.962	1.138	1.155	0.199
Cr	0.002	0.002	0.002	0.000	0.002	0.036	0.031	0.031	0.003	0.003	0.002	0.001	0.001
Mn	0.013	0.014	0.018	0.013	0.025	0.266	0.124	0.142	0.021	0.007	0.014	0.013	0.001
Fe^{2+}	0.817	0.876	0.876	0.708	0.214	11.444	10.828	7.608	0.048	0.000	0.036	0.089	0.135
Fe^{3+}	0.167	0.130	0.086	0.326	1.222	7.470	8.304	13.787	1.629	1.927	1.684	1.651	1.574
Cations	1.997	1.996	1.995	1.997	1.998	24.001	24.000	23.928	2.983	2.994	2.997	2.996	1.987

Table Hg-3. Representative electron microprobe analyses of coexisting oxides in samples DHL 295/69 (analyses 27-45) and DHL 296/69 (analyses 46-65). See caption to Table Hg-1 for explanation of abbreviations. T-adj = adjacent to trellis lamellae; center = between T and C ilmenites; C-C = between two C ilmenites; grains 1 through 4 are underlined, *e.g.*, grain 1 analyses are 27-31, etc.

	Titanomagnetites												
	T-adj	Center	C-adj	T-free	3 Trellis	Grain-edge C-adj		C-adj	T-adj	C-adj	T-adj	4 Trellis	C-C
	27	28	29	30	31	32	33	34	35	36	37	38	39
MgO	1.17	1.34	1.27	1.51	1.38	1.25	1.16	0.94	0.95	1.25	1.21	1.16	1.18
Al_2O_3	2.30	2.09	2.27	1.69	1.87	1.46	1.59	1.45	1.40	1.78	1.75	1.42	1.83
TiO_2	19.83	21.07	21.10	22.30	21.31	21.47	20.15	20.22	18.88	19.12	19.12	19.50	19.08
Cr_2O_3	0.02	0.02	0.05	0.01	0.00	0.04	0.05	0.06	0.08	0.04	0.02	0.02	0.00
MnO	0.77	0.67	0.54	0.59	0.72	0.58	0.58	0.83	0.45	0.43	0.47	0.51	0.55
FeO	47.56	48.53	48.24	48.93	48.11	48.55	47.40	47.30	46.97	46.68	46.62	46.98	46.44
Fe_2O_3	29.19	27.19	25.86	24.28	25.90	26.05	28.21	28.74	31.44	30.47	30.30	29.98	30.03
Total	100.84	100.91	99.33	99.31	99.29	99.40	99.14	99.54	100.17	99.77	99.49	99.57	99.11
Oxygens	32	32	32	32	32	32	32	32	32	32	32	32	32
Mg	0.511	0.586	0.563	0.670	0.610	0.555	0.519	0.422	0.421	0.555	0.538	0.517	0.525
Al	0.795	0.720	0.728	0.591	0.658	0.513	0.562	0.513	0.491	0.623	0.615	0.501	0.645
Ti	4.376	4.640	4.730	4.986	4.770	4.815	4.535	4.495	4.224	4.275	4.289	4.379	4.294
Cr	0.006	0.005	0.011	0.003	0.000	0.008	0.013	0.014	0.019	0.009	0.004	0.004	0.000
Mn	0.191	0.165	0.135	0.148	0.182	0.146	0.147	0.211	0.112	0.108	0.118	0.129	0.141
Fe^{2+}	11.674	11.889	12.031	12.168	11.978	12.114	11.869	11.861	11.691	11.611	11.633	11.733	11.629
Fe^{3+}	6.447	5.994	5.802	5.433	5.803	5.848	6.356	6.484	7.041	6.819	6.803	6.737	6.766
Cations	24.000	23.999	24.000	23.999	24.001	23.999	24.001	24.000	23.999	24.000	24.000	24.000	24.000

	Ilmenites												
	HI	T	T	T(Av)	C(Av)	C(Av)	I(+R)	I(+R)	Ī(+R+Pb)	Pb(+I+R)	Pb	Pb	Pb
	40	41	42	43	44	45	46	47	48	49	50	51	52
MgO	1.04	2.55	2.63	2.35	2.12	2.00	3.80	4.09	1.76	2.64	3.04	2.56	2.21
Al_2O_3	0.05	0.09	0.25	0.14	0.08	0.17	0.24	0.16	0.18	0.72	0.83	0.95	1.57
TiO_2	50.55	52.43	51.17	51.85	52.36	49.73	43.38	45.88	43.83	52.76	53.04	46.75	50.44
Cr_2O_3	0.03	0.00	0.02	0.01	0.02	0.00	0.00	0.00	0.00	0.00	0.00	0.01	0.00
MnO	0.78	0.84	0.58	0.69	0.68	0.60	0.85	1.08	0.67	0.25	0.26	0.26	0.27
FeO	42.79	41.74	40.71	41.73	42.60	40.54	31.35	32.87	35.58	11.24	10.77	6.16	9.73
Fe_2O_3	4.16	3.16	5.32	3.23	2.15	6.16	20.35	14.95	18.53	32.32	31.55	43.07	36.33
Total	99.40	100.81	100.68	100.00	100.01	99.20	99.97	99.03	100.55	99.93	99.49	99.76	100.55
Oxygens	3	3	3	3	3	3	3	3	3	5	5	5	5
Mg	0.039	0.093	0.097	0.087	0.078	0.075	0.140	0.152	0.066	0.151	0.174	0.147	0.126
Al	0.001	0.003	0.007	0.004	0.002	0.005	0.007	0.005	0.005	0.032	0.038	0.043	0.070
Ti	0.959	0.969	0.947	0.968	0.978	0.939	0.807	0.858	0.823	1.518	1.527	1.354	1.444
Cr	0.001	0.000	0.000	0.000	0.000	0.000	0.000	0.000	0.000	0.000	0.000	0.000	0.000
Mn	0.017	0.018	0.012	0.014	0.014	0.013	0.018	0.023	0.014	0.008	0.009	0.008	0.009
Fe^{2+}	0.904	0.858	0.838	0.867	0.886	0.852	0.649	0.683	0.743	0.360	0.345	0.198	0.310
Fe^{3+}	0.079	0.058	0.099	0.060	0.040	0.117	0.379	0.280	0.348	0.931	0.909	1.249	1.041
Cations	2.000	1.999	2.000	2.000	1.998	2.001	2.000	2.001	1.999	3.000	3.002	2.999	3.000

Table Hg-3 (continued).

	Ilmenites			Titanomagnetites									
	H+R 53	H 54	R 55	H+R 56	H 57	H 58	H 59	Mt(+Sp) 60	Pb(T) 61	Pb(C) 62	Pb 63	Pb 64	Pb 65
MgO	3.04	3.14	0.12	3.00	0.74	0.67	1.20	2.46	3.02	3.44	2.94	2.50	2.84
Al_2O_3	0.09	0.25	0.13	0.39	2.21	2.10	3.41	0.62	0.87	0.80	0.95	0.82	0.89
TiO_2	45.11	25.45	97.43	38.77	6.02	4.52	6.53	0.63	48.03	50.26	41.97	40.91	42.82
Cr_2O_3	0.00	0.00	0.00	0.00	0.02	0.03	0.02	0.03	0.00	0.00	0.00	0.00	0.00
MnO	1.19	1.53	0.00	0.62	0.14	0.11	0.26	1.11	0.30	0.34	0.24	0.14	0.19
FeO	2.62	15.73	1.84	0.00	3.96	2.77	3.47	26.66	6.33	7.45	1.10	1.23	2.25
Fe_2O_3	48.89	53.80	0.00	58.16	86.69	89.80	84.23	68.00	41.18	37.21	53.09	54.61	50.77
Total	100.94	99.90	99.52	100.94	99.78	100.00	99.12	99.51	99.73	99.50	100.29	100.21	99.76
Oxygens	5	3	2	5	3	3	3	32	5	5	5	5	5
Mg	0.173	0.119	0.002	0.171	0.029	0.026	0.046	1.111	0.173	0.196	0.168	0.144	0.163
Al	0.004	0.008	0.002	0.017	0.068	0.064	0.104	0.220	0.040	0.036	0.043	0.037	0.041
Ti	1.295	0.484	0.987	1.116	0.118	0.088	0.127	0.143	1.386	1.446	1.212	1.188	1.242
Cr	0.000	0.000	0.000	0.000	0.000	0.001	0.000	0.007	0.000	0.000	0.000	0.000	0.000
Mn	0.039	0.033	0.000	0.020	0.003	0.002	0.006	0.286	0.010	0.011	0.008	0.004	0.006
Fe^{2+}	0.084	0.333	0.021	0.000	0.086	0.060	0.075	6.746	0.203	0.239	0.035	0.040	0.073
Fe^{3+}	1.405	1.024	0.000	1.676	1.696	1.758	1.641	15.486	1.189	1.072	1.534	1.587	1.475
Cations	3.000	2.001	1.012	3.000	2.000	1.999	1.999	23.999	3.001	3.000	3.000	3.000	3.000

Table Hg-4. Representative electron microprobe analyses of coexisting oxides in sample H-5. See caption to Table Hg-1 for explanations of abbreviations. Pb (Av) = average of 5 analyses; H (Pb) = Hem_{ss} adjacent to Pb_{ss}; H (Sp) = Hem_{ss} adjacent to $magnesioferrite_{ss}$. Analysis 68 is for $pleonaste_{ss}$ lamellae in Mt_{ss} (see Fig. Hg-6b); analysis 78 is for $magnesioferrite_{ss}$ (see Fig. Hg-6f).

	Titanomagnetites												
	Mt(+Sp) 66	Sp 67	H 68	Pb 69	Pb 70	Pb(Av) 71	Pb(Av) 72	Sp 73	H 74	H(Pb) 75	H(Sp) 76	Pb 77	Sp 78
MgO	8.22	23.98	2.06	4.55	5.06	4.85	5.48	22.47	4.03	4.02	2.55	5.98	22.58
Al_2O_3	5.06	56.53	1.64	1.95	2.12	1.93	1.91	26.97	1.95	1.56	1.74	1.54	23.66
TiO_2	0.71	0.00	10.43	44.58	46.84	45.03	44.27	0.49	8.10	9.06	4.62	45.28	0.60
Cr_2O_3	0.05	0.00	0.03	0.00	0.02	0.01	0.00	0.00	0.11	0.86	0.24	0.00	0.14
MnO	2.45	0.69	0.22	0.25	0.23	0.26	0.71	2.66	1.29	1.07	1.25	0.67	2.56
FeO	18.13	4.07	5.49	0.29	0.88	0.05	0.00	0.00	0.00	0.00	0.00	0.00	0.00
Fe_2O_3	66.29	13.50	81.42	47.51	45.18	47.11	47.99	47.74	85.32	84.35	89.98	46.29	50.98
Total	100.91	98.77	101.29	99.13	100.33	99.24	100.36	100.33	100.80	100.92	100.38	99.76	100.52
Oxygens	32	32	3	5	5	5	5	32	3	3	3	5	32
Mg	3.384	7.169	0.078	0.258	0.282	0.274	0.305	7.742	0.151	0.178	0.097	0.334	7.871
Al	1.908	13.883	0.049	0.087	0.094	0.086	0.084	7.346	0.058	0.046	0.052	0.068	6.521
Ti	0.148	0.000	0.199	1.276	1.317	1.284	1.242	0.086	0.153	0.168	0.089	1.274	0.105
Cr	0.010	0.000	0.001	0.000	0.001	0.000	0.000	0.000	0.002	0.017	0.005	0.000	0.027
Mn	0.574	0.122	0.005	0.008	0.007	0.008	0.022	0.522	0.027	0.022	0.027	0.021	0.507
Fe^{2+}	4.190	0.709	0.116	0.009	0.027	0.002	0.000	0.000	0.000	0.000	0.000	0.000	0.000
Fe^{3+}	13.785	2.117	1.553	1.361	1.272	1.345	1.347	8.304	1.610	1.569	1.730	1.303	8.970
Cations	23.999	24.000	2.001	2.999	3.000	2.999	3.000	24.000	2.001	2.000	2.000	3.000	24.001

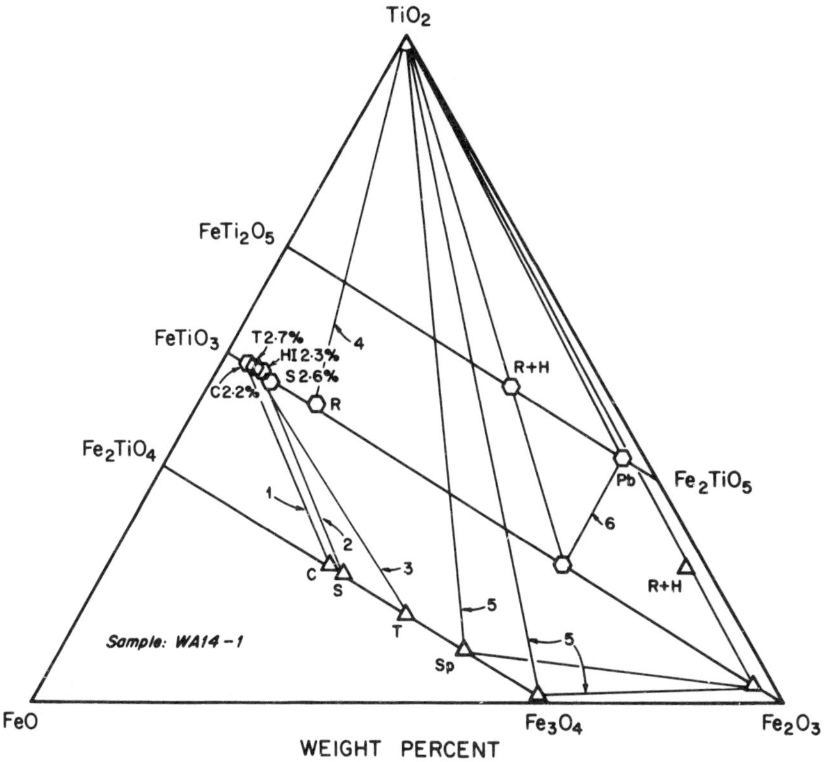

Figure Hg-10. Phase chemistry of coexisting assemblages in sample WA14-1. Hexagonals are ilmenites and oxidation products after ilmenite. Triangles are titanomagnetites and products after titanomagnetite. The following assemblages are present: (1) titanomagentite + composite ilmenite; (2) titanomagnetite + sandwich ilmenite; (3) titanomagnetite + trellis ilmenite; (4) ferrian ilmenite + ferrian rutile; (5) magnetite + R + Hem_{ss} (Hem_{ss} obtained from average analyses of R + Hem_{ss}; R + H on diagram--see text); and (6) Pb_{ss} + R + Hem_{ss} (obtained from average value of R + Hem_{ss}; R + H on diagram). The percentage values are for wt% MgO. Primary unoxidized ilmenite is marked as HI (homogeneous ilmenite). Analytical data in Table Hg-1.

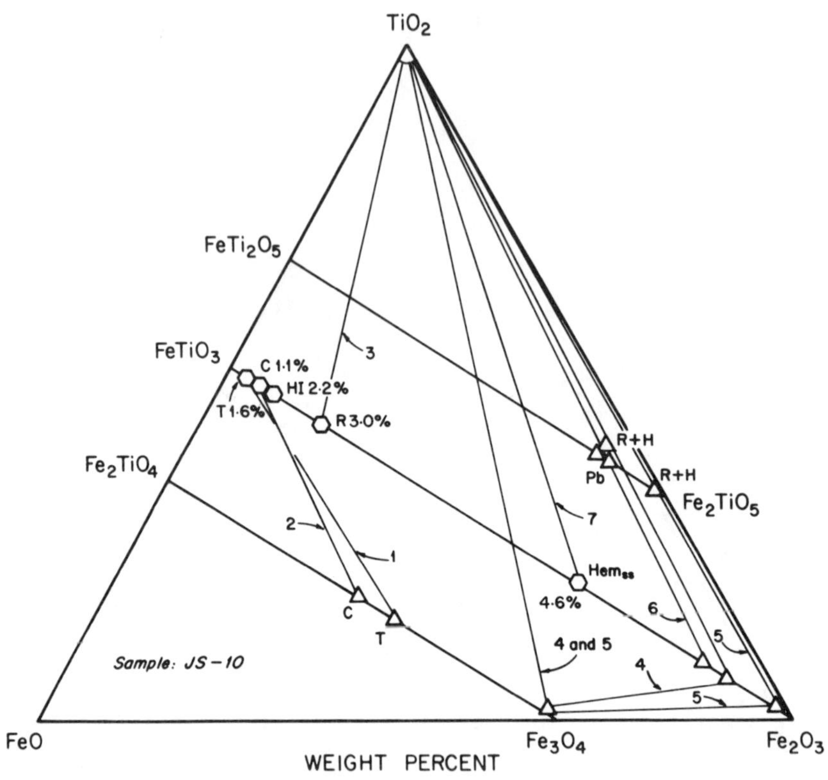

Figure Hg-11. Phase chemistry of coexisting assemblages in sample JS-10. Assemblages are as follows: titanomagnetite + trellis ilmenite; (2) titanomagnetite + composite ilmenite; (3) ferrian ilmenite + ferrian rutile; (4 & 5) Mt + R + Hem_{ss} (Hem_{ss} values obtained from R + H on diagram--see text); (6) Pb_{ss} + Hem_{ss} + R; and (7) R + Hem_{ss}. Primary ilmenite is HI. Percentage values are wt % MgO. Analytical data in Table Hg-2.

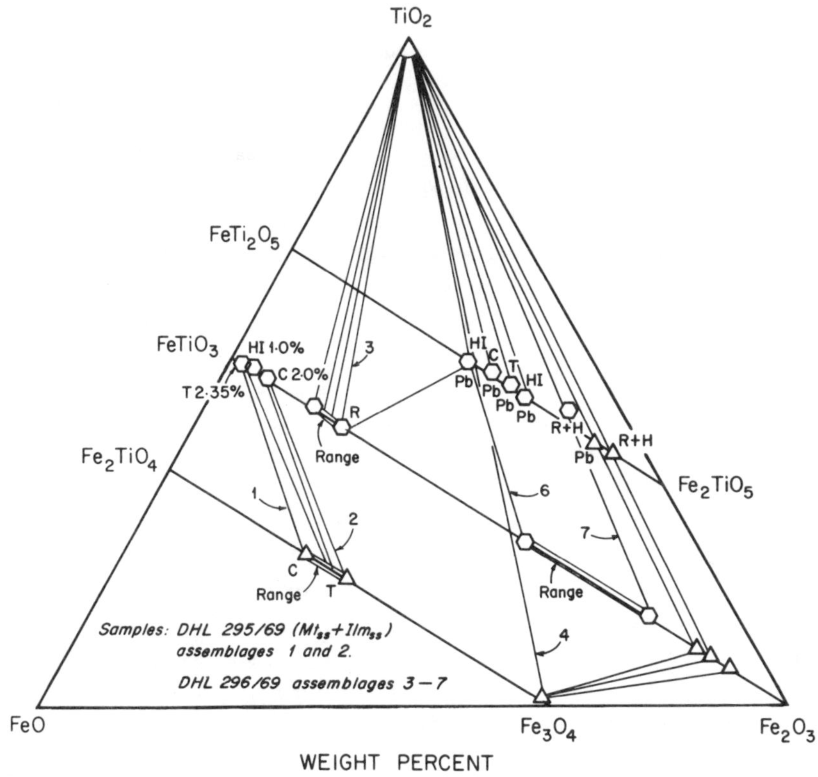

Figure Hg-12. Phase chemistry of coexisting assemblages in sample DHL 295/69 and DHL 296/69. The former sample is relatively unoxidized and consists of: (1) titanomagnetite + composite ilmenite; (2) titanomagnetite + trellis ilmenite; (3) ferrian ilmenite + ferrian rutile. The second sample is more highly oxidized and the assemblages are as follows: (4) Mt_{ss} + Hem_{ss} + Pb_{ss} + R; (5) Pb_{ss} + R after primary ilmenite (I), after composite ilmenite (C), and after trellis ilmenite (T); (6) ferrian ilmenite + titanohematite + Pb_{ss}; (7) R + Hem_{ss} after primary ilmenite. Analytical data are listed in Table Hg-3 for both samples which were collected approximately 80 cms apart across an oxidized joint selvage (see Fig. Hg-22).

Titanomagnetite-ilmenite assemblages

Homogeneous titanomagnetite is absent in all samples, and the compositions of discrete primary ilmenite are remarkably similar for the Iceland basalts and for the Oregon basalt. The latter have slightly lower initial MgO contents (1 vs. 2 wt % MgO) and the former have higher Hem_{ss} contents (8 vs. 4 wt % Fe_2O_3). The first textural intergrowths to be considered are titanomagnetite-ilmenite assemblages for ilmenite in trellis (T), sandwich (S) and composite (C) relationships. For the Iceland basalts the mean MgO contents of C, T, and S ilmenites in WA14-1 are 2.2, 2.7, and 2.6 wt % MgO respectively. In terms of Fe_2O_3 contents the C ilmenites contain the smallest proportion of Hem_{ss}, the S ilmenites the maximum, and the T ilmenites are intermediate between these two extremes. Minor element concentrations are virtually identical. For the second basalt (JS-10) the only significant difference between T and C ilmenites (sandwich laths are absent) is the lower Fe_2O_3 contents of trellis lamellae; this difference is the reverse of that shown for WA14-1 and is also more pronounced. The relatively unoxidized Oregon basalt (DHL 295/69) has a range of MgO contents in T lamellae between 2.4 and 2.6 wt % whereas the C inclusions are close to 2 wt % MgO. Ranges in minor element concentrations are similar for both T and C ilmenites. Trellis lamellae have an average Fe_2O_3 content of 3.2 wt % and a range between 3.2 and 5.3 wt %; the latter maximum value is clearly atypical and is restricted to a single grain. Significantly different Hem_{ss} contents are observed for the composite inclusions which fall into two populations: the first has a mean Fe_2O_3 of 2.2 wt % which is lower than the T lamellae and a second which has a mean Fe_2O_3 value of 6.2 wt %. For these three basalts two exhibit trellis compositions which are more oxidized than the coexisting composite inclusions and the Oregon basalt suggests that some C inclusions may have resulted by oxidation exsolution while others are primary precipitates. This complex distribution is not at all atypical, and it is noteworthy that the differences between discrete and intergrowth ilmenites and the differences among ilmenites in the second category are subtle. Expectedly wide variations in primary precipitated ilmenites from ilmenites which develop at subsolidus temperatures are in general not observed, except for MgO contents which tend to be higher in exsolved ilmenites.

Turning our attention now to the magnetites which host these ilmenites we are confronted with an equally complex distribution. Dealing firstly with WA14-1 which contains the three textural types of ilmenites, we find satisfactorily that the titanomagnetites which coexist with "exsolution" trellis lamellae (Tm = trellis magnetite) are more oxidized than those titanomagnetites with either S (Sm) or C (Cm) inclusions, which are almost identical. Note also that significant differences are present in the Al_2O_3 contents of Tm when compared with either of the Sm or the compositions of the Cm (2.5 wt % for Tm and 1.5-1.9 wt % for Cm and Sm). Magnesia is not as clearly defined: Trellis and sandwich magnetites are similar (0.9 wt %) whereas composite magnetites are enriched in MgO (1.3 wt %). For the second basalt, JS-10, the oxidation trend is repeated and is consistent with WA14-1; namely, that magnetites hosting T ilmenites are more highly oxidized and contain higher concentrations of Al_2O_3 than those hosting composite inclusions. MgO contents, although different by a factor of two, are <1 wt % (Cm = 0.44; Tm = 0.95 wt %). The Oregon sample (DHL 295/69) presents a more varied distribution, and examples typical of the rock are given for four titanomagnetite grains. Grain 1 contains abundant trellis lamellae and a composite inclusion; the least oxidized regions are in relatively trellis-free areas (24.2-25.9 wt % Fe_2O_3), and the most highly oxidized are immediately adjacent to lamellae (29.2 wt % Fe_2O_3). Titanomagnetites adjacent to composite ilmenites are comparable to the low oxidation areas (25.9 wt % Fe_2O_3) and become progressively more oxidizing as lamellae are approached, with values between 26 and 28 wt % Fe_2O_3 being typical. Other compositional differences are restricted to Al_2O_3 contents which are relatively high at Tm and Cm contacts (2.0-2.3 wt %), and low in trellis-free areas (1.7-1.9 wt %). For grain 2, grain boundary compositions are compared with compositions at Cm contacts; the homogeneous crystal edge is slightly less oxidized than the area adjacent to the composite inclusion, and all other values are comparable, except of course for TiO_2 which decreases with increasing Fe_2O_3 contents. Grain 3 is another Cm and Tm pair, and here once again titanomagnetites adjacent to trellis lamellae are more highly oxidized than magnetites adjacent to composite inclusions. Grain 4 is similar to grain 1 insofar as abundant trellis lamellae coexist with composite ilmenite, and compositions are all well within the experimental limits of the electron microprobe for the following associations: (a) titanomagnetite areas adjacent to C or T; (b) in areas bounded by four optically visible trellis lamellae; or (c) in areas between composite inclusions which are trellis free. The only consistent difference exists between Cm and Tm with the latter showing a slightly higher oxidation state (31.4 vs. 28.17 wt % Fe_2O_3).

The result common to these three rocks--and it is a result which is generally applicable--is that titanomagnetites which are closely associated with trellis lamellae are more highly oxidized than either the trellis-free areas within these titanomagnetites, or the areas associated with or adjacent to, composite or sandwich inclusions.

With respect to the primary or "exsolution" origin of these latter inclusions the minor element similarities among all textural types and their distinctions as a group when compared with discrete ilmenite would tend to suggest that these ilmenites have a common subsolidus origin which is distinct from primary ilmenite. The datum point for an evaluation of whether composite ilmenites are primary or secondary (*i.e.*, by subsolidus exsolution) must depend on the oxidation state of discrete primary ilmenite. For example, primary ilmenite in the Oregon basalt has the lowest Fe_2O_3 content (4.2 wt %), followed by WA14-1 (7.9 wt %) and JS-10 (8.9 wt %); the trellis lamellae for each example are consistently lower than these respective values, and composite ilmenites which approach the oxidation values of discrete ilmenite are therefore most probably primary. Using this criterion and in concert with the textural interpretations cited earlier, a clear case can be made for sandwich laths in WA14-1 as being primary, noting that MgO values have a limited range between 2.2 and 2.7 wt % for the four distinct ilmenites (discrete, trellis, sandwich and composite), and that the composite ilmenites in this sample are probably of "exsolution" origin (C = 6.1 and T = 7.0 wt % Fe_2O_3). Sample JS-10, which is perhaps most typical of basalts, shows that the oxidation stage of the composite ilmenite is intermediate between the trellis and the discrete types, but this example is atypical with respect to MgO. The MgO contents, although equivalent to the values for discrete ilmenite, are substantially larger than the MgO contents of the coexisting titanomagnetites which is consistent with the preferred partitioning of MgO between "exsolved" ilmenite and the titanomagnetite host. The Oregon sample is also typical of many basalts in which the Fe_2O_3 content criterion is either ambiguous or sufficiently far removed from the values of discrete ilmenite that no definitive evaluation is possible. The bimodal distribution which in many other samples can be related to the development of trellis lamellae, cannot be related here. These composite inclusions are both higher and lower in Hem_{ss} contents than discrete ilmenite and encompass the range for trellis lamellae, so that it is not possible to assign a specific origin to either of the two composite compositional types.

The process of oxidation exsolution induces a characteristic partitioning of minor element components. Distinctive distributions are noted for MgO, Al_2O_3 and Cr_2O_3, and for the examples noted here MgO tends to follow ilmenite and Al_2O_3 continues to reside in the spinel structure. In these low Cr_2O_3-bearing titanomagnetites the partitioning of chromium is not obvious but its tendency is to favor the titanomagnetite rather than the "exsolved" ilmenite. The case for MnO is not as clearly defined and it is most common to find approximately equal concentrations in both titanomagnetite and in ilmenite. These concentrations are uniformly low and rarely exceed values which are >1 wt % MnO.

Ferrian ilmenite and ferrian rutile

This assemblage is characteristic of oxidation stages R2 and R3 for discrete ilmenite, and for stage C4 in the titanomagnetite classification. In general the assemblage--which is restricted to trellis, composite and sandwich intergrowths in titanomagnetite--is extremely fine grained, and individual mineral components are beyond the resolution of the electron microprobe. A comparison, therefore, of the assemblage ferrian ilmenite-ferrian rutile in titanomagnetite intergrowths is not possible with similar assemblages in discrete ilmenite. Notwithstanding this limitation, the close proximities in mineral compositions of the four ilmenite textural types still permit the basic compositional trends which are established for discrete ilmenite to be related to those which develop in titanomagnetite. As noted previously, an optical comparison of ilmenite in titanomagnetite with discrete ilmenite shows that the trellis lamellae are initially more susceptible to oxidation than other textural types, and from the composition of these lamellae the explanation for this difference is most probably related to higher Hem_{ss} contents since all other elements are in approximately equal concentrations.

Assemblages of ferrian ilmenite + ferrian rutile are present in the two Iceland basalts, and the analyses quoted are respectively for grains with minor ferrian rutile (WA14-1) and for grains with abundent ferrian rutile (JS-10). The assemblage is also present in the second more highly oxidized Oregon basalt (DHL 296/69), and the data reported are for grains of approximately equivalent high ferrian rutile densities. The compositional characteristics to be noted are as follows: (1) For both samples the ferrian ilmenites are more highly oxidized and contain larger concentrations of MgO than their respective unoxidized counterparts; (2) grains containing lower densities of ferrian rutile (WA14-1) are less oxidized than grains with abundant ferrian rutile (JS-10), and the differences in MgO contents with respect to discrete ilmenite are proportionately larger for JS-10 (3.1 vs. 2.2 wt % MgO) and correspondingly smaller for WA14-1 (2.5 vs. 2.3 wt % MgO). This MgO enrichment is pronounced in the Oregon sample where the increase is a factor of four larger (1.0 vs. 3.8-4.1 wt % MgO) than that of discrete ilmenite;

(3) partitioning characteristics of other minor elements (Cr_2O_3, Al_2O_3, MnO) are ill defined for the Iceland samples, but noticeable increases in Al_2O_3 and MnO contents are evident for DHL 296/69. The compositions of the ferrian rutiles are variable and difficult to determine quantitatively because of x-ray excitation of the host ferrian ilmenite. Qualitative estimates of the FeO contents suggest a variation of between 5 and 10 wt % FeO, and minor element concentrations are equivalent to background levels. The essential features of ilmenite oxidation at this stage in the sequence are an increase in the hematite component, enrichments of MgO, MnO and Al_2O_3 in ferrian ilmenite, and a separation of the titanium-rich phase ferrian rutile.

Titanohematite, rutile, and magnetite

Titanohematite + rutile are characteristic of the R5 stage of ilmenite oxidation, and the assemblage titanohematite + rutile + accessory magnetite with exsolved aluminous magnesioferrite is characteristic of the C5 stage of titanomagnetite oxidation. These assemblages are present in the two Iceland basalts and in DHL 296/69.

Considering firstly the decomposition products of discrete ilmenite (R5) we observe that the TiO_2 contents of the hematite constituent are typically between 20 and 25 wt % TiO_2, that MgO varies between 3 and 4 wt % and that MnO is in the range 1-1.5 wt %. Compositions of rutile for this assemblage in the Iceland samples show that minor concentrations of FeO are present (<.1 wt %), but the more typical values observed are those given for the Oregon sample with FeO contents of between 1.5 and 2.5 wt % FeO, and with MgO and Al_2O_3 contents between 0.1 and 0.2 wt % for each oxide. In most cases the intimate nature of these intergrowths precludes the analyses of individual mineral constituents but broad beam electron microprobe results can be obtained for the entire assemblage. The proportion of rutile and titanohematite are approximately equal, and because this assemblage gives rise to and approaches the bulk compositions of members of the pseudobrookite solid solution series, the ferric iron component can be computed assuming that the assemblage is a Pb_{ss} rather than Hem_{ss} + R. This approach has been highly successful, and the analytical totals are consistently within the limits of experimental error. A comparison of WA14-1 and DHL 296/69 show that the compositions of the two assemblages are surprisingly similar, although the $Fe^{2+}:Fe^{3+}$ ratios and the MgO and MnO contents do differ by small amounts. These examples are typical of the values generally determined, and it is noteworthy that although the assemblage is at an advanced stage of oxidation major and minor element concentrations closely parallel those of unoxidized ilmenites in the same samples, indicating that decomposition is cationically isochemical.

Similar titanohematite + rutile assemblages develop in highly oxidized titanomagnetites, and these show a number of distinct compositional differences to the assemblages described above. The major differences result from the initially higher FeO and lower TiO_2 contents of titanomagnetites, but the mineral chemistry of the decomposition products is also controlled in part by the extent to which oxidation "exsolution" has proceeded before the onset of more advanced oxidation. If the "exsolution" of ilmenite is advanced, the spinel content is enriched in Mt_{ss} and consequently depleted in TiO_2. However, if oxidation is rapid and at high temperatures, oxidation "exsolution" is minimized and the spinel maintains its original Usp_{ss} content as a high Ti titanomagnetite. Both processes can in many cases be positively identified, and the dominantly developed {111} fabric is the genetic clue that ilmenite "exsolution" has taken place. As noted previously, the determination of individual mineral constituents in these intergrowths is beyond the resolving capabilities of the electron microprobe, and the data presented here are restricted to broad beam analyses of Hem_{ss} + R assemblages in the two Iceland basalts and in DHL 296/69, and to Hem_{ss} in the latter sample. The TiO_2 contents of the hematite component range between 4.5 and 6.5 wt % TiO_2, in contrast to the 20-25 wt % range determined for Hem_{ss} associated with discrete oxidized ilmenite. Another major distinction is in the Al_2O_3 contents of these hematites which are heterogeneous and typically vary between 2 and 4 wt % Al_2O_3. These variations are a simple reflection of the initial chemistry of the titanomagnetite and of the degree to which the analyzed area has incorporated products that have resulted from "exsolved" ilmenite with products of titanomagnetite decomposition. Broad beam analyses of R + Hem_{ss} assemblages confirm the heterogeneous nature of the bulk assemblage but show once again that TiO_2 contents are substantially reduced and that total Fe contents are increased with respect to analyses for equivalent assemblages in discrete oxidized ilmenite. In a broader sense, the compositions of such assemblages do, of course, approach those of discrete ilmenite for grains which have inherited a pre-existing {111} fabric, insofar as these compositions plot very close to the ideal Pb_{ss} join. This is the case for both JS-10 and DHL 296/69 but for WA14-1 a larger interaction with the titanomagnetite has developed, and compositions plot between the Hem_{ss} join and the Pb_{ss} join. On each of the ternary diagrams for these three rocks (Figs. Hg-10-Hg-12), average analyses can be used to determine the compositions of the titanohematite indirectly, by assuming that rutile contains on the order of <1-2 wt % FeO and that no other minerals are present. These compositions are obtained by extending the tie-line from the TiO_2 apex to the R + Hem_{ss} analysis (*i.e.*, the analysis which has been recalculated as Pb_{ss}) and continuing the tie-line to intersect the Hem_{ss} join. Compositions obtained in this way for the combined assemblage are very close to those obtained separately for titanohematite (see for example JS-10 and DHL 296/69). One further and final noteworthy point is that the bulk compositions depend, as noted above, on the degree of titanomagnetite decomposition, and an example is given for JS-10 and for two grains which have a small percentage and a large percentage of residual magnetite, respectively. The grain containing the larger concentration of unreacted magnetite has a higher TiO_2 content and a correspondingly lower total Fe content.

Magnetites which coexist with R + Hem_{ss} are a distinctive brown in reflected light oil immersion, and the presence of black aluminous spinels along {100} planes is ubiquitous. These magnetites are typically low in TiO_2 contents which rarely exceed 3 wt % TiO_2 and characterized by high MgO and Al_2O_3 contents (2-5 wt % total), but this depends on the degree of exsolution of the second spinel, and for the most part plot close to the end-member magnetite.

In summary, C6 and R6 assemblages yield actual and indirectly inferred titanohematite compositions which are enriched in the Hem_{ss} component for titanomagnetite decomposition and are relatively poorer in Fe_2O_3 for discrete ilmenite decomposition. In addition, there is evidence for the partial interaction of R + Hem_{ss} along pre-existing {111} planes, with R + Hem_{ss} assemblages in intra {111} planes as shown by the heterogeneous distribution of Al_2O_3 contents. Rutile + Hem_{ss} assemblages after discrete ilmenite are characteristically low in Al_2O_3, whereas the maximum Al_2O_3 contents are determined for residual, accessory magnetite after titanomagnetite.

Pseudobrookite, titanohematite, magnetite and Al-magnesioferrite

This assemblage is characteristic of C7 stages of titanomagnetite oxidation, and with accessory R + Hem_{ss} is characteristic of the R7 stage of ilmenite oxidation.

The compositions of the magnetites in the C7 assemblage are typically less magnesian and aluminous-rich than those associated with the C6 assemblage discussed above, and this is a reflection of the more complete exsolution of pleonaste-magnesioferrite. In general, however, magnetite is a minor mineral component at this advanced stage, but the resistant aluminous spinels do persist as disaggregated blebs throughout the associated titanohematite and occasionally within the pseudobrookite. Exceptions to this rule apply to initially high Mg-Al titanomagnetites which tend to resist oxidation because of the stabilizing effects of Mg and Al. Such an example is given for the picritic basalt (H-5) which has MgO and Al_2O_3 contents of 8.2 and 5.1 wt %, respectively. These magnetites contain relatively coarse lamellae of spinels which may be defined either as ferrian pleonaste or as aluminous magnesioferrite, and although precise quantitative analyses are difficult to obtain, it is clear that extensive variability exists within the three-component system $FeAl_2O_4$-$MgAl_2O_4$-$MgFe_2O_4$. Compositions typical of these lamellae and of spinels which have lost their magnetite hosts to oxidation are listed in Table Hg-4. The clear distinctions which exist for these two restricted examples is that the exsolved spinel is more nearly a ferrian pleonaste, whereas the more completely oxidized example (compare photomicrographs in Fig. Hg-6) is closer to an aluminous magnesioferrite.

The titanohematites associated with these magnetites and with pseudobrookite are more titaniferous than the Hem_{ss} associated with the C6 stage assemblage (Hem_{ss} + R), but less titaniferous than hematite associated with Pb_{ss} after discrete ilmenite oxidation. These distinctions are significant from the standpoint of R + Hem_{ss} reacting to form Pb_{ss} by continuous reaction and invite careful examination of the ternary diagrams in attempting to resolve an apparent incompatibility of phase relationships.

Pseudobrookite compositions are expectedly variable among decomposed ilmenites and oxidized titanomagnetites, which is a function of the compositional differences among parental phases but is a function also, particularly for titanomagnetite, of the extent to which equilibrium conditions are approached. If initial subsolidus oxidation has been extensive, then Ti enrichment is concentrated in those areas originally occupied by ilmenite, and in many instances the development of Pb_{ss} is restricted to lamellar zones which mimic the {111} subsolidus fabric. In other instances original composite inclusions can also be recognized, but their distinction from grains which have undergone very rapid oxidation can only be made if a comparison with unoxidized grains is also possible. Both types of grains yield identical textures of bulbous pseudobrookite in a matrix of titanohematite. Such a comparison is illustrated for DHL 296/69, for Pb_{ss} which have developed largely from coarse trellis lamellae, for pseudobrookites which have resulted from composite ilmenite decomposition, and for pseudobrookites which appear to have formed from oxidized titanomagnetite in which a very small percentage of subsolidus ilmenite was originally present. The trellis and composite pseudobrookites are distinctive in their MgO contents and specifically in their $FeO:Fe_2O_3$ ratios. These pseudobrookites have slightly higher proportions of $FeTi_2O_5$ contents when compared with those pseudobrookites which have developed from ilmenite-free titanomagnetite; these are enriched in Fe_2TiO_5 with lower MgO contents (2.5-2.9 vs. 3.0-3.4 wt %). It is noteworthy that Al_2O_3 is uniformly low, and that the Al content of the original titanomagnetite is now largely resident in coexisting titanohematite. A comparison of Pb_{ss} after discrete ilmenite show that the range of MgO contents (2.3-3.0 wt %) is almost equivalent to the range for pseudobrookites in oxidized titanomagnetite. A noticeable difference is that the $FeTi_2O_5$ contents of pseudobrookite after discrete ilmenite are equivalent to Pb_{ss} after trellis and composite Ilm_{ss}, and that these as a group have lower Fe_2TiO_5 concentrations than pseudobrookites developing after ilmenite-free titanomagnetites.

In the two more variably oxidized Iceland basalts, data are given for Pb_{ss} after discrete ilmenite (WA14-1) and for two optically different Pb_{ss} after trellis ilmenite in titanomagnetite (Tables Hg-1, Hg-2). The compositions of these three Pb_{ss} are virtually identical, at least within the latitude of the remaining differences which exist between the two samples, and are comparable in composition to the Pb_{ss} after ilmenite-free titanomagnetite in DHL 296/69. In WA14-1, Pb_{ss} are restricted to outermost mantles on R + Hem_{ss} cores after discrete ilmenite, and the broad beam composition of this latter assemblage, which was discussed earlier, is plotted in Figure Hg-10 with the deduced titanohematite composition and with the composition of coexisting Pb_{ss}. The discrepancies in composition are the result of incomplete reaction, but note that the R + Hem_{ss} composition is precisely within the range of Pb_{ss} for discrete, composite and trellis lamellae in DHL 296/69. These similarities are applicable to most basalts, a result perhaps of narrow limits for T°C, f_{O_2} and the composition of the oxidizing medium, with minor differences in initial chemistry playing a small but influential role in mineral stability.

Data for pseudobrookites and titanohematites in the picritic basalt (H-5) are listed for three grains of oxidized titanomagnetites in Table Hg-4. Grain 1 also contains Mg-Al magnetite + exsolved ferrian pleonaste, whereas in grains 2 and 3 the spinel assemblage has decomposed to form aluminous magnesioferrite. Differences in titanohematite compositions among these grains are as follows: The aluminum contents vary between 1.6 and 2 wt % Al_2O_3; magnesium contents in grain 1 (coexisting with magnetite) are similar to MgO contents in grain 3 adjacent to Al magnesioferrite (2-2.6 wt %), but these are a factor of two less than the MgO contents associated with Pb_{ss} (4 wt % MgO); TiO_2 values vary between 4.6 and 10.4 wt %, and if stoichiometry is assumed, grain 1 titanohematites contain on the order of 5 wt % FeO whereas those in grains 2 and 3 approach zero; MnO contents in the less reacted grain 1 are <0.5 wt % (with 2.5 wt % in the magnetite), in contrast to a maximum of 1.3 wt % MnO in the more completely reacted grains 2 and 3 (with 2.5-2.6 wt % MnO in coexisting Al-magnesioferrite). Coexisting pseudobrookites in these three grains reflect more complete reaction in grains 2 and 3 and are more highly oxidized with larger proportions of Fe_2TiO_5 than those Pb_{ss} coexisting with partially unreacted magnetite. The Al and Mg contents of these Pb_{ss} are uniformly high with sizable proportions of $MgTi_2O_5$ (karooite) and lesser tielite (Al_2TiO_5) in solid solution. The differences in oxide mineral chemistry of this picrite with the Iceland and Oregon basalts are due simply to initial differences in primary chemistry.

In summarizing the major compositional characteristics of R7 and C7 assemblages the following points should be noted: (1) Unreacted magnetites are less Mg and Al rich than those present at the C6 stage of titanomagnetite oxidation, although extreme examples do exist for picrites which have initially high Mg and Al contents; (2) titanohematites are Ti enriched when compared with the C6 assemblage and less titaniferous than Hem_{ss} after discrete oxidized ilmenites; (3) Pb_{ss} after discrete ilmenite contain less Fe_2TiO_5 and greater $FeTi_2O_5$, whereas the reverse is true for Pb_{ss} after titanomagnetite; and (4) Hem_{ss} coexisting with Pb_{ss} after titanomagnetite are more aluminous and less magnesian than the equivalent assemblage after ilmenite.

OXIDE SYSTEMATICS AND PHASE COMPATIBILITY RELATIONSHIPS

Introduction

Inherent in the discussions of textural relationships and of the mineral chemistry of the oxidation products of titanomagnetite and of ilmenite in the last two sections is that the deuteric process of oxidation at high temperatures in basalts is rarely an equilibrium process. Individual grains in many cases do approach equilibrium, but there are clearly wide variations in microenvironmental conditions which affect reaction rates and the degree to which cationic exchange processes will respond to rapidly changing physical and chemical parameters. The extremes exhibited in the examples quoted do, of course, provide a means of tracing the progressive stages of oxidation, but these extremes do not allow for a definitive estimate of T°C or of f_{O_2} for the entire rock; a grain-by-grain evaluation is the only reasonable but impractical approach.

The stages of oxidation as defined here are those distinctive assemblages which are not only widespread in oxidized basalts but for which some experimental data do exist which tend to support both the textural and compositional interpretations of coexisting pairs and of multicomponent mineral intergrowths. Although critical experimental data are lacking, particularly in the temperature range of interest (600-900°C), for unique interpretation of isolated assemblages an overall appraisal is possible in terms of idealized and equilibrium phase compatibility relationships.

The systematics of coexisting assemblages as a function of temperature for minerals in the pure end member system $FeO-Fe_2O_3-TiO_2$ are shown in Figures Hg-13 and Hg-14, and equations and reactions typical of each of the oxidation stages for ilmenite and for titanomagnetite are listed in Table Hg-5. Construction of these diagrams is based on the following experimental data: (1) $Usp-Mt_{ss}$ are complete above 600°C (Vincent *et al.*, 1957); (2) the "solvi delimiting" experimental curves by Lindsley (1973) were used for the $Ilm-Hem_{ss}$ series with a consulate point at approximately 800°C; (3) data for the join $Pb-FPb_{ss}$ are from Lindsley (1965), Haggerty and Lindsley (1969), and from Akimoto *et al.* (1957) which show that the series is complete above 1150°C; (4) the directions of conjugate tie-lines for coexisting assemblages are based on: (a) the 1200°C isothermal section by Webster and Bright (1961) shown in Figure Hg-13; (b) on data from Lindsley (1963) for coexisting $Mt-Usp_{ss}$ and $Ilm-Hem_{ss}$; and (c) as a general guideline on the mineral chemistry of natural coexisting assemblages.

The 1200°C isothermal section from Webster and Bright (1961) illustrates the variations of coexisting assemblages as a function of f_{O_2} with Hem_{ss} + Pb_{ss} + R as the stable assemblage at high f_{O_2} values (>$10^{-2.56}$ atms), and with FPb_{ss} + Ilm_{ss} + Usp_{ss} as the stable assemblage at correspondingly lower f_{O_2} values (<$10^{-10.79}$ atms). The oxidation of either titanomagnetite or of ilmenite is cationically isochemical, and in the decomposition of each of these minerals the bulk Fe:Ti ratio is kept constant. Progressive oxidation therefore may be followed along Fe:Ti isopleths from the ferrous to the ferric sides of the ternary diagrams. Such oxidation-reduction isopleths are shown for the end members Usp, Ilm and FPb, and if the case for Usp (Fe_2TiO_4) is considered, which is linked to Pb (Fe_2TiO_5), the constancy of Fe:Ti is clear, bearing in mind that iron in the former is Fe^{2+}, in the latter is Fe^{3+}, and that oxygen has been added to the system. Progressive oxidation of stoichiometric Fe_2TiO_4 should therefore result in stoichiometric Fe_2TiO_5 as the final end product, and at

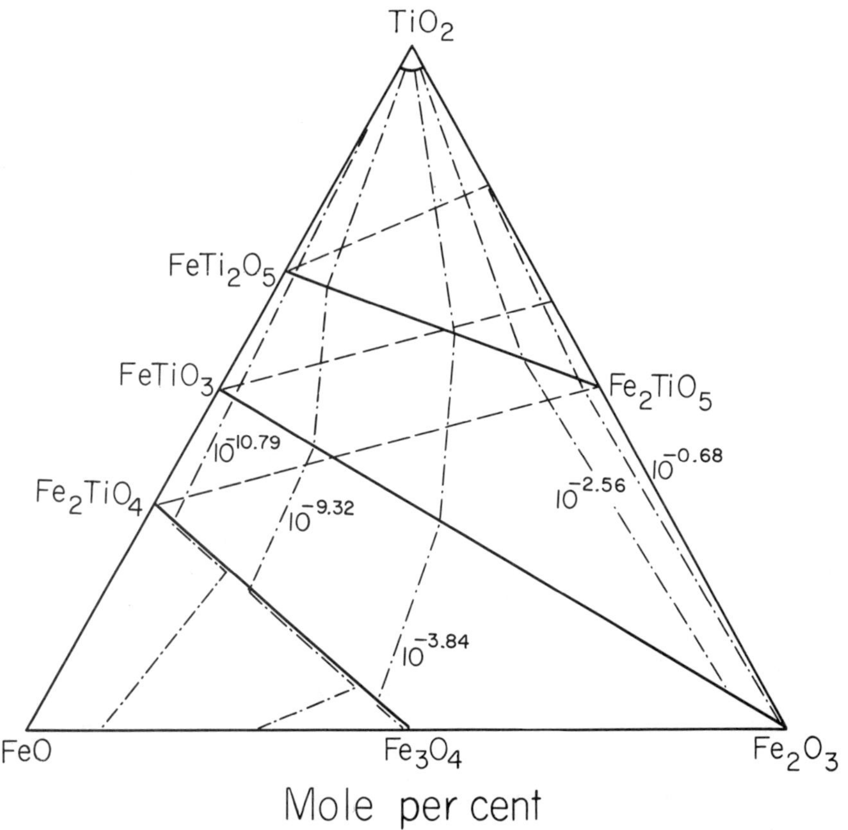

Figure Hg-13. $FeO-Fe_2O_3-TiO_2$ system at 1200°C (Webster and Bright, 1961). Continuous solid solution is shown between $Fe_2TiO_4-Fe_3O_4$ (ulvöspinel-magnetite), $FeTiO_3-Fe_2O_3$ (ilmenite-hematite), and $FeTi_2O_5-Fe_2TiO_5$ (pseudobrookite series). Oxidation-reaction lines (dashed) are lines of equal Fe:Ti. Oxygen isobars ($10^{-0.68}$ to $10^{-10.79}$) are illustrated for stable coexisting compositions at 1200°C.

Table Hg-5. Representative and possible equations for each of the oxidation stages for ilmenite (R1 to R7), equations 1 to 16, and for titanomagnetite (C1 to C7), equations 17 to 30. (S.S) signifies solid solution, *e.g.*, in equation 3 the Ilm_{ss} is represented by $9FeTiO_3 + Fe_2O_3$. Starting compositions are $Ilm_{90}Hem_{10}$ and $Usp_{60}Mt_{40}$.

Ilmenite	Assemblage	Typical Reactions	
R1	Homogeneous Ilm_{ss}	$Ilm_{90}Hem_{10}$	1
R2	Ilm_{ss}+R (minor)	$9FeTiO_3+Fe_2O_3+O_2 = 5FeTiO_3+3Fe_2O_3+4TiO_2$ (solid solution) (solid solution)	2
R3	Ilm_{ss}+R (major)		
R4	Ilm_{ss}+Hem_{ss}+R	$9FeTiO_3+Fe_2O_3+2O_2 = .9FeTiO_3+Fe_2O_3+.1FeTiO_3+4Fe_2O_3+8TiO_2$ (s. s.) (s. s.) (s. s.)	3
		$5FeTiO_3+3Fe_2O_3+4TiO_2+{}^5/_2O_2 = FeTiO_3+4Fe_2O_3+8TiO_2$ (s. s.) (s. s.)	4
		$5FeTiO_3+3Fe_2O_3+4TiO_2 = 3FeTiO_3+Fe_2O_3+2FeTiO_3+2Fe_2O_3+4TiO_2$ (s. s.) (s. s.)	5
R5	Hem_{ss}+R	$9FeTiO_3+Fe_2O_3+2O_2 = FeTiO_3+5Fe_2O_3+8TiO_2$ (s. s.) (s. s.)	6
		$5FeTiO_3+3Fe_2O_3+4TiO_2+2O_2 = FeTiO_3+5Fe_2O_3+8TiO_2$ (s. s.) (s. s.)	7
		$3FeTiO_3+Fe_2O_3+2FeTiO_3+2Fe_2O_3 = FeTiO_3+5Fe_2O_3+8TiO_2$ (s. s.) (s. s.) (s. s.)	8
R6	Hem_{ss}+R+Pb_{ss}	$9FeTiO_3+Fe_2O_3+O_2 = 4FeTiO_3+Fe_2O_3+2Fe_2TiO_5+FeTi_2O_5+TiO_2$ (s. s.) (s. s.) (s. s.)	9
		$9FeTiO_3+Fe_2O_3+{}^1/_2O_2 = FeTiO_3+Fe_2O_3+3Fe_2TiO_5+2FeTi_2O_5+TiO_2$ (s. s.) (s. s.) (s. s.)	10
		$5FeTiO_3+3Fe_2O_3+4TiO_2+{}^1/_2O_2 = 2FeTiO_3+Fe_2O_3+3Fe_2TiO_5+Fe_2TiO_5+2TiO_2$ (s. s.) (s. s.) (s. s.)	11
		$3FeTiO_3+2Fe_2O_3+2FeTiO_3+2Fe_2O_3+4TiO_2+{}^1/_2O_2 = 2FeTiO_3+2Fe_2O_3+2Fe_2TiO_5+FeTi_2O_5+3TiO_2$ (s. s.) (s. s.) (s. s.) (s. s.)	12

Table Hg-5 (continued).

Ilmenite	Assemblage	Typical Reactions	
R7	$Pb_{ss}(+Hem_{ss}+R)$	$\underset{(s.\ s.)}{9FeTiO_3+Fe_2O_3}+2O_2 = \underset{(s.\ s.)}{5Fe_2TiO_5+FeTi_2O_5}+2TiO_2$	13
		$\underset{(s.\ s.)}{9FeTiO_3+Fe_2O_3}+{}^3/{}_2O_2 = \underset{(s.\ s.)}{3Fe_2TiO_5+3FeTi_2O_5}+Fe_2O_3$	14
		$\underset{(s.\ s.)}{5FeTiO_3+3Fe_2O_3}+4TiO_2+O_2 = \underset{(s.\ s.)}{5Fe_2TiO_5+FeTi_2O_5}+2TiO_2$	15
		$\underset{(s.\ s.)}{5FeTiO_3+3Fe_2O_3}+4TiO_2+{}^1/{}_2O_2 = \underset{(s.\ s.)}{3Fe_2TiO_5+3FeTi_2O_5}+Fe_2O_3$	16
Titanomagnetite			
C1	$Usp_{60}Mt_{40}$		
C2	$Usp_{ss}+Ilm_{ss}$ (minor)	$\underset{(s.\ s.)}{6Fe_2TiO_4+4Fe_3O_4}+2O_2 = \underset{(s.\ s.)}{4Fe_2TiO_4+2Fe_3O_4}+\underset{(s.\ s.)}{4FeTiO_3+2Fe_2O_3}$	17
C3	$Mt_{ss}+Ilm_{ss}$ (major)	$\underset{(s.\ s.)}{6Fe_2TiO_4+4Fe_3O_4}+O_2 = \underset{(s.\ s.)}{3Fe_2TiO_4+3Fe_3O_4}+\underset{(s.\ s.)}{3FeTiO_3+3Fe_2O_3}$	18
C4	$Mt_{ss}+Ilm_{ss}+Hem_{ss}+R$	$\underset{(s.\ s.)}{6Fe_2TiO_4+4Fe_3O_4}+2O_2 = \underset{(s.\ s.)}{2Fe_2TiO_4+3Fe_3O_4}+\underset{(s.\ s.)}{FeTiO_3+5Fe_2O_3}+3TiO_2$	19
C5	$Hem_{ss}+R$	$\underset{(s.\ s.)}{6Fe_2TiO_4+4Fe_3O_4}+3O_2 = \underset{(s.\ s.)}{4FeTiO_3+10Fe_2O_3}+2TiO_2$	20
		$\underset{(s.\ s.)}{6Fe_2TiO_4+4Fe_3O_4}+3O_2 = \underset{(s.\ s.)}{2FeTiO_3+8Fe_2O_3}+4TiO_2+2Fe_3O_4$	21
		$\underset{(s.\ s.)}{6Fe_2TiO_4+4Fe_3O_4}+{}^7/{}_2O_2 = \underset{(s.\ s.)}{FeTiO_3+8Fe_2O_3}+5TiO_2+Fe_3O_4+2Fe_2O_3$	22
		$\underset{(s.\ s.)}{2Fe_2TiO_4+3Fe_3O_4}+\underset{(s.\ s.)}{FeTiO_3+5Fe_2O_3}+3TiO_2+{}^3/{}_2O_2 = \underset{(s.\ s.)}{FeTiO_3+9Fe_2O_3}+5TiO_2+Fe_3O_4+Fe_2O_3$	23

Table Hg-5 (continued).

Titanomagnetite		Typical Reactions	
C6	$Hem_{ss}+R+Pb_{ss}$	$\underset{(s.\,s.)}{6Fe_2TiO_4+4Fe_3O_4}+{}^7/_2O_2 = \underset{(s.\,s.)}{FeTiO_3+Fe_2O_3}+\underset{(s.\,s.)}{2Fe_2TiO_5+FeTi_2O_5}+TiO_2+2Fe_3O_4+5Fe_2O_3$	24
		$\underset{(s.\,s.)}{6Fe_2TiO_4+4Fe_3O_4}+{}^7/_2O_2 = \underset{(s.\,s.)}{FeTiO_3+10Fe_2O_3}+\underset{(s.\,s.)}{Fe_2TiO_5+FeTi_2O_5}+2TiO_2$	25
		$\underset{(s.\,s.)}{2Fe_2TiO_4+3Fe_3O_4}+\underset{(s.\,s.)}{FeTiO_3+5Fe_2O_3}+3TiO_2+{}^3/_2O_2 = \underset{(s.\,s.)}{2FeTiO_3+11Fe_2O_3}+4TiO_2$	26
		$\underset{(s.\,s.)}{2Fe_2TiO_4+3Fe_3O_4}+\underset{(s.\,s.)}{FeTiO_3+5Fe_2O_3}+3TiO_2+O_2 = \underset{(s.\,s.)}{FeTiO_3+5Fe_2O_3}+\underset{(s.\,s.)}{2Fe_2TiO_5+FeTi_2O_5}+TiO_2+Fe_2O_3+2Fe_3O_4$	27
C7	$Pb_{ss}+Hem_{ss}$	$\underset{(s.\,s.)}{6Fe_2TiO_4+4Fe_3O_4}+{}^7/_2O_2 = \underset{(s.\,s.)}{3Fe_2TiO_5+FeTi_2O_5}+FeTiO_3+8Fe_2O_3$	28
		$\underset{(s.\,s.)}{6Fe_2TiO_4+4Fe_3O_4}+3O_2 = \underset{(s.\,s.)}{2Fe_2TiO_5+FeTi_2O_5}+\underset{(s.\,s.)}{FeTiO_3+5Fe_2O_3}+2Fe_3O_4+TiO_2+Fe_2O_3$	29
		$\underset{(s.\,s.)}{2Fe_2TiO_4+3Fe_3O_4}+\underset{(s.\,s.)}{FeTiO_3+5Fe_2O_3}+3TiO_2+{}^3/_2O_2 = \underset{(s.\,s.)}{3Fe_2TiO_5+FeTi_2O_5}+\underset{(s.\,s.)}{FeTiO_3+8Fe_2O_3}$	30

1200°C the extremes in f_{O_2} which span this reaction are $<10^{-10.79}$ and $>10^{-0.68}$ atms; reduction of Fe_2TiO_5 should, conversely, result in Fe_2TiO_4.

For the temperature range (600-900°C) considered in the phase compatibility diagrams the $Usp\text{-}Mt_{ss}$ is complete, and an assymetric immiscibility gap is shown for $Ilm\text{-}Hem_{ss}$ at 600° and at 700°C. For the $Pb\text{-}FPb_{ss}$ series the end members decompose at 585° ± 15°C and 1140°± 10°C to R + Hem and to R + Ilm, respectively. Intermediate members break down at 800°C to Ilm_{ss} + Pb_{ss} + R and at lower temperatures (<750°C) to Ilm_{ss} or Hem_{ss} + R. The series is therefore incomplete for all temperatures between 600° and 900°C and solid solubility of $FeTi_2O_5$ (FBb) in Fe_2TiO_5 (Pb) increases with increasing temperature. Ranges of complete miscibility are shown as solid joins and immiscible regions are dashed.

Ilmenite systematics

The progressive oxidation of stoichiometric ilmenite at 900°C, proceeding along the oxidation reaction isopleth, will pass through the following sequences of assemblages according to the phase relationships shown in Figure Hg-14a: (1) Ilm_{ss} (*i.e.*, partially enriched in Hem_{ss}) + R; (2) Ilm_{ss} (now more highly enriched in Hem_{ss}) + R + Pb_{ss}; (3) Pb_{ss} + $IlmHem_{ss}$; and (4) Pb_{ss} + R. At 800°C the two-phase field R + Ilm_{ss} and the three-phase field Ilm_{ss} + R + Pb_{ss} are expanded proportionately and the assemblage sequence is now: (1) Ilm_{ss} + R; (2) Ilm_{ss} + R + Pb_{ss}; and (3) Pb_{ss} + R. These expansions continue at 700° and at 600°C because of the rapidly contracting Pb_{ss} limit and because the $Ilm\text{-}Hem_{ss}$ miscibility gap is encountered. This marked change splits the original three-phase regions and the oxidation sequence is: (1) Ilm_{ss} + R; (2) Ilm_{ss} + Hem_{ss} + R; (3) Hem_{ss} + R; (4) Pb_{ss} + Hem_{ss} + R; and (5) Pb_{ss} + R. For ilmenites of basaltic composition the only likely changes in the reaction sequence are for oxidation at 900°C where the onset of oxidation will yield Pb_{ss} coexisting with R + Ilm_{ss} (see Table Hg-2). Stabilizing minor element components or ilmenites with initially larger proportions of Fe_2O_3 in solid solution will influence the rate at which oxidation is initialized, but the progressive sequence is unlikely to differ in overall format.

Titanomagnetite systematics

Spinel oxidation is abundantly more complex than that of ilmenite oxidation because of the associated production of "exsolved" ilmenite at subsolidus temperatures. However, if chosen to be viewed separately, the ilmenite component of the assemblage will oxidize in parallel to and will conform to the

Figure Hg-14. (a-d) Phase compatibility diagrams constructed from the solvus data for Ilm-Hem_{ss} (Lindsley, 1973) and the decomposition data for the Pb_{ss} series (Haggerty and Lindsley, 1970) at 600°, 700°, 800°, and 900°C. Tie lines indicate the compositional limits of possible coexisting phases. The Fe_3O_4-Fe_2TiO_4 series is complete at these temperatures; Ilm-Hem_{ss} is complete at approximately 800°C and is taken as complete in (c); the Pb_{ss} is incomplete for the temperatures considered and the small but distinct limit of solid solubility of $FeTi_2O_5$ in Fe_2TiO_5 should be noted for the 600°C section (a). The sequence of assemblages discussed in the text are obtained by moving across each of the ternary diagrams along oxidation-reaction lines of constant Fe:Ti (dashed) from the TiO_2-FeO join towards the TiO_2-Fe_2O_3 join (see Fig. Hg-13). (e) Oxide assemblages relevant to each of the appropriate phase regions for each of the stages of oxide oxidation.

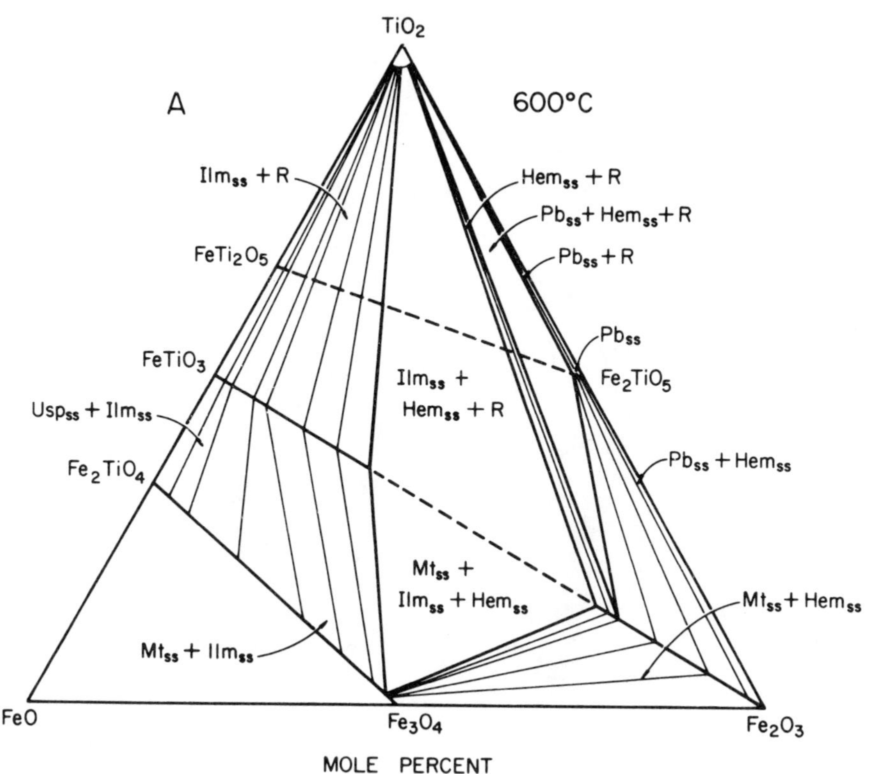

Figure Hg-14(a).

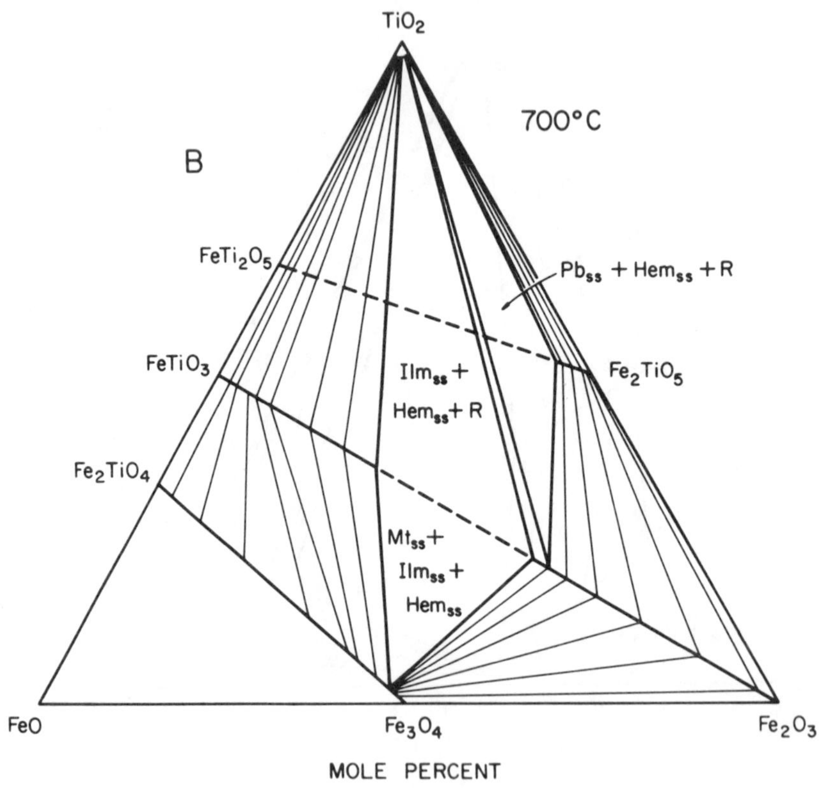

Figure Hg-14(b).

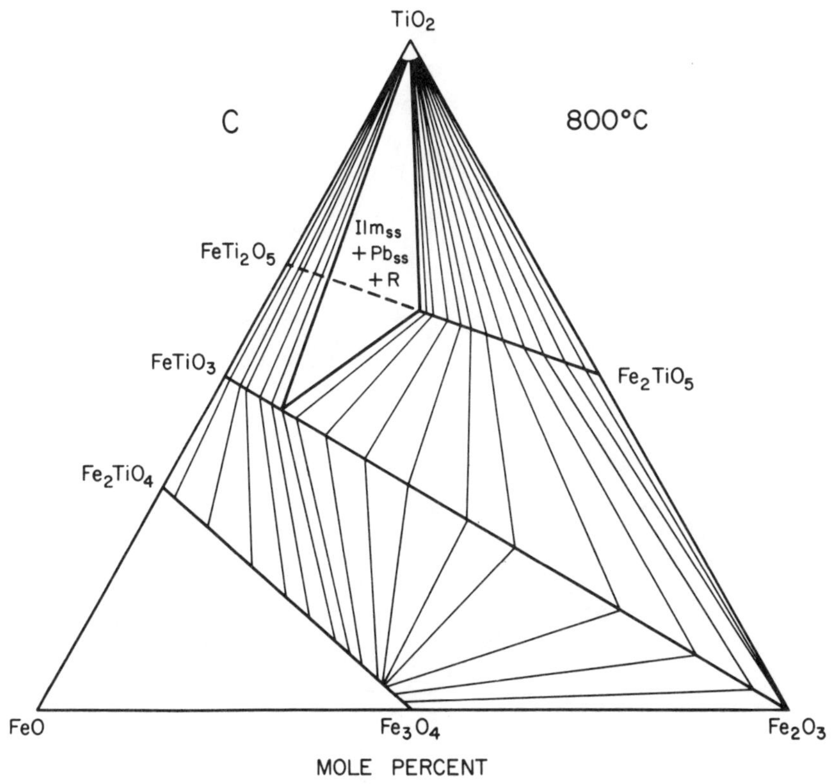

Figure Hg-14(c).

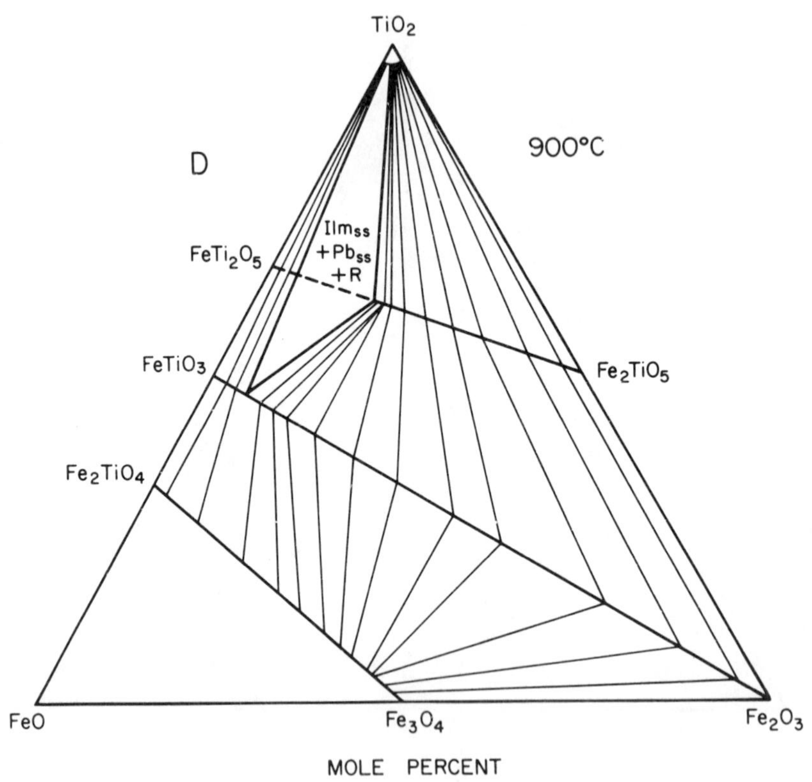

Figure Hg-14(d).

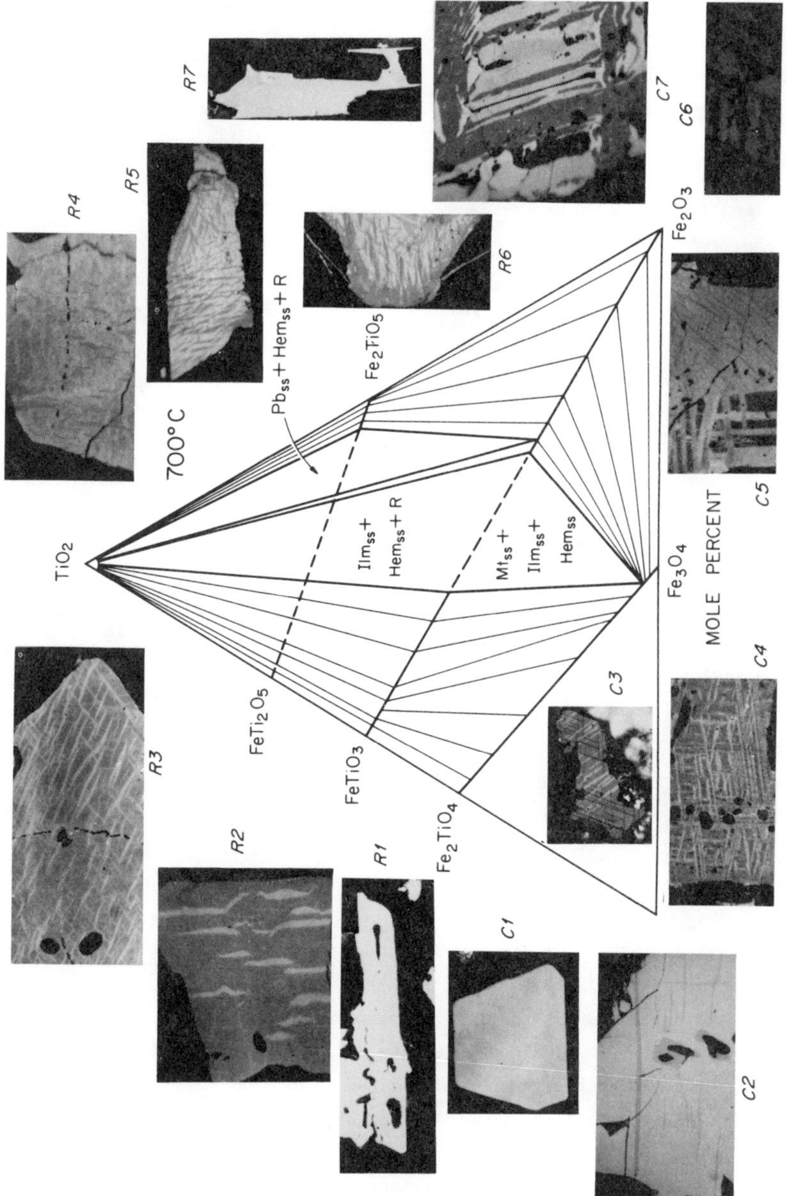

Figure Hg-14(e).

sequence systematics defined by discrete ilmenite, while the residual and Ti-depleted titanomagnetite will oxidize largely to titanohematite with perhaps minor associated R + Pb_{ss}. On the other hand, if subsolidus "exsolution" is limited then relatively large proportions of Usp_{ss} will remain in the spinel and a sequence of assemblages which are not too far removed from that of ilmenite will result. Both processes can be evaluated from the phase compatibility diagrams and both types of reaction sequences are considered in the equations listed in Table Hg-5.

For the more typical case in which oxidation exsolution does take place, the sequence of reactions at 900°C for an ilmenite with >90 mole % $FeTiO_3$ is: (1) Ilm_{ss} + R; (2) Ilm_{ss} + R + Pb_{ss}; (3) Pb_{ss} + Ilm-Hem_{ss}; and (4) Pb_{ss} + R. For values of <Ilm_{90}, the two-phase R + Ilm_{ss} assemblage is likely to be omitted but the remaining sequence will hold. With progressively decreasing temperatures the oxidation reaction trends will assure the formation of R + Ilm_{ss}, with progressively larger concentrations of Fe_2O_3 being possible at least to the limit of the Ilm-Hem_{ss} solvi limb. For the most part the remaining assemblages do not differ from those of discrete ilmenite at lower temperatures. The spinel upon "exsolution" becomes enriched in Fe_3O_4, and at 800° and at 900°C the assemblages most likely to result are: (1) Mt_{ss} + Hem_{ss} followed by (2) Hem_{ss} (with larger Fe_2O_3) + Pb_{ss}. At lower temperatures (600-700°C however, the solvi-interrupted Ilm-Hem_{ss} series produces two large three-phase fields in which the stable assemblages are Ilm_{ss} + Hem_{ss} + R or Mt_{ss} + Hem_{ss} + Ilm_{ss}. For Usp_{ss}-rich compositions (*i.e.*, Fe_2O_3 enriched) compositions the latter assemblage will result. Depending on the initial composition of Mt_{ss} (*i.e.*, on the proportion of Usp in solid solution), oxidation beyond the three-phase assemblage Mt_{ss} + Ilm_{ss} + Hem_{ss} may result in: (1) Mt_{ss} + Hem_{ss} or Hem_{ss} + R; followed by (2) Pb_{ss} + Hem_{ss} + R; and (3) Pb_{ss} + Hem_{ss}. For the more Usp-rich starting product, as noted above, the assemblage sequence most likely to result is: (1) Mt_{ss} + Hem_{ss} + Ilm_{ss}; (2) Ilm_{ss} + Hem_{ss} + R; (3) Hem_{ss} + R; (4) Pb_{ss} + Hem_{ss} + R; and (5) Pb_{ss} + Hem_{ss}. Note that the end products are similar and that the assemblage is Pb_{ss} + Hem_{ss}.

If the alternative situation is considered for kinetically rapid oxidation in which ilmenite "exsolution" is either retarded or totally inhibited, the resultant end products at high temperatures (*e.g.*, 900°C) will be Pb_{ss} + Hem_{ss}. Such reactions appear to be possible even at 800°C, but below this temperature the fields of Ilm_{ss} + Hem_{ss} + R, Hem_{ss} + R, and Pb_{ss} + Hem_{ss} + R are encountered prior to completion of the reaction which is finalized by Pb_{ss} + Hem_{ss}.

Applications

It is important to note that in deducing the most likely temperature at which oxidation occurred or in applying the compositions of an assemblage to these phase compatibility diagrams that the following points be borne in mind and rigorously adhered to: (1) These diagrams are based on the best estimates of currently available experimental data; (2) that the assemblages are based on equilibrium relationships; (3) initial variations in mineral chemistry, or in the influences of minor element concentrations or of atypical partitioning characteristics totally precludes the use of the diagrams; (4) that astutely determined independent estimates of reaction rates are important; and (5) that conjugate tie-lines do not link the compositions of actually determined phases nor do these represent experimentally determined points of coexistence--these are merely best estimates based on available data which require experimental confirmation.

Within the limitations of these five constraints a clearly defined case can be made for the affirmation of the oxide decomposition sequences of titanomagnetite in stages C1 to C7, and of ilmenite in stages R1 to R7. An equally well-justified case can also be made for the two Iceland and the two Oregon basalts quoted earlier which show that the mineral chemistry of the oxide components are consistent with oxidation between 700° and 800°C. Oxidation was almost certainly continuous down to at least 600°C, and undoubtedly below this temperature, although because these reactions are kinetically sluggish no evidence (for example, Pb_{ss} decomposition) is available defining the absolute lower limit of oxidation quenching. Given the prerequisite of a chemically reactive and oxidizing gaseous environment, a very large percentage of subaerially extruded basalts appear to have received and reflect oxidation temperatures which are very close to 700-750°C. Many basalts are only partially oxidized, variably oxidized, or totally unoxidized, and these comments pertain only to those basalts which have undergone intense oxidation.

In summary, the results of this and previous sections are perhaps best gleaned from a careful inspection of the montage illustrated in Figure Hg-14e, and a clearer understanding of the relevancies and relationships among the mineral oxidation classification, mineral chemistry and phase compatibility considerations should now be possible.

Figure Hg-15. Chromian spinels. (a) Euhedral spinel core mantled by chromian-titanomagnetite. The rounded crystal to the left is partially zoned and is enclosed in olivine. (b) An irregularly shaped chromian-spinel mantled by chromian-titanomagnetite, and titanomagnetite. Electron microprobe data for this spinel complex are listed in Table Hg-6 and the analytical profile (approximately E-W) is shown in Figure Hg-16b. (c) Several adjoining spinels with diffuse reaction contacts against chromian-titanomagnetite mantles. (d) A zoned chromian-spinel core mantled by the assemblage titanomagnetite + trellis ilmenite. Both mantle phases show partial oxidation as evidenced by the irregular mottled appearance. The white stringers are hematite. (e) An oxidized chromian-spinel inclusion (black) mantled by an original titanomagnetite crystal which has decomposed to Mt_{ss} + $pleonaste_{ss}$; associated trellis and external composite assemblages, after ilmenite, are now R + Hem_{ss}. The chromian spinel has hematite lamellae along {111} planes which are parallel to the {111} planes of the mantle assemblage. (f) A discrete chromian spinel with oxidation hematite lamellae along {111} and a diffusion rim of hematite along the spinel grain boundary. The spinel is enclosed in a highly oxidized olivine crystal, and the bright white flecks are magnetite + hematite.

Scale bar = 25 μm.

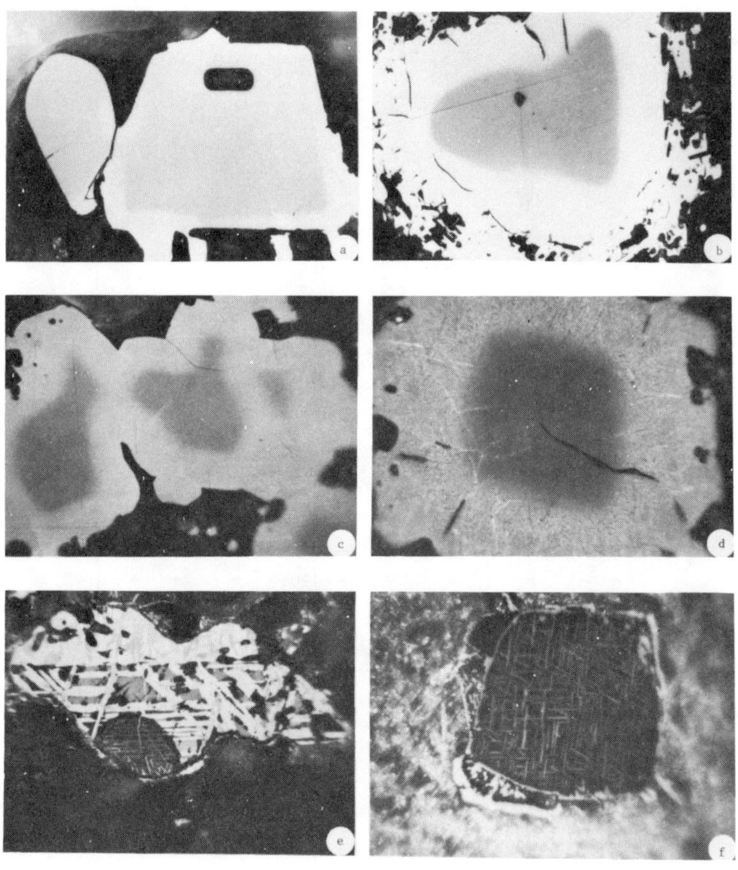
a
b
c
d
e
f

Table Hg-6. Electron microprobe analyses for a zoned spinel in a traverse from crystal edge (analysis 1) to crystal core (analysis 13). See Figure Hg-16 for the complete distribution and Figure Hg-15b for a photomicrograph of the grain. Sample D6-2. Analyses are at approximately 5 μm intervals.

	1	2	3	4	5	6	7	8	9	10	11	12	13
MgO	2.39	2.41	2.33	2.58	2.72	2.82	3.46	5.00	6.37	7.29	9.54	10.17	10.46
Al_2O_3	3.98	4.00	4.71	5.42	6.82	8.30	13.19	21.37	22.83	23.11	23.68	23.43	23.72
TiO_2	16.87	16.63	14.89	13.63	11.06	8.94	5.17	1.60	1.10	1.17	1.11	1.09	0.98
Cr_2O_3	11.37	12.46	14.57	17.62	22.29	25.33	29.72	31.63	32.41	33.11	33.47	34.25	34.60
MnO	0.48	0.55	0.63	0.47	0.50	0.62	0.58	0.49	0.51	0.43	0.40	0.43	0.41
FeO	43.27	42.93	41.72	40.59	38.51	36.21	32.67	28.54	25.86	24.87	21.43	20.58	20.26
Fe_2O_3	21.13	20.40	21.17	20.01	19.22	17.71	15.12	11.24	9.64	9.47	9.65	9.95	9.94
Total	99.42	99.38	100.02	100.32	101.12	99.93	99.91	99.87	98.72	99.45	99.28	99.90	100.37
Oxygens	32	32	32	32	32	32	32	32	32	32	32	32	32
Mg	1.036	1.047	1.002	1.100	1.143	1.189	1.419	1.952	2.471	2.790	3.594	3.801	3.880
Al	1.365	1.374	1.604	1.830	2.266	2.766	4.274	6.600	7.008	6.994	7.055	6.923	6.959
Ti	3.693	3.642	3.234	2.935	2.344	1.901	1.068	0.316	0.216	0.226	0.210	0.206	0.184
Cr	2.618	2.869	3.328	3.988	4.968	5.663	6.461	6.553	6.672	6.723	6.690	6.789	6.811
Mn	0.118	0.135	0.154	0.114	0.119	0.149	0.136	0.109	0.113	0.095	0.085	0.092	0.087
Fe^{2+}	10.540	10.461	10.078	9.720	9.082	8.563	7.513	6.254	5.632	5.342	4.531	4.313	4.218
Fe^{3+}	4.632	4.472	4.601	4.312	4.077	3.768	3.128	2.216	1.889	1.831	1.835	1.878	1.862
Cations	24.002	24.000	24.001	23.999	23.999	23.999	23.999	24.000	24.001	24.001	24.000	24.002	24.001

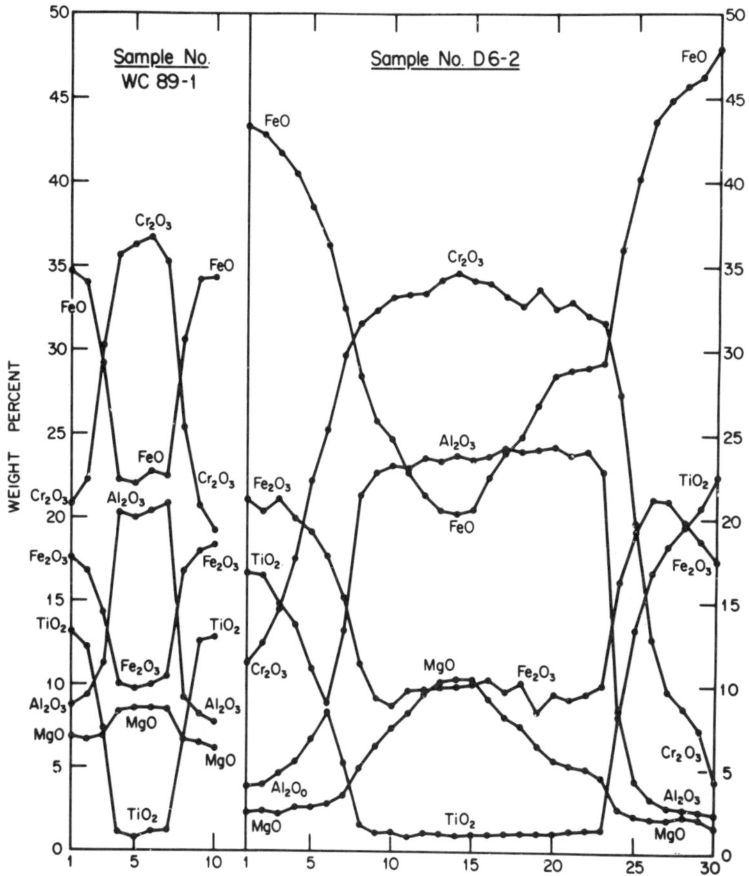

Figure Hg-16. Electron microprobe profiles across two zoned spinel crystals. WC89-1 is from western Iceland, and D6-2, which is illustrated in Figure Hg-16b, is from eastern Iceland. Both spinels are groundmass crystals in basalts. Data points are at approximately 5 μm intervals. The distribution of major elements for both crystals are almost identical. Note that the cores are TiO_2, FeO, and Fe_2O_3 poor, and MgO, Al_2O_3, and Cr_2O_3 rich. Analytical data for one half of the profile across D6-2 are listed in Table Hg-6.

ASSOCIATED MINERAL OXIDATION

Introduction

The high temperature oxide assemblages discussed in previous sections can be demonstrated to result by deuteric cooling; that is, oxidation as a direct consequence of volatile reaction at subsolidus temperatures during the initial cooling of a basalt lava. Other minerals commonly occur in place of the high temperature assemblage, and examples of these are titanomaghemite replacement of titanomagnetite, and of sphene or aenigmatite replacement of either ilmenite or of titanomagnetite. Titanomaghemite is a low temperature process mineral with a probable thermal stability range between 250° and 550°C, and although it may develop during the final stages of cooling, its occurrence can also be related to supergene weathering or to burial metamorphism. For sphene and aenigmatite the reaction with residual magmatic fluids are suggested, although the most prominent development of sphene is more typically related to low grade metamorphism or to reactions associated with the slow cooling of plutons. More directly associated with intense high temperature oxidation is the decomposition of chromian-bearing spinel and the oxidation of olivine, and within the context of this chapter the following brief discussion is limited to these two minerals.

Chromian spinel oxidation

Chromian-bearing spinels are compositionally complex, are accessory minerals in basalts and have two modes of occurrence: (1) as inclusions in olivine, but sometimes also in pyroxene; and (2) as discrete groundmass crystals. The inclusions crystals are generally homogeneous, although zoning and Fe-Ti enrichment mantles are also observed; the groundmass crystals on the other hand, are always spectacularly zoned, and examples are illustrated in Figure Hg-15. The zonal trends are typical of reactions with the liquid and in most cases the extreme outer mantles are compositionally similar to discrete titanomagnetite. Major element electron microprobe profiles across two such crystals are shown in Figure Hg-16, and selected analyses are listed in Table Hg-6 High temperature oxidation of the spinel mantles parallels those of titanomagnetite with similar textures and assemblages, but the cores of these spinels only decompose under extreme and intensely oxidizing conditions. For the most part decomposition is restricted to the oxidation of Fe^{2+}, and this is expressed by the development of finely textured hematite along {111} planes

(Fig. Hg-15e-f), which is similar in every respect to oxidation "exsolved" ilmenite in titanomagnetite. This partitioning of Fe^{3+} enriches the core spinel in Mg, Al and Cr, and under reflection oil-immersion optics changes the color from pale gray to black with an accompanying reduction in reflectivity.

Olivine oxidation

In all olivine basalts which have approached or have gone beyond the C5, R5 (R + Hem_{ss}) stage of oxide oxidation, evidence for the decomposition of the fayalitic component of olivine can be demonstrated. The oxide components which are generated are magnetite and hematite as illustrated in Figure Hg-17. Magnetite is most commonly restricted to the interior of olivine crystals and hematite to the exterior, although intense oxidation will transform all Fe_3O_4 to Fe_2O_3. Olivine oxidation is particularly pertinent to magnetic studies because these secondarily produced magnetites are fine grained and in many cases approach either single domain size or small multidomain clusters which have high coercivities and are magnetically stable. In the presence of Pb_{ss} + Hm_{ss} or R + Hem_{ss} assemblages, the residual magnetites which may persist are small in concentration when compared with unoxidized samples, and the mainstay of the magnetic signature therefore is now largely dependent on the chemically produced magnetite from olivine, which in itself may carry a thermoremanent magnetization if oxidation occurred above the Curie point of Fe_3O_4. The potential significance of olivine oxidation to rock magnetism and to paleomagnetism, and details of textural intergrowths and assemblages are to be found in the following sources: Haggerty and Baker (1967), Baker and Haggerty (1967), Champness and Gay (1968), Champness (1970), Riding (1969), and Hoye and O'Reilly (1972,1973). The stability field of the olivine series with respect to oxidation has been determined by Nitsan (1974), who has shown that compositions between Fa_{10} and Fa_{100} at 700°C are stable between 10^{-14} and 10^{-18} atms, respectively. A new and novel technique for decorating dislocations in olivine by oxidation and iron oxide precipitation is discussed for Kohlstedt and Vander Sande (1975), by Kohlstedt *et al.* (1976), and by Zuench and Green (1976); and in fact the textural distribution of oxides shown in Figure Hg-17f-g is dominantly controlled by such dislocations which act as the preferred sites of nucleation.

Figure Hg-17. Olivine oxidation. (a) A central core of partially oxidized olivine with symplectic magnetite which gradually increases in abundance towards the crystal grain boundary. The hematite diffusion rim is less prominently developed than the example shown in grain (b). (b) An oxidized olivine with a well-developed hematite diffusion rim and with symplectic magnetite + hematite throughout a large proportion of the crystal. (c-e) These three crystals display varying degrees of symplectic magnetite growth. Grains (c) and (d) have marginally-developed hematite whereas grain (d) is entirely magnetite with the central core of the olivine unoxidized. (f-g) These olivine crystals were oxidized in air at 800° and 950°C, respectively. Grain (f) contains hematite which has precipitated along olivine dislocations, and grain (g) shows fine symplectic magnetite in subspherical units.

Scale bar = 25 μm.

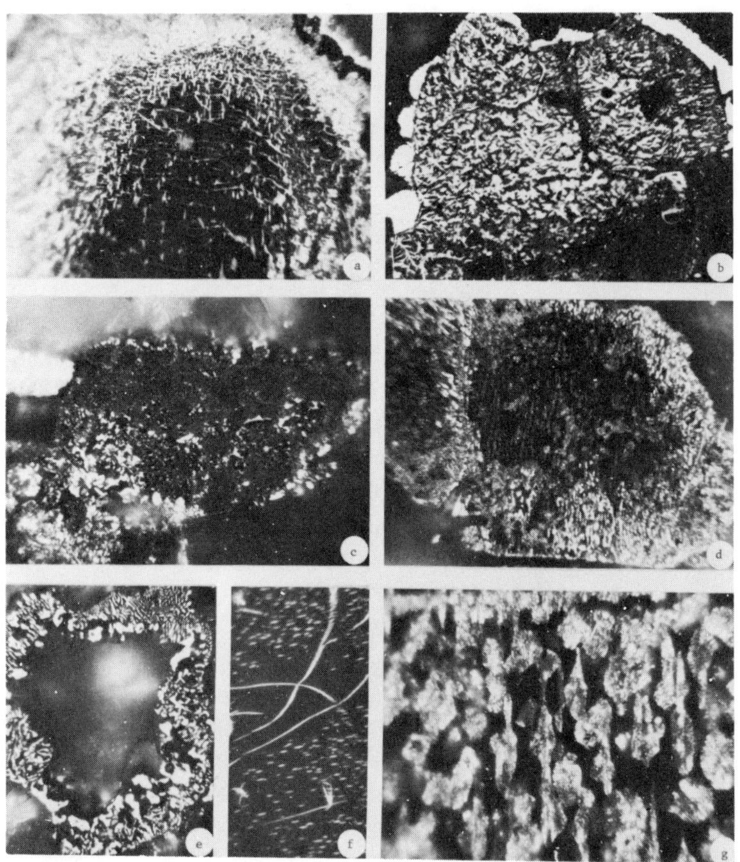
a
b
c
d
e
f
g

OXIDE DISTRIBUTIONS IN BASALT PROFILES

Introduction

The evidence for deuteric oxidation is based on two factors: (1) the high temperatures associated with the lower thermal stability limits of the oxides, referring specifically to members of the Pb_{ss} series; and (2) on the distribution of oxidation zones in profiles across single lava flows which have previously been shown to reach maximum levels of oxidation towards the central-most regions of lavas; these high oxidation zones can be related to migrating isotherms as a function of cooling rate (Sato and Wright, 1966; Watkins and Haggerty, 1967).

Although the general tenor of discussion in previous sections has been on the non-equilibrium nature of deuteric oxidation, all studies published to date show that, although variations are present in the degree of oxidation on a grain-to-grain basis, when considered as a unit marked variations in the average state of oxidation are present in vertical traverses between the upper and lower cooling surfaces of either basalt lavas or of ponded lava lakes.

In earlier studies estimates of the mean oxidation level were established by evaluating the dominant oxide assemblage which was present as determined microscopically. For example, a sample containing >80% of titanomagnetite in its unoxidized state (*i.e.*, optically homogeneous) was expressed as a Class I sample and in the revised classification as C1. In all instances ilmenite will also be homogeneous at the R1 stage, and the sample is therefore classified as C1R1. It soon became obvious that average evaluations were insensitive in attempts to correlate magnetic property data with oxide mineralogy. Rationalizing bulk chemical data for extreme variations in $FeO:Fe_2O_3$ ratios was also ample demonstration that a more quantitative approach was necessary, since both ferric iron and magnetic property variations are almost entirely dependent on the variations expressed by the state of oxidation of the Fe-Ti oxides. For extreme assemblages at either very high or very low oxidation states, chemical, mineralogical and magnetic correlations are good but for intermediate stages or for samples containing a variety of assemblages, correlations are ill defined or depart from the expected norm. Three advances have been made to correct, or at least quantify the mineralogy: The first is the advent of automated electron microprobes; the second is the refined oxide classification discussed in previous sections which is now more acceptable in terms of phase compatibility relationships due to the revised experimental data by Lindsley;

and the third is the modal determination of assemblages using classical point counting techniques.

Mean oxidation numbers

In modal determinations of the distribution of oxide assemblages, estimates to within $\pm$ 5% are possible for counts of between 200-500 oxide grains in 2.5 cm diameter sections for crystals with a mean grain size of 50 μm and a concentration of 5-10% by volume (statistics based on determinations by 12 observers and 22 observations of the same sample). The number of grains counted depends on variations in grain size, on the ratio or abundance of titanomagnetite and of ilmenite, and on whether a sample is uniformly oxidized or displays a wide disparity in assemblages. With experience a decision can be made quickly for additional grain counts or for multiple observations. The technique is rapid and straightforward if the oxide classification is followed and if the following criteria are employed in distinguishing between original titanomagnetite and original ilmenite: (1) grain morphology and the distinctions in titanomagnetite and ilmenite skeletal crystal habits; (2) the ratios of $R{:}Hem_{ss}$ and of $Pb{:}Hem_{ss}$ (these ratios are larger in oxidized ilmenite); (3) relic {111} planes; and (4) residual rods or blebs of black spinels (pleonaste-magnesioferrite) are indicative of original titanomagnetite.

Mean titanomagnetite and ilmenite oxidation numbers are determined by the percentage of grains which fall into each of respective stages of oxidation. This value is expressed either as MC (for titanomagnetite) or as MR (for ilmenite) according to:

$$MC = \frac{(C1 \times n_1) + (C2 \times n_2) + (C3 \times n_3) \ldots (C7 \times n_7)}{n_1 + n_2 + n_3 \ldots n_7}$$

where MC or MR = mean oxidation number

C1 to C7 = titanomagnetite oxidation stages

R1 to R7 = ilmenite oxidation stages

n_1 to n_7 = % of grains in each oxidation class.

Apart from the decision of whether an oxidized grain was originally titanomagnetite or originally ilmenite, the only additional precautionary measures which should be exercised are evaluations which depend on the abundances of ilmenite lamellae in titanomagnetite (C2 or C3), on whether composite or sandwich intergrowths are primary or oxidation products, and on the abundance levels once again of ferrian rutile in ilmenite (R2 or R3). As a

Figure Hg-18. Oxidation profiles across 14 single lava flows from Iceland based on the assemblages of oxidized titanomagnetite. The thinner flows AS to DS are flow units and the remaining flows have clearly-defined upper and lower chilled margins. Profiles S and MS are from the same flow sampled at approximately 100 m apart, and are shown in greater detail in Figure Hg-19. Flow HS which is the thickest flow is discussed in Figure Hg-20. The oxidation indices are listed from 1 to 6, where 6 is equivalent to C7 as defined in the text. Note that the maximum levels of oxidation vary from one flow to another, and vary also as a function of flow thickness. Those flows exhibiting Pb_{ss} + Hem_{ss} (6 or C7) show preferred maxima towards the central portions of flows or maxima at one third or two thirds from the base.

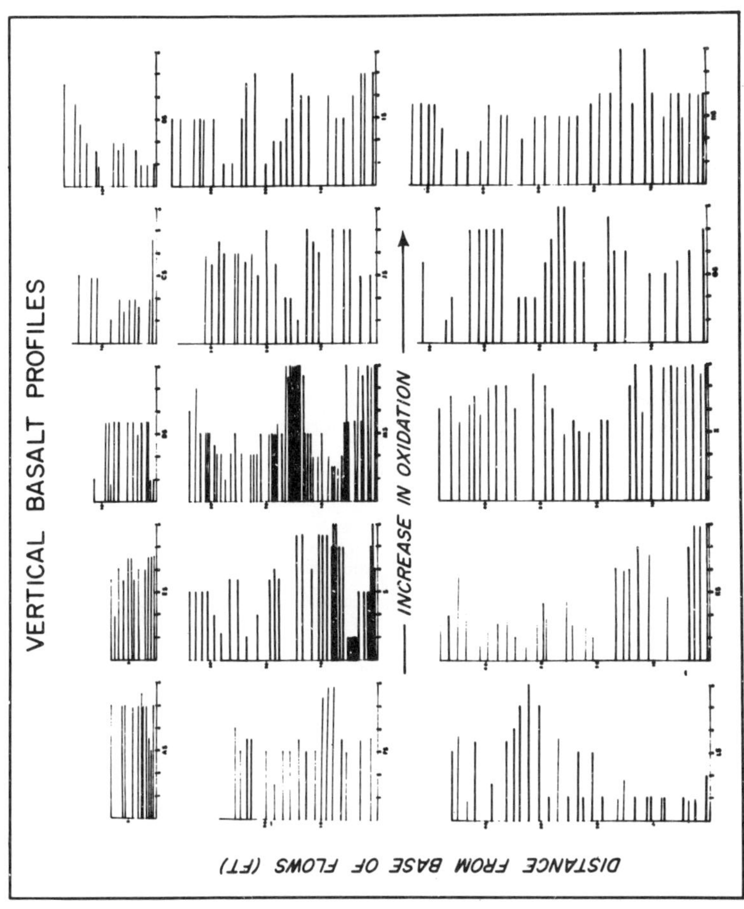
VERTICAL BASALT PROFILES
INCREASE IN OXIDATION
DISTANCE FROM BASE OF FLOWS (FT)

general rule, incipient or small numbers of lamellae are C2 and R2 and large numbers (>25% by area) R3 and C3. Composite intergrowths require independent compositional information, and these grains should be classified as C1, C2 or C3 with the appropriate suffixes denoting composite and sandwich types, for example as $C2_c$ (composite), $C2_s$ (sandwich), or $C2_{cs}$ if both are present in association with a small number of trellis ilmenite lamellae. For magnetic studies it is imperative also that some estimates be made of the level of oxidation of olivine and close attention given to the ratios of unoxidized olivines to oxidized olivines and of the ratios of Fe_3O_4:Fe_2O_3 in crystals which display oxidation (see Fig. Hg-17).

Oxide distributions

Three examples are given to illustrate the variations in oxide assemblages within single cooling units. The examples are for Iceland basalts (Watkins and Haggerty, 1967; Wilson *et al.*, 1968), a drill core from the Makaopuhi lava lake (Sato and Wright, 1966; Haggerty, 1971), and for zones of oxidation adjacent to joint selvages from the Oregon Picture Gorge basalts (Lindsley, 1960; Lindsley and Haggerty, 1971). The examples are restricted to extrusive basalts because wide variations in the oxide assemblages of plutons and dikes are known to be limited. The discussion is necessarily brief because the major attention should be given to the accompanying diagrams and figure captions which are self explanatory.

The oxide distributions across 14 single lavas, with two profiles across one lava (S and MS), are shown in Figure Hg-18. These distributions and those that follow are for titanomagnetite based on a scale of 1 to 6 where 1-5 are equivalent to C1 to C5, 6 is equivalent to C7, and the distinctions between C6 (incipient Pb_{ss}) and C7 (Pb_{ss}+Hem_{ss}) as defined here are not shown. This subtle discrimination is important but does not affect the overall trend of oxide distributions or the positions of maximum oxidation. Oxide distributions and intensity of magnetization are shown in Figures Hg-19a-b and Hg-19c-d, respectively, for profiles S and MS which were samples at approximately 100m apart. The central zone of high oxidation at approximately 2m above the base of the flow are shown in Figures Hg-19e and Hg-19f as functions of the concentration of titanomagnetite at stage C7, and of ilmenite at stages R6 and R7. The oxidation of Ilm_{ss} and Mt_{ss} closely parallel each other, and if these distributions are now compared with the intensity of magnetization profile (Fig. Hg-19c), a good correlation among parameters is apparent. Two points should be noticed: (1) The rise and final peak in maximum oxidation is relatively sharp when approached from the base of the flow but falls off progressively on the upper side, a situation which is reversed in profile MS (Fig. Hg-19d); and (2) the maxima in oxide oxidation at the base of the flow is not reflected in the intensity of magnetization profile, a result of Mt_{ss} decomposition but of only partial olivine decomposition. In Figure Hg-19g the temperature distribution profile is shown for traverse S after a period of 58 weeks assuming an extrusion temperature of 1000°C and deuteric cooling of the flow on an underlying basalt and a free-air upper face (Jaeger, 1961; Watkins and Haggerty, 1967). The phase compatibility ternary at 650°C closely models the distribution of oxidation assemblages with two and three phase assemblages on the Fe_2O_3-rich portion of the diagram being typical of the central high oxidation region, and with two phase assemblages on the FeO-rich portion of the diagram being typical of adjacent relatively unoxidized zones.

Magnetic and oxide parameters for a second flow (HS), which is the thickest flow shown in Figure Hg-18, are given in Figure Hg-20a-c. These data are from Wilson *et al.* (1968), and in this example the peak in oxidation is at approximately 5m from the base of the flow which is 16m thick. Here, once again, intensity of magnetization correlates with high oxide oxidation and with the oxidation of olivine. Other features of interest are the remarkable constancy of Curie temperatures and of the relative positions of maximum grain sizes for ilmenite which correlates with the zone of high oxidation, and for titanomagnetite which is at a maximum at 9m above the flow base.

The major points to notice for the Iceland basalts are as follows: (1) The state of oxidation is highly variable throughout the flows; (2) maximum states of oxidation are present towards the central portions of the flows although in many cases maximum values are either one third or two thirds from the base; (3) the Fe-Ti oxidation index correlates positively with $FeO:Fe_2O_3$ ratios; (4) grain size variations parallel the distribution of oxide assemblages; (5) traverses from the same lava show differences in the positions of maximum oxidation; and (6) magnetic property measurements show that highly oxidized zones are more intensely magnetic and more stable magnetically than unoxidized zones.

The example from the Makaopuhi lava lake is shown in Figure Hg-21a for the distribution of an oxidized zone between 550° and 750°C. Temperatures and values of f_{O_2} were measured *in situ*, and the data are from Sato and Wright (1966). Oxide parameters as a function of depth and of $Fe_2O_3:FeO$ for

Figure Hg-19. (a-b) Oxide distributions from S and MS as a function of thickness. The profiles are approximately 100 m apart. Oxidation indices I-V are equivalent to C1 to C5, and index VI is equivalent to C7. Data from Watkins and Haggerty (1967). (c-d) Intensity of natural magnetization J (emu/gm x 10^{-3}) for the same suite of samples shown in Figure Hg-19a-b with a close correspondence between high intensities and high oxidation. The only discrepancy is at the base of the flow which is a reflection of the oxidation of the oxides but not of olivine. (e-f) Relative percentages of oxides at each of the oxidation stages C7 and R6, R7 for samples from the zone of maximum oxidation in profile S. These distributions bear a much stronger relationship to the intensity of magnetization profile, and it is of interest to note that C7 and R7 (Pb_{ss} + Hem_{ss} after titanomagnetite and after primary ilmenite, respectively) are virtually identical. (g) The curve is the polytherm after approximately one year for a lava of 10 m in thickness extruded at an initial temperature of 1000°C on an underlying basalt and with a free-air upper face. The assymetry of the polytherm which is modeled ideally by profile S results from the relative rates of convective heat loss which is rapid at the free-air surface but relatively slower at the base because of the insulating effects of the underlying lava. The polytherm is calculated from the data by Jaeger (1961) and is discussed in Watkins and Haggerty (1967). The ternary shows the expected phase compatibility relationships for the oxides at approximately 650°C. The upper portions of the flow are characterized by Mt_{ss} + Ilm_{ss}, below this by R + Hem_{ss} and within the zone of maximum oxidation by Pb_{ss} + Hem_{ss}. The polytherm shows clearly that the lower central third of the flow cools at the slowest rate, and hence it is within this region that volatile accumulation takes place and high oxidation ensues.

PROFILE S

10
7.5
Meters 5
2.5

I II III IV V VI
Oxidation Index

(a)

PROFILE MS

10
7.5
Meters 5
2.5

I II III IV V VI
Oxidation Index

(b)

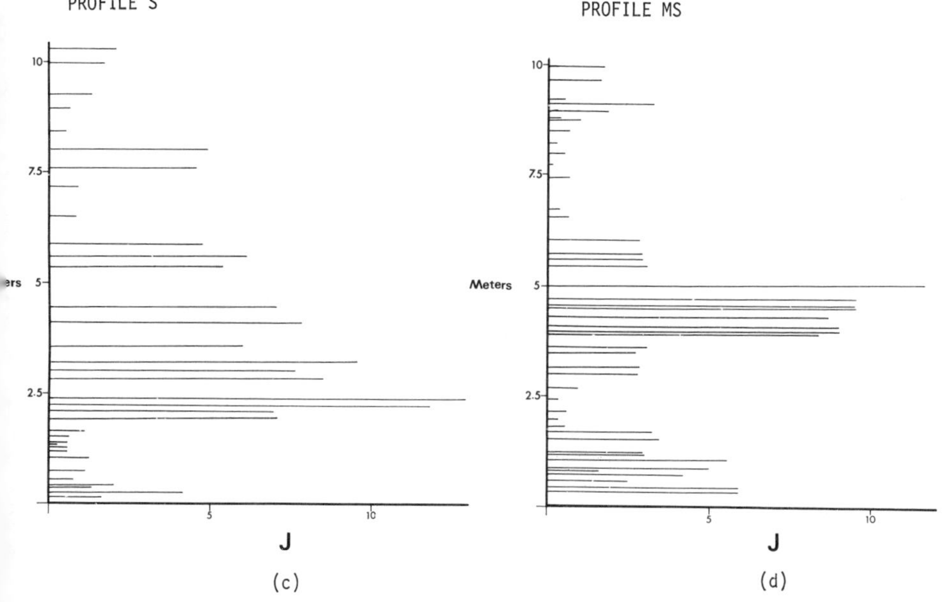

(c)

(d)

Figure Hg-19(a-d).

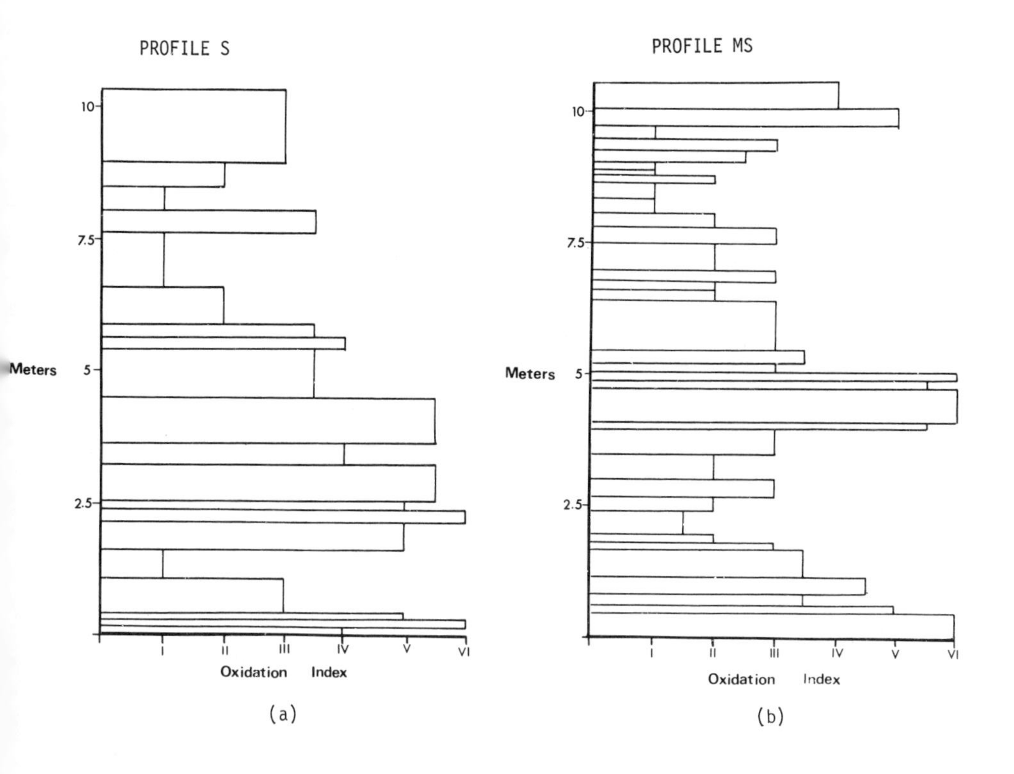

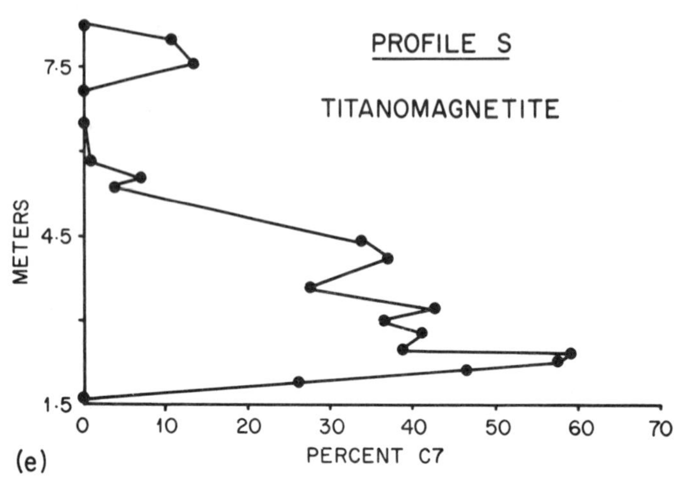

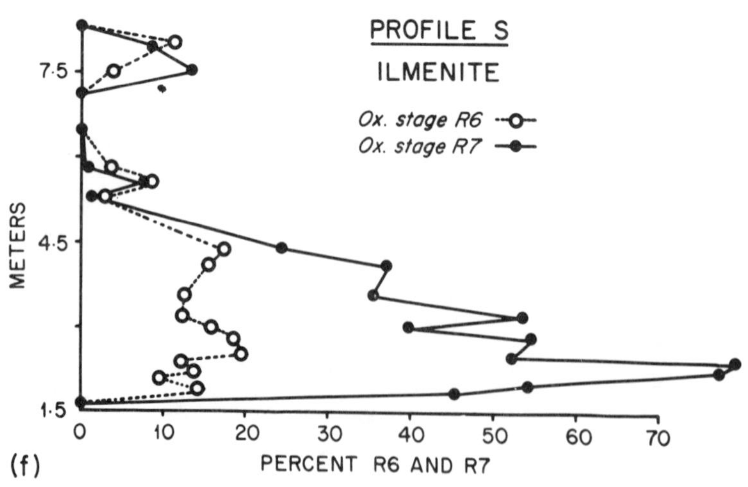

Figure Hg-19(e-f).

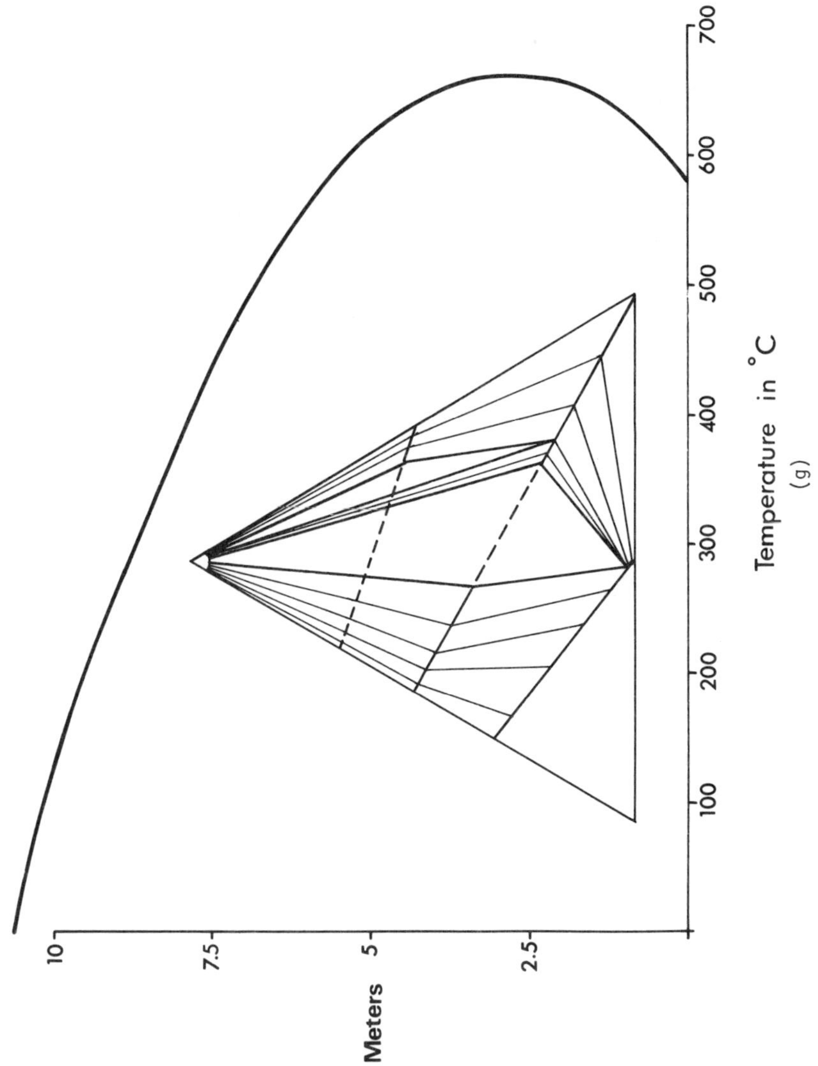

Figure Hg-19(g).

Figure Hg-20. (a-b) Magnetic and oxide parameters for 30 samples from profile HS. Sample 1 is at the base of the flow and sample 30 is at the top. Data are from Wilson *et al.* (1968). (c) Fine-scale definition of the zone of maximum oxidation showing a close correlation among parameters except for high Fe_2O_3:FeO ratios at the base of the flow. Note the surprising correlation of ilmenite grain size and that the maximum grain size for titanomagnetite is 3 m higher in the flow (refer also to Fig. Hg-19b).

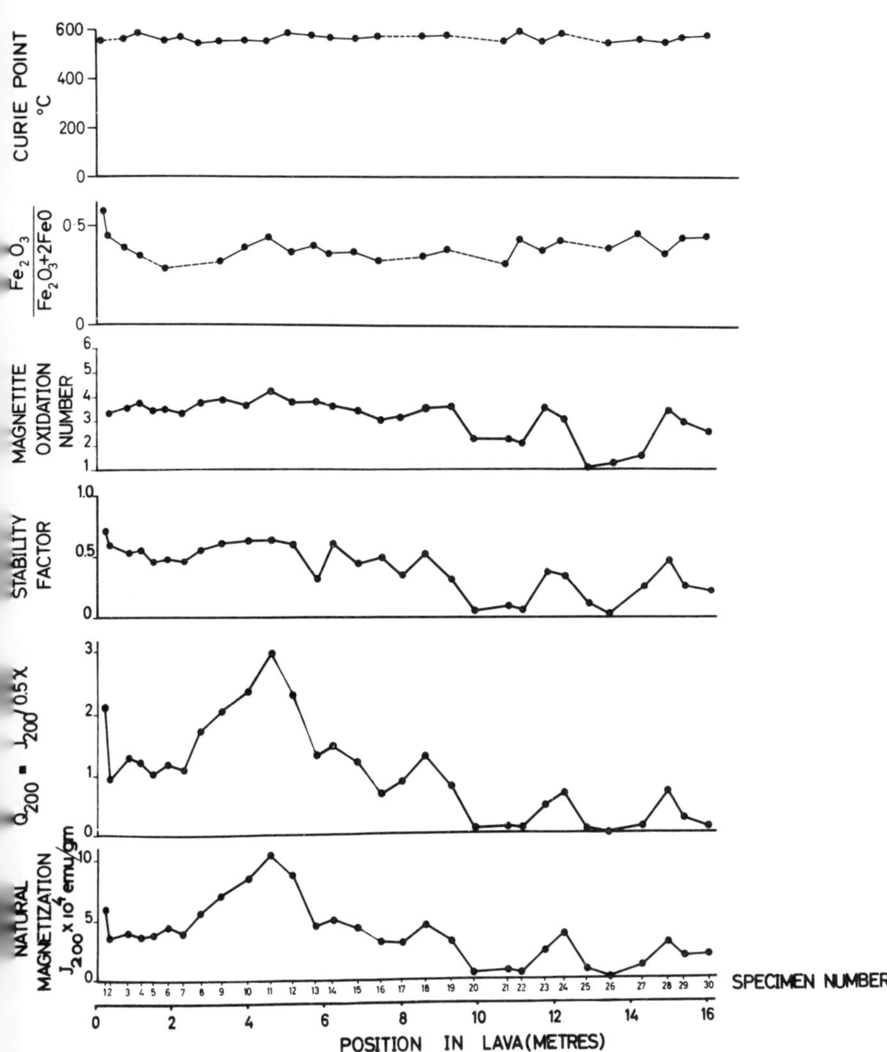

(a)

Figure Hg-20(a).

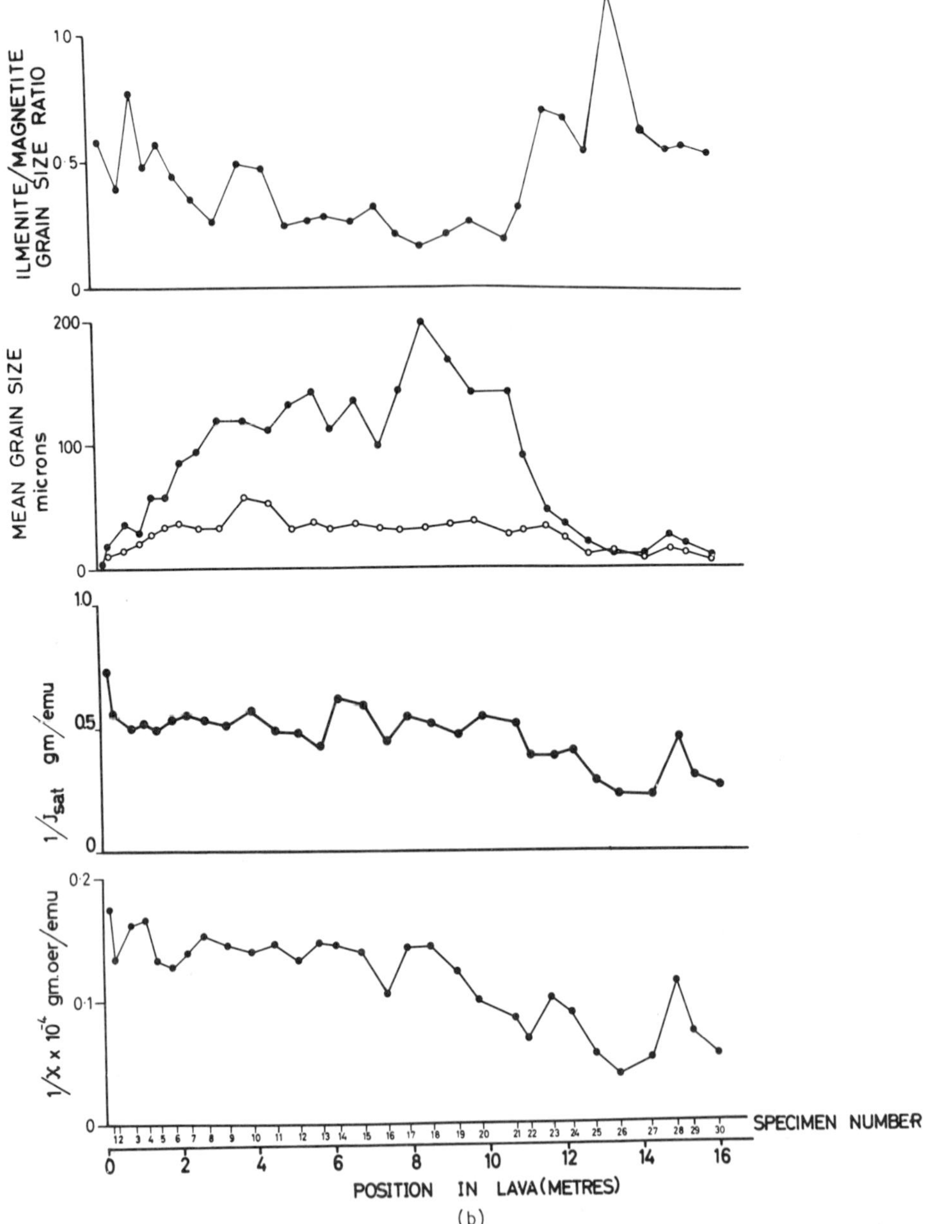

(b)

Figure Hg-20(b).

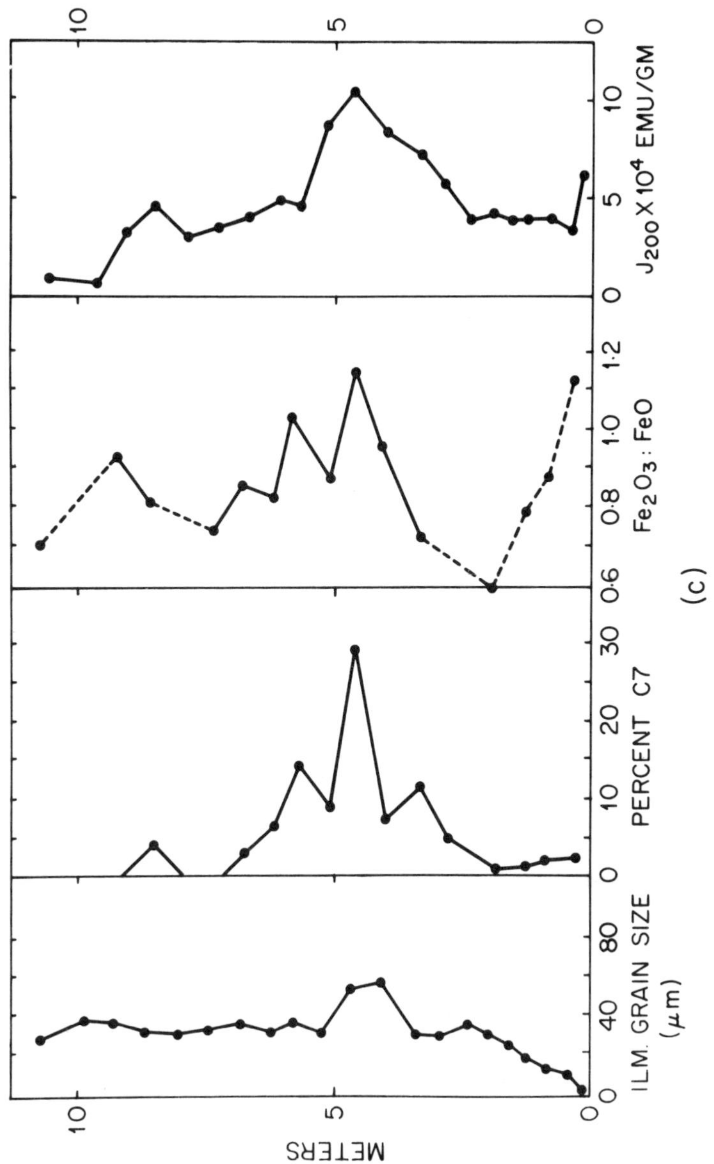

Figure Hg-20(c).

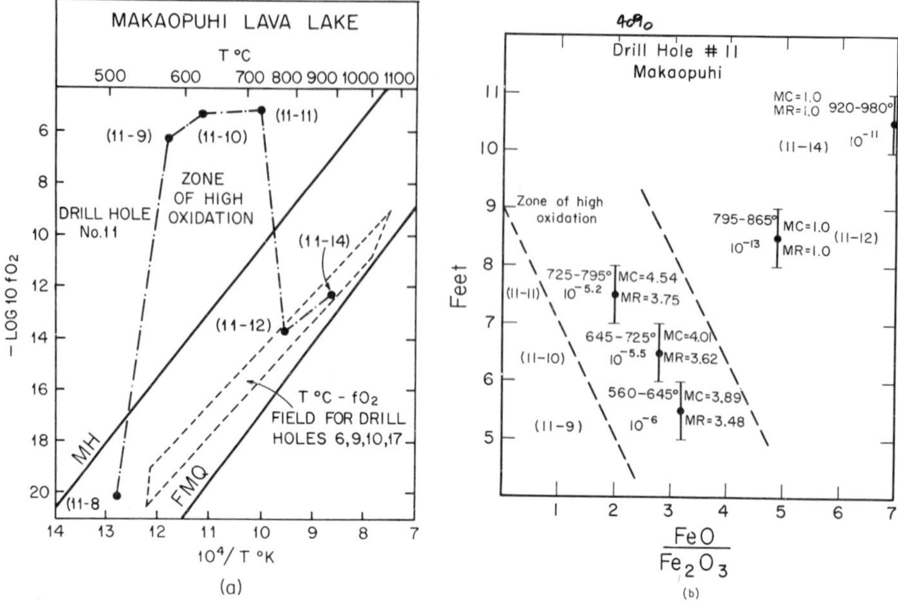

Figure Hg-21. T°C-f_{O_2} relationships for *in situ* measurements of drill holes in the cooling Makaopuhi lava lake, Hawaii. MH and FMQ are the magnetite-hematite, and the fayalite-magnetite-quartz buffers, respectively. Data are from Sato and Wright (1966). (b) Oxide data of samples for drill hole #11 as a function of the depth of collection, $FeO:Fe_2O_3$, T°C and f_{O_2}. Data are from Haggerty (1971).

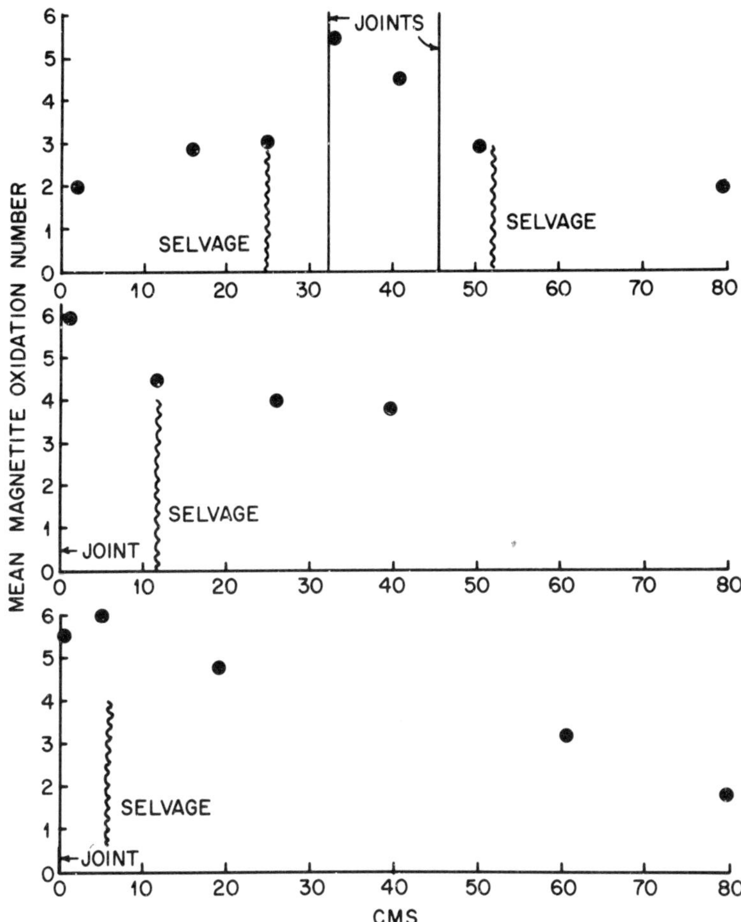

Figure Hg-22. (a-c) Variation in mean magnetite oxidation numbers as a function of distance from vertical joints in basalts from Picture Gorge, Oregon. The selvages are toughened annealed margins adjacent to the joints. DHL 296/69 and DHL 295/69 are shown in (c) at 1 cm and at 80 cm, respectively, and the mineral chemistry of these samples is discussed in Figure Hg-12.

samples from drill hole #11 are shown in Figure Hg-21b. The samples fall into two groups: Those within the oxidized zone were collected between 560° and 795°C, and those outside of the zone of maximum oxidation were sampled between 795° and 890°C. Values of f_{O_2} for the high temperature and low oxidation suite are between 10^{-11} and 10^{-13} atms whereas those samples within the oxidized zone fall between $10^{-5.2}$ and 10^{-6} atms. Mean oxidation numbers correlate with these distinctions, are related to $FeO:Fe_2O_3$, and demonstrate that the oxides are sufficiently responsive to record an estimate of the quenched environmental conditions.

The distribution of oxidation stages for the joint selvage examples (Fig. Hg-22) are illustrative of the effects of escaping comagmatic volatiles during deuteric cooling, and although the sample array is on a very much smaller scale than either the Iceland or the Hawaii examples the oxide assemblages are similar and the sequential stages of oxidation are identical to those present in thick cooling units.

Mechanism of oxidation

The preferred location of high temperature oxidation maxima towards the central portions of lava flows demonstrates that volatiles were trapped or tended to accumulate in zones of prolonged cooling. For the joint selvages the volatiles were vented and oxidation was hence restricted to parallel walls lining these joints. Lindsley (1960) suggested that the high oxidation zones adjacent to joints was the result of hydrogen loss and similar mechanisms have been proposed by Sato and Wright (1966) and by Watkins and Haggerty (1967). The Sato and Wright model is elegantly expressed as the semipermeable membrane mechanism in which oxygen and water molecules can no longer diffuse through the basalt freely, while hydrogen continues to escape because of higher diffusivity constants. This situation occurs increasingly with falling temperature, and for the Makaopuhi lava lake the temperature interval most conducive to oxidation is between 550° and 750°C. There is uniform agreement that high temperature oxidation is a consequence of the dissociation of water with the retention of oxygen and an ensuing loss of hydrogen. The compositions of volcanic gasses (see Anderson, 1976 for an extensive review) support the contention for the dominantly oxidizing agent, insofar as water vapor is the major component of gasses liberated from active volcanic vents. The relatively unoxidized portions of the Makaopuhi lava lake, between 500° and 1100°C, fall close to the FMQ buffer curve (Fig. Hg-21a) whereas the oxidized zone is well above the MH buffer curve. The preferred hydrogen loss then does not reduce the oxide assemblage, in the sense that metallic Fe should form, but increases the H_2:H_2O ratio by a small percentage in the basalt layers which overlie the oxidized zone. Since H_2O is many orders of magnitude larger than H_2 the environment is still highly oxidized, and although there is a small decrease in the relative state of oxidation it is insufficient to affect the oxides.

In summary, application of the oxidation classification to studies of single cooling units show that maximum high temperature oxidation develops towards the interior of lava flows rather than at the upper and lower cooling flanks. The mechanism of oxidation is related to volatile accumulation, H_2O dissociation and preferred hydrogen loss. Oxidation zones adjacent to joint surfaces result from volatile venting, and the mechanisms of oxidation as well as the oxide assemblages are similar to those of larger bodies.

IMPLICATIONS FOR ROCK MAGNETISM AND OXIDE PETROGENESIS

Introduction

The complexity of assemblages and the correspondingly wide variations in mineral chemistry which are exhibited by the opaque oxide mineral group for the spectrum of oxidation discussed in previous sections is at a maximum for basalts. This is due in part to the relatively high TiO_2 contents of basalts; Ti concentrations decrease proportionately with increasing acidity resulting in small modal percentages and in either an ilmenite-free assemblage or the assemblage Ti-poor magnetite $\pm$ ilmenite. If the maximum complexity exists for basalts, there is at least some consolation in the fact that other systems are simpler and also that the implications of oxidation to rock magnetism and to oxide petrogenesis is relatively straightforward. However, it is pertinent that the following points be borne in mind in the summary which follows: (1) The oxides are the major magnetic mineral carries, and modifications to these oxides have a direct influence on magnetic property measurements; (2) oxidation is widespread and the oxides are highly susceptible to changing environmental conditions; (3) oxidation is highly variable within single cooling units and differences in assemblages exist between cooling units; (4) oxidation is related to the mode of emplacement, it is at a maximum for subaerially extruded basalts, intermediate for hypabyssal and plutonic suites and approaches minimum variability in deep sea environments; (5) the modifying effects of mutual exsolution between $Usp\text{-}Mt_{ss}$ and between $Ilm\text{-}Hem_{ss}$ are restricted to plutons; exsolution in basalts is limited to the formation of pleonaste-magnesioferrite$_{ss}$ from Mt_{ss} which is induced by oxidation and results in Mg-Al-Fe^{3+} supersaturation; and (6) the oxides which do not undergo high temperature oxidation become prime candidates for decomposition during supergene weathering or cycles of burial metamorphism.

Oxidation

The influences of oxidation are twofold: (1) Primary crystallization oxidation controls the compositions of members of each of the solid solution series Usp-Mt_{ss} and Ilm-Hem_{ss}; and (2) deuteric oxidation controls the degree to which the primary assemblages are modified beyond the limits of the compositions initially defined by crystallization oxidation. Magmas with low values of f_{O_2} yield Usp-rich and Ilm-rich solid solution members upon primary crystallization whereas the extremes of high f_{O_2} deuteric oxidation may result in Pb_{ss} + Hem_{ss}, Pb_{ss} + R, or R + Hem_{ss}.

Primary crystallization oxidation in basalts produces initial compositions which are typically between Usp_{50} and Usp_{80} coexisting with ilmenite solid solutions which are typically between Ilm_{90} and Ilm_{95}. The compositions of coexisting assemblages are dependent on T°C and f_{O_2} and equilibrium crystallization closely parallels the FMQ buffer curve. From the magnetic standpoint the end member Fe_2TiO_4 (Usp_{100}) is antiferromagnetic and has a Curie temperature of -150°C; the Curie temperature of Usp_{70} is 100°C and there is an approximately linear increase to Mt_{100} which has a Curie temperature of 585° and is ferrimagnetic. Members of the Ilm-Hem_{ss} series are antiferromagnetic and the Néel temperature of the end members and Curie temperatures for intermediate members defined approximately by the limbs of the solvi are as follows: Ilm_{100} = -190°C; Ilm_{60} = 230°C; Ilm_{14} = 570°C; Hem_{100} = 675°C (Hargraves and Banerjee, 1973). The compositions of the primary components of oxides in basalts suggest therefore that the major magnetic carrier is titanomagnetite for compositions $\leq Usp_{70}$.

The effects of deuteric oxidation may be divided into three classes of oxidation related to oxidation intensity: (1) cationic oxidation; (2) exsolution oxidation; and (3) pseudomorphic oxidation.

In cationic oxidation a small percentage of Fe^{2+} is transformed to Fe^{3+} creating either a slight departure from ideal solid solution stoichiometry, or minor enrichments of Fe_3O_4 in Usp_{ss} and Fe_2O_3 in Ilm_{ss}. These increases in Fe^{3+} contents are magnetically sensitive and although Ilm_{ss} will not contribute to the overall magnetism of the rock any change of an initially rich Usp_{ss} to higher values of Mt_{ss} will increase both Curie temperature and all related magnetic property measurements. It is important to note that no new phases develop in cationic oxidation although both distortions and defects to the cell may be incurred.

Exsolution oxidation best describes the development of crystallographically oriented ilmenite lamellae along {111} planes in Usp-Mt_{ss}. The segregation of a Ti-rich component depletes the spinel in Fe_2TiO_4 and drives the composition of the host towards higher values of Fe_3O_4 and towards a state of ferrimagnetism. Exsolution oxidation therefore has a profound magnetic response to susceptibility, coercivity and intensity of magnetization. A related property of considerable interest is magnetic stability which in this context is a function of the decrease in "effective magnetic grain size" (Strangway *et al.*, 1968); noting that the finer grain size and the more nearly the sizes of particles approach single domain dimensions (Fe_3O_4 = .03 μm diam.) the more strongly magnetic is the particle and the more stable is the magnetism of the rock (see review by Hargraves and Banerjee, 1973). Fine-grained particles are normally associated with quench products but particle size reduction can also be accomplished by the multiple-division of a large crystal into discrete magnetic volumes. The forms of these volumes in titanomagnetite with {111} ilmenite trellis lamellae are octahedrons, or more accurately spherical octahedrons if the curvilinear shape of lamellae are considered; depending on the section through the crystal these appear as spherical triangles or as concave spherical rectangles when viewed in two dimension. The critical factor is that large unoxidized crystals have a very large number of interacting domains, but on exsolution oxidation there is a redistribution of domains in the original crystal to a multitude of new domain sectors with each volume behaving as a magnetic entity separated from adjacent volumes by ilmenite lamellae barriers. Thus a single large crystal with {111} ilmenite would have the magnetic appearance of a large number of non-interacting discrete crystals and this would assume the property of increased stability.

The final class of oxidation referred to previously as pseudomorphic oxidation is the process in which the primary mineralogy is pseudomorphed by an assemblage of higher oxides which are largely antiferromagnetic or paramagnetic. Closely coupled with the destruction of the primary ferrimagnetic character is the allied chemical event of olivine oxidation and the secondary formation of magnetite and of hematite. The major significance of this latter process is that the newly formed oxides are extremely fine grained, are present in great abundance, develop at high temperatures and may impart to the rock an intensity of magnetization and a level of stability which supercedes that of the primary unoxidized assemblage.

Single cooling units

Studies of the distribution of oxide assemblages between the upper and lower cooling faces of basalt lava flows show that the degree of oxidation is variable, that the magnetic mineralogy is accordingly variable, and that magnetic property measurements closely correlate with variations expressed by the Fe-Ti oxides and the bulk chemical environment of oxidation. Therefore, the position of sampling relative to these cooling faces is critical to rock magnetic studies because of assemblage variations; to paleomagnetic studies because of the uncertainties associated with post-cooling events; and to paleointensity measurements of the geomagnetic field which depend on the basic assumptions of a primary TRM (thermal remanent magnetization) acquisition, and on non-interacting magnetic subunits (*e.g.*, between discrete crystals, or within intergrowths). Unless multiple samples from the same cooling unit are measured the credibility of the data base may be assumed to be in question and the conclusions as a consequence open-ended.

A second significant factor related to cooling is grain-size distribution and the magnetic stability of fine particles relative to coarse particles. Primary distributions are controlled by (1) quenching at the margins of lava flows; (2) rapid cooling in zones adjacent to quenched layers; and (3) relatively slower cooling rates towards the center. In the absence of extreme deuteric oxidation (*i.e.*, in the absence of olivine decomposition) the most magnetically stable regions of a flow are therefore at the edges of the lava flow. However, if extrusion is rapid and a thick succession of thin flow units evolve, this sequence will cool in a manner which is closely similar to that of an originally thick flow and the grain size of oxides at contacts between units may be no finer than those towards the center of a single thick flow. It is important to note finally that the upper parts of lava flows are more highly susceptible to reheating by later extrusives than either the interior or the base of the flow so that the choice for sample collection requires careful field observation.

SUMMARY AND CONCLUSIONS

This review on the oxidation of opaque mineral oxides in basalts has attempted to characterize the sequences of oxidation assemblages which develop during high-temperature oxidation and deuteric cooling. Classes of assemblages for ilmenite and for titanomagnetite have been established in terms of oxidation stages defined as C1 to C7 for the cubic Fe-Ti spinel series ($Usp\text{-}Mt_{ss}$) and R1-R7 for the rhombohedral ilmenite-hematite solid solution series ($Ilm\text{-}Hem_{ss}$). The phase chemistry of examples typical of each of the oxidation stages when integrated into ideal phase compatibility relationships demonstrates that deuteric oxidation is consistent with an estimated temperature of approximately 750°C and f_{O_2} values which range between 10^{-5} and 10^{-6} atms. The evidence for deuteric oxidation is based primarily on the distribution of oxidation zones in profiles across single lava flows, but closely related to this factor are the experimentally-determined lower thermal stability limits for pseudobrookite solid solution members in the series $Fe_2TiO_5\text{-}FeTi_2O_5$ which show a maximum development towards the central interiors of lava flows. Correlations with basic magnetic property data are ample demonstration of the close interdependencies which exist between magnetism and mineralogy. The preferred accumulation of volatiles in the more slowly cooled portions of lava flows and the sensitivity of the oxide mineral group to oxidation places severe constraints on the minimum number of samples required for a reliable data base and unambiguous magnetic characterization.

The foregoing should be regarded as an initial framework for future studies and clearly many new experimental data points are required. These include specifically: (1) The statistical correlations which have been demonstrated between reversed directions of magnetization and high oxidation states should be re-examined; (2) extensions of single lava studies to lavas of other compositions; (3) experimental confirmation of phase compatibility relationships in the temperature range 600-900°C as a function of oxygen fugacity; (4) detailed mineral chemistry and magnetic property correlative studies; (5) refinement of oxidation classes based on experimental data and on compositions of other lavas; (6) refined quantitative mineral parameters that result in absolute numbers comparable to those of magnetic property measurements; (7) mineralogical discrimination techniques should be developed for identifying magnetic source materials to determine the relative contributions that complex assemblages impart to the overall magnetism of a rock; (8) more extensive sampling and mineral-magnetic studies of modern extrusives undergoing active cooling for which T°C and f_{O_2} are known; (9) more detailed studies of the mineral and magnetic effects associated with volatile venting along fractures and joints; and (10) studies to determine the lower limits of deuteric cooling, the effects of high-temperature reheating, and a determination of the kinetic stability of the oxide mineral group in post-deuteric hydrothermal and burial metamorphic cycles.

REFERENCES

References for Chapter 4 are included at the end of Chapter 8.

Ahmed El Goresy

Chapter 5

INTRODUCTION

The recovery of lunar rocks from various landing sites on the moon brought into the hands of geologists rocks which have not been subjected to terrestrial-type weathering. Lunar minerals and rocks have not suffered the chemical and textural changes experienced by their terrestrial counterparts. The importance of opaque oxides became evident after the recovery of the Apollo 11 basalts, which were found to be exceptionally enriched in opaque minerals. The opaque oxide content of lunar rocks from the various landing sites may vary from a few percent up to 20 percent by volume. Another important phenomenon of lunar rocks in comparison with terrestrial equivalents is the ubiquitous occurrence of metallic iron thus indicating formation under extremely low oxygen fugacities. At such low oxygen fugacities several opaque oxide assemblages may help in approximately reconstructing the conditions that prevailed during the formation of these rocks. The content of ferric iron is extremely low in lunar opaque oxides, thus eliminating many complications in recalculating concentrations of end members from electron microprobe analyses. Opaque oxide assemblages co-existing with metallic iron in lunar rocks mainly belong to the following solid solution series: (a) chromite-ulvöspinel series; (b) ilmenite-geikielite series; (c) armalcolite-anosovite series; and (d) rutile.

MINERALOGY

Chromite-ulvöspinel series

Due to the reducing conditions that prevailed during formation of lunar rocks the amount of trivalent iron in various oxide minerals is negligible and hence magnetite is nonexistent as an end member in the spinel series. Chromite, $FeCr_2O_4$, and ulvöspinel, Fe_2TiO_4, are the most abundant end members in spinel solid solutions encountered in lunar rocks. Other end members reported are hercynite, $FeAl_2O_4$, spinel, $MgAl_2O_4$, and magnesiochromite, $MgCr_2O_4$ (Haggerty, 1972; El Goresy *et al.*, 1972). In highland breccias and anorthosites members of the hercynite-spinel solid solution series ($FeAl_2O_4$-$MgAl_2O_4$) dominate, reflecting the high Al_2O_3 content of these rocks (Haggerty, 1972).

In mare-type basalts recovered from Apollo 12, Apollo 15, and Luna 16 landing sites, the solid solution and textural relationships are much more complex; however, the main end members are magnesiochromite, chromite, and ulvöspinel. The paragenetic sequence during crystallization of many mare basalts indicates precipitation of spinels enriched in $MgCr_2O_4$ followed by members enriched in $FeCr_2O_4$ and at last members enriched in Fe_2TiO_4. In many mare basalts, numerous spinel grains were found to display this sequence in the same grain with the $MgCr_2O_4$-rich spinel in the core and the Fe_2TiO_4-rich spinel comprising the outermost boundary with $FeCr_2O_4$-rich spinel in between. Reflectivity and color of $MgCr_2O_4$ and $FeCr_2O_4$ end members are so similar that such zoning from magnesiochromite to chromite can only be traced with the electron microprobe. The chemical zoning between these two spinels is always gradational. Spinels enriched in Fe_2TiO_4 are brownish or tan in color and hence can be readily distinguished in reflected light from chromite spinels. The boundary between the ulvöspinel mantle and the chromite core can be gradational or abrupt. Gradational zoning from chromite-rich to ulvöspinel-rich zones indicates continuous precipitation at temperatures at which solid solution between chromite and ulvöspinel was still complete (Nehru *et al.*, 1974, El Goresy *et al.*, 1976). On the other hand, abrupt zoning between chromite and ulvöspinel is suggestive of crystallization at lower temperature at which the solid solution between the two end members was interrupted by a solvus (Nehru *et al.*, 1974, El Goresy *et al.*, 1976). In the majority of the vitrophyres from Apollo 12 and 15 landing sites this sharp chromite-ulvöspinel zoning reflects the original relationship indicating crystallization of ulvöspinel-rich members late in the sequence long after chromite crystallization terminated. Relationships in coarse-grained basalts with both sharp and gradational zoning indicate multiple precipitation and resorption reactions before the late stage ulvöspinel crystallized (El Goresy *et al.*, 1976). Due to the complicated zoning relationships and to the complex solid solutions lunar spinels could be used as indicators for crystallization sequences of mare basalts (El Goresy *et al.*, 1976). Figure EG-1 displays the chemical variations of lunar spinels reported from the Apollo 15 lunar rocks plotted in the spinel prism with the six end members $MgCr_2O_4$-$MgAl_2O_4$-$FeCr_2O_4$-$FeAl_2O_4$-Mg_2TiO_4-Fe_2TiO_4. Lamellar intergrowth of spinels along (100), a well-known phenomenon in terrestrial rocks, has never been encountered in lunar spinels. Several coarse-grained basalts from Apollo 12, Apollo 14, and Apollo 15 sites contain only ulvöspinel-rich members as late stage, but idiomorphic grains.

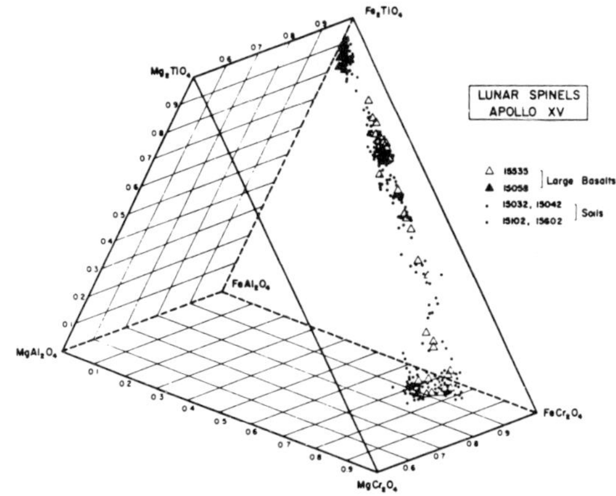

Figure EG-1. Lunar spinel compositions in 50 basalt samples collected from the Apollo 15 site (from Haggerty, 1972).

Basalts rich in TiO_2 collected from Apollo 11 and 17 sites contain ulvöspinel-rich members also, however, as an early quench phase, usually as idiomorphic clusters included in olivine. Lunar chromites contain appreciable amounts of ulvöspinel in solid solution and hence are designated Ti-chromites. Ulvöspinels on the other hand may contain appreciable amounts of $FeCr_2O_4$ and are designated Cr-ulvöspinel.

Ilmenite-geikielite series

Ilmenite is present in almost every lunar sample recovered from the moon. In many mare basalts it is the dominant opaque oxide. In TiO_2-rich basalts ilmenite content may reach up to 18 percent by volume of the total rock. Due to the extremely low ferric iron content of lunar rocks hematite is non-existent as an end member in lunar ilmenite. Manganese content is also low (El Goresy *et al.*, 1971,1972,1974,1975) so that pyrophanite, $MnTiO_3$, is negligible in lunar ilmenites. The lunar mineral is primarily a member of the ilmenite, $FeTiO_3$, -geikielite, $MgTiO_3$, solid solution series. Members of the ilmenite-geikielite series in TiO_2-poor basalts occur either as primary minerals precipitated from the cooling liquid or as exsolution lamellae in ulvöspinel-rich spinels oriented parallel to (111) directions of the host spinels. The content of $MgTiO_3$ is by no means a function of temperature of crystallization since it was found that ilmenite continually equilibrates with the coexisting silicates and thus changes its composition continuously upon cooling (Usselman, 1975). The MgO content of lunar ilmenites may vary from a fraction of 1 wt. percent to 9 wt. percent. In TiO_2-rich basalts ilmenite relationships are much more complex than in other lunar rocks. A major part of the ilmenite in TiO_2-rich basalts is formed due to reactions involving armalcolite, ulvöspinel, metallic iron, or the cooling basaltic liquid. Many Apollo 14 breccias and some highland rocks from the Apollo 16 landing site are characterized by the assemblage ilmenite-baddeleyite, ZrO_2 (El Goresy *et al.*, 1972).

Armalcolite-anosovite series

Armalcolite was first discovered as a new mineral in many TiO_2-rich basalts from the Apollo 11 landing site (Anderson *et al.*, 1970). Armalcolite is a member of the pseudobrookite series. However, due to the reducing conditions under which the mineral was formed, it does not contain trivalent iron. The majority of terrestrial pseudobrookites always contain trivalent iron and usually lie along the join ferripseudobrookite, Fe_2TiO_5, -ferropseudobrookite, $FeTi_2O_5$, or the join ferripseudobrookite-karrooite, $MgTi_2O_5$. Ternary members are called kennedyite (von Knorring and Cox, 1961). Lunar armalcolites in comparison were believed to lie along the join ferropseudobrookite-karrooite (Anderson *et al.*, 1970). Recent detailed studies, however, indicated that the majority of lunar armalcolites are members of the system ferropseudobrookite-karrooite-anosovite, Ti_3O_5 (El Goresy *et al.*, 1974, Wechsler *et al.*, 1975). Lunar armalcolites are hence characterized by the presence of trivalent titanium. The anosovite content may be as high as 10 mole percent (Wechsler *et al.*, 1975). Armalcolite is one of the major opaque oxide phases in TiO_2-rich basalts recovered from Apollo 11 and Apollo 17 landing sites. Zirconium-rich analogues were also reported from some highland rocks (Haggerty, 1973).

Rutile

Rutile is the only TiO_2 polymorph reported from the lunar samples (Ramdohr and El Goresy, 1970; Haggerty *et al.*, 1970; El Goresy *et al.*, 1975). It is usually present as fine lamellae intergrown with ilmenite either formed by exsolution or by complex subsolidus reactions. Lunar rutiles are characterized by their high Nb and Zr contents (Marvin, 1975; El Goresy *et al.*, 1975). In TiO_2-rich basalts rutile occurs frequently in the assemblage armalcolite-ilmenite-rutile or ilmenite-spinel-rutile-metallic iron.

OXIDE RELATIONS IN DIFFERENT ROCK TYPES RECOVERED FROM THE MOON

Opaque oxides in TiO_2-poor basalts

Basalts poor in TiO_2 were recovered from Apollo 12 and Apollo 15 landing sites. Low TiO_2 basalts were also collected from the Luna 16 site; however, compared to Apollo 12 and Apollo 15 basalts their Al_2O_3 content is much higher (16 wt% versus 8 wt%). The opaque oxide assemblage of these mare basalts is quite similar. Ilmenite, members of the normal and inverse spinel series (magnesiochromite, chromite, hercynite, ulvöspinel), and rutile constitute the dominant assemblage in all basalts of the three sites.

Spinels

Of the oxide phases reported from the low-TiO_2 basalts, members of the spinel series exhibit various zoning relationships reflecting the chemical changes which took place in the cooling liquid during growth of the spinels. Members of the spinel group are sensitive indicators to compositional changes in the basaltic liquids since: (1) Spinels are able to incorporate appreciable amounts of numerous major elements (except Si and Ca) due to the complex solid solution series; and (2) textural relationships indicate that spinel crystallization may span a major part of the cooling period of a basalt. Changes in abundances of major and some minor elements, reflecting the precipitation of a silicate phase or the build-up of some oxides during cooling of the basaltic liquid, may be stored in zoned individual spinel grains whose growth spanned a great deal of the cooling period of a basalt. Two principal spinel groups occur in basalts of the three landing sites: (a) blue-to-bluish grey Ti-chromites; and (b) pink or tan to brownish Cr-ulvöspinel. In the majority of the basalts the two spinel types most commonly occur together. The harder chromite is always rimmed by the ulvöspinel. Euhedral crystals of Ti-chromite commonly occur alone as inclusions in olivine and in pyroxene. In a few basalts Cr-ulvöspinel without chromite cores was encountered in the groundmass. The contacts between Ti-chromite cores and Cr-ulvöspinel rims are sharp or gradational. Sharp contacts between Ti-chromite cores and Cr-ulvöspinel rims were formerly explained as due to a compositional gap between the normal and inverse spinel series (Haggerty and Meyer, 1971; Haggerty, 1972a) or due to peritectic reaction of early chromite with the cooling liquid (Kushiro *et al.*, 1970). On the other hand, relationships among spinels with diffuse or gradational contacts between core and rim were not fully understood partially due to assumptions that chromite cores should be enriched in Cr, Al, V and Mg, whereas the ulvöspinel rims should contain high concentrations of Ti and Fe but be depleted in Mg and trivalent cations. Any chemical zoning deviating from these "rules" was classified as non-systematic and atypical. The Cr-ulvöspinel rims with progressively zoned contacts were explained as having been formed by continuous growth of spinels during crystallization with compositions changing gradually from chromite to ulvöspinel.

Of the several hypotheses proposed to explain the compositional gap between chromite cores and ulvöspinel rims with sharp contacts, two seem to agree in part with the textural and chemical evidence (Nehru *et al.*, 1974). The first involves a solvus (Muan *et al.*, 1972) in the spinel series, where a continuous solid-liquid loop in the chromite-ulvöspinel system is required, descending from chromite to ulvöspinel, with a solvus whose crest is below the solidus. If continuous spinel crystallization commenced at sufficiently high temperature the solvus would not be intersected and spinel compositions could vary almost continuously from chromite to ulvöspinel. In other cases, however, crystallization temperature may be reduced so that the solvus may be intersected thus producing a compositional gap between chromite and ulvöspinel. The relationship in this hypothesis involves, at equilibrium, a peritectic reaction of chromite with liquid to form ulvöspinel. If crystallization was sufficiently rapid that there was no actual reaction, ulvöspinel could begin to form rims around chromite crystals. The second hypothesis involves a reaction (under equilibrium conditions) of chromite, and probably olivine, to form a chromian pyroxene (Nehru *et al.*, 1974). After this reaction, the liquid would crystallize chromian pyroxene without spinel for a time, and would then recommence spinel-group mineral crystallization, this time ulvöspinel, because Cr would have been depleted in the melt by pyroxene crystallization, and Ti in the melt would have continuously increased. The proposed chromite-pyroxene reactions would probably occur simultaneously with the olivine-pyroxene reaction (the chromian pyroxene end member could not exist by itself) and the later ulvöspinel would also appear simultaneously with the later fayalite (Nehru *et al.*, 1974).

Neither of the two hypotheses contradict the textural and compositional evidence of sharp chromite-ulvöspinel zoning in Apollo 12 and Apollo 15 fine-grained pigeonite (quartz normative) basalts. A distinct textural feature characterizes the sharp chromite-ulvöspinel zoning in the fine-grained pigeonite basalts in comparison with sharp zoning in coarse-grained pigeonite and olivine basalts. In the former basalt types the chromite cores display throughout *idiomorphic* sharp boundaries to the ulvöspinel rim without any sign of reaction prior to ulvöspinel precipitation (Fig. EG-2). In contrast, coarse-grained olivine basalts (*e.g.*, 15555), pigeonite basalts (*e.g.*, 15065), or ilmenite basalts (12051) contain corroded and rounded chromite cores as inclusions in late chromian ulvöspinel (Fig. EG-3). It appears that considerable reaction between basaltic liquid and Ti-chromite took place thus removing the outer rims of chromite prior to precipitation of late chromian ulvöspinel. In the same basalts Ti-chromites with gradational zoning to tan chromian ulvöspinel also occur. Several grains were encountered with

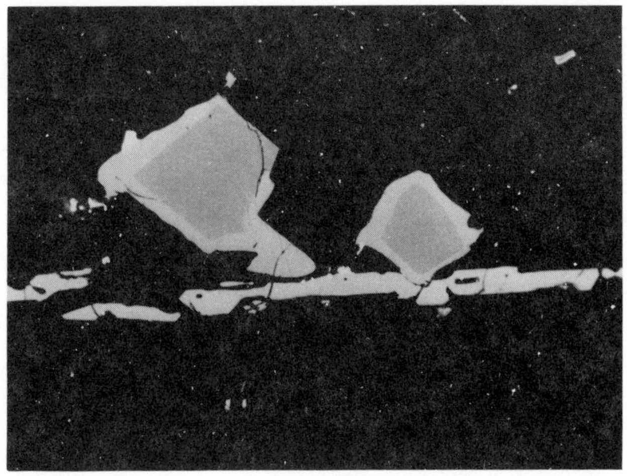

Figure EG-2. Idiomorphic Ti-chromite (gray) with sharp zoning to chromian ulvöspinel (white). Long grain below ulvöspinel is ilmenite, length of photograph 400 microns.

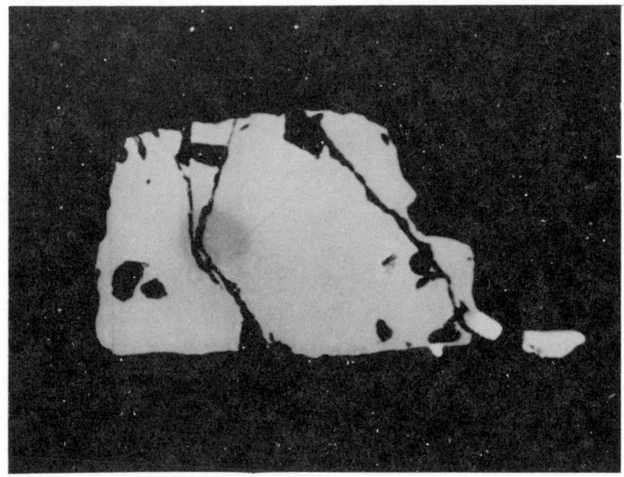

Figure EG-3. Rounded and corroded Ti-chromite core (gray) in sharp zoning relationship with chromian ulvöspinel (light gray); white grain in lower right is FeNi. Length of photograph 300 microns.

gradational zoning and idiomorphic boundaries on one side and corroded chromite cores with sharp boundaries to chromian ulvöspinel on the other side (Fig. EG-4). Presence of gradational and sharp chromite-ulvöspinel zoning in the same basalt and even in the same spinel grain indicate that a different mechanism other than fast cooling is responsible for these features. Textural evidence indicates that zoned spinels showing both gradational and sharp contacts to chromian ulvöspinel are shielded with pyroxene on the side with gradational zoning (Fig. EG-5). Neither of the two hypotheses mentioned above can explain these textural features.

A direct way of understanding the coexistence of both abrupt and gradational zoning in the same basalt and in the same spinel grain is the detailed study of the zoning trends in individual grains with the electron microprobe. Detailed study of zoning trends in numerous individual abrupt and gradational zoned spinels in several pigeonite basalts, olivine basalts, and one ilmenite basalt (El Goresy *et al.*, 1976) indicate the presence of three distinct coupled chemical trends:

1. Slight increase in $TiO_2/(TiO_2+Cr_2O_3+Al_2O_3)$ ratio, decrease in $Cr_2O_3/(Cr_2O_3+Al_2O_3)$ ratio, increase in V_2O_3, decrease in FeO/(FeO+MgO) ratio; FeO/(FeO+MgO) (FFM) initial ratio of the spinel core ≤ 0.9.
2. Sharp increase in $TiO_2/(TiO_2+Cr_2O_3+Al_2O_3)$ ratio, increase in $Cr_2O_3/(Cr_2O_3+Al_2O_3)$ ratio, decrease in V_2O_3, increase in FeO/(FeO+MgO) ratio; FFM initial ratio of the spinel core ≥ 0.9.
3. Constant or slightly increasing $TiO_2/(TiO_2+Cr_2O_3+Al_2O_3)$ ratio, increase in $Cr_2O_3/(Cr_2O_3+Al_2O_3)$ ratio, decrease in V_2O_3, sharp increase in FeO/(FeO+MgO) ratio; FFM initial ratio of the spinel core ≤ 0.85.

Thus the zoning relationships are much more complex than proposed by the simple scheme of the sequence magnesiochromite-chromite-ulvöspinel. The first two chemical trends were encountered in many coarse-grained basalts. The third trend, however, occurs both in coarse-grained and fine-grained rocks. The first zoning trend is restricted to early spinels enclosed in olivine in coarse pigeonite basalts and olivine basalts. This trend is unique as it indicates an increase in the concentrations of Al_2O_3, V_2O_3, and MgO during the growth of the chromite from the basaltic liquid (Figs. EG-6a, EG-7a). These chemical zonings may indicate precipitation of chromite from a silicate liquid with continuous build-up in Al_2O_3, V_2O_3, and MgO contents. Crystallization of this Ti-chromite evidently took place before crystallization of

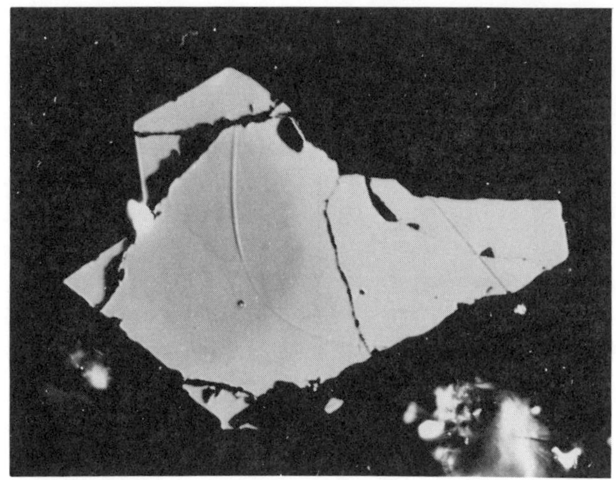

Figure EG-4. A spinel with gradational zoning on the left-hand side and abrupt and corroded boundary to late Cr-ulvospinel on the right-hand side. Length of photograph 300 microns.

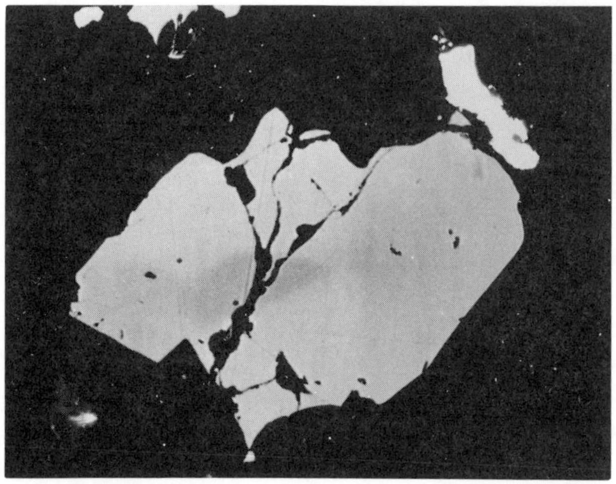

Figure EG-5. Spinel shielded by pyroxene on the left-hand side with gradational zoning. Corroded core with sharp contacts to ulvöspinel on the right-hand side. Length of photograph 300 microns.

plagioclase and pyroxene and probably either before or during olivine precipitation. Decrease in the FeO/(FeO+MgO) ratios of those spinels may, however, be also due in part to subsolidus equilibration between Ti-chromite and host olivine, thus allowing considerable Mg diffusion from the olivine to the chromite. Spinels displaying both gradational and sharp zoning and occurring in the same basalts with early chromites are characterized by the second chemical trend (Figs. EG-6a, EG-7a). Comparison between early spinels and this spinel type indicates a distinct difference in textural features and chemical trends and thus puts important constraints on the behavior of basaltic liquids during cooling and crystallization. The FFM initial ratio is usually higher than that of the early chromites, indicating: (a) Precipitation from a liquid with higher FFM ratio and thus later in the crystallization sequence; and (b) the cooling basaltic liquid did indeed subsequently precipitate various chromite generations with various FFM ratios due to changes in its composition and at a certain cooling rate (El Goresy *et al.*, 1976). On the sides of the grains with gradational zoning there is an increase in the $TiO_2/(TiO_2+Cr_2O_3+Al_2O_3)$ ratio from core to rim from 0.02 up to 0.48 (Figs. EG-6a, EG-7a, EG-8a). On the other side of the grains with abrupt zoning, evidently, reaction took place between basaltic liquid and the Ti-rich outer rims thus corroding the spinels on this side down to the chromite core prior to new precipitation of the chromian ulvöspinel. Compositional variation of chromian ulvöspinel rim on the other side of the chromite with sharp contacts is usually chemically continuous with the outermost rims of the uncorroded side (Figs. EG-6b, EG-7b, EG-8b). The process of precipitation of Ti-chromite with gradational zoning followed by resorption of the Ti-rich layers could have taken place several times in the course of crystallization (*e.g.* Fig. EG-7b) and as much as three generations of Ti-chromites with various initial FFM ratio could be encountered in the same basalt. Increase in the $TiO_2/(TiO_2+Cr_2O_3+Al_2O_3)$ ratio, decrease in V_2O_3 content, and increase in FFM ratio of these spinels may be indicative of co-precipitation with pyroxene but before massive crystallization of plagioclase took place. Resorption of the Ti-rich zones of the same spinel should result in an increase in the Cr concentration and to a lesser extent Ti-concentration of the basaltic liquid. This slight but sudden increase in both Cr and Ti concentrations may be reflected in the change of chemistry of the co-precipitating zoned pyroxene.

The third chemical trend was encountered in olivine basalts and coarse- and fine-grained pigeonite basalts. The Ti-chromite cores usually have a low initial FFM ratio (0.80-0.85). This ratio increases sharply from core to rim

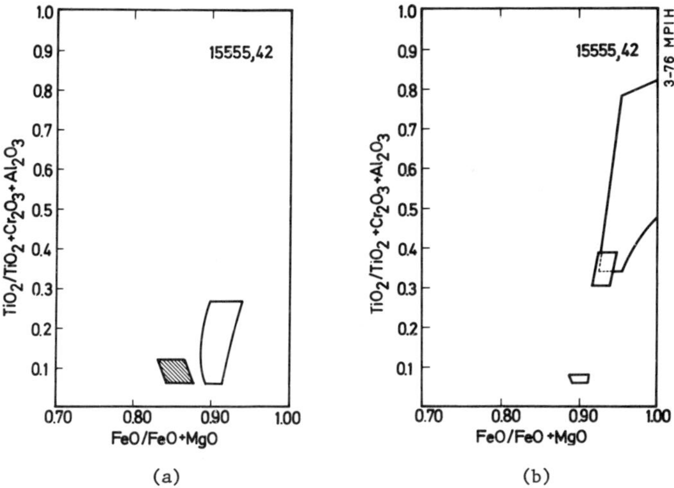

Figure EG-6. Projections of the rectangular face of the spinel prism displaying the first and second zoning trends in an olivine basalt. (a) (left) Compositional variation of early spinel (hachured area) and gradationally zoned spinels on their shielded side with the zoning filling the Apollo 12 gap. (b) (right) Compositional variations of the chromite corroded side and the late ulvöspinel demonstrating the gap after removal of the gradationally zoned equivalents.

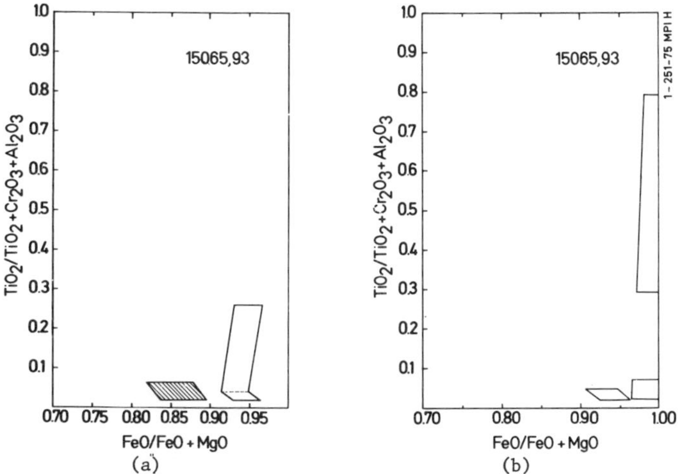

Figure EG-7. Projections of the rectangular face of the spinel prism with the first and second zoning trends in a pigeonite basalt. (a) (left) Compositional variation in early spinels (hachured area) and gradational zoning on the shielded sides of spinels of the second trend. (b) (right) Compositional variation on the corroded side and the late ulvöspinel demonstrating the gap after removal of the gradationally zoned layers.

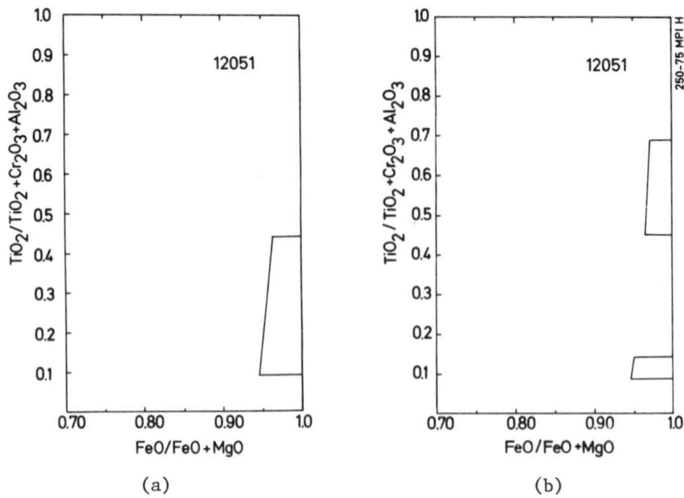

(a) (b)

Figure EG-8. Projections of the rectangular face of the spinel prism displaying (a) (left) the second zoning trend on the uncorroded and (b) (right) on the corroded side of spinels in an Apollo 12 ilmenite basalt.

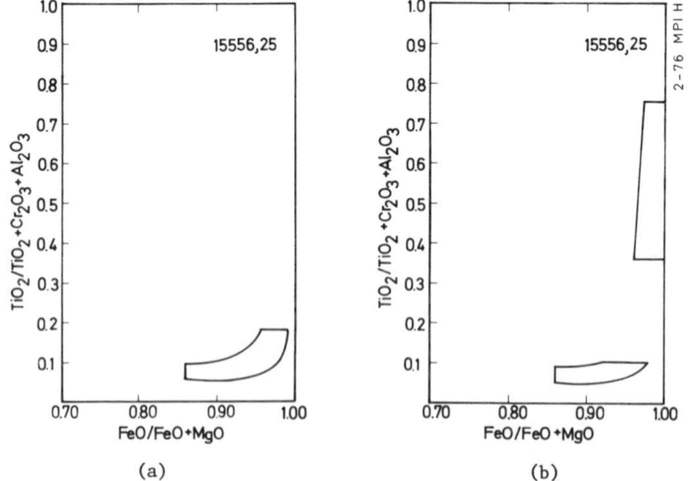

(a) (b)

Figure EG-9. Projections of the rectangular face of the spinel prism displaying the third zoning trend in an olivine basalt. (a) (left) Sharp increase in the FFM ratio and sharp rise in the $TiO_2/(TiO_2+Cr_2O_3+Al_2O_3)$ only in the terminating stage. (b) (right) Corrosion of the TiO_2-rich zones and compositional variation of the late stage ulvöspinel.

(Figs. EG-9a,b) with slight increase in the $TiO_2/(TiO_2+Cr_2O_3+Al_2O_3)$ ratio and decrease in V_2O_3 content thus demonstrating crystallization from a liquid continuously enriched in FeO content. In the coarse-grained olivine and pigeonite basalts the Ti-chromite cores are rounded and corroded indicating that likewise the second trend reaction between basaltic liquid and chromite took place. In olivine basalts with spinels indicative of this trend numerous spinels display both gradational and abrupt zoning. The chemical trend on the continuously zoned side is indicative of enrichment in the Fe_2TiO_4 end member just before termination of the overgrowth, yet the trend is distinct from the second type discussed above. The compositional variation of late stage ulvöspinel is almost identical in all basalts regardless of grain size or type: sharp increase in $TiO_2/(TiO_2+Cr_2O_3+Al_2O_3)$ ratio, decrease in $Cr_2O_3/(Cr_2O_3+Al_2O_3)$ ratio, decrease in V_2O_3, and increase in FeO/(FeO+MgO) ratio (Figs. EG-6,7,8,9,10).

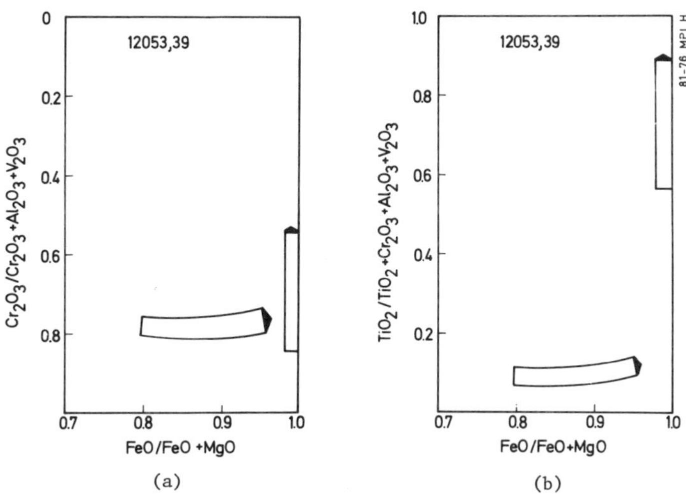

Figure EG-10. (a) (left) Base of the spinel prism showing the third compositional variation trend of idiomorphic chromite core and late ulvöspinel in an Apollo 12 pigeonite basalt. (b) (right) Rectangular face of the prism displaying the third zoning with the sharp increase in FFM ratio but without any significant increase in $TiO_2/(TiO_2+Cr_2O_3+Al_2O_3)$.

Cationic relationships and substitutions

The coupled oxide relationships of the three chemical trends discussed above substantiate the application of zoning trends in spinels for a better understanding of the crystallization histories of mare basalts. It is not our intention to present a detailed treatment of distribution of the various cations among the tetrahedral and octahedral sites of the spinel structure. Rather important is the substitutional relationship in zoned individual spinel grains along with the zoning as a direct function of change in chemistry of the cooling basaltic liquid due to crystallization of silicate phases.

Chromite, spinel and hercynite are structurally normal spinels with the trivalent cations (Cr,Al,V) octahedrally coordinated in the B site and the divalent cations (Fe,Mg) tetrahedrally coordinated in the A site. In contrast, ulvöspinel is an inverse spinel with half of the B sites occupied by tetravalent Ti and the divalent cations distributed among the A site and the other half of the B site.

Cationic substitutions in a zoned spinel in the B site involving the trivalent cations Cr, Al and V are extremely sensitive to changes in the concentration of these elements in the liquid. Changes in the substitutional trends may signal the crystallization of a major phase like pyroxene (Cr,V) or plagioclase (Al). Changes in the substitutional trends for the divalent cations Fe and Mg are also dependent on changes in the relative concentration of these elements in the cooling liquid due to crystallization of olivine and/or pyroxene. However, subsolidus equilibration between chromite inclusion and olivine host may have slightly modified or even obscured the initial trends and such an effect cannot be ruled out. Substitutional trends for the pairs Cr-Al, Cr-V, Al-V, and Fe-Mg offer an accurate way to trace the crystallization sequence (El Goresy *et al.*, 1976). However, the substitutional trends of these pairs cannot be treated independently since precipitation of a major phase should be documented in the majority of trends involving the different pairs. The Cr-Al substitutional ratio is a function of this ratio in the liquid whereby decrease in Al reflects decrease of the concentration of this element in the liquid due to crystallization of plagioclase. Decrease in the Cr and V ratios, on the other hand, indicates crystallization of pyroxene.

Cr-Al substitutional trends

In olivine and pigeonite basalts with spinel generations displaying the first and second coupled chemical trends, three distinct Cr-Al substitutional

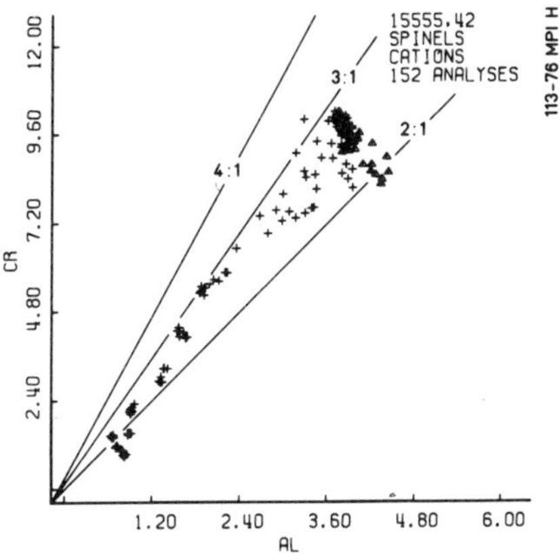

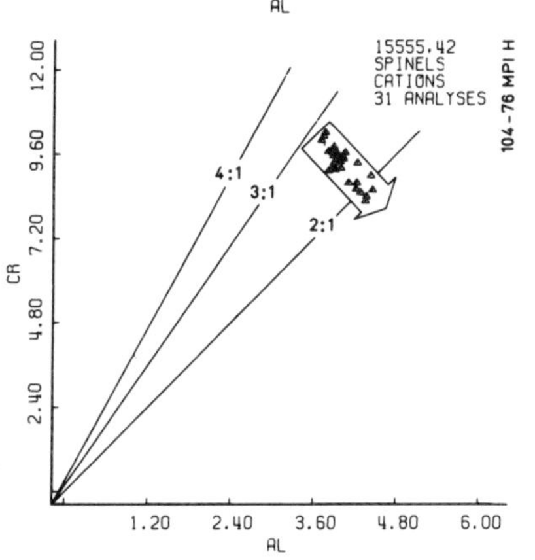

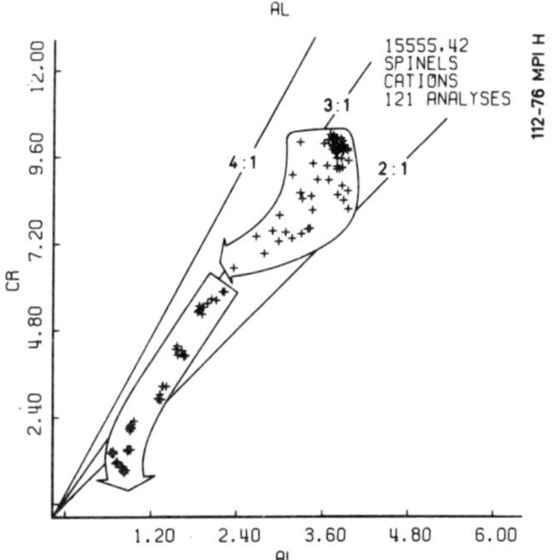

Figure EG-11. Cr-Al substitutional trends in various spinels in an Apollo 15 olivine basalt. (a) (top) Cr-Al substitutions for all spinels. Note that the relationships are obscured. (b) (middle) Cr-Al substitutional relationship for early spinels displaying the negative substitutional trend. (c) (bottom) Cr-Al substitutional trends of the second Ti-chromite type and late ulvöspinel (lower branch).

---

trends were encountered (Figs. EG-11,12). The substitutional trends become obscured if all the Cr and Al cations of both spinel generations, along with that of late ulvöspinel, are plotted altogether in the same diagram (Figs. EG-11a, EG-12a). Early chromites in a coarse pigeonite basalt (Fig. EG-11b) and an olivine basalt (Fig. EG-12b) show a unique trend that varies from core to rim from Cr/Al ratios of 4:1 to 2:1 and of 3:1 to 2:1, respectively (Figs. EG-11b, EG-12b). These trends demonstrate that early chromites in these two basalts grew from a liquid with continuous build-up of Al and thus before crystallization of plagioclase. The Cr/Al substitutional trend for early chromites in both rocks is entirely different from the trends of other chromite generations and ulvöspinel. However, it is quite similar to the substitutional trend reported for $MgAl_2O_4$-rich spinels in Apollo 14 samples (Haggerty, 1972a). Spinels with the second chemical trend display a curved substitutional trend starting for cores at a Cr/Al ratio of 4:1 with a steep positive slope to 2:1 and with a sharp turn back to ratios higher than 4:1. This curvature was interpreted as indicative of change in the activities of

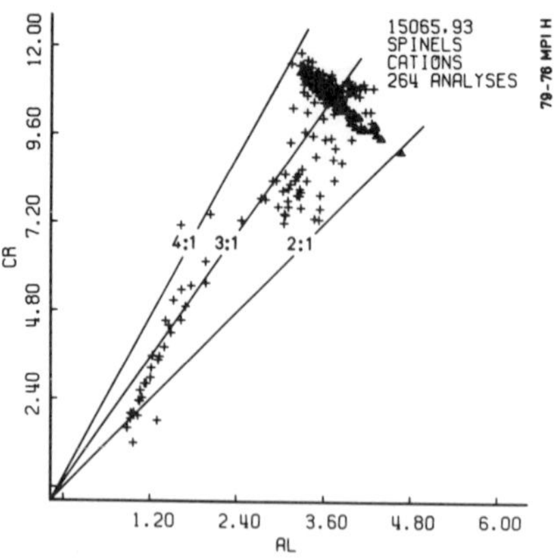

Figure EG-12. Cr-Al substitutional trends in spinels in an Apollo 15 pigeonite basalt. (a) (top) Cr-Al substitutions for all spinels. (b) (middle) Cr-Al substitutions for early spinels. Note the slight difference in slope from Fig. EG-11b. (c) (bottom) Cr-Al substitutional trends for the second chromite type and late ulvöspinel.

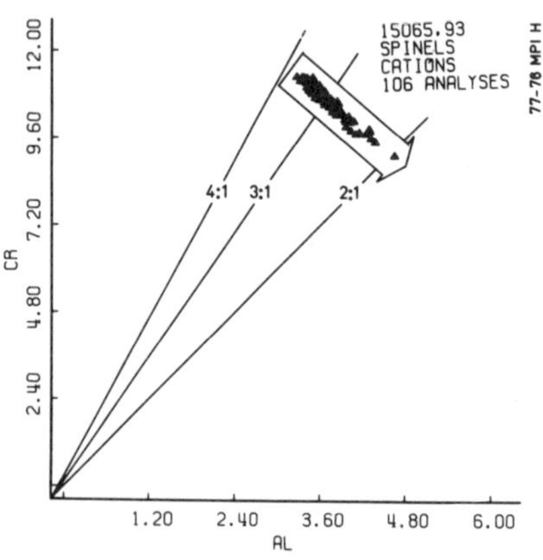

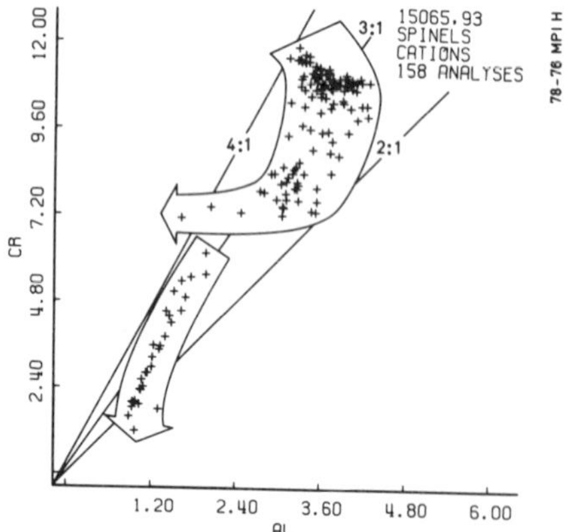

Cr_2O_3 and Al_2O_3 due to subsequent crystallization of pyroxene and plagioclase (El Goresy *et al.*, 1976). Along the first branch of this trend simultaneous crystallization of pyroxene and plagioclase is likely; however, the steep drop in Cr is indicative that at this stage pyroxene along with Ti-chromite was the main crystallizing phase. The sharp turn back to much lower Al concentration signals the entry of plagioclase as a major crystallizing phase (Figs. EG-11c,EG-12c). The Cr/Al substitutional trend for chromian ulvöspinel is indicative of crystallization from a liquid continuously depleted in both Cr and Al with the trend beginning at a Cr:Al ratio of 3:1, but changing continuously to a ratio lower than 2:1 for the outer rims of the ulvöspinels. Probably at that stage of crystallization both pyroxene and plagioclase coprecipitated. The crystallization sequence deduced from the Cr-Al substitutional trends is also consistent with V/Cr, V/Al, and Fe/Mg substitutional trends for both rocks thus substantiating the application of these trends in zoned spinels as an indicator of the crystallization histories. The conclusions drawn from the analyses of the substitutional trends are in excellent agreement with experimental results obtained by Kesson (1975).

V-Cr and V-Al substitutional trends

Though a trace element in spinels, vanadium is a valuable element because of its partitioning between chromite and ulvöspinel and its preference for

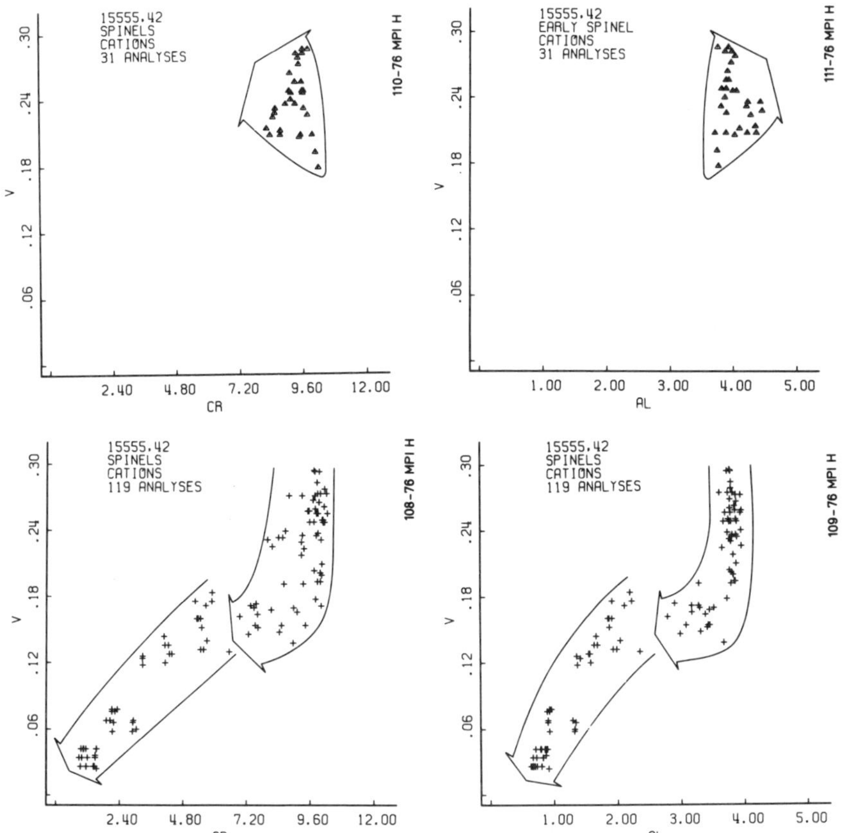

Figure EG-13. V-Al and V-Cr substitutional relationships in an olivine basalt. (a) (upper left) and (b) (upper right) substitutional trends of early spinels indicating increase in V and Al from core to rim. (c) (lower left) and (d) (lower right) substitutional trends for later chromite and ulvöspinel.

pyroxene (Laul and Schmitt, 1973) and hence it provides an additional check for pyroxene entry as a crystallizing phase. V-Cr and V-Al substitutional trends provide a good control for the position of both pyroxene and plagioclase in the crystallization sequence. Figures EG-13 and EG-14 document the substitutional trends in two different rocks. The unique substitutional trend of the early spinels (Figs. EG-13a,c, Fig. EG-14a,c) is also well developed thus indicating crystallization before pyroxene and plagioclase as evidenced by the positive V-Al sympathetic trend and negative antipathetic V-Cr relationship. The trends for chromites with the second chemical trend and the late ulvöspinels are antipathetic both for V-Cr and V-Al.

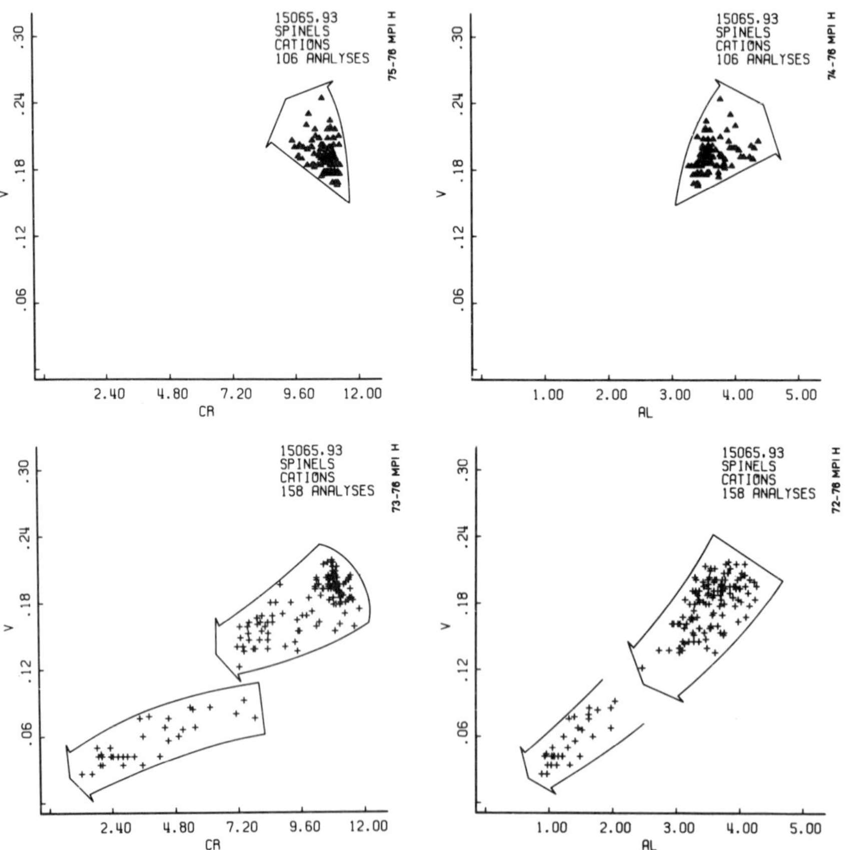

Figure EG-14. V-Al and V-Cr substitutional relationships in a pigeonite basalt. (a) (upper left) and (b) (upper right) Substitutional trends of early spinels indicating increase in V and Al from core to rim. (c) (lower left) and (d) (lower right) Substitutional trends for later chromite and ulvöspinel.

Fe-Mg substitutional trends

Several features can be recognized in the Fe-Mg substitutional trends shown in Figures EG-15 and EG-16. All the substitutional trends observed in the Cr-Al, V-Cr, V-Al diagrams are also encountered in the Fe-Mg relationships. All Ti-chromite and ulvöspinel generations with various initial FFM ratios emerge. Early spinels again display a unique antipathetic trend with negative slope with increasing Mg substitutions for Fe from core to rim. This trend very probably indicates growth from a liquid with decreasing FFM ratio. Haggerty argued (1972a,b,c) that the divalent Fe versus Mg relationship is poor to totally incoherent and that distinct linear slopes do emerge if the Ti content of the spinels is considered. Each slope proposed by Haggerty would correlate spinel compositions with similar $TiO_2/(TiO_2+Cr_2O_3+Al_2O_3)$ ratios. This attempt does not reflect the real substitutional trend and even obscures the Fe-Mg relationship in the zoned spinels. This conclusion is demonstrated in Figures EG-15 and EG-16, where vertical trends with sharply increasing Fe substitution for Mg from core to rim for the later Ti-chromite and chromian ulvöspinel generations are evident. In fact, these steep trends cross the several slope lines for compositions with various Ti-contents proposed by Haggerty. Furthermore, slopes constructed for compositions with similar Ti-content connect spinels of the various generations crystallized at different times and the relationship observed for early spinels would then completely disappear. The Fe-Mg substitutional trends of the second zoning trend demonstrate a continuous increase in the FFM ratio of the liquid after precipitation of olivine (El Goresy *et al.*, 1976).

Ti-(V+Cr+Al) substitutions

The Ti-(V+Cr+Al) substitutional ratio displays the occupancy in the B site in the normal-inverse solid solution series of tetravalent and trivalent cations. Non-stoichiometry of the spinels should cause a departure of the data points from the 8 (Ti) to the 16 (V+Cr+Al) ratio. No evidence of departure from the 8:16 ratio was found and hence cation deficiency is questionable (Nehru *et al.*, 1974; El Goresy *et al.*, 1976) (Fig. EG-17). Neglecting to include V and Si in the spinel analyses is responsible for speculation that either the B site is deficient or there is increase in the octahedral site occupancy for divalent cations.

The Luna 16 mare type basalts are characterized by their high Al_2O_3 content (∿16%). Spinel analysis in these rocks indicate similar compositional variation trends to Apollo 12 and Apollo 15 basalts (Haggerty, 1972). However,

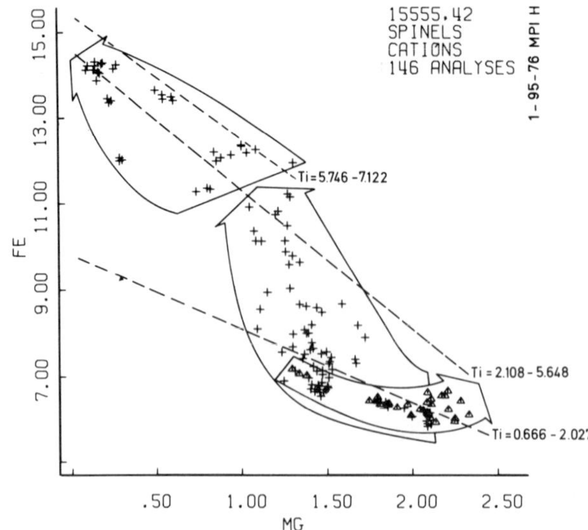

Figure EG-15. Fe-Mg substitutional trends for various zoned spinels in an olivine basalt. Note antipathetic negative trend for early spinels towards higher Mg substitutions. Zoning trends of later chromites are steep and indicate sharp Fe substitution for Mg. Substitutional trends of zoned grains cross several of the slope lines by Haggerty (1972a,b,c).

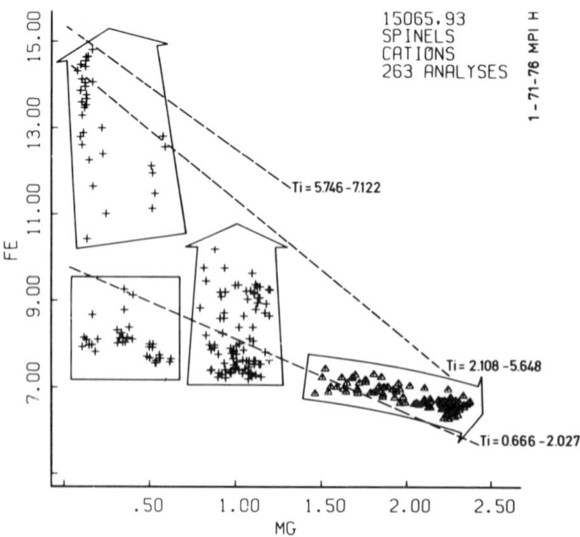

Figure EG-16. Fe-Mg substitutional trends for various zoned spinels in a pigeonite basalt. Antipathetic negative trends for early spinels is more pronounced than in Fig. EG-15. All chromite generations with various FFM ratios emerge.

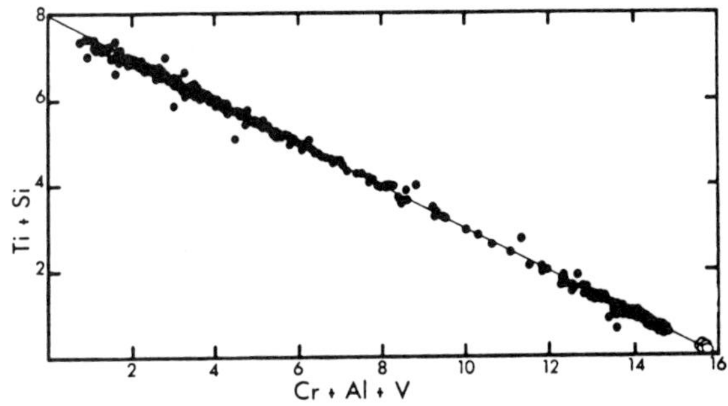

Figure EG-17. Plot of Ti+Si against Cr+Al+V cations for spinels from Apollo 15 rake samples. Open circles, spinels; filled circles, chromite and ulvöspinel (from Nehru *et al.*, 1974).

---

they are characterized by considerable solid solution towards $FeAl_2O_4$ and $MgAl_2O_4$ (Fig. EG-18). Evidently, these spinels crystallized from a liquid with high but continuously decreasing Al concentrations (Fig. EG-19) as documented by the sharp decrease in the Al/Cr ratio. This may indicate that in comparison to Apollo 12 and 15 basalts, anorthitic plagioclase and pyroxene co-precipitated from the Luna 16 magmas (Bence *et al.*, 1972). Continuous crystallization of anorthitic plagioclase would explain the continuous and drastic decrease in the Al/Cr ratio. Spinel compositions from various basalt types from Apollo 12, Apollo 15 and Luna 16 sites are shown in Table EG-1.

Ilmenite textures in basalts of the three landing sites are quite similar (Haggerty, 1971; El Goresy *et al.*, 1971). The concentration of geikielite is usually in direct relationship with the total MgO content of the rock, *e.g.*, high geikielite concentrations in rocks with high MgO content. Ilmenite in the majority of the TiO_2-poor basalts is usually late in the crystallization sequence probably after the late Cr-ulvöspinel.

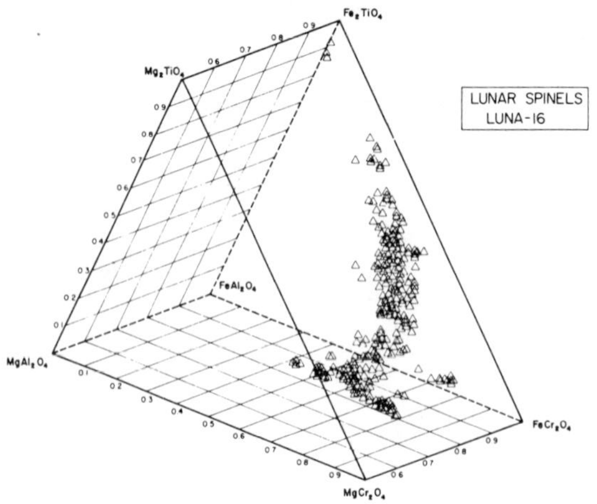

Figure EG-18. Compositions of Luna 16 spinels in the modified spinel prism (from Haggerty, 1972b).

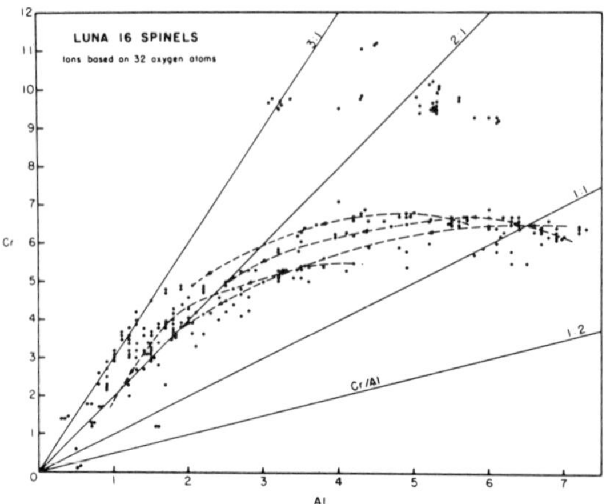

Figure EG-19. Atomic proportions of Cr as a function of Al for Luna 16 spinels. The dashed lines indicate the zonal trends (from Haggerty, 1972b).

Table EG-1. Chemical analyses of spinels.

	1	2	3	4	5	6	7
SiO_2	-	-	0.52	0.11	0.17	0.76	0.35
TiO_2	6.74	25.80	3.60	32.60	7.09	0.92	29.70
Cr_2O_3	42.20	16.30	47.10	1.28	30.96	51.49	5.54
Al_2O_3	11.10	4.37	10.70	2.58	21.51	14.48	1.89
V_2O_3	0.97	0.62	0.87	0.01	-	-	-
FeO	36.60	51.30	32.80	62.70	34.42	25.85	61.05
MgO	1.89	1.70	2.94	0.52	4.60	6.67	0.17
MnO	0.40	0.36	0.35	0.31	0.36	0.46	0.46
CaO	0.05	0.05	0.34	0.26	0.08	0.38	0.17
Total	99.50	99.80	99.22	100.36	99.19	101.01	99.33

1: Titanian chromite, Apollo 12 (Taylor et al.1971, Table 5, p.865)
2: Chromian ulvöspinel, Apollo 12 (Taylor et al. 1971, Table 5, p.865)
3: Magnesian-aluminian chromite, Apollo 15 (Nehru et al.1974, Table 2, p. 1225)
4: Chromian ulvöspinel, Apollo 15 (Nehru et al.1974, Table 2, p.1225)
5: Magnesian-aluminian chromite, Luna 16 (Haggerty, 1972 b, Table 1, p. 335)
6: Cr-rich aluminian-magnesian chromite, Luna 16 (Haggerty, 1972 b, Table 1, p. 334)
7: Chromian ulvöspinel, Luna 16 (Haggerty, 1972 b, Table 1, p. 335)

Opaque oxides in TiO_2-rich basalts

TiO_2-rich basalts were collected during the Apollo 11 and 17 flights from Mare Tranquillitatis and the Taurus-Littrow Site, respectively. These high-titanium basalts were formed in the period from 3.82 to 3.55 G.Y. and are mainly confined to the eastern half of the side of the moon facing the earth. Textural variations and compositional similarity among the Taurus-Littrow high-titanium basalts are suggestive of a similar source for both landing sites (LSPET, 1974). The TiO_2-rich basalts can be classified into two major types: (1) plagioclase poikilitic ilmenite basalts; and (2) olivine porphyritic ilmenite basalts. So far, the first basalt type was not encountered in the Apollo 11 site and is restricted to the Apollo 17 samples. The opaque oxides encountered in the studied basalts are: ilmenite, armalcolite, chromian ulvöspinel, secondary titanian chromite, and rutile. Of special interest are textural relationships between armalcolite and the coexisting silicates and opaque oxides in the different rocks as well as variations of armalcolite chemistry and opaque oxides in the different rocks. Armalcolite in the two different major rock types show differences in their Cr and Fe/Mg ratios.

Armalcolite relationships

Two optically different armalcolite types were reported in several Apollo 17 basalts (Haggerty, 1973; El Goresy *et al.*, 1974): (a) a grey variety usually mantled by Mg-rich ilmenite present in olivine porphyritic ilmenite basalts; and (b) a tan variety encountered in medium- and coarse-grained plagioclase poikilitic ilmenite basalts. Preliminary investigations indicated that the grey armalcolite shows higher MgO and Cr_2O_3 contents than the tan variety and this may be responsible for the difference in color (El Goresy *et al.*, 1973). Smyth and Brett (1973) demonstrated that the two armalcolite types are indistinguishable in terms of crystal structure ruling out the possibility that the types are different polymorphs. Their study, however, showed similar differences in MgO and Cr_2O_3 contents as reported by El Goresy *et al.* (1974). Textural relationships of armalcolite-bearing assemblages in Apollo 11 TiO_2-rich basalts indicate that only the grey variety mantled by Mg-rich ilmenite is present. Apollo 17 plagioclase poikilitic ilmenite basalts do not have equivalents in the Apollo 11 landing site. Papike *et al.* (1974) report that armalcolite morphology in olivine porphyritic and plagioclase poikilitic ilmenite basalts is due to local variation in silicate crystallization rather than to major paragenetic differences. According to El Goresy *et al.* (1974) there are indeed differences in the paragenetic sequence between plagioclase poikilitic and olivine porphyritic basalts. These textural and paragenetic differences are outlined below.

1. *Plagioclase poikilitic ilmenite basalts:* This rock type is characterized by the presence of two pyroxenes: (a) titanaugite with sectoral zoning and (b) pigeonite as single crystal overgrowths on augite (Hodges and Kushiro, 1974). Olivine and Cr-ulvöspinel were among the first minerals to crystallize followed by tan armalcolite followed by titanaugite and pigeonite. Ilmenite, plagioclase and then cristobalite were the last minerals to crystallize. Armalcolite occurring in these rocks *is only of the tan variety* and is exclusively present as inclusions in the titanaugite (Fig. EG-20); occasionally, it is present with ilmenite in sealed grain boundaries with armalcolite morphology (Haggerty, 1973). The dominant feature, however, is the idiomorphic blocky appearance of armalcolite crystal clusters occurring only in the cores of titanaugite. In these rocks massive ilmenite precipitation started after the majority of the pyroxenes crystallized. In none of the studied fragments, regardless of their grain size, were ilmenite reactions rims around armalcolite observed, although Papike *et al.* (1974) report ilmenite reaction rims in sample

Figure EG-20. Cluster of idiomorphic tan armalcolite crystals enclosed in a clinopyroxene. Apollo 17 plagioclase poikilitic ilmenite basalt. Length of field 400 microns.

70035. The above-described crystallization path is not dependent on the cooling rate. The few exceptions reported by Papike *et al.* do not negate the crystallization sequence described above.

2. *Olivine porphyritic ilmenite basalts:* These rocks are characterized by the presence of olivine phenocrysts (partially as a quench phase) with Cr-ulvöspinel inclusions. These two minerals, as in the plagioclase poikilitic basalts, were the first phases to crystallize (El Goresy *et al.*, 1974). Both were then followed by grey armalcolite, then ilmenite and at last augite, plagioclase and then tridymite. In coarse-grained rocks olivine and armalcolite were not encountered. The textures in both Apollo 11 and Apollo 17 samples are identical.

Two main features differentiate the two rock types.

a. Olivine porphyritic ilmenite basalts contain *only* the grey armalcolite variety regardless if armalcolite is mantled by ilmenite or not.

b. In olivine porphyritic ilmenite basalts, ilmenite precipitated directly after armalcolite, whereas in plagioclase poikilitic ilmenite basalts, ilmenite crystallized after the major part of titanaugite and pigeonite precipitated.

In olivine porphyritic ilmenite basalts the majority of the armalcolite grains are surrounded by ilmenite rims. Shape and width of the ilmenite mantle vary from grain to grain. Usually, the armalcolite grains are surrounded by continuous ilmenite mantles (Fig. EG-21) and the composite grain still displays armalcolite morphology. The origin and shape of the mantling ilmenite will be discussed separately in detail in a later section.

The above-described differences between the two rock types, especially the presence of two pyroxenes in plagioclase poikilitic basalts and the inverted pyroxene-ilmenite crystallization sequence in the olivine porphyritic basalts, regardless of the grain size of the rock, is suggestive of different mehcanisms other than cooling rate to explain these features.

Origin of ilmenite rims around armalcolite in olivine porphyritic basalts

Studies in reflected light on numerous basalt samples (El Goresy *et al.*, 1974; Papike *et al.*, 1974) indicate that the ilmenite mantles around grey armalcolite grains are formed according to one or a combination of the following processes:

1. Reaction between the cooling basaltic liquid and early crystallized armalcolite

$$FeTi_2O_5 + FeO \text{ (from melt)} \rightarrow 2FeTiO_3$$

Figure EG-21. Several idiomorphic armalcolite crystals displaying bireflection and ilmenite reaction rims. Olivine prophyritic ilmenite basalt. Length of field 200 microns.

---

2. Reaction between chromian ulvöspinel and armalcolite according to the idealized reaction:

$$Fe_2TiO_4 + FeTi_2O_5 \rightarrow 3FeTiO_3$$

3. Reaction between metallic Fe^o and armalcolite as suggested by Harzman and Lindsley (1973)

$$4FeTi_2O_5 + Fe^o \rightarrow 5FeTiO_3 + \text{"}Ti_3O_5\text{"} \text{ (solid solution in armalcolite)}$$

4. Breakdown of armalcolite to ilmenite and rutile
5. Simple overgrowth of ilmenite around armalcolite.

The majority of the above-described processes may be present in the same basalt sample in Apollo 11 and Apollo 17 material.

1. The reaction between the cooling basaltic liquid and armalcolite is evidently the major process responsible for the formation of ilmenite rims around armalcolite in the TiO_2-rich basalts of the Taurus-Littrow and Apollo 11 sites (Lindsley *et al.*, 1974; El Goresy *et al.*, 1974). Ilmenite formed due to this reaction was apparently enriched in TiO_2 in contrast to primary

ilmenites precipitated directly from the basaltic liquid, since the ilmenite mantles usually show numerous rutile inclusions which probably exsolved on cooling. Many armalcolite grains show reactions only on certain sides, namely where the basaltic liquid had a free path to the armalcolite crystal (El Goresy *et al.*, 1974; Papike *et al.*, 1974). This feature is indeed strong evidence that reaction (1) is responsible for the formation of the ilmenite rims around armalcolite. In the fine-grained vitrophyres a few armalcolite grains display no or little reaction although these armalcolites were not protected by other silicates. This could be due to the very fast cooling of the basaltic liquid and the deposition of pyroxene and plagioclase quench crystals before the reaction started.

2. Textures strongly suggestive of reaction (2) were observed in a few lithic fragments and large basalts from the Apollo 17 landing site. In an ideal case pure ulvöspinel would react with armalcolite to form ilmenite according to the equation

$$Fe_2TiO_4 + FeTi_2O_5 \rightarrow 3FeTiO_3$$

Since ulvöspinel in the Apollo 17 basalts is in a broad sense a member of ulvöspinel-chromite solid solution series, secondary titanian chromite will precipitate in addition to ilmenite as a result of this reaction. Thus, the chromian ulvöspinel will change its composition according to the degree of the reaction. Normally, the boundaries of the original ulvöspinel grain are still visible (Fig. EG-22) whereby the newly formed chromite deposited between ulvöspinel and ilmenite. In advanced stages the reaction is also accompanied by exsolution of ilmenite from the host ulvöspinel (Fig. EG-23). The chromite is confined to ulvöspinel boundaries where the reaction with armalcolite took place. Drastic enrichment in $MgAl_2O_4$ in the secondary spinel took place due to this reaction (20.5 wt. % Al_2O_3 versus 9.3% in the original ulvöspinel). Ilmenites formed due to this reaction were also probably rich in TiO_2, since rutile exsolved from the ilmenites.

3. Harzman and Lindsley (1973) and Lindsley *et al.* (1974) report that armalcolite heated within its stability field with metallic iron yields ilmenite$_{ss}$ + a different armalcolite in which part of Ti^{4+} is reduced to oxidize Fe^o and the Ti^{3+} produced enters the armalcolite as Ti_3O_5 component. El Goresy *et al.* (1974) observed in many lithic fragments textures strongly suggestive of this reaction. Several armalcolite grains mantled by ilmenite were observed with small iron globules at the boundary between the armalcolite core and the ilmenite mantle. Electron microprobe analyses of these armalcolites indicate a significant enrichment of Ti compared to the coexisting armalcolite in the

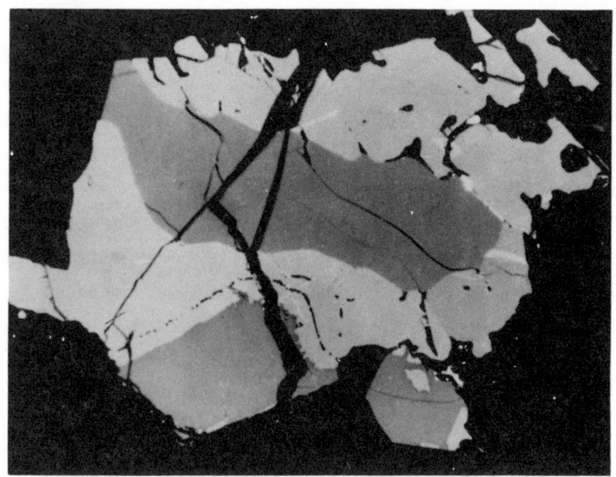

Figure EG-22. Armalcolite (center, gray) which reacted in its lower part with ulvöspinel to form ilmenite and secondary titanian chromite deposited between ilmenite and ulvöspinel. Original boundaries of ulvöspinel are still visible as marked by small aligned silicate inclusions above the ulvöspinel-ilmenite boundary. Length of field 200 microns.

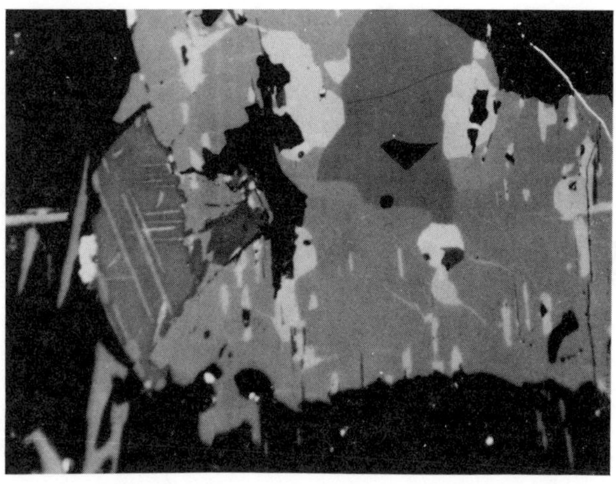

Figure EG-23. A very advanced stage of reaction 2, gray at top is armalcolite surrounded by secondary ilmenite which exsolved rutile (light gray). Gray at left is ulvöspinel with ilmenite exsolutions. Big patches of secondary titanian chromite (dark gray) are located between ulvöspinel and armalcolite. Length of field 150 microns.

same lithic fragment. The total cations (Ti is calculated as Ti^{4+}) are also slightly lower than those of coexisting armalcolites.

4. Pure ferropseudobrookite ($FeTi_2O_5$) decomposes to ilmenite and rutile at and below 1140 $\pm$ 10°C (Harzman and Lindsley, 1973). Harzman and Lindsley also report that armalcolite of a given Fe/Mg ratio decomposes first to Mg-enriched armalcolite + ilmenite + rutile. This breakdown requires the presence of ilmenite and rutile in almost 1:1 ratio. Apollo 11 and 17 basalts were inspected for this reaction and only very few grains of armalcolite in the 17 material were found mantled by ilmenite and rutile satisfying this reaction. The breakdown of armalcolite to rutile + ilmenite is, however, a common phenomenon in Apollo 11 TiO_2-rich basalts.

5. In olivine porphyritic ilmenite basalts, ilmenite precipitates after armalcolite and before titanaugite. According to this crystallization sequence simple overgrowths of ilmenite around pre-existing armalcolite should be expected in these rocks. An important criterion to recognize this texture is that the composite ilmenite-armalcolite grain does not show any resemblance to the armalcolite morphology. However, this kind of overgrowth is quite rare compared to reactions 1, 2, and 3.

Study of equilibrium phase relations of synthetic TiO_2-rich basalts as a function of oxygen fugacity (f_{O_2}) indicates that there is a direct relationship between the crystallization sequence and f_{O_2} for basaltic liquids with the same composition (Usselman *et al.*, 1975). The observed difference in the composition of armalcolites and the reversal of ilmenite and pyroxene in the crystallization sequence in plagioclase poikilitic and olivine porphyritic basalts was found to be a function of f_{O_2} (Usselman *et al.*, 1975) (Fig. EG-24).

Chemistry of armalcolite

Haggerty (1973) reports that tan armalcolite and gray armalcolite are compositionally indistinguishable in terms of major element abundances. More than 400 complete analyses (El Goresy *et al.*, 1974) strongly suggest that both tan and gray armalcolite are cation deficient since the number of cations calculated on the basis of 5 oxygens never totalled 3. The total number of cations for all armalcolites analyzed range from 2.91 to 2.97. According to Lind and Housley (1972) and Smyth (1973), armalcolite crystallizes in the space group Bbmm with the cations strongly ordered whereby Ti^{4+} cations occupy the $8f(M_2)$ and Fe^{2+} and Mg^{2+} cations are randomly distributed among the $4C(M_1)$ sites. Following this model, the majority of the analyses revealed that Fe^{2+} and Mg^{2+} do not satisfy the 4C site occupancy since they never totalled 1,

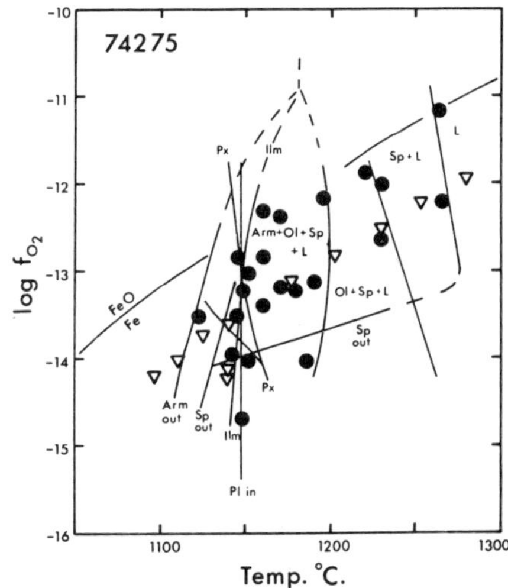

Figure EG-24. Melting relations of Apollo 17 sample 74275. Triangular points are those of O'Hara and Humphries (1975) at their stated oxygen fugacities. The iron-wüstite (Fe-FeO) curve is shown as reference.

although there is a full complement of almost two Ti cations per five oxygens (El Goresy *et al.*, 1974). Smyth suggested that Ti may be present as Ti^{3+} and perhaps Cr as Cr^{2+}. Wechsler *et al.* (1975) calculated 4-10% $Ti_2^{3+}TiO_5$ component for many lunar armalcolites, thus supporting the presence of Ti^{3+} rather than cation deficiency of the armalcolite structure.

The gray armalcolite variety is characterized by relatively higher Cr_2O_3 and MgO contents than the tan variety (El Goresy *et al.*, 1974). However, Papike *et al.* (1975) indicate that many gray armalcolites are zoned. Papike *et al.* report a decrease in Cr_2O_3 and increase in FeO content from the core to the rim of an armalcolite grain. The compositional variation of a zoned crystal was found to overlap a major part of the separate fields assigned for tan and gray armalcolite.

The Mg versus Fe and Mg versus Cr cationic distributions for tan and gray armalcolites are shown in Figures EG-25 and EG-26, respectively. Two important features are recognized: (a) There is indeed a compositional bimodal distribution of tan and gray armalcolites with slight overlap of the fields; (b) the Mg-Fe substitutional relationship for the tan armalcolite variety is almost coherent with a negative slope, indicating that Fe is substituting for Mg. Probably, in analogy to olivine, armalcolites which crystallized earlier from the silicate melt have a higher Mg/Fe ratio than those crystallized later. The gray armalcolites, regardless if they are mantled by ilmenite or not, show generally higher Mg concentrations than tan armalcolite. Compared

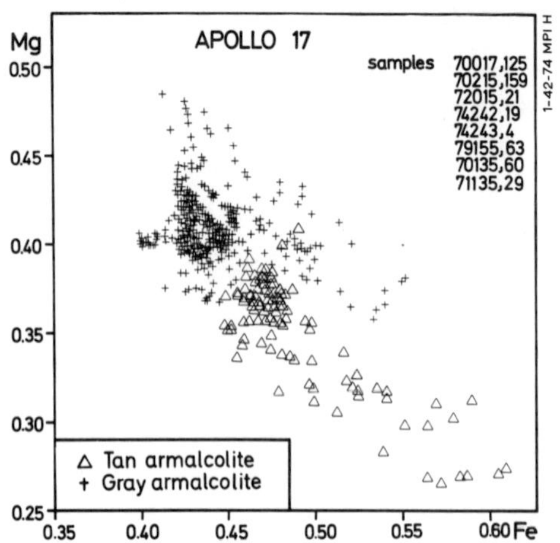

Figure EG-25. Mg-Fe cationic substitutional relationship (based on 5 oxygens) for tan and gray armalcolite.

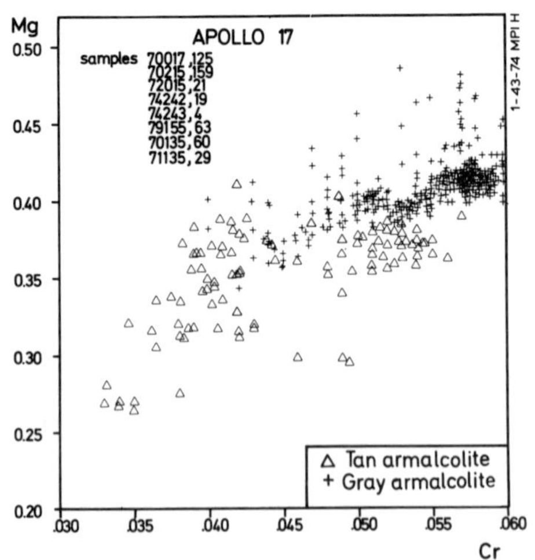

Figure EG-26. Mg-Cr cationic substitutional relationship for tan and gray armalcolites.

to tan armalcolite the Mg-Fe substitutional relationship of the gray armalcolite is not coherent. This scatter was interpreted (El Goresy *et al.*, 1974) as due to enrichment of armalcolite in Mg resulting from reactions 1, 2, 3, or 4 described above. Electron microprobe analyses indicate such strong preference of Mg for armalcolite rather than for mantling ilmenite (El Goresy *et al.*, 1974).

The compositional bimodality is also demonstrated in Figure EG-26 which shows the Mg-Cr substitutional relationship for both armalcolite types. Despite the scatter in the data points for tan armalcolite, this figure is suggestive of a positive relationship between Mg and Cr. The Mg-Cr substitutional relationship for gray armalcolite is coherent, indicating that these two elements have similar partitioning behavior. Armalcolites in olivine porphyritic basalts have higher MgO and Cr_2O_3 contents than tan armalcolites. The data presented here are also strongly suggestive of a partitioning of Mg and Cr between armalcolite and mantling ilmenite with strong preference of these elements for armalcolite. Ilmenite rims around gray armalcolite cores were also analyzed with the microprobe (El Goresy *et al.*, 1974), and the suggested partitioning is confirmed in the plot of

$$\frac{\text{MgO in armalcolite}}{\text{MgO in mantling ilmenite}} \quad \text{versus} \quad \frac{Cr_2O_3 \text{ in armalcolite}}{Cr_2O_3 \text{ in mantling ilmenite}}$$

(Fig. EG-27). The coherent positive slope of the data points indicates a preference of both Mg and Cr for armalcolite. This slope, however, may not represent the actual partitioning relationship between armalcolite and the mantling ilmenite, since the majority of the ilmenite mantles exsolved rutile after their formation which may have caused an additional redistribution of

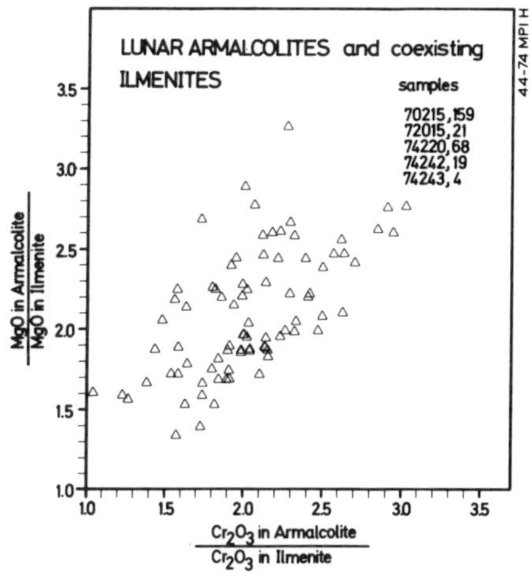

Figure EG-27. MgO Arm./MgO Ilm. versus Cr_2O_3 Arm./Cr_2O_3 Ilm. relationship for gray armalcolites and mantling ilmenites.

both Cr and Mg between ilmenite and rutile. The partitioning of Cr and Mg between ilmenite and rutile does not necessarily need to be similar to the partitioning of these two elements between armalcolite and ilmenite. Thus, gray armalcolite shows higher Mg and Cr concentrations than tan armalcolite. Reflection measurements on numerous armalcolite types from several Apollo 17 basalts indicate that the reflection curves of tan armalcolite show a continuous increase above 600 nm compared to a flat curve for gray armalcolite. This is responsible for the difference in color. Armalcolite compositions from various rocks are displayed in Table EG-2.

Table EG-2. Chemical analyses of armalcolite.

	1	2	3	4	5	6	7
SiO_2	0.33	0.30	0.08	0.08	-	-	-
TiO_2	70.44	71.61	74.13	73.91	74.30	73.00	72.50
Cr_2O_3	1.63	1.69	2.00	2.01	2.17	1.72	1.43
Al_2O_3	1.72	1.77	2.10	1.99	1.93	1.96	1.91
V_2O_3	0.18	0.27	0.10	0.05	-	-	-
FeO	18.43	16.30	14.08	14.44	13.4	16.30	17.60
MgO	5.47	6.63	7.86	7.75	7.95	6.27	5.32
MnO	0.08	0.09	0.14	0.17	0.00	0.00	0.00
CaO	0.57	0.44	0.04	0.03	-	-	-
Total	98.85	99.10	100.53	100.43	99.80	99.20	98.80

1: Tan armalcolite, Apollo 17 (El Goresy et al. 1974, Table 2, p. 641)
2: Tan armalcolite, Apollo 17 (El Goresy et al. 1974, Table 2, p. 641)
3: Gray armalcolite, Apollo 17 (El Goresy et al. 1974, Table 2, p.641)
4: Gray armalcolite, Apollo 17 (El Goresy et al. 1974, Table 2, p. 641)
5: Core of a gray armalcolite, Apollo 17 (Papike et al.1974, Table 6, p. 492)
6: Same gray armalcolite few microns away from the core, Apollo 17 (Papike et al. 1974, Table 6, p. 492)
7: Same gray armalcolite just at the ilmenite rim, Apollo 17 (Papike et al. 1974, Table 6, p. 492)

Chromian ulvöspinel

Chromian ulvöspinel occurs in both basalt types not only as idiomorphic crystals but also as clusters of grains enclosed in olivine or pyroxene. In both rock types ulvöspinel is probably an early quench phase. In coarse-grained ilmenite basalts this early crystallized ulvöspinel was not encountered; instead, tiny discrete late-stage ulvöspinel grains were observed in the mesostasis. The early crystallized ulvöspinels in Apollo 17 basalts resemble the Apollo 11 ulvöspinel since their composition is intermediate between chromite and ulvöspinel. The Apollo 17 ulvöspinels, however, show considerable solid solution towards $MgAl_2O_4$ (Fig. EG-28). Figure EG-28 displays the composition of ulvöspinels and chromite exsolution lamellae in ilmenite in the spinel prism. It is noteworthy to mention that ulvöspinel shows systematically higher MnO and V_2O_3 contents than the coexisting armalcolite.

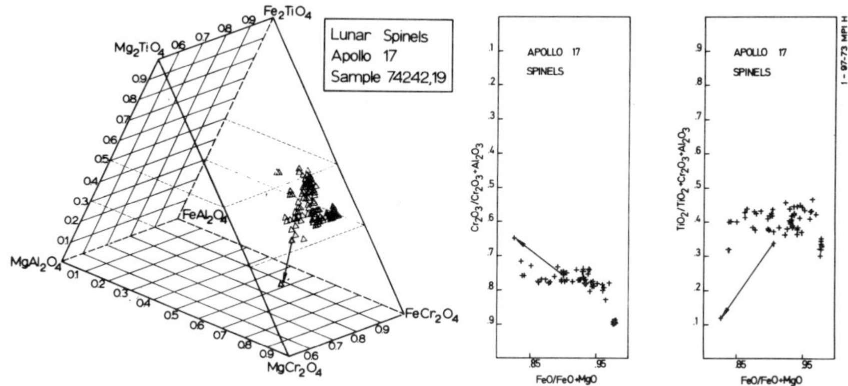

Figure EG-28. Composition of chromian ulvöspinels in Apollo 17 basalts as projected in the multicomponent spinel prism (left); projection of the spinel prism base (middle); and projection of front rectangular face of spinel prism (right).

Ilmenite

Ilmenite is the most abundant opaque mineral both in Taurus-Littrow and Apollo 11 basalts. The amount of ilmenite in the studied samples range from 15% to 20% by volume. Ilmenite occurs in both rock types as primary blocky crystals and in olivine porphyritic basalts as reaction rims around armalcolite formed according to processes 1, 2, 3 or 4 as described above. This latter ilmenite type is designated secondary ilmenite to distinguish it from the former primary ilmenite which precipitated directly from the silicate melt. The majority of ilmenite formed by processes 1 and 2 show numerous rutile inclusions exsolved from the TiO_2-rich secondary ilmenite on cooling. Analyses of primary and secondary ilmenites formed by processes 1 and 2 indicate a systematic difference in the composition of both ilmenites (El Goresy *et al.*, 1974). Secondary ilmenites were found to contain significantly higher Cr_2O_3 contents (0.93-1.05%) than coexisting primary ilmenites (0.39-0.65%).

Rutile

Rutile occurs mainly as exsolution lamellae, as a product of reaction 4 described above, or is formed due to subsolidus reduction of magnesian chromian ilmenite to rutile + spinel + metallic Fe (El Goresy *et al.*, 1975). Many of these rutile grains show anomalous optical properties with prominent bireflection and anisotropism and a typical blue reflection color (El Goresy *et al.*, 1975). These features are characteristic of oxygen-deficient rutile. The majority of rutile grains were found to be highly enriched in ZrO_2 (Table EG-3) regardless if they were formed by exsolution or subsolidus reduction reaction (El Goresy *et al.*, 1975).

Table EG-3. Chemical analyses of rutile.

	1	2	3
SiO_2	0.42	0.11	0.11
TiO_2	97.80	98.60	98.30
Cr_2O_3	0.23	0.23	1.54
Al_2O_3	0.00	0.00	0.26
V_2O_3	0.00	0.00	0.05
FeO	0.93	0.59	0.86
MgO	0.04	0.06	0.00
MnO	0.00	0.00	0.14
CaO	0.05	0.07	0.04
ZrO_2	0.82	0.32	1.31
Total	100.29	99.98	102.61

1: Stoichiometric rutile, Apollo 17 (El Goresy et al. 1975, Table 4, p. 744)
2: Stoichiometric rutile, Apollo 17 (El Goresy et al. 1975, Table 4, p. 744)
3: Blue oxygen deficient rutile, Apollo 17 (El Goresy et al.1975, Table 4, p. 744)

Opaque oxides in anorthositic rocks and highland breccias

On the basis of the occurrence of a few anorthositic lithic fragments in Apollo 11 fines, Wood *et al.* (1970) proposed that the Highlands of the moon are largely composed of anorthosite. Prinz *et al.* (1973) demonstrated that the anorthositic rocks range in composition and mineralogy from anorthositic to noritic and troctolitic. These rocks very probably constitute a major part of the lunar crust. Due to continuous bombardment of the lunar surface in the past, the original textures of the rocks of this group are obscured and altered. Rocks belonging to this group were collected at the Luna 20 landing site (in the area of Apollonius C crater), Apollo 16 at the Descartes site, and at the

Apollo 14 (Fra Mauro) site. Two major rock suites were recognized in the samples collected from the three landing sites (Prinz *et al.*, 1973b).

1. ANT (anorthositic-noritic-troctolitic) suite which could be subdivided to anorthosites, norites, and (spinel) troctolite).

2. High-alumina basalt suite (> 45% $\lessdot$ 60% plagioclase).

Opaque oxide minerals are rare and generally fine grained. They also occur as discrete rounded to angular grains in microbreccias and agglutinates and as rare euhedral to anhedral grains in crystalline fragments (El Goresy *et al.*, 1973; Brett *et al.*, 1973; Haggerty, 1973). In the ANT suite members of the spinel series are the most abundant opaque oxides followed by ilmenite and rutile. In the high-alumina basalt suite, ilmenite is quite frequent, usually occurring in the assemblages ilmenite-rutile-baddeleyite (ZrO_2) and ilmenite-rutile-baddeleyite-chromite (El Goresy *et al.*, 1973).

Spinels

Two distinct spinel types occur in the highland rock suites: (a) Members of the spinel ($MgAl_2O_4$)-hercynite ($FeAl_2O_4$) series mainly encountered in the ANT suite, especially in spinel troctolites (Brett *et al.*, 1973). Their composition ranges from 93 to 56 mole % $MgAl_2O_4$ (Brett *et al.*, 1973). They are characterized by a low TiO_2-content (0.02-1.0 wt %). A minor amount of the chromite component is also present but their compositions are clearly separated from the chromite-ulvöspinel solid solution series. (b) Members of the chromite-ulvöspinel series occurring as a minor component in the ANT suite but more abundant in the high-alumina basalt. Although no optical zoning was observed in these spinels, quantitative electron microprobe analyses indicate a zonational trend similar to the third trend present in mare-type basalts (El Goresy *et al.*, 1973). The zoning trend of these spinels shows, however, a sharp increase in the $Cr_2O_3/(Cr_2O_3+Al_2O_3)$ ratio from core to rim (El Goresy *et al.*, 1973) (Fig. EG-29). The Cr/Al cationic relationship is similar to that found in Luna 16 rocks (Haggerty, 1973).

Ilmenite

Ilmenites encountered in highland rocks contain appreciable amounts of the geikielite component just as do those from mare basalts. The MgO-content of ilmenite varies between 0.02 and 8 wt % (Brett *et al.*, 1973).

Rutile

Rutile is usually confined to oriented lamellae and irregular areas in

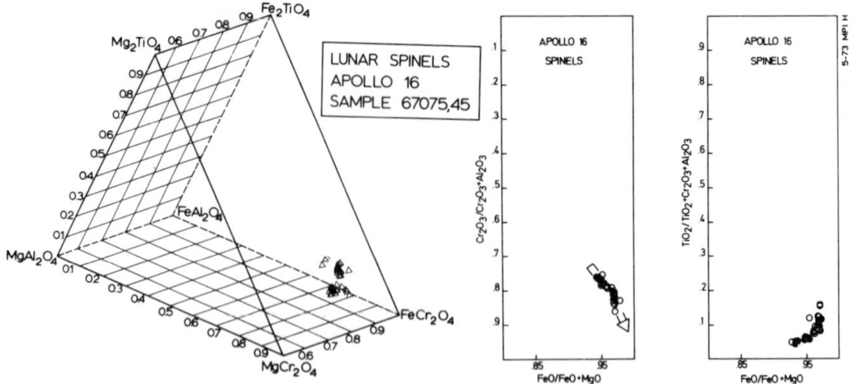

Figure EG-29. Composition of spinels in an anorthosite sample as projected in the multicomponent spinel prism (left); projection of the spinel prism base (middle) showing the crystallization trend of the spinels; and front rectangular face of the spinel prism (right) also depicting the trend with increase of TiO_2-content.

---

ilmenites (Haggerty, 1973). The mineral contains between 2.3 and 2.9 wt % FeO and between 0.4 and 0.5 wt % Cr_2O_3.

SUBSOLIDUS REACTIONS

Lunar rocks differ from the majority of terrestrial counterparts mainly in their extremely low oxygen fugacities. This feature is manifested by the ubiquitous occurrence of metallic iron as a primary phase, the presence of textures suggestive of subsolidus reduction, and the extremely low amount of Fe^{3+} in silicate and oxide minerals. Very probably, many of the lunar magmas were already in a reduced state at the time of extrusion (Sato *et al.*, 1973). Considerable variation in the intensity of reduction has been reported among rocks of the same landing site (El Goresy *et al.*, 1972; Haggerty, 1972a) reflecting the complexity of the formation history of the lunar rocks and supporting the existence of various mechanisms for reduction. An attempt to understand the processes responsible for reduction reactions is due to Mao *et al.* (1974), who demonstrated that trapped solar wind hydrogen is probably an important reducing agent on the surface of the moon, especially during the flow of basaltic liquid on hydrogen-enriched regolith.

Subsolidus reduction reactions in the lunar rocks

Evidence for five subsolidus reactions for which experimentally-determined buffer curves exist was reported from the lunar samples of all landing sites. The experimentally-determined buffer curves involve the following reactions (Sato *et al.*, 1973; Williams, 1971; Taylor *et al.*, 1972; Lindsley *et al.*, 1974):

1. $Fe_2TiO_4 \rightarrow FeTiO_3 + Fe + \frac{1}{2}O_2$
2. $FeTiO_3 \rightarrow Fe + TiO_2 + \frac{1}{2}O_2$
3. $Fe_2SiO_4 \rightarrow 2Fe + SiO_2 + O_2$
4. $2FeTiO_3 \rightarrow FeTi_2O_5 + Fe + \frac{1}{2}O_2$
5. $FeTi_2O_5 \rightarrow Fe + 2TiO_2 + \frac{1}{2}O_2$

In applying these reactions, only *estimations* of the degree of reduction and temperature can be made, since the reduced lunar minerals are by no means pure compounds as compared to the experimentally-investigated synthetic phases. Furthermore, the host minerals subjected to reduction change their composition continuously as the reaction proceeds to the right-hand side of the above equations. These compositional variations were reported by El Goresy *et al.* (1972) and Haggerty (1972a) who demonstrated that chromian ulvöspinel changes its composition continuously upon subsolidus reduction to become enriched in the chromite component. In comparison to equation 2, lunar magnesian ilmenites break down to rutile + spinel + Fe (Haggerty, 1973a). The activity of Fe_2SiO_4 in lunar fayalite decreases continuously upon reduction to Fe + SiO_2. Host ilmenite also changes its composition to become enriched in the geikielite component. Reactions 4 and 5 were found to be induced upon flow of basaltic liquid on regolith loaded with solar wind hydrogen (Mao *et al.*, 1974) and hence are regarded as exogenic reactions. The reduction described by Mao *et al.* (1974) indicates that extensive reduction of lunar basalts probably first took place on the surface of the moon during the eruption process.

Figure EG-30 displays schematically the oxygen fugacity curves of the univariant reactions listed above, after Sato *et al.* (1973) and Taylor *et al.* (1972), and extrapolated down to 700°C. Among the univariant reactions described, the I-Q-F curve holds a key position. The quartz-fayalite-iron (I-Q-F) univariant curve intersects the ulvöspinel-ilmenite-iron (Fe-il-uv) curve at 950°C and the ilmenite-rutile-iron (Fe-ru-il) curve at roughly 830°C (Taylor *et al.*, 1972). Any understanding of the temperature and reduction history of the lunar samples requires the study of *all assemblages* involved in the first three reactions. Apollo 14 samples 14053 and 14072 were previously

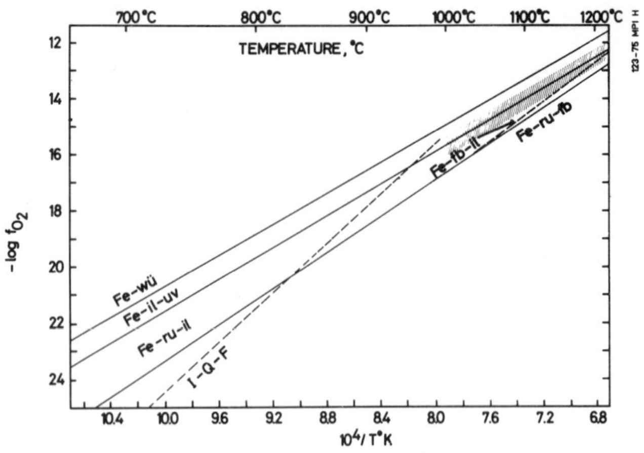

Figure EG-30. Oxygen fugacity curves for the reduction reactions 1-5. Data taken from Sato *et al.* (1973) and Taylor *et al.* (1972). Path of fugacity curves below 830°C is extrapolated and hence is uncertain. Symbols are: Fe or I = iron, Wü = wüstite, il = ilmenite, fb = ferropseudobrookite, and ru = rutile.

---

considered to be the most reduced basalts, due to the presence of the breakdown of fayalite to silica + Fe in addition to the extensive reduction of chromian ulvöspinel to Ti-chromite + ilmenite + Fe (El Goresy *et al.*, 1972; Haggerty, 1972a). Taylor *et al.* (1972) conclude from their experimental investigations that the presence of fayalite reduction in these two samples is *direct* evidence that these two rocks have undergone more reducing conditions than any other rocks, terrestrial or lunar reported from Apollo 11 through Apollo 14. A comparison between the breakdown assemblages in these two rocks and in several TiO_2-rich basalts from the Taurus-Littrow region as well as the application of the first three reactions negates a severe reduction. Table EG-4 displays the assemblages involved in the univariant reactions 1 to 3 in sample 14053 as compared to assemblages in Apollo 17 samples 70017, 70035, and 70135. Comparison

Table EG-4. Assemblages in reduced lunar basalts.

14053	70017; 70035; 70135
1. Ulvöspinel ⟶ ilmenite+Fe	1. Ulvöspinel ⟶ ilmenite+Fe
2. Fayalite ⟶ Fe+silica	2. Ilmenite ⟶ spinel+rutile+Fe
3. Incipient breakdown of ilmenite (only very few grains)	3. No fayalite breakdown

between the assemblages in 14053 and in the three Apollo 17 basalts strongly suggests that the rocks of the two landing sites had different thermal histories. Very probably, samples 14053 and 14072 were subjected to reduction during a heating event to a temperature in excess of 830°C and/or close to 950°C. A mechanism of reduction upon heating to these temperatures at the given oxygen fugacity explains best the simultaneous breakdown of fayalite to Fe + silica and ulvöspinel to ilmenite + Fe. Incipient breakdown of ilmenite to rutile + Fe may suggest that the opaque fayalite assemblages were placed below the Fe-ru-il fugacity curve above 830°C and very probably close to 950°C in order to account for the extensive subsolidus reduction of both fayalite and chromian ulvöspinel followed by rapid cooling, thus preventing further ilmenite breakdown. Textures and mineralogy of sample 14053 do not support reduction during an initial cooling process. Fayalite in this sample always occurs in a mesostasis assemblage, which is widely accepted to be the last assemblage to crystallize from the silicate melt. Reduction of fayalite must have taken place *after* solidification of the mesostasis assemblage, since the breakdown texture still displays the morphology of fayalite grains (El Goresy *et al.*, 1972).

Apollo 17 samples 70017, 70035, and 70135 display extensive subsolidus reduction of chromian ulvöspinel to Al-Ti chromite + ilmenite + Fe and of Mg ilmenite to Al-Ti chromite + rutile + Fe. Accessory fayalite in these samples does not show any sign of reduction to Fe + silica. On the basis of the pervasive reduction of ilmenite alone, and neglecting fayalite (El Goresy *et al.*, 1972; Taylor *et al.*, 1972; Haggerty, 1972a), one may conclude that these samples were severely reduced. Such a conclusion is unrealistic. The three assemblages shown in Table EG-4 suggest that the reduction of these rocks took place below 830°C, probably during cooling. These assemblages may have been reduced to a f_{O_2} *below* the Fe-ru-il curve, but above the I-Q-F curve. This is only possible below 830°C, where the fayalite buffer curve intersects the ilmenite buffer curve.

It is worthwhile to mention that the above-given temperatures are subject to some correction since the activities of Fe_2TiO_4 and $FeTiO_3$ in the lunar ulvöspinel and ilmenite, respectively, are by no means equal to 1.

Haselton and Nash (1975) discussed a model (formerly proposed by Lindsley *et al.*, 1974) of isochemical reaction of opaque oxides with part of Ti present as Ti^{3+} in the initial phases to explain the subsolidus reactions observed. An important aspect of this model is the fact that the components of the assemblages are indeed not in their standard states. Such a system would display deviation from ideality. This deviation from stoichiometry indicates that the f_{O_2} curve will not be a straight line. Experimental studies in the system Fe-Ti-O with assemblages coexisting with metallic Fe^o (Simons, 1973) indeed indicate a compositional variation as a function of temperature. The mechanism presented by Haselton and Nash (1975) could explain the textures in the assemblage armalcolite-ilmenite-rutile-Fe^o. However, it could hardly explain the breakdown features of chromian ulvöspinel since Ti^{3+} was not detected in an ulvöspinel-magnetite$_{ss}$ coexisting with metallic Fe^o between 1000 and 1300°C (Simons, 1974).

Nature of reducing agent in Apollo 17 basalts

The causes for the reduction of lunar basalts have been a subject of debate since recovery of the Apollo 11 samples. Several mechanisms were proposed: (1) internal redistribution of valence of states of Fe, Ti, and Cr by assuming initial presence of Ti^{3+} and Cr^{2+} in the melt (Brett *et al.*, 1971,1972); (2) vacuum pumping of oxygen (O'Hara *et al.*, 1970; Biggar *et al.*, 1971; Ford *et al.*, 1972; Haggerty, 1972a); (3) sulfur loss from lunar magmas as S^{2-} (Brett, 1975); (4) alkali volatilization from lunar magmas during extrusion (O'Hara *et al.*, 1970; Biggar *et al.*, 1971,1972, Ford *et al.*, 1972; (5) reduction by carbon or carbon monoxide at a depth shallower than a few kilometers (Sato *et al.*, 1973).

Vacuum pumping of oxygen could never be a plausible cause of reduction because of the preferential escape of Fe vapor at low oxygen fugacities (Sato *et al.*, 1973). Nor could alkali loss explain the reduced state of the lunar basalts since escape of alkalis as elements would cause oxidation, whereas volatilization of alkalis as oxides (Na_2O) would not change the oxidation state of lunar basalts (Sato *et al.*, 1973). The mechanisms of sulfur loss from lunar magmas or of redistribution of the valence of states of Fe, Ti, and Cr may well explain the precipitation of metallic iron from the lunar basaltic liquids, but they are unable to account for subsolidus reduction reactions of iron-titanium oxide assemblages.

El Goresy *et al.* (1975) propose that gaseous activity took place after solidification of the lunar basalts. They found evidence for genetic relationship between this gaseous activity and subsolidus reactions of the opaque oxides. Many of the open cracks penetrating silicate and oxide minerals as well as cleavage in pyroxenes in several Apollo 12, Apollo 15, and Apollo 17 samples were found to be filled with metallic iron. The metallic iron forms a network system of veins across several mineral grains along several hundred microns. El Goresy *et al.* (1975) believe that these features exclude the possibility that iron liquid has been injected in the crack system. These observations can be explained as due to deposition of iron from a gaseous phase which permeated these rocks after crystallization. Ilmenite grains penetrated by such iron-filled cracks display textures typical of subsolidus reduction to spinel + rutile + Fe^o. These reduction products always originate from the cracks radiating into the ilmenite grains. El Goresy *et al.* (1975) consider these features as strongly suggestive that a single event is responsible for deposition of iron in the cracks and reduction of opaque oxides. A gaseous mixture of CO and carbonyle iron compounds ($Fe(CO)_5$) was proposed to have initiated these reactions. Upon cooling, many of those complex gaseous compounds become unstable and break down at different temperatures to Fe^o+ CO. Upon breakdown, metallic iron would fill many of the open cracks, whereas carbon monoxide released would account for the reduction processes. The breakdown and the reduction processes may have continued during cooling down to 200°C. It is also plausible that these gases may have been released from the same magma chamber from which the basalts originated and permeated those rocks after solidification and formation of tension cracks. The reduction process is probably not the result of a release of solar wind hydrogen during an impact process since no evidence of reheating or shock features in the plagioclase or ilmenite were found. All these observations do indeed support the reduction mechanism proposed by Sato *et al.* (1973) and emphasize the possibility of endogenic gaseous activity during eruption of lunar lavas.

REFERENCES

Anderson, A. T., T. E. Bunch, E. N. Cameron, S. E. Haggerty, F. R. Boyd, L. W. Finger, O. B. James, K. Keil, M. Prinz, P. Ramdohr, and A. El Goresy (1970) Armalcolite: A new mineral from the Apollo 11 samples. *Proc. Apollo 11 Lunar Sci. Conf., Geochim. Cosmochim. Acta, Suppl. 1, 1*, 55.

Bence, A. E., W. Holzwarth, and J. J. Papike (1972) Petrology of basaltic and monomineralic soil fragments from the Sea of Fertility. *Earth Planet. Sci. Lett. 13*, 299.

Biggar, G. M., M. J. O'Hara, A. Peckett, and D. J. Humphries (1971) Lunar lavas and the achondrites: Petrogenesis of protohypersthene basalt in the mare lava lakes. *Proc. Lunar Sci. Conf. 2nd*, 617.

Brett, R. (1975) Reduction of mare basalts by sulfur loss (abstr.) In, *Lunar Science VI*, 89. The Lunar Science Institute, Houston.

_________, P. Butler, Jr., C. Meyer, Jr., A. M. Reid, H. Takeda, and R. J. Williams (1971) Apollo 12 igneous rocks 12004, 12008, 120022: A mineralogical and petrological study. *Proc. Lunar Sci. Conf. 2nd*, 301.

El Goresy, A., P. Ramdohr, and L. A. Taylor (1971) The opaque minerals in the lunar rocks from Oceanus Procellarum. *Proc. Lunar Sci. Conf. 2nd*, 219.

El Goresy, A., P. Ramdohr, and L. A. Taylor (1972) Fra Mauro crystalline rocks: Mineralogy, geochemistry and subsolidus reaction of opaque minerals. *Proc. Lunar Sci. Conf. 3rd*, 333.

________, ________, and O. Medenbach (1973) Lunar samples from the Descartes site: Opaque mineralogy and geochemistry. *Proc. Lunar Sci. Conf. 4th*, 733.

________, ________, ________, and H.-J. Bernhardt (1974) Taurus-Littrow TiO_2-rich basalts: Opaque mineralogy and geochemistry. *Proc. Lunar Sci. Conf. 5th*, 627.

________ and P. Ramdohr (1975) Subsolidus reduction of lunar opaque oxides: Textures, assemblages, geochemistry, and evidence for a late-stage endogenic gaseous mixture. *Proc. Lunar Sci. Conf. 6th*, 729.

________, M. Prinz, and P. Ramdohr (1976) Zoning in spinels as an indicator of the crystallization histories of mare basalts. *Proc. Lunar Sci. Conf. 7th*, in press.

Ford, C. E., G. M. Biggar, D. J. Humphries, G. Wilson, D. Dixon, and M. J. O'Hara (1972) Role of water in the evolution of lunar crust; an experimental study of sample 14310; an indication of lunar calc-alkaline volcanism. *Proc. Lunar Sci. Conf. 3rd*, 207.

Haggerty, S. E., F. R. Boyd, P. M. Bell, L. W. Finger, and W. B. Bryan (1970) Opaque minerals and olivine in lavas and breccias from Mare Tranquillitatis. *Proc. Apollo 11 Lunar Sci. Conf.*, 513.

________ and H. O. A. Meyer (1970) Apollo 12: Opaque oxides. *Earth Planet. Sci. Lett. 9*, 379.

________ (1972a) Apollo 14: Subsolidus reduction and compositional variations of spinels. *Proc. Lunar Sci. Conf. 3rd*, 305.

________ (1972b) Luna 16: An opaque mineral study and systematic examination of compositional variations of spinels from Mare Fecunditatis. *Earth Planet. Sci. Lett. 13*, 328.

________ (1972c) Solid solutions, subsolidus reduction and compositional characteristics of spinels in some Apollo 15 basalts. *Meteoritics 7*, 353.

________ (1973a) Apollo 17: Armalcolite paragenesis and subsolidus reduction of chromian-ulvöspinel and chromian-picroilmenite (abstr.). *E⊕S (Trans. Am. Geophys. U.) 54*, 593.

________ (1973b) Armalcolite and genetically associated minerals in the lunar samples. *Proc. Lunar Sci. Conf. 4th*, 777.

________ (1973c) Luna 20: Mineral chemistry of spinel, pleonast, chromite, ulvöspinel, ilmenite, and rutile. *Geochim. Cosmochim. Acta 37*, 857.

Harzman, M. J. and D. H. Lindsley (1973) The armalcolite join ($FeTi_2O_5$-$MgTi_2O_5$) with and without excess Fe^o: Indirect evidence of Ti^{3+} on the moon (abstr). *Ann. Meeting Geol. Soc. Am. 5*, 593.

Haselton, J. D. and W. P. Nash (1975) Observations on titanium in luna oxides and silicates (abstr.). In, *Lunar Science VI*, 343. The Lunar Science Institute, Houston.

Hodges, F. N. and I. Kushiro (1974) Apollo 17 petrology and experimental determination of differentiation sequences in model moon compositions. *Proc. Lunar Sci. Conf. 5th, 1*, 505.

Kesson, S. E. (1975) Mare basalts: Melting experiments and petrogenetic interpretations. *Proc. Lunar Sci. Conf. 6th*, 921.

Knorring, O. V. and K. G. Cox (1961) Kennedyite, a new mineral of the pseudobrookite series. *Mineral. Mag. 32*, 672.

Kushiro, I., Y. Nakamura, and S. Akimoto (1970) Crystallization of Cr-Ti spinel solid solutions in an Apollo 12 rock, and source rock of magmas of Apollo 12 rocks (abstr.). *Am. Geophys. U. Ann. Meeting*, 64.

Laul, J. C. and R. A. Schmitt (1973) Chemical composition of Apollo 15, 16, and 17 samples. *Proc. Lunar Sci. Conf. 4th*, 1349.

Lindsley, D. H., S. E. Kesson, M. J. Hartzman, and M. K. Cushman (1974) The stability of armalcolite: Experimental studies in the system MgO-Fe-Ti-O. *Proc. Lunar Sci. Conf. 5th, 1*, 521.

LSPET (Lunar Sample Preliminary Examination Team) (1974) *Preliminary Examination of Lunar Samples*, pp. 7.1-7.46. L. B. Johnson Space Center.

Mao, H. K., A. El Goresy, and P. M. Bell (1974) Evidence of extensive chemical reduction in lunar regolith samples from the Apollo 11 site. *Proc. Lunar Sci. Conf. 5th*, 673.

Marvin, U. (1975) The perplexing behavior of Niobium in meteorites and lunar samples. *Meteoritics 10*, 452.

Muan, A., J. Hauck, and T. Löfall (1972) Equilibrium studies with a bearing on lunar rocks. *Proc. Lunar Sci. Conf. 3rd, 1*, 185.

Nehru, C. E., M. Prinz, E. Dowty, and K. Keil (1974) Spinel-group minerals and ilmenite in Apollo 15 rake samples. *Am. Mineral. 59*, 1220.

O'Hara, J. M., G. M. Biggar, S. W. Richardson, and C. E. Ford (1970) The nature of seas, mascons, and the lunar interior in the light of experimental studies. *Proc. Apollo 11 Lunar Sci. Conf.*, 695.

________ and D. J. Humphries (1975) Armalcolite crystallization, phenocryst assemblages, eruption conditions and origin of eleven high titanium basalts from Taurus-Littrow (abstr.). In, *Lunar Science VI*, p. 616-618. The Lunar Science Institute, Houston.

Papike, J. J., A. E. Bence, and D. H. Lindsley (1974) Mare basalts from the Taurus-Littrow region of the moon. *Proc. Lunar Sci. Conf. 5th, 1*, 471.

Prinz, M., E. Dowty, K. Keil, and T. E. Bunch (1973a) Mineralogy, petrology and chemistry of lithic fragments from Luna 20 fines: Origin of the cumulative ANT suite and its relationship to high-alumina and mare basalts. *Geochim. Cosmochim. Acta 37*, 979.

________, ________, ________, and ________ (1973b) Spinel troctolite and anorthosite in Apollo 16 samples. *Science 179*, 74.

Ramdohr, P. and A. El Goresy (1970) Opaque minerals in the lunar rocks and dust from mare Tranquillitatis. *Science 167*, 615.

Sato, M., N. L. Hickling, and J. E. McLane (1973) Oxygen fugacity values of Apollo 12, 14 and 15 lunar samples and reduced states of lunar magmas. *Proc. Lunar Sci. Conf. 4th*, 1061.

Simons, B. (1974) Zusammensetzung und Phasenbreiten der Fe-Ti-Oxyde in Gleichgewicht mit metallischem Eisen. *Diplomarbeit, Technische Hochschule, Aachen*, 104.

Smyth, J. R. and P. R. Brett (1973) The crystal structure of armalcolites from Apollo 17 (abstr.) *Ann. Meeting Geol. Soc. Am. 5 (7)*, 814.

________ (1974) The crystal chemistry of armalcolites from Apollo 17. *Earth Planet. Sci. Lett. 24*, 262.

Taylor, L. A., G. Kullerud, and W. B. Bryan (1971) Opaque mineralogy and textural features of Apollo 12 samples and a comparison with Apollo 11 rocks. *Proc. Lunar Sci. Conf. 2nd*, 855.

________, R. J. Williams, and R. H. McCallister (1972) Stability relations of ilmenite and ulvöspinel in the Fe-Ti-O system and applications of these data to lunar mineral assemblages. *Earth Planet. Sci. Lett. 16*, 282.

Usselman, T. M. (1975) Ilmenite chemistry in mare basalts, an experimental study. Origin of mare basalts and their implications for lunar evaluation (abstr.). In, *Lunar Science*, 164. The Lunar Science Institute, Houston.

________ and G. E. Lofgren (1976) Phase relations of high-titanium rare basalts as a function of oxygen fugacity (abstr.). *Lunar Science VII*, 888. The Lunar Science Institute, Houston.

Wechsler, B. A., C. T. Prewitt, and J. J. Papike (1975) Structure and chemistry of lunar and synthetic armalcolite (abstr.) In, *Lunar Science VI*, 860. The Lunar Science Institute, Houston.

Williams, R. J. (1971) Reaction constants in the system Fe-MgO-SiO_2-O_2 at 1 atmosphere between 900°C and 1300°C: Experimental results. *Am. J. Sci. 270*, 334.

OPAQUE OXIDE MINERALS in METEORITES

Ahmed El Goresy

Chapter 6

INTRODUCTION

Records of stones falling from the sky can be traced back to classical Chinese and ancient Greek or Latin literature. The stone of Ensisheim in Alsace, France, is the oldest preserved meteorite fall. A stony meteorite weighing 127 kg fell on November 16, 1492 in Ensisheim and the event was recorded by chroniclers of the town in a detailed illustrated description. However, it was only after the meteorite shower of L'Aigle, France which fell on April 26, 1803 that the majority of the scientific community finally accepted the extraterrestrial origin of meteorites, due to the convincing demonstration by J. B. Biot to the members of the Academie Française that the L'Aigle stones fell from the sky. Ever since that time the scientific community all over the world has continuously and carefully recorded meteorite falls. The increasing interest in the study of meteorites, especially in the last 100 years, is mainly due to the fact that these objects document the wide variety of solar system processes, *e.g.*, (1) processes in the solar nebula prior to formation of planets; (2) processes in planet-like bodies, analogous to the earth; and (3) those resulting from collisional events between interplanetary objects creating shock and fragmentation. Little is known about the exact source of meteorites, although it is widely accepted that they probably come from the asteroid belt.

The wide variation in the compositions of the primary objects from which meteorites were derived is well demonstrated by the variations in meteorite compositions. The most commonly used classification of meteorites among petrologists is the scheme of Mason (1962) which is based on that of Prior (1920). This classification is shown in Table EG-5. Chondrites and achondrites could be grouped together as stony meteorites since silicate and oxide minerals comprise the major part of these meteorites. The distinction between the two categories of stony meteorites is straightforward: "chondritic" meteorites contain spherical silicate or oxide objects called chondrules whereas these spherical objects are absent in achondrites. However, a few stones considered as chondrites actually contain no chondrules but are so classed because they are chemically and mineralogically similar to the chondrule-bearing stones of the same type (Mason, 1962). The chondrites are the most abundant of all

Table EG-5. Classification of meteorites.*

Class	Subclass
I. Chondrites	A. Enstatite chondrites B. Olivine-bronzite chondrites C. Olivine-hypersthene chondrites D. Carbonaceous chondrites
II. Achondrites	A. Calcium-poor achondrites 1. Enstatite achondrites (aubrites) 2. Hypersthene achondrites (diogenites) 3. Olivine achondrites (chassignites) 4. Olivine-pigeonite achondrites (ureilites) B. Calcium rich achondrites 1. Augite achondrites (angrites) 2. Diopside-olivine achondrites (nakhlites) 3. Pyroxene-plagioclase achondrites a) Eucrites b) Howardites
III. Stony irons	A. Olivine stony irons (pallasites) B. Bronzite-trydimite stony irons (siderophyres) C. Bronzite-olivine stony irons (lodranites) D. Pyroxene-plagioclase stony irons (mesosiderites)
IV. Irons	A. Hexahedrites B. Octahedrites 1. Coarsest octahedrites 2. Coarse octahedrite 3. Medium octahedrites 4. Fine octahedrites 5. Finest octahedrite C. Nickel-rich ataxites

**From Mason, 1962, 1967*

meteorites. A metal phase occurs in the majority of meteorites; it consists of two nickel-iron alloys: (a) *Alpha iron or kamacite* is an iron-nickel alloy with constant composition of about 5.5% Ni; it crystallizes in a body-centered cubic lattice. (b) *Gamma iron, or taenite* is a nickel-iron alloy of variable composition ranging from about 27 to about 65% Ni; it crystallizes in a face-centered cubic lattice. Most iron meteorites are mixtures of both kamacite and taenite. The classification of iron meteorites into hexahedrites, octahedrites, and Ni-rich ataxites is mainly based on the configuration of kamacite and taenite in a polished etched surface. Hexahedrites consist of large crystals of kamacite with cleavage parallel to the faces of a cube. In octahedrites the orientation of kamacite and taenite bands are parallel to octahedral planes. This structure is also known as "Widmanstätten pattern" after its discoveror, Baron Alois von Widmanstätten. The texture of the pattern is in direct but inverse relationship with the Ni content of the meteorite. The higher the Ni content the finer the octahedral structure. As the nickel content of octahedrites increases, the bands of kamacite become narrower. At 12-14% Ni, they become extremely narrow and discontinuous, and the Widmanstätten pattern disappears. Meteorites of this type are classified as nickel-rich ataxites.

In stony meteorites, chondrites and achondrites, opaque oxides occur as accessory minerals usually evenly dispersed in the silicate groundmass. In chondrites the abundance, assemblages, and textures of opaque oxide minerals in chondrules (*e.g.*, in many ordinary chondrites) or in Ca, Al-rich inclusions (*e.g.*, carbonaceous chondrites) can vary drastically from those in the groudmass of the same meteorite. In iron meteorites, opaque oxides are extremely rare compared to the mass of meteorites in which they are encountered. The opaque oxides occur intergrown with silicates in silicate-rich inclusions or with troilite (stoichiometric FeS) and graphite which also occurs as macroscopic inclusions in the NiFe alloy groundmass.

Members of the spinel group (chromite, spinel, and to a lesser extent magnetite) are the most abundant opaque oxides in meteorites. Ilmenite and rutile constitute only a small fraction of the oxides. Members of the pseudobrookite series have never been encountered in meteorites. Enstatite chondrites and enstatite achondrites appear to be barren of opaque oxides.

MINERALOGY

Spinel group minerals

Chromite is by far the most abundant end member of the spinel series in meteorites. Chromite occurs in various quantities in more than 90% of all stony meteorites (chondrites and achondrites) and is the most abundant oxide accessory in mesosiderites, palasites, and iron meteorites (El Goresy, 1965). However, it is rare or almost absent in carbonaceous chondrites. Instead, the groundmass of carbonaceous chondrites is especially enriched in magnetite (Jedwab, 1971; Ramdohr, 1973). In a few calcium-rich achondrites (Nakhlites) titanomagnetite is the dominant spinel (Bunch and Reid, 1975; Boctor *et al.*, 1976). Spinel ($MgAl_2O_4$) occurs as a minor component especially in a few chondrules in ordinary chondrites (Ramdohr, 1973); however, the mineral is a major constituent of Ca and Al-rich inclusions in some carbonaceous chondrites (Sztrokay, 1960; Christophe Michel-Lévy, 1969; Marvin *et al.*, 1970). Many of these Ca- and Al-rich inclusions enriched in melilite, spinel, pyroxene, and perovskite are widely accepted as high-temperature condensates from the primordial solar nebula (Grossman, 1975).

Based on textures and assemblages, Ramdohr (1967,1973) recognized the following various types of chromites in ordinary chondrites: (1) coarse chromite; (2) clusters of chromite aggregates; (3) pseudomorphous chromite; (4) exsolution chromite; (5) chromite chondrules, and (6) myrmekitic (or symplectitic) chromite. The number of these groups would increase markedly, if subtle morphological details were taken into account.

1. Coarse chromite: This chromite type occurs in the overwhelming majority of ordinary chondrites, achondrites, stony irons, and iron meteorites. It is present as coarse euhedral to subhedral grains interlocked in the silicate matrix. Coarse chromite exhibits idiomorphic features only if in contact with troilite or metallic iron (Ramdohr, 1973). Its boundaries to silicates are usually anhedral. According to Ramdohr (1973), coarse chromite is later than olivine and pyroxene in the crystallization sequence. Planimetric integration of 73 chondrites (Keil, 1962) indicates that the chromite content of chondrites varies between 0.01 and 0.61 wt %. No apparent relationship was found between the chromite content and the chondrite type (L = low iron, or H = high iron groups). Exsolution lamellae of ilmenite along (111) planes of coarse chromite is a rare feature in chondrites, but not rare in chromites in mesosiderites or pallasites (Fig. EG-31).

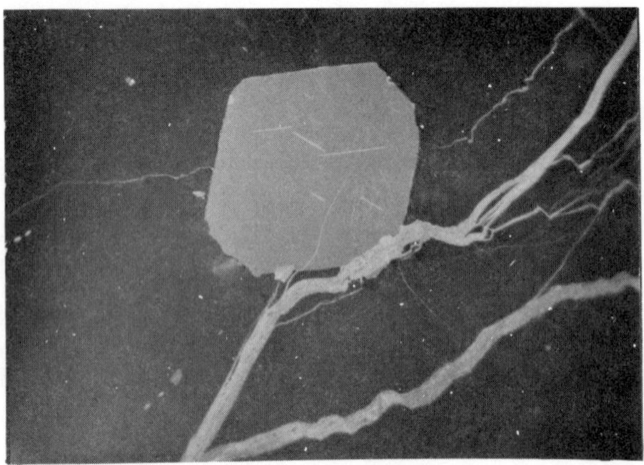

Figure EG-31. Mount Padbury (stony iron), Australia. A grain of coarse chromite in pyroxene with rutile exsolution lamellae (white); length of photograph 150 microns (from Ramdohr, unpublished).

2. Clusters of chromite aggregates: This type seems to be restricted to ordinary chondrites. The cluster consists of medium- to fine-grained idiomorphic to subhedral grains of chromite usually embedded in plagioclase (Fig. EG-32). This chromite type is also later in the crystallization sequence than olivine and pyroxene.

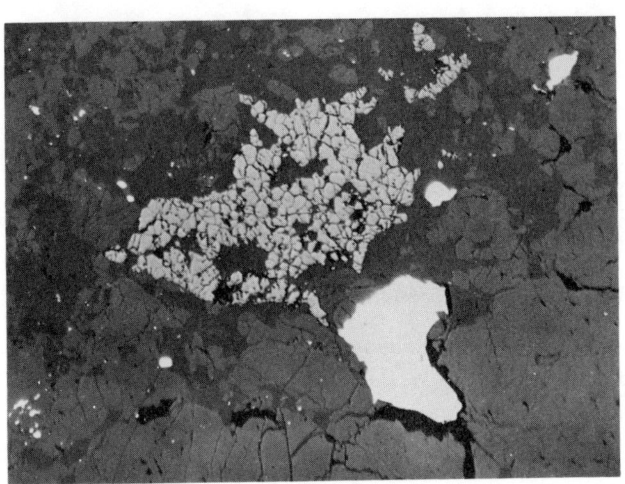

Figure EG-32. Nardoo (olivine-bronzite chondrite), New South Wales. Cluster of aggregate chromite, white is kamacite; length of photograph 350 microns (from Ramdohr, 1973).

---

3. Pseudomorphous chromite: This chromite type appears to be restricted to chondrules in ordinary chondrites. It is absent in achondrites, stony irons, and iron meteorites. The assemblage in which this type occurs consists chiefly of albitic plagioclase and chromite, and sometimes minor clinopyroxene (Ramdohr, 1967,1973). The chromite content of such chondrules may be as high as 40% by volume. This chromite occurs as fine grains (less than 10 microns in diameter) frequently oriented in fan-shaped laths of a former unknown mineral which broke down to albite + chromite (Fig. EG-33) (Ramdohr, 1973). The texture is indeed characteristic for breakdown. Ramdohr (1976) suggested the precursor is very probably ureyite (kosmochlor), $NaCrSi_2O_6$, to account for the 1:3 ratio of chromite to albite. Yoder and Kullerud (1971) report that at 700°C and 2 kb albite + chromite, as well as albite + eskolaite (Cr_2O_3), were found to be stable. However, a mixture of kosmochlor + anorthite + enstatite reacted at the same conditions to form an albitic plagioclase + eskolaite + chromite + clinopyroxene.

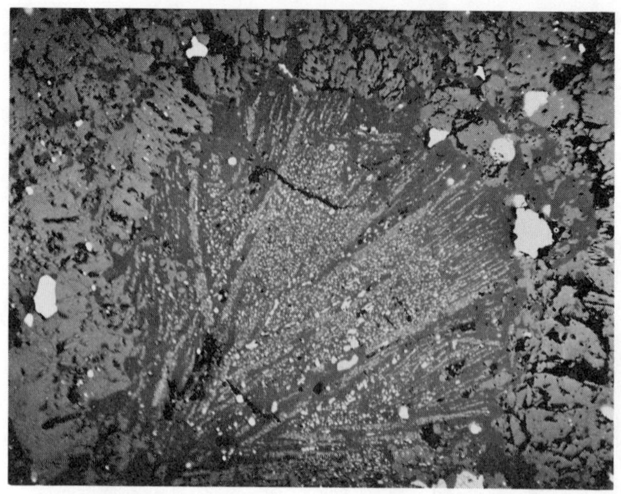

Figure EG-33. Bachmut (olivine-hypersthene chondrite), Ukraine. A chondrule with pseudomorphous chromite in lath-shaped radiating pseudomorphs in albite (dark gray) matrix; length of photograph 1.2 mm (from Ramdohr, 1967).

Yoder and Kullerud (1971) thus propose that an explanation for the assemblage chromite + albite (+ minor clinopyroxene) could be that kosmochlor is a stable product or metastable quench product that reacts with anorthite as well as bronzite and possibly olivine to form albitic plagioclase, chromite, and a clinopyroxene.

4. Exsolution chromite: This type is encountered in clinopyroxene-rich chondrules as fine-grained clusters of chromite at the boundaries between clinopyroxene and plagioclase. Ramdohr (1973) proposes exsolution from a chromium-rich silicate to account for this assemblage.

5. Chromite chondrules: Though uncommon in chondrites, the chromite type is spectacular with regard to its possible origin and the chemical variation of chondrules. This type of chondrule was first discovered by Ramdohr (1967). The chondrules consist usually of a two-phase assemblage: chromite and plagioclase. The chromite:plagioclase ratio can vary drastically, and in some cases almost pure chromite chondrules are encountered, *e.g.*, Loot and Burdette meteorites (Ramdohr, 1967). The chromite occurs in concentric alternating layers of different grain size, compactness, and plagioclase content. The outermost shell of the chondrule is usually surrounded by a thin feldspar layer of fairly

uniform thickness, overgrown by FeNi alloy and troilite. Some chondrules exhibit an atoll-like feature with a plagioclase core and a uniform chromite shell, *e.g.*, Harrisonville chondrite. A very common type of chromite chondrule is characterized by the presence of large idiomorphic chromite crystals in a plagioclase groundmass which, in turn, is loaded with numerous small anhedral chromite grains dispersed throughout the plagioclase (Fig. EG-34). The chondrule is sealed by an almost continuous shell of chromite, which in turn is separated from the meteorite groundmass by a continuous layer of plagioclase (Fig. EG-34).

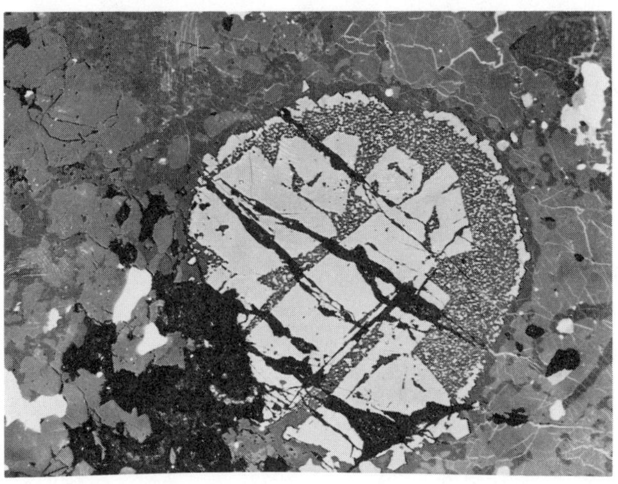

Figure EG-34. Plainview (polymict olivine-bronzite chondrite), Hale County, Texas. Chromite chondrule with idiomorphic chromite crystals, fine-grained chromite-plagioclase intergrowth, and chromite rim; length of photograph 450 microns (from Ramdohr, 1967).

The amount of chromite usually exceeds 50% by volume of the chondrule. A less common type is shown in Figure EG-35. Although chromite and plagioclase are the major constituents, clinopyroxene occurs in minor amounts. A major part of the chromite occurs as skeletal crystals, indicative of quenching due to rapid cooling. These chondrules also contain pseudomorphous chormite, indicating a genetic link between the two chromite types (Fig. EG-35). The predominance of chromite and plagioclase in these chondrules document the unusual composition of the liquid from which the chondrules were derived.

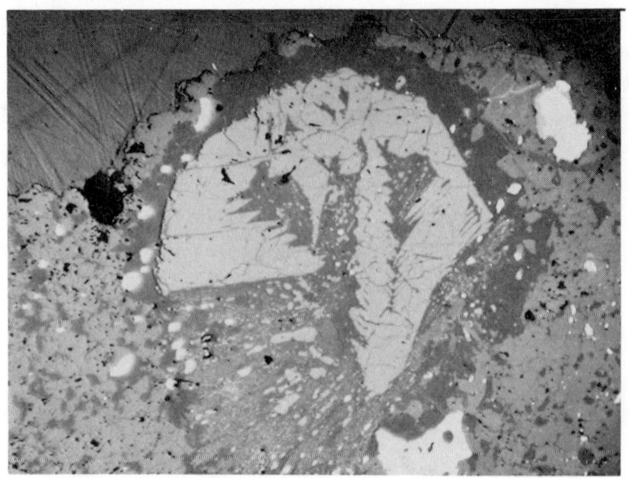

Figure EG-35. Mangwendi (polymict olivine-hypersthene chondrite), Rhodesia. A chromite chondrule with skeletal crystals of chromite and pseudomorphous chromite; length of photograph 450 microns (from Ramdohr, 1973).

---

6. Myrmekitic (symplectitic) chromite: This type seems to be restricted to mesosiderites and iron meteorites. The assemblage consists of major olivine and chromite (Fig. EG-36). So far, neither plagioclase nor pyroxene has been observed (Ramdohr, 1973). Texture and assemblage are indicative of eutectic intergrowth of olivine and chromite.

Magnetite is by far the major opaque oxide in the groundmass of numerous carbonaceous chondrites. Usually, it occurs as reaction rims around metallic FeNi alloy or troilite (Ramdohr, 1973). However, the main mass is present as spherules of various sizes dispersed in the groundmass (Fig. EG-37). A detailed study with the scanning electron microscope (Jedwab, 1971) indicates that these magnetite spherules are in fact composed of stacking platelets or spirals suggestive of condensation from the vapor phase. This interpretation would indicate that these magnetites are one of the latest components to condense directly from the cooling solar nebula.

Titanomagnetites occasionally occur in Ca-rich achondrites (Bunch and Reid, 1975; Boctor *et al.*, 1976); they usually exhibit ilmenite lamellae parallel to (111) of the magnetite and very fine lamellae of ulvöspinel parallel to (100) of the magnetite host. These features document the close similarity between nakhlites and many terrestrial igneous rocks (Boctor *et al.*, 1976).

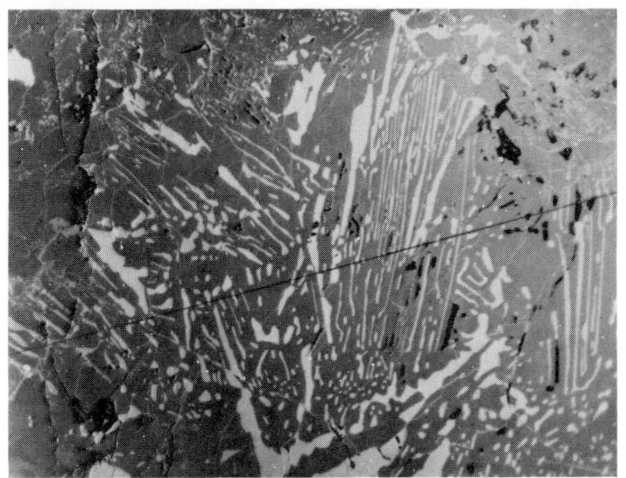

Figure EG-36. Vaca Muerta (mesosiderite), Chile. Myrmekitic (symplectic) intergrowth between chromite and olivine; length of photograph 800 microns (from Ramdohr, unpublished).

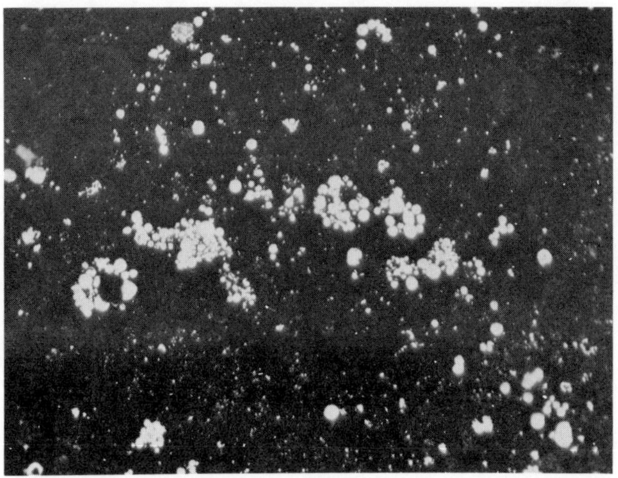

Figure EG-37. Esebi (carbonaceous chondrites), Zaire. Magnetite spherules in carbonaceous and silicate groundmass; length of photograph 300 microns (from Ramdohr, unpublished).

Ilmenite-geikielite-pyrophanite series

Members of this series were reported from ordinary chondrites, stony irons, and iron meteorites (Ramdohr, 1963,1973; Snetsinger and Keil, 1969; El Goresy, 1965; Bunch and Keil, 1971). They seem to be absent in carbonaceous chondrites and are very rare in achondrites. In nakhlites, $ilmenite_{ss}$ usually occurs as lamellae along with ulvöspinel in titanomagnetite (Boctor *et al.*, 1976). Only in very few iron meteorites is it also present as exsolution lamellae in coarse chromite. In contrast to terrestrial igneous and metamorphic rocks, meteoritic ilmenite never displays titanohematite lamellae. Rutile lamellae are very rare but reported to be frequent in some mesosiderites, *e.g.*, Vacu Muerta (Ramdohr, 1973). Many of the ilmenites reported from meteorites were found to exhibit twin lamellae presumably formed by shock due to meteorite collisions in space (Fig. EG-38) (Ramdohr, 1973).

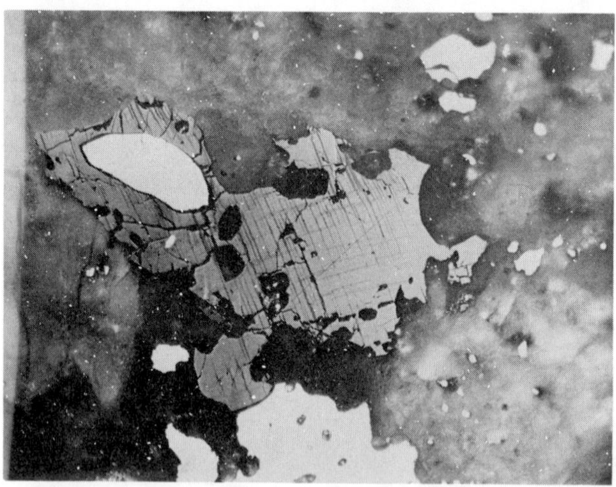

Figure EG-38. Adlie Land (olivine-hypersthene chondrite), Antarctica. Xenomorphic ilmenite with shock-induced twin lamellae, white is troilite; length of photograph 600 microns (from Ramdohr, 1973).

The amount of the components geikielite ($MgTiO_3$) and pyrophanite ($MnTiO_3$) in ilmenite is usually higher in chondritic ilmenites than in nonchondritic (achondritic, stony irons) ilmenites (Snetsinger and Keil, 1969; Bunch and Keil, 1971). The FeO content of ilmenite in ordinary chondrites tends to increase with increasing FeO/(FeO+MgO) in coexisting olivine, but MgO tends to decrease (Snetsinger and Keil, 1969).

Rutile

Rutile is quite rare in ordinary chondrites and achondrites and seems to be absent in carbonaceous chondrites. However, it is a frequent accessory mineral in the majority of mesosiderites (Ramdohr, 1965; Marvin and Klein, 1964) and many pallasites. The mineral is also not rare in silicate inclusions in iron meteorites (El Goresy, 1965,1971). Busek and Keil (1966) report that among chondrites, rutile is most abundant in the Farmington chondrite. In chondrites and stony irons the mineral has a colorless to pale blueish appearance in reflected light with frequent internal reflections. Both in the Farmington chondrite and Vaca Muerta mesosiderite it occurs as lamellae in ilmenite (Ramdohr, 1965; Busek and Keil, 1966). However, in several mesosiderites rutile lamellae in chromite were also reported (Ramdohr, 1965; Busek and Keil, 1966). Ramdohr (1964) suggested that rutile occurring in the assemblage rutile-ilmenite-αFe^o was probably formed due to the subsolidus reaction

$$2FeTiO_3 \rightarrow 2TiO_2 + 2Fe^o + O_2$$

El Goresy (1965,1971) reports that rutile present in iron meteorites has anomalous optical properties with a greenish color in reflected light. These rutiles are characterized by a relatively high Cr_2O_3 (1.23%) and NbO_2 (2.93%) contents.

OXIDE ASSEMBLAGES IN VARIOUS METEORITE GROUPS

Chondrites

The classification of chondritic meteorites in the four subclasses given in Table EG-5 is based mainly on the mineralogical composition of the meteorites. The abundance of Fe, Mg, Si, and O altogether in chondrites is very high and may total 90% in the majority of chondrites (Wasson, 1974). As mentioned before, in contrast to terrestrial rocks the majority of meteorites contain metallic FeNi alloy. Prior (1916) noted that "the less the amount of Ni-Fe in chondritic stones, the richer it is in Ni, and the richer in Fe are the magnesium silicates." These two relationships are known as Prior's rules. From the petrological point of view, Prior's rules are expressions of the f_{O_2} conditions under which meteorites were formed, since the prevailing oxygen fugacity at a given temperature will control the partitioning of iron between the coexisting silicates and the metal phase. In reduced chondrites, *e.g.*, enstatite chondrites, most of the iron is present in the metal phase, and the Ni content of the metal is correspondingly low. With increasing degree of oxidation more of the iron is oxidized; the iron content of the silicates increases, and since the amount of metal decreases while the amount of Ni remains constant, the Ni content of the metal increases. Prior's rules should be considered as qualitative and hold only for ordinary chondrites (but not for enstatite or carbonaceous chondrites) because the fractionation in the Fe/Si ratio discovered by Urey and Craig (1953) demonstrated that the ordinary chondrites could not be understood in terms of such a simple model. Urey and Craig (1953) demonstrated the existence of two groups of chondrites, one having a low total iron content (22.33 wt %), the other a high content (28.58 wt %) which they designated the *L group* and the *H group*, respectively. Determination of the fayalite content of olivine in some 800 chondrites (Mason, 1963) and the iron and magnesium distribution in coexisting olivines and rhombic pyroxenes (Keil and Fredriksson, 1964) established the existence of three major groups of ordinary chondrites: (1) high iron (H)-group (olivine-bronzite chondrites); (2) low iron (L)-group (olivine-hypersthene chondrites); (3) low iron-low metal (LL)-group (olivine-hypersthene chondrites). Figures EG-39 and EG-40 demonstrate the existence of these three groups as

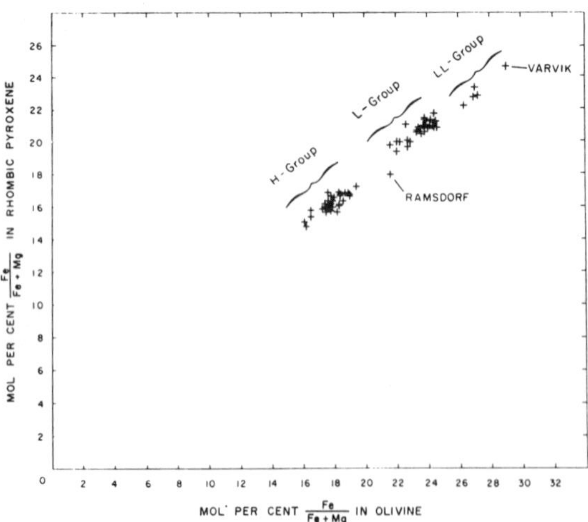

Figure EG-39. Ratios of Fe/(Fe+Mg) in olivine plotted against ratios of Fe/(Fe+Mg) in rhombic pyroxene for 86 chondrites (from Keil and Fredriksson, 1964).

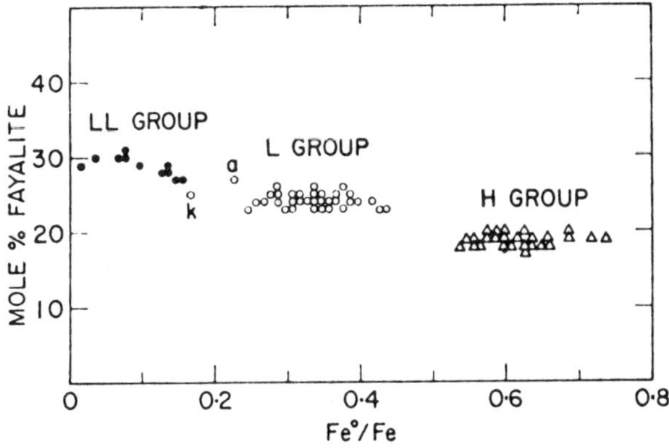

Figure EG-40. Plot of mole percent of fayalite in olivine versus Fe^{O}/Fe ratios for ordinary chondrites (from Van Schmus and Wood, 1967).

---

obtained from the distribution of Fe and Mg among coexisting olivines and pyroxenes (Fig. EG-39) and the mole percent fayalite in olivine versus Fe^{O}/Fe ratios in the meteorite. This classification, however, does not account for the degree of equilibration among silicates. Van Schmus and Wood (1967) introduced a simplified classification using two parameters: chemical and petrologic (degree of equilibration). Figures EG-41 and EG-42 summarize this two parameter classification. Increasing number indicates increase in the degree of equilibration, *e.g.*, LL3 means low to medium equilibrated LL chondrite, whereas LL6 is a highly equilibrated LL chondrite.

Spinels. Bunch *et al.* (1967) carried out a detailed electron microprobe survey of chromite chemistry in ordinary chondrites as a function of chemistry and texture (equilibration) of the chondrites. Their study, however, is restricted to the coarse chromite, and hence the variability in chemistry of the other chromite types described by Ramdohr (1967,1973) is not known. The compositional variability of chromites in equilibrated chondrites (subgroups 5 and 6) was found to be within the precision of the method except for Al and Ti. The largest compositional variabilities were observed in chromite in unequilibrated chondrites (subgroups 3). Zoning in chromites in unequilibrated chondrites is also a frequent feature (Bunch *et al.*, 1967). These results indeed support the classification introduced by Van Schmus and Wood (1967). A direct correlation exists between chromite composition and the classification of

Petrologic type

Chemical group	1	2	3	4	5	6
E	E1 —	E2 —	E3 **1***	E4 **4**	E5 **2**	E6 **6**
C	C1 **4**	C2 **16**	C3 **8**	C4 **2**	C5 —	C6 —
H	H1 —	H2 —	H3 **7**	H4 **35**	H5 **74**	H6 **44**
L	L1 —	L2 —	L3 **9**	L4 **18**	L5 **43**	L6 **152**
LL	LL1 —	LL2 —	LL3 **4**	LL4 **3**	LL5 **7**	LL6 **21**

* Number of examples of each meteorite type now known is given in its box.

Figure EG-41. Classification of chondrites. Number of each meteorite type now known is given in its box (from Van Schmus and Wood, 1967).

Petrologic type

	1	2	3	4	5	6	
E	*	*	Enstatite chondrites				
C	Carbonaceous chondrites			C4	*	*	
H	*	*	H3‡	Bronzite chondrites			†
L	*	*	L3‡	Hypersthene chondrites			†
LL	*	*	LL3‡	Amphoteric chondrites			†
				†			

* Unpopulated
† Ordinary chondrites.
‡ Unequilibrated ordinary chondrites.

Figure EG-42. Location of meteorite type in the petrological chemical classification (from Van Schmus and Wood, 1967).

---

equilibrated chondrites into H, L, and LL groups. From Table EG-6 it is apparent that FeO and TiO_2 increase and Cr_2O_3, MgO, and MnO decrease from H to L to LL groups. Bunch *et al.* (1967) did not include chromites from subgroups 3 and 4 since they were found to vary in composition and are apparently not in equilibrium with the silicates. Correlation between chromite composition and classification of equilibrated chondrites becomes more evident if major oxide contents are compared to the iron oxide content of coexisting olivine expressed as mole percent FeO/(FeO+MgO) (Fig. EG-43). In most cases, H, L, and LL groups are well separated. Both FeO and TiO_2 in chromite increase from subgroups 3 to

Table EG-6

Composition of chromite from equilibrated chondrites.*

	H group (10)**	L group (7)	LL group (6)
Cr_2O_3	56.9	56.1	54.4
Al_2O_3	5.9	5.3	5.7
V_2O_3	0.68	0.72	0.73
TiO_2	2.33	2.81	3.23
FeO	31.2	33.0	34.5
MgO	2.66	1.99	1.62
MnO	0.94	0.74	0.63
Total	100.61	100.66	100.81

* *Analyses for subgroups 5 and 6; all from Bunch et al. 1967, Table 1, p.1571*

** *Numbers in parentheses indicate number of meteorites analysed*

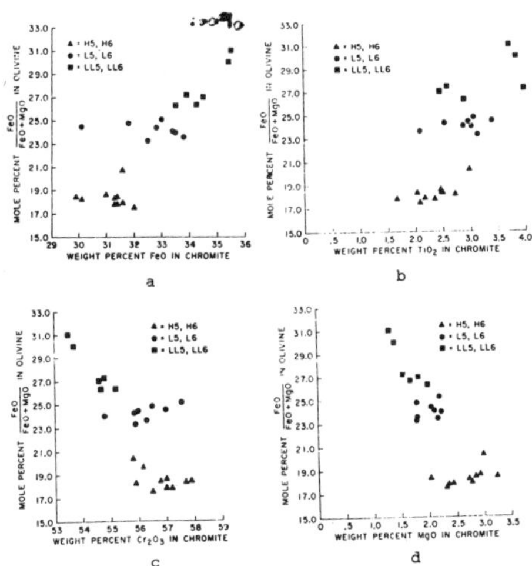

Figure EG-43. Correlation between chromite composition and classification of equilibrated H-, L-, and LL-group chondrites (subgroups 5 and 6). With increase of FeO/(FeO+MgO) in coexisting olivine, FeO and TiO_2 of chromite increase (a and b) and Cr_2O_3 and MgO decrease (c and d) (from Bunch *et al.*, 1967).

5 in H and L groups. However, FeO in chromite decreases slightly from subgroups 5 to 6, whereas TiO_2 continues increasing (Bunch *et al.*, 1967). The classification of Van Schmus and Wood (1967) implies that their petrographic subgroups within a major chondritic group (*e.g.*, H3 through H6) represent metamorphic series, *i.e.*, mineral compositions in H6 chondrites were established by metamorphism of H3-type assemblages after chondrite formation. Keil and Fredriksson (1964), Ramdohr (1967), and many others indicate, however, that chemical and mineralogical equilibration may well be attained during a primary process (*e.g.*, by slow cooling). However, Bunch *et al.* (1967) indicate that relations between chromite compositions and subgroup classification are consistent with either a metamorphic or a primary origin of chondrites.

Spinels found in Ca-, Al-rich inclusions in carbonaceous chondrites coexist with a Ti-rich fassaite, gehlenite, anorthite and perovskite (sometimes with hibonite, $Ca_2((Al,Ti)_{24}O_{38})$ (Marvin *et al.*, 1970; Keil and Fuchs, 1971). Two types of coarse-grained Ca-, Al-rich inclusions with spinel as a major constituent were reported from the Allende meteorite (Grossman, 1975). Type A contains 80-85 percent melilite, 15-20 percent spinel and 1-2 percent perovskite. Clinopyroxene, if present, is usually restricted to thin rims around inclusions or surrounding cavities in the interior (Grossman, 1975). Type B contains 35-60 percent clinopyroxene, 15-30 percent spinel, 5-20 percent plagioclase, and 5-20 percent melilite. The main differences between the two types of inclusions are abundance and composition of clinopyroxene. Thermodynamic calculations for phases condensing at 10^{-4} atmospheres in the solar nebula indicate the following sequence: corundum, hibonite, perovskite, gehlenite, spinel, Fe-metal, diopside, forsterite, Ti_3O_5, anorthite, enstatite, rutile, albite, and nepheline (Grossman, 1972).

The spinel-bearing Ca-, Al-rich inclusions are considered to be early condensates from the solar nebula. Spinels occurring in these inclusions in Allende are almost pure $MgAl_2O_4$ but contain traces of FeO, Cr_2O_3, TiO_2, and CaO (Marvin *et al.*, 1970; Grossman, 1975; El Goresy, unpublished). The Cr_2O_3 and most FeO, CaO, and TiO_2 contents are well below 1 percent. No systematic difference in composition was found between spinels enclosed in melilite and those enclosed in pyroxene. Figure EG-44 displays the Cr_2O_3 content versus TiO_2 content of spinels in type A and type B inclusions (Grossman, 1975).

Ilmenite. Snetsinger and Keil (1969) indicate that the average composition of ilmenite from equilibrated ordinary chondrites is, like chromite, related to chondrite classification: FeO increases and MgO and MnO decrease from H to L to LL groups (Table EG-7). These relationships are displayed in Figure EG-45

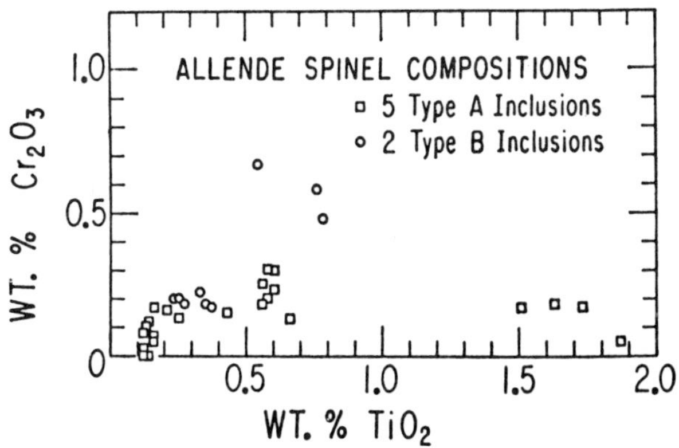

Figure EG-44. Cr_2O_3 and TiO_2 contents of spinels in Allende inclusions. They seldom exceed 1 percent (from Grossman, 1975).

Table EG-7. Ilmenite from equilibrated ordinary chondrites.*

	H 5,6	L 5,6	LL 6
FeO	40.9	41.9	44.2
MgO	4.10	3.30	1.80
MnO	3.20	1.50	1.10
TiO_2	51.7	52.7	51.7
Cr_2O_3	0.09	0.31	0.31
Total	99.99	99.71	99.11

* *From Snetsinger and Keil, 1969, Table 3, p. 784*

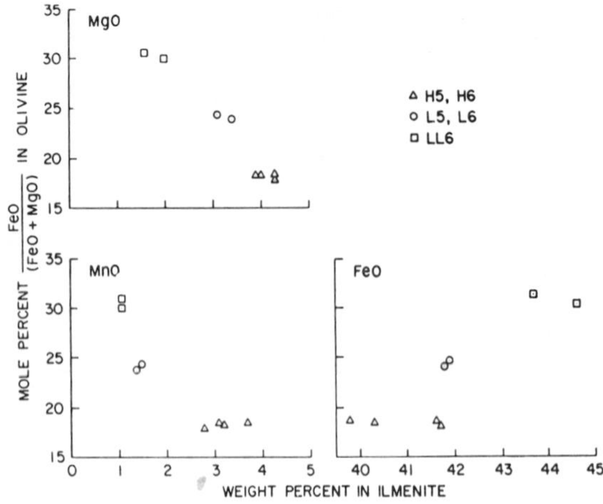

Figure EG-45. Mole percent FeO/(Fe+MgO) in olivine versus weight percent of MgO, MnO and FeO in ilmenite of equilibrated chondrites (from Snetsinger and Keil, 1969).

---

where FeO/(FeO+MgO) in olivine is plotted against MgO, MnO, and FeO in coexisting ilmenite. Although the statistics are poor due to the rarity of ilmenite, compositional trends similar to those observed in chromites can be recognized. The correlation indicates that the four phases olivine, orthopyroxene, chromite, and ilmenite did indeed equilibrate in these chondrites.

Achondrites

Chromite is the most abundant opaque oxide in achondritic meteorites, followed by ilmenite and then rutile (Ramdohr, 1973). Chromite compositions in diogenites, chassignites, and several eucrites have been published by Ramdohr and El Goresy (1969), Bunch and Keil (1971), and Lovering (1975). Spinel compositions in Angra dos Reis (angrite) were recently reported by Keil *et al.* (1976). Chromian titanomagnetites from nakhlites were also reported by Bunch and Reid (1975), and Boctor *et al.* (1976). A detailed study of mineral assemblages in different clasts in the Kapoeta Howardite was recently published by Dymek *et al.* (1976). Bunch and Keil (1971) report compositional variability (particularly for Ti and Al) of chromites from achondrites (without specifying the meteorite group), although Lovering (1975) indicates a complete chemical homogeneity of all chromites he analyzed in the Moama eucrite. Bunch and Keil

(1971) indicate some distinctions between chromites in diogenites and chromites in eucrites: Chromites in diogenites are higher in Al_2O_3 and MgO, whereas chromites in eucrites are higher in TiO_2, FeO, and V_2O_3. Spinel in Angra dos Reis is a magnesian, chromian hercynite (Keil *et al.*, 1976). The howardite Kapoeta was found to contain various basaltic clasts which could be grouped into two broad lithologic types on the basis of modal mineralogy--(a) basaltic (pyroxene- and plagioclase-bearing) and (b) pyroxenitic (pyroxene-bearing) (Dymek *et al.*, 1976). Type b clasts are unusually enriched in opaque oxides (chromite and ilmenite). Spinel compositions in various types of achondrites (other than nakhlites) are given in Table EG-8. Two important features can be recognized in Table EG-8: (1) Chromites in achondrites do not show any correlation due to the strong variability of mineralogy and history of the various subclasses; and (2) comprehensive studies similar to those presented by Dymek *et al.* (1976) of additional achondrites are needed for better understanding of phase petrology.

The nakhlites are characterized, compared to other meteorites, by the presence of titanomagnetite with ilmenite and ulvöspinel exsolution lamellae (Bunch and Reid, 1975; Boctor *et al.*, 1976). Ilmenite in the Lafayette meteorite was found to have broken down to rutile + hematite. Boctor *et al.* (1976) report that the maximum calculated hematite content corresponds to a temperature of about 740°C and f_{O_2} of 10^{-17}. Such values are similar to those obtained for the Skaergaard gabbro. Boctor *et al.* conclude that the phases they reported from the Lafayette meteorite represent the late stages of differentiation of the original magmas and that the parent body from which Lafayette was derived may have undergone major primary differentiation of a nature similar to that believed operative in the early crystallization history of the lunar crust.

Ilmenite. Zoning and minor grain-to-grain compositional variability of ilmenite in several achondrites was reported by Bunch and Keil (1971). Achondritic ilmenite is usually depleted in MgO and MnO compared to ilmenite in ordinary chondrites.

Stony irons

Chromites in pallasites tend to occur as coarse to very coarse grains (*e.g.*, up to a few centimeters in the Brenham pallasite) either coexisting with olivine or completely embedded in the iron mass. They are usually characterized by their low TiO_2 content (0.18% average) but variable Cr_2O_3 (60.5 to 69.0%) and Al_2O_3 (1.5 to 9.1%) contents (Bunch and Keil, 1971). Compositional variability of chromites in mesosiderites is more pronounced than in pallasites. The TiO_2

Table EG-8. Composition of spinels from various achondrites.

	1	2	3	4	5	6	7	8
Cr_2O_3	55.50	46.10	48.90	3.30	38.87	40.25	47.35	46.18
Al_2O_3	10.10	9.80	9.30	54.50	4.92	5.33	7.49	9.19
V_2O_3	0.43	0.28	0.75	0.07	0.58	0.38	0.34	0.24
TiO_2	1.09	3.70	4.10	0.65	10.86	11.01	3.12	5.57
FeO	28.80	36.50	35.70	28.40	40.57	41.45	36.04	36.61
Fe_2O_3	-	-	-	3.40	-	-	-	-
MgO	3.90	2.86	0.59	8.00	0.43	0.71	2.78	1.02
MnO	0.68	0.54	0.59	0.18	1.04	0.95	1.06	0.98
Total	100.50	99.97	99.93	98.50	97.26	100.08	98.26	99.79

1: *Average chromite composition in diogenites (Bunch and Keil, 1971, Table 4, p. 149)*

2: *Chromite composition in Chassigny (chassignites) (Bunch and Keil, 1971, Table 5, p. 150)*

3: *Average chromite composition in eucrites (Bunch and Keil, 1971, Table 4, p. 149)*

4: *Spinel composition in Angra dos Reis (Angrites) (Keil et al., 1976, Table 1, p. 444)*

5: *Chromite composition in basaltic clast A in Kapoeta Howardite (Dymek et al., 1976, Table 1, p. 1117)*

6: *Chromite composition in pyroxenitic clast B in Kapoeta Howardite (Dymek et al., 1976, Table 2, p. 1118)*

7: *Chromite composition in fine grained pyroxenite clast C in Kapoeta Howardite (Dymek et al., 1976, Table 3, p. 1120)*

8: *Chromite composition in fine grained porphyritic basalt clast P in Kapoeta Howardite (Dymek et al ., 1976, Table 4, p. 1121)*

content of chromites in mesosiderites is relatively higher than that of chromites in pallasites (Bunch and Keil, 1971). Table EG-9 shows a comparison between chromite compositions in pallasites, mesosiderites and lodranites. Compared to chromites in other stony irons, chromites present in lodranites are unique due to their high ZnO content (0.78-1.68%). In this respect, they are analogous to chromites in iron meteorites.

Compositions of ilmenites in mesosiderites are generally similar to ilmenite compositions in achondrites. However, ilmenites in mesosiderites tend to be higher in MgO, MnO, and Cr_2O_3 than ilmenites in achondrites (Bunch and Keil, 1971).

Iron meteorites

Chromites in iron meteorites occur both in the metal groundmass and together with silicates in troilite and graphite inclusions (El Goresy, 1965). In many troilite inclusions it coexists with the thiospinel daubreelite, $FeCr_2S_4$ (El Goresy, 1965). This documents the chalcophile and lithophile behavior of chromium in the same meteorite. In the metal groundmass chromites *always* form perfect euhedral crystals with sharp crystal faces to the iron. In troilite inclusions with silicates, chromite exhibits idiomorphic boundaries only against

Table EG-9. Composition of chromites in pallasites, mesosiderites, and lodranites.

	1	2	3
Cr_2O_3	64.00	52.00	61.83
Al_2O_3	5.60	11.50	4.43
V_2O_3	0.54	0.54	0.46
TiO_2	0.18	1.84	0.91
FeO	23.20	31.00	22.28
MgO	5.80	2.29	6.47
MnO	0.65	0.77	1.10
ZnO	-	-	1.27
Total	100.13	99.94	98.75

1: Average chromite composition in pallasites (Bunch and Keil, 1971, Table 7, p. 152)
2: Average chromite composition in mesosiderites (Bunch and Keil, 1971, Table 7, p. 152)
3: Average chromite composition in Lodran (lodranite) (Bild, unpublished analyses).

troilite but not against silicates (Fig. EG-46). This feature may indicate equilibration between chromite and coexisting rhombic pyroxene and olivine in the early history of the meteorite. Bunch *et al.* (1970) classified silicate inclusions in iron meteorites into several groups: (1) Odessa type; (2) Copiapo type; (3) Weekero Station type; and three other "miscellaneous" types: Enon, Kendall County, and Netschaevo. The classification introduced by Bunch *et al.* (1970) is based on mineralogical composition, mineral abundance, texture, and shape of inclusion. Due to the problem of small sample populations, such classifications are difficult and somewhat ambiguous. Chromite is common in Odessa and Toluca (El Goresy, 1965; Bunch *et al.*, 1970) but was not found in several members of the Copiapo-type inclusions. Chemistry of chromites in the majority of iron meteorites is unique compared to chondritic, achondritic, pallasitic, and mesosideritic (except for Lodran) chromites. Usually, they contain appreciable amounts of MnO (up to 4.2%) and ZnO (up to 2.31%). This again puts some constraints on the geochemical behavior of both Mn and Zn in iron meteorites. In the same inclusions these chromites coexist with alabandite, MnS, and sphalerite, ZnS (El Goresy, 1965). This establishes both the chalcophile and lithophile behavior of these two elements in iron meteorites. Compositions of chromites from various iron meteorites are given in Table EG-10.

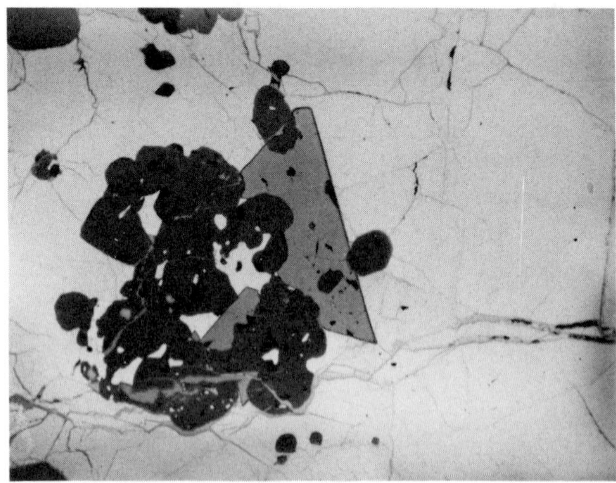

Figure EG-46. Mundrabilla (coarse octahedrite), Australia. Chromite in silicate-bearing troilite nodules. Note sharp boundaries to troilite and subhedral features towards silicates; length of photograph 200 microns (from Ramdohr, unpublished).

Table EG-10. Composition of chromite in various iron meteorites.

	1	2	3	4
Cr_2O_3	69.40	71.90	71.70	68.40
Al_2O_3	2.51	1.24	0.42	10.90
V_2O_3	0.31	0.26	0.57	0.68
SiO_2	0.21	-	-	-
TiO_2	1.01	0.40	0.48	0.02
FeO	7.00	12.60	15.10	2.50
MgO	16.00	10.20	7.10	14.20
MnO	2.22	2.28	3.40	4.20
ZnO	1.37	1.39	1.70	0.02
Total	100.03	100.27	100.47	100.88

1: Average of 25 grains in Mundrabilla (Ramdohr et al., unpublished data).
2: Odessa chromite (Bunch et al., 1970, Table 7, p. 314)
3: Copiapo chromite (Bunch et al., 1070, Table 7, p. 314)
4: Kendall County chromite (Bunch et al., 1970, Table 7, p. 314)

The very high Cr_2O_3 content (except in Kendall County) is striking. In Weekero Station, Colomera, Kodiakanal, Enon, Kendall County, and Netschaevo, chromites are characterized by their relatively high Al_2O_3 content (between 2.68 and 17.6%). Figure EG-47 displays the $Fe^{+2}/(Fe^{+2}+Mg^{+2})$ in olivine versus $Fe^{+2}/$

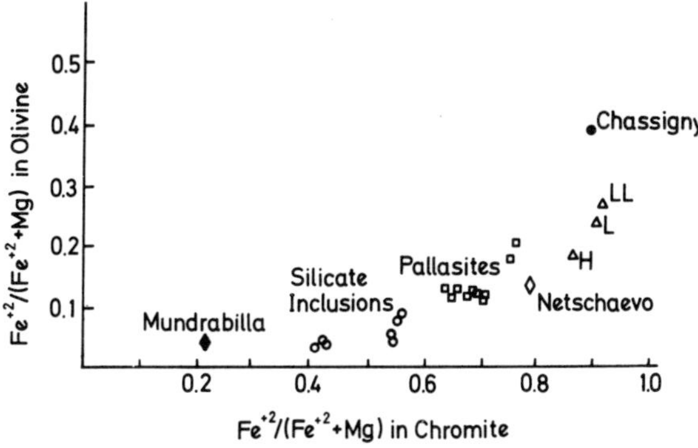

Figure EG-47. $Fe^{+2}/(Fe^{+2}+Mg)$ in olivine versus $Fe^{+2}/(Fe^{+2}+Mg)$ in chromite in Chassigny, LL-, L-, H-chondrites, pallasites, silicate inclusions in iron meteorites (modified from Bunch *et al.*, 1970).

(Fe^{+2}+Mg^{+2}) in coexisting chromite with chondrites, pallasites, Chassigny, and silicate inclusions in iron meteorites. Both chromite and olivine in iron meteorites show the lowest Fe/(Fe+Mg) ratios compared to other meteorite groups. Also evident is the decreasing Fe/(Fe+Mg) trend from Chassigny (achondrite) to LL, L, H chondrites, pallasites, and iron meteorites.

Rutile in iron meteorites occurs as separate grains in silicate-rich troilite inclusions or with chromite in silicate-free nodules (El Goresy, 1965). Rutile in iron meteorites is unique compared to rutiles in chondrites (Busek and Keil, 1966), since El Goresy (1971) found that rutile in iron meteorites is usually enriched in Cr_2O_3 and NbO_2 (Table EG-11).

Table EG-11. Composition of rutile in iron meteorites and Vaca Muerta mesosiderite.*

	1	2	3	4
TiO_2	95.06	95.10	92.48	95.58
NbO_2	2.93	2.89	1.63	0.38
FeO	0.73	0.10	1.00	2.27
MgO	<0.05	<0.05	0.06	0.20
MnO	-	-	<0.05	0.48
Cr_2O_3	1.16	1.15	0.75	0.18
Al_2O_3	-	<0.05	<0.05	<0.05
V_2O_3	<0.05	0.10	0.08	0.06
Total	99.88	99.34	96.00	99.15

**All analyses from El Goresy, 1971, Table 1, p. 360*

REFERENCES

Boctor, N. Z., H. O. Meyer, and G. Kullerud (1976) Lafayette meteorite: Petrology and opaque mineralogy. *Earth Planet. Sci. Lett.*, in press.

Bunch, T. E., K. Keil, and K. G. Snetsinger (1967) Chromite composition in relation to chemistry and texture of ordinary chondrites. *Geochim. Cosmochim. Acta 31*, 1569.

________, ________, and E. Olsen (1970) Mineralogy and petrology of silicate inclusions in iron meteorites. *Contrib. Mineral. Petrol. 25*, 297.

________, ________ (1971) Chromite and ilmenite in non-chondritic meteorites. *Am. Mineral. 56*, 146.

________, and A. M. Reid (1975) The nakhlites--part I: Petrology and mineral chemistry. *Meteoritics 10*, 303.

Busek, P. R. and K. Keil (1966) Meteoritic rutile. *Am. Mineral. 51*, 1506.

Christophe, M. (1969) Etude minéralogique de la chondrite CIII de Lancé. In, P. M. Millman, Ed., *Meteorite Research*, Reidel, 492.

Dymek, R. F., A. L. Albee, A. A. Chodos, and G. J. Wasserburg (1976) Petrography of isotopically-dated clasts in the Kapoeta howardite and petrologic constraints on the evolution of its parent body. *Geochim. Cosmochim. Acta 40*, 1115.

El Goresy, A. (1965) Mineralbestand und Strukturen der Graphit- und Sulfideinschlüsse in Eisenmeteorite. *Geochim. Cosmochim. Acta 29*, 1131.

________ (1971) Meteoritic rutile: A niobium bearing mineral. *Earth Planet. Sci. Lett. 11*, 359.

Grossman, L. (1972) Condensation in the primitive solar nebula. *Geochim. Cosmochim. Acta 36*, 597.

________ (1975) Petrography and mineral chemistry of Ca-rich inclusions in the Allende Meteorite. *Geochim. Cosmochim. Acta 39*, 433.

Jedwab, J. (1971) La magnétite de la metéorite d'Orgueil vue en microscope électronique à Balayage. *Icarus 15*, 319.

Keil, K. (1962) Quantitiv-erzmikroskopische Integrationsanalyse der Chondrite. *Chemie der Erde XXII*, 281.

________, and K. Fredriksson (1964) The iron, magnesium and calcium distribution in coexisting olivines and rhombic pyroxenes of chondrites. *J. Geophys. Res. 69*, 3486.

________, and L. H. Fuchs (1971) Hibonite $Ca_2(Al,Ti)_{24}O_{38}$ from the Leoville and Allende chondritic meteorites. *Earth Planet. Sci. Lett. 12*, 184.

________, M. Prinz, P. F. Hlava, C. B. Gomes, W. S. Curvello, G. J. Wasserburg, F. Tera, D. A. Papanastassiou, J. C. Huneke, A. V. Murali, M. S. Ma, R. A. Schmitt, G. W. Lugmair, K. Marti, B. Scheinin, and R. N. Clayton (1976) Progress by the consorts of Angra dos Reis (abstr). In, *Lunar Science VII*, Lunar Science Institute, Houston, 443.

Lovering, J. F. (1975) The Moama eucrite--a pyroxene-plagioclase adcumulate. *Meteoritics 10*, 101.

Marvin, U. B. and C. Klein (1964) Meteoritic zircon. *Science 146*, 919.

________, J. A. Wood, and J. S. Dickey, Jr. (1970) Ca-Al rich phases in the Allende meteorite. *Earth Planet. Sci. Lett. 7*, 346.

Mason, B. (1962) *Meteorites.* John Wiley and Sons, Inc.

________ (1963) Olivine compositions in chondrites. *Geochim. Cosmochim. Acta 27*, 1011.

________ (1967) Meteorites. *Am. J. Sci. 55*, 429.

Prior, G. T. (1916) On the genetic relationship and classification of meteorites. *Mineral. Mag. 18*, 26.

________ (1920) The classification of meteorites. *Mineral. Mag. 19*, 51.

Ramdohr, P. (1964) Einiges über Opakerze in Achondriten und Enstatitachondriten. *Abhandl. deutsch. Akad. Wiss. Berlin 5*, 1.

________ (1965) Über den Mineralbestand von Mesosideriten und Pallasiten und einige genetische Überlegungen. *Monatsber. deutsch. Akad. Wiss. Berlin I*, 923.

________ (1967) Chromite and chromite chondrules in meteorites--I. *Geochim. Cosmochim. Acta 31*, 1961.

________ (1973) *The Opaque Minerals in Stony Meteorites.* Elsevier Press.

________, and A. El Goresy (1969) Peckelsheim. A new bronzite achondrite. *Naturwiss. 56*, 512.

________, M. Prinz, and A. El Goresy (1975) Silicate inclusions in the Mundrabilla meteorite. *Meteoritics 10*, 477.

Snetsinger, K. G. and K. Keil (1969) Ilmenite in ordinary chondrites. *Am. Mineral. 54*, 780.

Sztrokay, K. (1960) Über einige Meteroitenmineralien des kohlenwasserstoffhaltigen Chondriten von Kaba, Ungarn. *Neues Jahrb. Mineral. Abh. 94*, 1284.

Urey, H. C. and H. Craig (1953) The composition of the stone meteorites and the origin of the meteorites. *Geochim. Cosmochim. Acta 4*, 36.

Van Schmus, W. R. and J. A. Wood (1967) A chemical-petrologic classification for the chondritic meteorites. *Geochim. Cosmochim. Acta 31*, 747.

Wasson, J. T. (1974) *Meteorites.* Springer-Verlag, Berlin-Heidelberg-New York.

Yoder, H. S., Jr. and G. Kullerud (1971) Kosmochlore and the chromite-plagioclase association. *Annu. Rep. Dir. Geophys. Lab., Year Book 69*, 155.

The MANGANESE OXIDES - A BIBLIOGRAPHIC COMMENTARY

J. Stephen Huebner

Chapter 7

INTRODUCTION

The rock- and ore-forming manganese oxides form large deposits which have considerable economic importance as sources of manganese for batteries and steel, as well as numerous and widespread small concentrations of manganese that are not now economic. For these reasons--the economic importance and the almost ubiquitous occurrences--it is appropriate that manganese oxides be treated in a short course on oxides. The bibliography which follows will introduce the reader to the mineralogy of the simple oxides of manganese. This review is divided into sections on Crystal Structures and Chemistry; Geologic Occurrences; Thermochemistry and Phase Relations; and Petrology of Manganese Oxides. It is an objective of this review to reveal gaps in the knowledge of manganese oxides and to stimulate further research. For instance, the phase relations and geologic conditions of formation of the compositionally variable and complex "higher oxides" of manganese are not reviewed here because, despite their great value as ores, virtually nothing is known about their relative stabilities.

Crystal structures and chemistry

Tetravalent oxides. The structural relationships of the tetravalent manganese oxides are reviewed by Burns and Burns (1975), who point out that, as in the silicates, these various structures can be derived by connecting basic units [$(MnO_6)^{-8}$ octahedra] in different ways. Pyrolusite, a polymorph of the compositionally simple MnO_2, has the simplest structure (TiO_2 type) with single chains of $(MnO_6)^{-8}$ octahedra; ramsdellite, another polymorph of MnO_2, has double chains. The other tetravalent oxides have more complicated and open structures permitting non-stoichiometry and the substitution of a variety of cations which are commonly large, *e.g.*, hollandite, $BaMn_8O_{16}\cdot xH_2O$; cryptomelane, $KMn_8O_{16}\cdot xH_2O$; coronadite, $PbMn_8O_{16}\cdot xH_2O$; chalcophanite, $ZnMn_3O_7\cdot 3H_2O$; romanechite (psilomelane), $(Ca,Mn)Mn_4O_9\cdot 3H_2O$; birnessite, $(Ca,Na)Mn_7O_{14}\cdot 3H_2O$.

The chemistry of the naturally occurring tetravalent oxides reflects their crystal structures. The structurally simple and compact pyrolusite (Strunz, 1943) and ramsdellite (Bystrom, 1949) have compositions close to MnO_2. Nsutite, another form of "MnO_2," is a random structural intergrowth of pyrolusite and ramsdellite units (de Wolff, 1959; Giovanoli *et al.*, 1967), the structural disorder permitting the observed nonstoichiometry and variety of x-ray patterns (Zwicker *et al.*, 1962). Two forms of MnOOH, manganite (Dachs, 1963) and groutite (Glasser and Ingram, 1968) are similar to pyrolusite and ramsdellite, respectively. The relationships between composition and structure of the more complex tetravalent manganese oxides are reviewed by Burns *et al.* (1975).

Trivalent oxides. Partridgeite (α-Mn_2O_3) and bixbyite (α-$(Fe,Mn)_2O_3$) were originally believed to have the rare-earth, C-type cubic structure related to that of fluorite (Zachariasen, 1928; Wells, 1950). Hase *et al.* (1967) pointed out that bonding in pure α-Mn_2O_3 was unlike other oxides in the group, and Norrestam (1967) showed that α-Mn_2O_3 was actually orthorhombic. Geller *et al.* (1967), Hase *et al.* (1969), and Geller (1971) have shown clearly that whereas Mn_2O_3 (ideal partridgeite) is orthorhombic, the presence of small quantities of Fe_2O_3 in solid solution to form $(Fe,Mn)_2O_3$ stabilizes the cubic structure. Naturally occurring partridgeite contains some iron and is probably cubic; if so, the mineral name partridgeite could be dropped in favor of bixbyite. The solubility limit of Fe_2O_3 in bixbyite is very dependent upon temperature (Mason, 1943); natural bixbyite can contain up to 60% (molecular) Fe_2O_3, a value consistent with the maximum solubility limit (Muan and Somiya, 1962; Schmal and Hennings, 1969b; Ono *et al.*, 1971). The substitution of scandium up to 0.35% by weight Sc_2O_3 in natural bixbyite is described by Frondel (1970). Bixbyite is not known to deviate from ideal stoichiometry of 2 cations per 3 oxygens; stoichiometric Mn_2O_3 has been synthesized several times (Rode, 1949; Moore *et al.*, 1950).

The basic structure of natural braunite (approximate composition Mn_7SiO_{12}) was determined by Byström and Mason (1943). In an investigation of the phase relations of the system MnO-SiO_2-O_2, Muan (1959a,b) found that analogues of braunite can be synthesized which have 0-40% by weight SiO_2. A recent crystal structure determination of braunite (deVilliers, 1975) indicates a double Mn_2O_3 cell and provides a basis for a structural model in which the silica content varies from 0% (orthorhombic Mn_2O_3) to the theoretical 44% (tetragonal braunite). DeVilliers and Herbstein (1967) report a silica-deficient phase, braunite II, which, deVilliers (1975) proposes, consists of a regular stacking arrangement of one braunite and two Mn_2O_3 cells. Mason (1943) reviews the chemistry of braunite and notes that deviations from ideal composition (MnSi = 7:1) are small. Braunite commonly contains appreciable quantities of boron, 0.3-1.2% by weight B_2O_3 (Wasserstein, 1943). In view of the relationship between the braunite and bixbyite structures and the association of braunite with more iron-rich manganese-oxide minerals, it is surprising that natural braunites contain little iron (commonly <5% as FeO; maximum 15%).

The phase Mn_5O_8 has been reported to have formed during the oxidation of Mn_3O_4 to Mn_2O_3 in the laboratory (Feitknecht, 1964), but is not known to have formed in nature. Oswald and Wampetich (1967) demonstrated a Mn_5O_8 structure based on sheets of Mn and of oxygen ions.

"Spinel-type" oxides. The structure of hausmannite, Mn_3O_4, is tetragonal ($I4_1$/amd) at room temperature (Aminoff, 1926). If a face-centered cell is chosen, the hausmannite structure is analogous to that of the Fe_3O_4 spinel, magnetite. The axial ratio of Mn_3O_4 is 1.16, indicating considerable distortion from a cubic lattice. Both Fe_3O_4 and Mn_3O_4 invert from cubic to lower symmetry as temperature decreases, but whereas the former inversion is at 119°K, the latter is near 1150°C (Southard and Moore, 1942; McMurdie and Golovato, 1948; Van Hook and Keith, 1958). Driessens (1967) found little dependence of oxygen fugacity (f_{O_2}) on the tetragonal to cubic inversion temperature in hausmannite, suggesting little effect of f_{O_2} on the Mn/O ratio in hausmannite. The effects of solid solution with iron, nickel, cobalt, and zinc end-member spinel components is considered by Brabers (1971), Finch *et al.* (1957), and Dreissens (1967).

A phase designated γ-Mn_2O_3, which can be described as a cation-deficient hausmannite (much as maghemite is a cation-deficient magnetite) has been synthesized at low temperature in aqueous solution by Verwey (1935), Verwey and de Boer (1936), and Moore *et al.* (1950); Bricker (1965) synthesized a related, hydrated form of α-Mn_2O_3. In an investigation of $Fe_{3-x}Mn_xO_{4+y}$ synthesized at high temperature in air, Wickham (1969) found that manganese-rich members of the series were essentially stoichiometric (y = 0). The compound γ-Mn_2O_3 is not known to occur naturally, but in view of its possible low temperature laboratory occurrence and powder pattern (which is nearly identical to that of hausmannite), γ-Mn_2O_3 may exist, unrecognized, in nature.

Natural hausmannite is commonly quite pure; the most common substitution, Fe_3O_4 in solid solution, is not known to exceed 7% by weight (Mason, 1943). This observation agrees generally with the experimental observations of

Van Hook and Keith (1958), Mason (1943), and Ono *et al.* (1971) who found a two-phase region of hausmannite solid solution + $(Fe,Mn)_3O_4$ spinel (jacobsite) extending from 91 to 54 molecular percent Mn_3O_4. Naturally occurring oriented intergrowths of hausmannite + jacobsite are termed vredenburgite. Rarely, ZnO substitutes in natural Mn_3O_4 to form a solid solution with hataerolite, $ZnMn_2O_4$ (Mason, 1947; Ogawa, 1967). Recent commercial interest in the magnetic properties of manganates and manganiferous spinels has stimulated many syntheses and structural investigations of compositions which do not occur in nature: Cervinka and Vetterkind (1968); Rieck and Driessens (1966); Finch *et al.* (1957); Schmahl and Hennings (1969b); Miller (1968); MacChesney *et al.* (1967); Makram (1967); Yamada and Iida (1968); Holba *et al.* (1973). Driessens (1967) reviews literature describing 18 possible substitutions in Mn_3O_4.

Wüstite-type oxide. Manganosite, $Mn_{1-x}O$, has the NaCl structure (Levi, 1924) and in the laboratory forms solid solutions with FeO, MnO, NiO, MgO, etc. In nature, the mineral is of such limited occurrence that little is known of its geological crystal chemistry. Most natural manganosites are quite pure; some contain a little FeO. Laboratory investigations of the compositional range of "MnO" have included its oxygen content (Hed and Tannhauser, 1967a; Bransky and Tallan, 1971; Foster and Welch, 1956; Schmahl and Hennings, 1969a; Towhidi and Neuschutz, 1972; Davies and Richardson, 1959); the maximum departure from stoichiometric composition appears to be $Mn_{0.89}O$. Solid solutions of "MnO" with other monoxides have been synthesized and measured by various high-temperature techniques: $(Fe,Mn)_{1-x}O$ by Foster and Welch (1956), Voeltzel and Manenc (1967), and Goodwin *et al.* (1975); Mn_xO-MgO by Jones and Cutler (1966, 1971); NiO-MnO by Cameron and Unger (1970); and CaO-MnO-FeO by Tiberg and Muan (1970).

Manganese monoxide has been subjected to a wide variety of physical property measurements: reflectivity (Kinney and O'Keefe, 1969); covalency parameters (Fender *et al.*, 1968); electrical conductivity (Eror and Wagner, 1971; Hed and Tannhauser, 1967b; Price and Wagner, 1970; Bocquet *et al.*, 1967). Chemical diffusivity values and vacancy concentrations were derived in the above studies.

Geologic occurrences

Different genetic types of oxide-bearing manganese deposits are reviewed and summarized by Hewett and Fleischer (1960) and Roy (1968).

Stratified manganese oxides which have been subjected to alteration processes no more intense than diagenesis are reviewed by Varentsov (1964).

Sokolova (1964), and Hewett (1966). Many of the sedimentary manganese oxide deposits contain little iron; Krauskopf (1957) explains this efficient geochemical separation with Eh-pH diagrams, and Handa (1970) investigated the importance of the kinetics of precipitation. Interesting depositional or diagenetic textures appear in published photographs, but have never been investigated in detail. The association of manganese oxide strata or layers with layers of manganese carbonate and sheet silicates, sometimes on a very fine scale (Huebner, 1967; Hewett *et al.*, 1961), suggests that there are facies of manganese formation similar to the oxide, carbonate, and silicate facies of iron formation proposed by James (1954).

Unmetamorphosed manganese-rich oxides also are found as nodules and crusts in marine (Hubred, 1975; Moore and Vogt, 1976; Scott *et al.*, 1974; Manheim, 1965) and freshwater environments (Varentsov, 1972; Schoettle and Friedman, 1971; Cronan and Thomas, 1972). Manganese oxides have also been thought to precipitate in tidal pools (Mart and Sass, 1972). Tetravalent manganese oxides form crusts associated with terrestrial hot springs in volcanic zones (Hewett *et al.*, 1963; Hewett and Fleischer, 1960; White, 1955). Some sedimentary hausmannite and braunite is probably related to submarine volcanism (Sorem and Gunn, 1967; Hewett *et al.*, 1961, p. 48).

Hypogene manganese-oxide veins are usually epithermal and are related to volcanic activity. They also share a feature with many stratified deposits--the manganese concentrations are high, indicating effective separation of manganese and iron. Hewett (1964) reviews the occurrences of manganese oxide veins in the southwest United States and northern Mexico.

Only two attempts have been made to compare systematically different kinds of metamorphosed manganese deposits; the first emphasizes the nature of the manganese formation or protore (Roy, 1965) whereas the other stresses differences in the mineral assemblages and deduces differences in conditions of formation (Huebner, 1967). Most authors have described individual manganese deposits (and districts); some have confined themselves geographically to a country, region or state: Japan (Lee, 1955), California (Hewett *et al.*, 1961), central India (Fuchs, 1970a,b), southern India (Roy, 1973).

Distinct kinds of metamorphic manganese deposits that should be included in any comprehensive list include the following:

Buckeye of the California Coast Ranges; hausmannite, braunite, rhodochrosite (Huebner, 1967; Trask, 1950).

Noda Tamagawa, Japan; hausmannite, braunite, pyrochroite (previously manganosite), carbonate, silicates (Watanabe, 1959; Watanabe *et al.*, 1970).

Tamworth, N.S.W.; hausmannite, silicates (Segnit, 1962).

Langban, Sweden; manganosite, braunite, hausmannite (Magnusson, 1930; Ridge, 1959; Moore, 1971).

Gondite ores, India; hausmannite, braunite, bixbyite and silicates (Roy, 1965,1973; Narayanaswami, 1963; Rao, 1963; Fuchs, 1970a,b).

Merid Mine, Brazil; hausmannite, carbonate, silicates (Horen, 1953).

Postmasburg, South Africa; hausmannite, bixbyite, braunite (deVilliers, 1960).

Thermochemistry and phase relations

Thermochemical data for simple manganese oxides is summarized by Mah (1960) and by Robie and Waldbaum (1968). In addition, numerous electrochemical and phase-equilibria studies have yielded Gibbs free energies and enthalpies of reaction. Comparison of sets of thermochemical data, and application of such data to natural assemblages, is hindered by uncertainties in the compositions of the phases considered. Measurements on the three-phase assemblage $Mn_{1-x}O$-Mn_3O_4-Gas illustrate these difficulties (Huebner and Sato, 1970, p. 944-948).

The oxygen fugacity of the metal plus oxide assemblage Mn-$Mn_{1-x}O$ at one atm pressure can be calculated from the electrochemical measurements of Alcock and Zador (1967) and is log atm $f_{O_2} = -\frac{40600}{T°K} + 7.97$; presumably this manganosite is nearly stoichiometric. Oxygen fugacity data (Fig. SH-1) for the assemblages $Mn_{1-x}O$-Mn_3O_4-Gas, Mn_3O_4-Mn_2O_3-Gas, and Mn_2O_3-MnO_2-Gas are presented and reviewed by Huebner and Sato (1970) and Fukunaga *et al.* (1969). Stability field boundaries for γ-Mn_2O_3, the MnO_2 "polymorphs" ramsdellite and nsutite, and the complicated tetravalent oxides are not known, but in some cases conditions of synthesis have been reported (Endo *et al.*, 1974; McKenzie, 1971; Wadsley, 1950). The relationship between "silica-rich" braunite, Mn_3O_4, and $MnSiO_3$ at high temperature is given by Muan (1959a,b); Huebner (1967) gives the reaction sequences for both silica-rich and silica-poor (natural) braunite under low pressure as a function of temperature and f_{O_2}. Manganese oxide stabilities are further limited by reaction with CO_2 gas to form rhodochrosite, $MnCO_3$ (Fig. SH-2) and with H_2O to form the hydroxides pyrochroite, $Mn(OH)_2$ and manganite, γ-MnOOH (Bricker, 1965; Burns and Brown, 1972; Giovanoli and Leuenberger, 1969; Glasser and Smith, 1968; Klingsberg and Roy, 1959; Klingsberg, 1958).

The activity-composition relationships of manganosite at high temperature and low pressure have been specifically analyzed by Schmahl and Hennings (1969a) and by Davies and Richardson (1959). Activity data are also available for solid

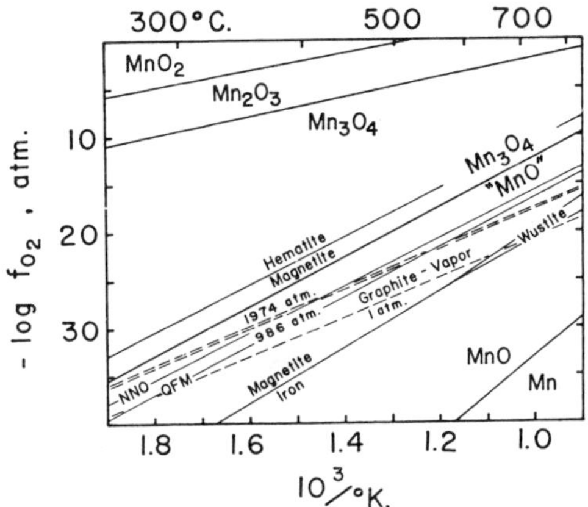

Figure SH-1. Reactions in the systems Mn-O, Fe-O, and C-O plotted as a function of temperature and oxygen fugacity. The change of f_{O_2} with pressure is significant only for the graphite-vapor assemblage. Note that the field of manganosite is extensive when compared with the wüstite field, and that manganese oxides are stable at much higher oxygen fugacity values than corresponding iron oxides. Source: Huebner, 1969, fig. 3.

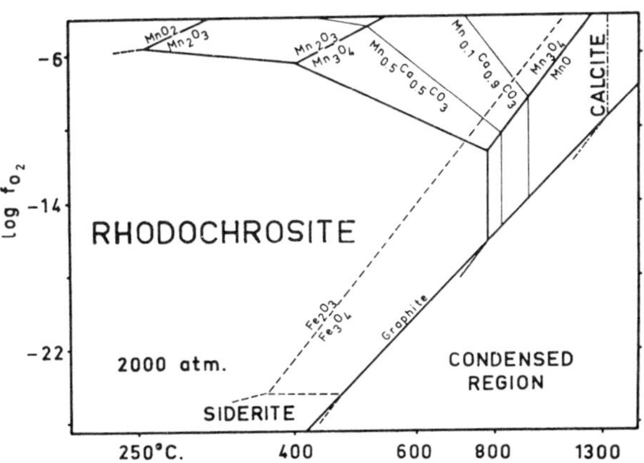

Figure SH-2. T-log f_{O_2} section at 2000 atm C-O gas showing that in the presence of carbon-dioxide rhodochrosite ($MnCO_3$) replaces the manganese oxides. For comparison, the iron oxide fields for hematite and magnetite extend to lower values of oxygen fugacity and temperature than the manganese oxide fields. The addition of $CaCO_3$ component stabilizes rhodochrosite at the expense of manganese oxides (which do not take a calcium oxide component into solid solution). Source: Huebner, 1969, fig. 9 .

solutions of $Mn_{1-x}O$ with NiO (Cameron and Unger, 1970), CaO (Brežný *et al.*, 1970), MgO (Woermann and Muan, 1970), and $Fe_{1-x}O$ (Balesdent *et al.*, 1972). Schmahl and Hennings (1969b) investigated the activity-composition relationships of Mn_3O_4 and Mn_2O_3 in $(Mn,Fe)_3O_4$ solid solutions (both tetragonal and cubic) and $(Mn,Fe)_2O_3$ solutions.

Petrology of manganese oxides

Attempts to classify manganese oxide deposits have traditionally been hindered by imperfect knowledge of the manganese oxides present in any particular location. Older descriptions and petrogenetic theories did not have available recent understandings of the structures and crystal chemistry of the tetravalent oxides. The last attempts to synthesize the mineralogy and petrogenesis of continental manganese oxide deposits were led by Foster Hewett. At this time attempts were made to formulate criteria by which supergene or hypogene oxides could be distinguished (Hewett and Fleischer, 1960; Hewett *et al.*, 1963; and Hewett, 1972). Simultaneously attempts were made to understand and classify stratified manganese oxides (Hewett, 1966; Varentsov, 1964) and their metamorphosed products (Roy, 1965; Roy and Purkait, 1965). Studies of the relationship between the manganese mineral assemblages present and the physico-chemical conditions of metamorphism (Huebner, 1967; Roy, 1973) were hindered by inadequate descriptions of (equilibrium) assemblages in the literature. Most of the recent investigations of manganese oxides have considered both marine (Calvert and Price, 1970; Glasby, 1972; Ostwald and Frazer, 1973; Burns and Burns, 1976) and nonmarine (Schoettle and Friedman, 1971; Cronan and Thomas, 1972; Varentsov, 1972) manganese and ferromanganese oxide nodules. These manganese nodules contain minerals which are metastable relative to hausmannite and manganite in the marine environment, yet exist because of catalytic and kinetic processes during precipitation (Crerar and Barnes, 1974).

In a study of the metamorphism of manganiferous sediments to form oxide-bearing deposits, Huebner (1967) constructed a petrogenetic grid (Fig. SH-3) for the system Mn-Si-C-O showing the phase assemblages that occur in nature. The stability fields of the manganese oxides are bounded by reactions to form rhodochrosite ($MnCO_3$), pyroxenoid (pyroxmangite or rhodonite, $MnSiO_3$), and tephroite (Mn_2SiO_4). The grid can be positioned in P-T-f_{O_2} space by reference to the manganese oxide reactions (Fig. SH-1) and the manganese carbonate stability field (Fig. SH-2 and Huebner, 1969). One such section, for 2000 bars total pressure and a CO_2 + CO gas, is shown (Fig. SH-4). Several conclusions can be drawn from sections such as these. Only one manganese oxide, mangano-

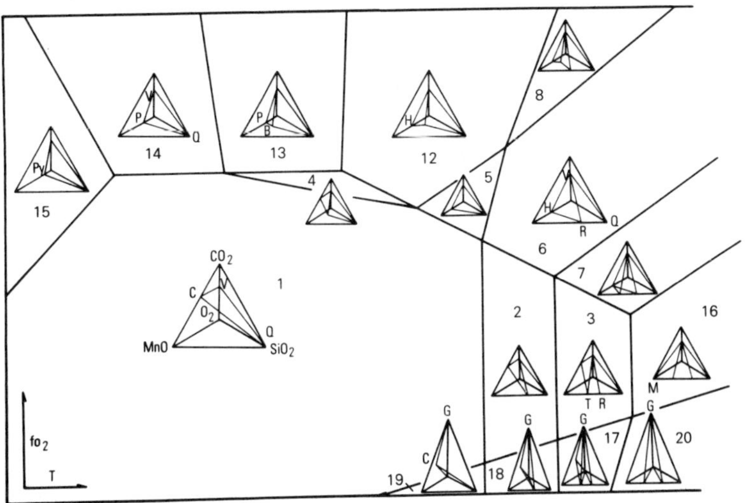

Figure SH-3. Petrogenetic grid in isobaric f_{O_2}-T space showing topology of stable reactions and phase assemblages in the system $MnO-SiO_2-C-O_2$. Phase designations are as follows:

M	manganosite	C	rhodochrosite	Q	quartz
H	hausmannite	G	graphite	R	pyroxenoid
P	partridgeite	V	vapor or gas	T	tephroite
Py	pyrolusite	B	braunite (silica-poor)		

Numbers in Figures SH-3 and SH-4 refer to possible assemblages, as follows; an asterisk indicates a condensed vapor phase.

1	CQV	12	BHQ, HBV
2	CRV, RQV	13	BQV, PBV
3	CTV, TRV, RQV	14	PQV
4	CBV, BQV	15	PyQV
5	HQV	16	MTV, TRV, RQV
6	HRV, RQV	17	CTGV*, TRGV*, RQGV*
7	HTV, TRV, RQV	18	CRGV*, RQGV*
8	HBV, BRV, RQV	19	CQGV*
		20	MTGV*, TRGV*, RQGV*

Source: after Huebner, 1967, fig. V-2.

site, is stable at the oxygen fugacity values characteristic of common ferruginous rocks in which magnetite is the characteristic oxide. The occurrence of hausmannite, braunite, or bixbyite in metasedimentary manganese ore bodies indicates the existence of an oxygen fugacity that was orders of magnitude greater than the f_{O_2} of surrounding ferruginous rocks. The high f_{O_2} values necessary to stabilize bixbyite or braunite deposits during metamorphism cannot be imposed by the more reduced country rocks, and hence must be inherent in the original manganiferous sediment (Huebner, 1967,1969). At lower f_{O_2} values, hausmannite and manganosite can occur only if the f_{CO_2} is low, thereby suppressing rhodochrosite. Also at lower f_{O_2}, manganese oxides will react with silica to form pyroxenoids and the manganese olivine, tephroite. It appears, then, that in metamorphic deposits, manganese oxides will occur only when the manganese was in an oxidized state prior to metamorphism.

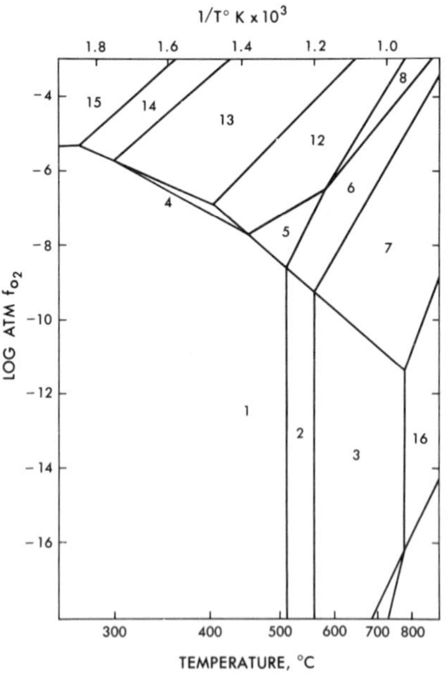

Figure SH-4. Petrogenetic grid of Figure SH-3 placed on f_{O_2} and T coordinates at $P_T = P_{CO_2} + P_{CO} = 2000$ atm. Numbers in fields have the same meaning as in the previous figure. Source: after Huebner, 1967, fig. V-1.

REFERENCES

Alcock, C. B. and S. Zador (1967) Thermodynamic study of the manganese/manganous-oxide system by the use of solid oxide electrolytes. *Electrochim. Acta 12*, 673-677.

Aminoff, G. (1926) Uber die Kristallstruktur von Hausmannit ($MnMn_2O_4$). *Z. Kristallogr. 64*, 475-490.

Balesdent, D., O. Evrard, and C. Gleitzer (1972) Extension de la methode de Darken aux systemes heterogenes pour le calcul des activites des constituants des manganowüstites et magnesiowüstites. *Rev. Chim. Mineral. 9*, 233-244.

Bocquet, J.-P., M. Kawahara, and P. Lacombe (1967) Conductibilite electronique, conductibilite ionique et diffusion thermique dans MnO a haute temperature (de 900 a 1150°C). *C. R. Acad. Paris 265*, 1318-1321.

Brabers, V. A. M. (1971) Cation migration valencies and the cubic-tetragonal transition in $Mn_xFe_{3-x}O_4$. *J. Phys. Chem. Solids 32*, 2181-2191.

Bransky, I. and N. M. Tallan (1971) A gravimetric study of nonstoichiometric MnO. *J. Electrochem. Soc. 118*, 788-793.

Brežný, B., W. R. Ryall, and A. Muan (1970) Activity-composition relations in CaO-MnO solid solutions at 1100-1300°C. *Mater. Res. Bull. 5*, 481-488.

Bricker, O. P. (1965) Some stability relations in the system $Mn-O_2-H_2O$ at 25° and one atmosphere total pressure. *Am. Mineral. 50*, 1296-1354.

Burns, V. M., R. G. Burns, and W. K. Zwicker (1975) Classification of natural manganese dioxide minerals. In, A. Kozawa and R. J. Brodd, Eds., *Proc. Int. Symp. Manganese Dioxide*, Electrochemical Soc., pp. 288-305.

Burns, R. G. and B. A. Brown (1972) Nucleation and mineralogical controls on the composition of manganese nodules. In, D. R. Horn, Ed., *Conference on Ferromanganese Deposits on the Ocean Floor*, pp. 51-61.

________ and V. M. Burns (1975) Structural relationships between the manganese (IV) oxides. In, A. Kozawa and R. J. Brodd, Eds., *Proc. Int. Symp. Manganese Dioxide*, Electrochemical Soc., pp. 306-327.

________, ________ (1976) Understanding the mineralogy and crystal chemistry of deep-sea manganese nodules, a polymetallic resource of the twenty-first century. Presented at the Discussion Meeting: *Mineralogy, Towards the Twenty-First Century*. Organized jointly by the Royal Society and the Mineralogical Society, April 7-8, 1976, London.

Bystrom, A. and B. Mason (1943) The crystal structure of braunite, $3Mn_2O_3 \cdot MnSiO_3$. *Arkiv for Kemi, Mineralogi och Geologi Utgivet av K. Svenska Vetenskapsakademien 16*, 1-8.

Bystrom, A. M. (1949) The crystal structure of ramsdellite, an orthorhombic modification of MnO_2. *Acta Chem. Scandinavica 3*, 163-173.

Calvert, S. E., and N. B. Price (1970) Composition of manganese nodules and manganese carbonates from Loch Fyne, Scotland. *Contrib. Mineral. Petrol. 29*, 215-233.

Cameron, D. J., and A. E. Unger (1970) The measurement of the thermodynamic properties of NiO-MnO solid solutions by a solid electrolyte cell technique. *Met. Trans. 1*, 2615-2621.

Cervinka, L., and D. Vetterkind (1968) The influence of low temperature (293°K-4.5°K) on the structure of manganese-rich manganese ferrite. *J. Phys. Chem. Solids 29*, 171-179.

Crerar, D. A., and H. L. Barnes (1974) Deposition of deep-sea manganese nodules. *Geochim. Cosmochim. Acta 38*, 279-300.

Cronan, D. S., and R. J. Thomas (1972) Geochemistry of ferromanganese oxide concretions and associated deposits in Lake Ontario. *Geol. Soc. Am. Bull. 83*, 1493-1502.

Dachs, H. (1963) Neutronen und Rontgenuntersuchungen am Manganit, MnOOH. *Z. Kristallogr. 118*, 303-326.

Davies, M. W., and F. D. Richardson (1959) The non-stoichiometry of manganous oxide. *Trans. Faraday Soc. 55*, 604-610.

de Villiers, J. P. R. (1975) The crystal structure of braunite with reference to its solid-solution behavior. *Am. Mineral. 60*, 1098-1104.

de Villiers, J. P. R. (1960) The manganese deposits of the Union of South Africa. *Geol. Surv. South Africa, Handbook 2*, 271p.

de Villiers, P. R., and F. H. Herbstein (1967) Distinction between two members of the braunite group. *Am. Mineral. 52*, 20-30.

de Wolff, P. M. (1959) Interpretation of some γ-MnO_2 diffraction patterns. *Acta Crystallogr. 12*, 341-345.

Driessens, F. C. M. (1967) Place and valence of the cations in Mn_3O_4 and some related manganates. *Inorg. Chim. Acta 1*, 193-201.

Endo, T., S. Kume, M. Shimada, and M. Koizumi (1974) Synthesis of potassium manganese oxides under hydrothermal conditions. *Mineral. Mag. 39*, 559-563.

Eror, N. G., and J. B. Wagner, Jr. (1971) Nonstoichiometric disorder in single crystalline MnO. *J. Electrochem. Soc. 118*, 1665-1670.

Feitknecht, W. (1964) Einfluss der Teilchengrosse auf den Mechanismus von Festkorperreaktionen. *Pure and Appl. Chem. 9*, 423-440.

Fender, B. E. F., A. J. Jacobson, and F. A. Wedgwood (1968) Covalency parameters in MnO, α-MnS, and NiO. *J. Chem. Phys. 48*, 990-994.

Finch, G. I., A. P. B. Sinha, and K. P. Sinha (1957) Crystal distortion in ferrite-manganites. *Proc. Roy. Soc., Ser. A, 242*, 28-35.

Foster, P. K., and A. J. E. Welch (1956) Metal-oxide solid solutions--Part 1.--Lattice-constant and phase relationships in ferrous oxide (wustite) and in solid solutions of ferrous oxide and manganous oxide. *Trans. Faraday Soc. 52*, 1626-1635.

Frondel, C. (1970) Scandium-rich minerals from rhyolite in the Thomas Range, Utah. *Am. Mineral. 55*, 1058-1060.

Fuchs, H. D. (1970a) Metamorphose und Genese zentralindischer Manganerze. *Clausthaler Hefte 9*, 66-84.

________ (1970b) Stratigraphische und petrographische Untersuchungen im Kristallin des zentral-indischen Manganerz-Gürtels. Das Gebiet des südlichen Bawanthari-Flusses. *Neues Jahrb. Geol. Palaont. Abh. 136*, 262-302.

Fukunaga, O., K. Takahashi, T. Fujita, and J. Yoshimoto (1969) Phase equilibrium between MnO_2 and Mn_2O_3. *Mater. Res. Bull. 4*, 315-322.

Geller, S. (1971) Structures of α-Mn_2O_3, $(Mn_{0.983}Fe_{0.017})_2O_3$ and $(Mn_{0.37}$-$Fe_{0.63})_2O_3$ and relation to magnetic ordering. *Acta Crystallogr. B27*, 821-828.

________, J. A. Cape, R. W. Grant, and G. P. Espinosa (1967) Distortion in the crystal structure of α-Mn_2O_3. *Phys. Letters 24A*, 369-371.

Giovanoli, R., and U. Leuenberger (1969) Über die Oxydation von Manganoxidhydroxid. *Helv. Chim. Acta 52*, 2333-2347.

________, R. Mauer, and W. Feitknecht (1967) Zur Struktur des γ-MnO_2. *Helv. Chim. Acta 50*, 1072-1080.

Glasby, G. P. (1972) The mineralogy of manganese nodules from a range of marine environments. *Marine Geol. 13*, 57-72.

Glasser, L. S. D., and L. Ingram (1968) Refinement of the crystal structure of groutite, α-MnOOH. *Acta Crystallogr. B24*, 1233-1236.

________, and I. B. Smith (1968) Oriented transformations in the system MnO-O-H_2O. *Mineral. Mag. 36*, 976-987.

Goodwin, C. A., H. K. Bowen, and W. D. Kingery (1975) Phase separation in the system (Fe,Mn)O. *J. Am. Ceram. Soc. 58*, 317-320.

Handa, B. K. (1970) Chemistry of manganese in natural waters. *Chem. Geol. 5*, 161-165.

Hase, W., K. Kleinstuck, and G. E. R. Schulze (1967) Kristallstrukturuntersuchungen an Sesquioxiden mit α-Mn_2O_3-Struktur. *Z. Kristallogr. 124*, 428-451.

________, W. Bruckner, J. Tobisch, H.-J. Ullrich, and G. Wegerer (1969) Untersuchungen zur Kristallstruktur von $(Mn_{1-x}Fe_x)_2O_3$. *Z. Kristallogr. 129*, 360-364.

Hed, A. Z., and D. S. Tannhauser (1967a) Contribution to the Mn-O phase diagram at high temperature. *J. Electrochem. Soc. 114*, 314-318.

________, ________ (1967b) High-temperature electrical properties of manganese monoxide. *J. Chem. Phys. 47*, 2090-2103.

Hewett, D. F. (1964) Veins of hypogene manganese oxide minerals in the southwestern United States. *Econ. Geol. 59*, 1429-1472.

________ (1966) Stratified deposits of the oxides and carbonates of manganese. *Econ. Geol. 61*, 431-461.

________ (1972) Manganite, hausmannite, braunite: features, modes of origin. *Econ. Geol. 67*, 83-102.

________, and M. Fleischer (1960) Deposits of the manganese oxides. *Econ. Geol. 55*, 1-55.

________, C. W. Chesterman, and B. W. Troxel (1961) Tephroite in California manganese deposits. *Econ. Geol. 56*, 39-58.

________, M. Fleischer and N. Conklin (1963) Deposits of the manganese oxides: Supplement. *Econ. Geol. 58*, 1-51.

Holba, P., M. A. Khilla, and S. Krupicka (1973) On the miscibility gap of spinels $Mn_xFe_{3-x}O_4$ + γ. *J. Phys. Chem. Solids 34*, 387-395.

Horen, A. (1953) The manganese mineralization at the Merid Mine, Minas Gerais, Brazil. Ph.D. Dissertation, Harvard University, Cambridge, Massachusetts.

Hubred, G. (1975) Deep-sea manganese nodules: A review of the literature. *Minerals Sci. Eng. 7*, 71-85.

Huebner, J. S. (1967) Stability relations of minerals in the system Mn-Si-C-O. Ph.D. Thesis, The Johns Hopkins University, Baltimore, Maryland.

________ (1969) Stability relations of rhodochrosite in the system manganese-carbon-oxygen. *Am. Mineral.* *54*, 457-481.

________, and M. Sato (1970) The oxygen fugacity-temperature relationships of manganese oxide and nickel oxide buffers. *Am. Mineral.* *55*, 934-952.

James, H. L. (1954) Sedimentary facies of iron formation. *Econ. Geol.* *29*, 253-293.

Jones, J. T., and I. B. Cutler (1971) Interdiffusion in the system M_xO-MgO. *J. Am. Ceramic Soc.* *54*, 335-338.

________, and I. B. Cutler (1966) Stoichiometry of MgO-Mn_xO solid solutions in air at high temperature. *J. Am. Ceramic Soc.* *49*, 572-573.

Kinney, T. B., and M. O'Keefe (1969) The dielectric and Reststrahlen parameters of MnO. *Solid State Comm.* *7*, 977-978.

Klingsberg, C. (1958) The system Mn-O-OH. Ph.D. Thesis, The Pennsylvania State University, University Park, Pennsylvania.

________, and R. Roy (1959) Stability and interconvertibility of phases in the system Mn-O-OH. *Am. Mineral.* *44*, 819-838.

Krauskopf, K. B. (1957) Separation of iron from manganese in sedimentary processes. *Geochim. Cosmochim. Acta* *12*, 63.

Lee, D. E. (1955) Mineralogy of some Japanese manganese ores. *Stanford Univ. Publ., Univ. Ser., Geol. Sci.* *5*, 64p.

Levi, G. R. (1924) Il reticolo cristallino dell'ossido manganoso. *Gazz. Chim. Ital.* *54*, 704-708.

MacChesney, J. B., H. J. Williams, J. F. Potter, and R. C. Sherwood (1967) Magnetic study of the manganate phases: $CaMnO_3$, $Ca_4Mn_3O_{10}$, $Ca_3Mn_2O_7$, Ca_2MnO_4. *Phys. Rev.* *164*, 779-785.

Magnusson, N. H. (1930) Långbans malmtrakt. *Sver. Geol. Undersok., Ser. ca* *23*, 111p.

Mah, A. D. (1960) Thermodynamic properties of manganese and its compounds. *Bur. Mines Rep. Investigations* *5600*, 1-34.

Makram, H. (1967) Growth of nickel manganite single crystals. *J. Crystal Growth* *1*, 325-326.

Manheim, F. T. (1965) Manganese-iron accumulations in the shallow marine environment. In, D. R. Schink and J. T. Corless, Eds., *Symposium on Marine Geochemistry.* The University of Rhode Island, Publ. No. 3, pp. 217-276.

Mart, J., and E. Sass (1972) Geology and origin of the manganese ore of Um Bogma, Sinai. *Econ. Geol.* *67*, 145-155.

Mason, B. (1943) Mineralogical aspects of the system FeO-Fe_2O_3-MnO-Mn_2O_3. *Geol. Foren. Stockholm Forhandlingar* *65*, 97-180.

Mason, B. (1947) Mineralogical aspects of the system Fe_3O_4-Mn_3O_4-$ZnMn_2O_4$-$ZnFe_2O_4$. *Am. Mineral.* *32*, 426-441.

McKenzie, R. M. (1971) The synthesis of birnessite, cryptomelane, and some other oxides and hydroxides of manganese. *Mineral. Mag.* *38*, 493-502.

McMurdie, H. F., and E. Golovato (1948) Study of the modifications of manganese dioxide. *J. Res. Natl. Bur. Stand. Res. Pap. RP1941,* *41*, 589-600.

Miller, A. (1968) Determination of the valence state of copper in cubic $CuMn_2O_4$ spinel by X-ray absorption edge measurements. *J. Phys. Chem. Solids 29*, 633-639.

Moore, P. B. (1971) Mineralogy and chemistry of Langban-type deposits in Bergslagen, Sweden. *Mineral. Record 1*, 154-172.

Moore, T. E., M. Ellis, and P. W. Selwood (1950) Solid oxides and hydroxides of manganese. *J. Am. Chem. Soc. 72*, 856-866.

Moore, W. S., and P. R. Vogt (1976) Hydrothermal manganese crusts from two sites near the Galapagos Spreading Axis. *Earth Planet. Sci. Letters 29*, 249-356.

Muan, A. (1959a) Phase equilibria in the system manganese oxide--SiO_2 in air. *Am. J. Sci. 257*, 297-315.

_______ (1959b) Stability relations among some manganese minerals. *Am. Mineral. 44*, 946-960.

_______, and S. Somiya (1962) The system iron oxide-manganese oxide in air. *Am. J. Sci. 260*, 230-240.

Narayanaswami, S. *et al.* (1963) The geology and manganese ore-deposits of the manganese belt in Madhya Pradesh and adjoining parts of Maharashtra. *Geol. Surv. India Bull., Ser. A 22*, 69p.

Norrestam, R. (1967) α-manganese (III) oxide--a C-type sesquioxide of orthorhombic symmetry. *Acta Chem. Scandinavica 21*, 2871-2884.

Ogawa, S. (1967) Oxidation rate and phase diagram of Mn-Zn ferrite determined from the cation diffusion coefficient. *Jpn. J. Appl. Phys. 6*, 1427-1433.

Ono, K., T. Ueda, T. Ozaki, Y. Ueda, A. Yamagushi, and J. Moriyama (1971) Thermodynamic study of the iron-manganese-oxygen system. *Nippon Kinzoku Gakkaishi 35*, 757-763.

Ostwald, J., and F. W. Frazer (1973) Chemical and mineralogical investigations on deep sea manganese nodules from the Southern Ocean. *Mineral. Deposita 8*, 303-311.

Oswald, H. R., and M. J. Wampetich (1967) Die Kristallstrukturen von Mn_5O_8 und $Cd_2Mn_3O_8$. *Helv. Chim. Acta 50*, 2023-2034.

Price, J. B., and J. B. Wagner, Jr. (1970) Diffusion of manganese in single crystalline manganous oxide. *J. Electrochem. Soc. 117*, 242-247.

Rao, J. S. R. K. (1963) Microscopic examination of manganese ores of Srikakulam and Visakhapatnam (Vizagapatam) Districts, Andhra Pradesh, India. *Econ. Geol. 58*, 434-440.

Ridge, J. D. (1959) The unusual manganese-iron deposits of Langban, in Sweden. *Mineral. Indus. 29-3*, 1-5.

Rieck, G. D., and F. C. M. Driessens (1966) The structure of manganese-iron-oxygen spinels. *Acta Crystallogr. 20*, 521-525.

Robie, R. A., and D. R. Waldbaum (1968) Thermodynamic properties of minerals and related substances at 298.15°K (25.0°C) and one atmosphere (1.013 bars) pressure and at higher temperatures. *U. S. Geol. Surv. Bull. 1259*, 256p.

Rode, E. Ya. (1949) Physicochemical study of manganese oxides: Izvest. Sektora Fiz.-Khim. Anal., Inst. Obshchei i Neorg. Khim., Akad. Nauk SSSR *19*, 58-68. Cited in *Chem. Abstr. 44*, 9228.

Roy, S. (1965) Comparative study of the metamorphosed manganese protores of the world. *Econ. Geol. 60*, 1238-1260.

Roy, S. (1973) Genetic studies on the Precambrian manganese formations of India with particular reference to the effects of metamorphism. UNESCO, Earth Sci. Ser., No. 9, 229-242. In, *Genesis of Precambrian Iron and Manganese Deposits. Proc. Kiev Symp.*, UNESCO, Paris.

_______ (1968) Mineralogy of the different genetic types of manganese deposits. *Econ. Geol. 63*, 760-786.

_______, and P. K. Purkait (1965) Stability relations of manganese oxide minerals in metamorphic orebodies corresponding to sillimanite grade in Gowari Wadhona Mine Area, Chhindwara District, Madhya Pradesh, India. *Econ. Geol. 60*, 601-613.

Schmahl, N. G., and D. Hennings (1969a) Die Nichtstöchiometrie des Mangan (II)--Oxids im thermischen Gleichgewicht. *Z. Physik. Chemie Neue Folge 63*, 111-124.

________, and D. Hennings (1969b) Phasengleichgewichte und Kryptomodifikationen im Spinell-und Sesquioxidbereich des stabilen Systems Mn-Fe-O. *Z. Physik. Chemie Neue Folge 64*, 313-332.

Schoettle, M., and G. M. Friedman (1971) Fresh water iron-manganese nodules in Lake George, New York. *Geol. Soc. Am. Bull. 82*, 101-110.

Scott, M. R., R. B. Scott, P. A. Rona, L. W. Butler, and A. J. Nalwalk (1974) Rapidly accumulating manganese deposit from the Mid-Atlantic Ridge. *Geophys. Res. Letters 1*, 355-358.

Segnit, E. R. (1962) Manganese deposits in the neighborhood of Tamworth, New South Wales. *Proc. Australasian Inst. Min. Metal. 202*, 47-61.

Sokolova, E. I. (1964) *Physicochemical Investigations of Sedimentary Iron and Manganese Ores and Associated Rocks*. Daniel Davey and Co., Inc., New York, 220p. (transl. *Fiziko-Khimicheskie Issledovaniya Osadochnykh Zheleznykh I Margantsevykh Rud I Vmeshchayushchikh Ikh Porod.*)

Sorem, R. K., and D. W. Gunn (1967) Mineralogy of manganese deposits, Olympic Peninsula, Washington. *Econ. Geol. 62*, 22-56.

Southard, J. C., and G. E. Moore (1942) High temperature heat content of Mn_3O_4, $MnSiO_3$, and Mn_3C. *J. Am. Chem. Soc. 64*, 1769-1770.

Strunz, H. (1943) Beitreg zum Pyrolusitproblem. *Naturwiss. 31*, 89-91.

Tiberg, N., and A. Muan (1970) Activity-composition relations at 1100°C in the system calcium oxide-iron oxide-manganese oxide in contact with metallic iron. *Metal. Trans. 1*, 435-439.

Towhidi, N., and D. Neuschutz (1972) Kinetik der Reduktion und elektrische Leitfahigkeit von Mangan (II)--oxid. *Archiv fur das Eisenhuttenwesen 3*, 219-227.

Trask, P. D. (1950) *Geologic Description of the Manganese Deposits of California*. California Div. Mines Bull. *152*, 378p.

Van Hook, H. J., and M. L. Keith (1958) The system Fe_3O_4-Mn_3O_4. *Am. Mineral. 43*, 69-83.

Varentsov, I. M. (1972) Geochemical studies on the formation of iron-manganese nodules and crusts in Recent basins. I Eningi-Lampi Lake, central Karelia. *Acta Mineral.-Petrogr. 20*, 363-381.

________ (1964) *Sedimentary Manganese Ores*. Elsevier, New York, 119p.

Verwey, E. J. W. (1935) The crystal structures of γ-Fe_2O_3 and γ-Al_2O_3. *Z. Kristallogr. 91*, 65-69.

Verwey, E. J. W., and J. H. DeBoer (1936) Cation arrangement in a few oxides with crystal structures of the spinel type. *Rec. Trav. Chim. 55*, 531-540.

Voeltzel, J., and J. Manenc (1967) Etude des solutions solides $(Fe,Mn)_{1-y}O$. *Mem. Sci. Rev. Metal. 64*, 191-194.

Wadsley, A. D. (1950) Synthesis of some hydrated manganese minerals. *Am. Mineral. 35*, 485-499.

Wasserstein, B. (1943) On the presence of boron in braunite and manganese ores. *Econ. Geol. 38*, 389-398.

Watanabe, T. (1959) The minerals of the Noda-Tamagawa Mine, Iwate Prefecture, Japan. *Mineral. J. 2*, 408-421.

________, S. Yui, and A. Kato (1970) Metamorphosed bedded manganese deposits of the Noda-Tamagawa Mine. In, T. Tatsumi, Ed., *Volcanism and Ore Genesis*, pp. 143-152.

Wells, A. F. (1950) *Structural Inorganic Chemistry*. Clarendon Press, Oxford, 590p.

White, D. F. (1955) Thermal springs and epithermal ore deposits. In, A. M. Bateman, Ed., *50th Anniv. Vol., Econ. Geol.*, pp. 99-154.

Wickham, D. G. (1969) The chemical composition of spinels in the system Fe_3O_4-Mn_3O_4. *J. Inorg. Nucl. Chem. 31*, 313-320.

Woermann, E., and A. Muan (1970) Derivation of approximate activity-composition relations in MgO-MnO solid solutions. *Mater. Res. Bull. 5*, 779-788.

Yamada, N., and S. Iida (1968) Details of the Richter type magnetic relaxations in Mn $Fe_{3-x}O_{4+\gamma}$. *J. Phys. Soc. Jpn. 24*, 952.

Zachariasen, W. (1928) Uber die Kristallstructur von Bixbyite, sowie vom kunstlichen Mn_2O_3. *Z. Kristallogr. 67*, 455-464.

Zwicker, W. K., W. O. J. G. Meijer, and H. W. Jaffe (1962) Nsutite-A widespread manganese oxide mineral. *Am. Mineral. 47*, 246-266.

Stephen E. Haggerty

Chapter 8

GENERAL INTRODUCTION

The opaque mineral oxides form a small but ubiquitous constituent of all igneous rocks and are an important mineral group from the standpoints of their yielding data on geo-hermometry and oxygen geobarometry, and because of their unique geophysical property of agnetism. This chapter is divided into the following four review topics: (I) systematic xide mineralogy; (II) the distribution of oxides among igneous rocks; (III) temperatures nd oxygen fugacities of coequilibrated oxide pairs; and (IV) a detailed tabulation of paque mineral oxides in igneous rocks.

The first section is structured to provide an overview of the mineralogy of igneous ocks and observed reactions, with an emphasis on reflection microscopy. The number of ptical parameters available for unambiguous identification are limited and hence the ection is covered largely in the form of photomicrographic plates for hands-on application hich is intended to substitute for the crux of the text. In the second section, the mphasis is on oxide distributions as a function of rock type considered both within the ontext of bulk chemistry and with respect to modes of emplacement. This review is a syn-esis of data which is covered largely in tabulated form in the final section and includes ta obtained by both classical mineral-chemical techniques as well as magnetic property chniques. The third section is a compilation of T°C and f_{O_2} values obtained for co-uilibrated oxide pairs and it includes an overall evaluation of the conditions of oxide rmation for a variety of rock types. The tabulated section is perhaps best described as ort notes for a course but this is, in essence, the body of the text for detailed study d the source for a majority of the literature referrals. While the intention, in part, that each of these four sections could stand alone, it is important that the reader oss-reference each of the sections for a final synthesis; the assumption in later sections ll be that the first section at least, has been thoroughly digested and understood.

SYSTEMATIC MINERALOGY

nenclature

Spinel series

The limits of solid solution of most direct interest to igneous rocks for minerals in spinel series fall into four categories (Figs. Hg-23-24).

(1) Solid solution between Fe_2TiO_4 (ulvöspinel) and Fe_3O_4 (magnetite). Members of the series are referred to as titanomagnetites and are abbreviated as Usp-Mt_{ss} (Fig. Hg-23a-b).

(2) Cationic deficient spinels lie along the join Fe_3O_4 and γFe_2O_3 (maghemite), and within the Fe^{3+}-rich field defined by Fe_3O_4-Fe_2TiO_4-$FeTiO_3$-Fe_2O_3. These latter spinels are titanomaghemites (Fig. Hg-23b).

(3) Titanomagnetite solid solutions incorporating $FeCr_2O_4$ (chromite), $MgCr_2O_4$ (picrochromite), $FeAl_2O_4$ (hercynite), $MgAl_2O_4$ (spinel), $MgFe_2O_4$ (magnesio-

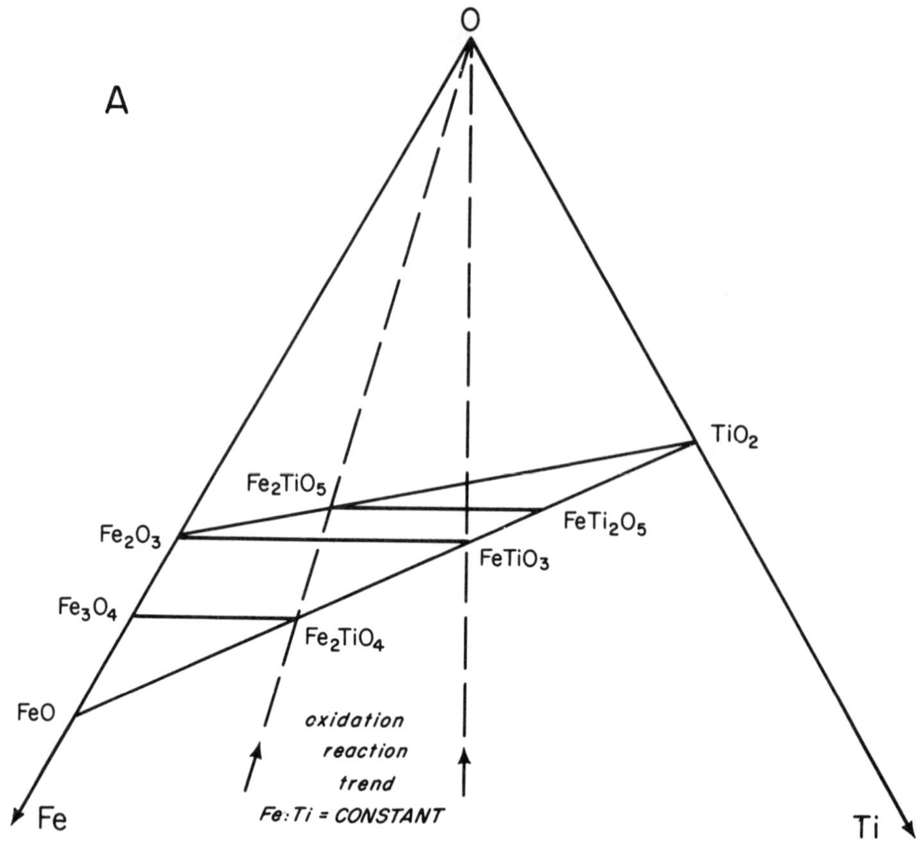

Figure Hg-23a. A portion of the system Fe-Ti-O containing the subsystem $FeO-Fe_2O_3-TiO_2$ Horizontal tie-lines are solid solution joins and dashed lines are oxidation and reduction trends for which Fe:Ti is constant.

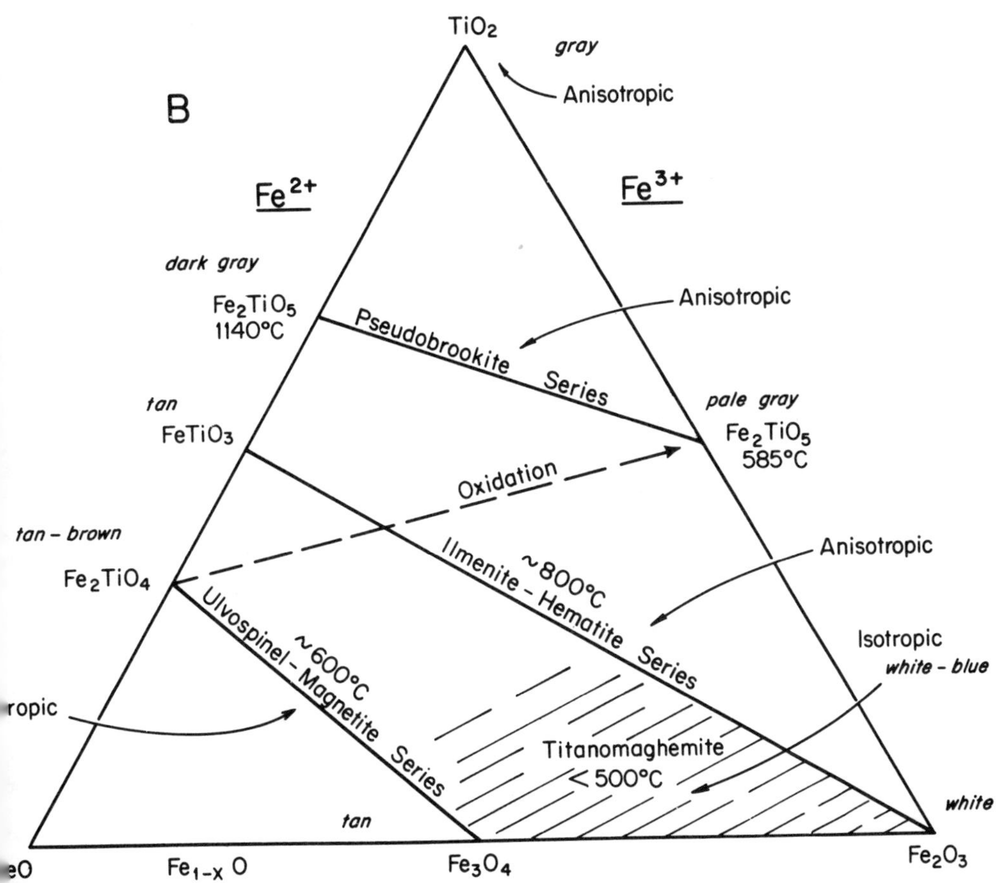

Figure Hg-23b. The subsystem $FeO-Fe_2O_3-TiO_2$ showing the major solid solution joins for the cubic, rhombohedral and orthorhombic series. Temperatures along joins are consolute temperatures (critical points for the solvi); temperatures for the orthorhombic end-members are lower thermal stability limits. Isotropic, anisotropic, and the color variations are the optical properties as viewed in reflected light oil-immersion.

ferrite) and Mg_2TiO_4 (magnesian-titanate) are referred to collectively as magnesian, aluminian and chromian titanomagnetites depending on the major dilutant (Fig. Hg-24a).

(4) Spinels listed in category 3 form extensive mutual solid solubility, are rarely present as pure end-members and are denoted either by solid solution subscripts, *e.g.*, $chromite_{ss}$ (Chr_{ss}) or by the varietal names as shown in Figure Hg-24a. Compositions within this multicomponent system may be conveniently plotted within an orthorhombic prism or onto projections of the spinel prism as illustrated in Figure Hg-24. Nickel, Zn, Mn, and V are rarely present in substantial proportions, although compositions approaching the end members trevorite ($NiFe_2O_4$), franklinite ($ZnFe_2O_4$), gahnite ($ZnAl_2O_4$), jacobsite ($MnFe_2^{3+}O_4$) and coulsonite (FeV_2O_4) are known.

Ilmenite series

The four members that comprise the ilmenite series in igneous rocks are ilmenite ($FeTiO_3$), hematite (αFe_2O_3), geikielite ($MgTiO_3$), and pyrophanite ($MnTiO_3$). Important solid solutions among these end-members are: (1) between ilmenite and hematite ($Ilm\text{-}Hem_{ss}$ where ilmenite-rich solid solutions are referred to as ferrian ilmenite, and hematite-rich solid solutions are termed titanohematite; (2) between ilmenite and geikielite ($Ilm\text{-}Geik_{ss}$ and (3) between ilmenite and pyrophanite ($Ilm\text{-}Pyh_{ss}$). Magnesium and Mn-rich ilmenites are termed picroilmenites and manganoan ilmenites, respectively. Zinc, Al, Cr and V are commonly present in minor concentrations but do not reach end-member proportions.

Pseudobrookite series

The pseudobrookite series of interest in igneous rock petrogenesis is defined by the end-members $FeTi_2O_5$ (ferropseudobrookite), pseudobrookite (Fe_2TiO_5) and karrooite ($MgTi_2O_5$ and by minor concentrations of tielite (Al_2TiO_5) in solid solution. Mutual solid solution are present among all end-members but those that dominate are between $FeTi_2O_5$ and Fe_2TiO_5 ($Fpb\text{-}Pb_{ss}$); between Fe_2TiO_5 and Mg_2TiO_5 ($Pb\text{-}Kar_{ss}$), where the intermediate member is kennedyite; and between $FeTi_2O_5$ and $MgTi_2O_5$ ($Fpb\text{-}Kar_{ss}$), where the intermediate member is armalcolite.

Apart from Al, chromium is the only other minor dilutant and does not reach end-member proportions.

TiO_2 polymorphs

The three polymorphs recognized are rutile, anatase and brookite. Minor concentrations of Zr, Nb, Ta and Fe^{3+} are commonly present.

Exsolution

The term "exsolution" is defined here as the separation of a one-phase solid solution into a two-phase intergrowth of solid solutions that is caused by cooling the one phase solid solution into a P-T region where a solvus is stable.

Exsolution within the spinel multicomponent system is most abundant between $Usp\text{-}Mt_{ss}$ and between titanomagnetites and the ternary spinels $FeAl_2O_4$-$MgAl_2O_4$-$MgFe_2O_4$. Exsolution within both subsystems is along {100} cubic spinel planes.

Exsolution among members of the ilmenite series is restricted to $Ilm\text{-}Hem_{ss}$ ($FeTiO_3$-Fe_2O_3) with exsolution developing along basal {0001} rhombohedral planes.

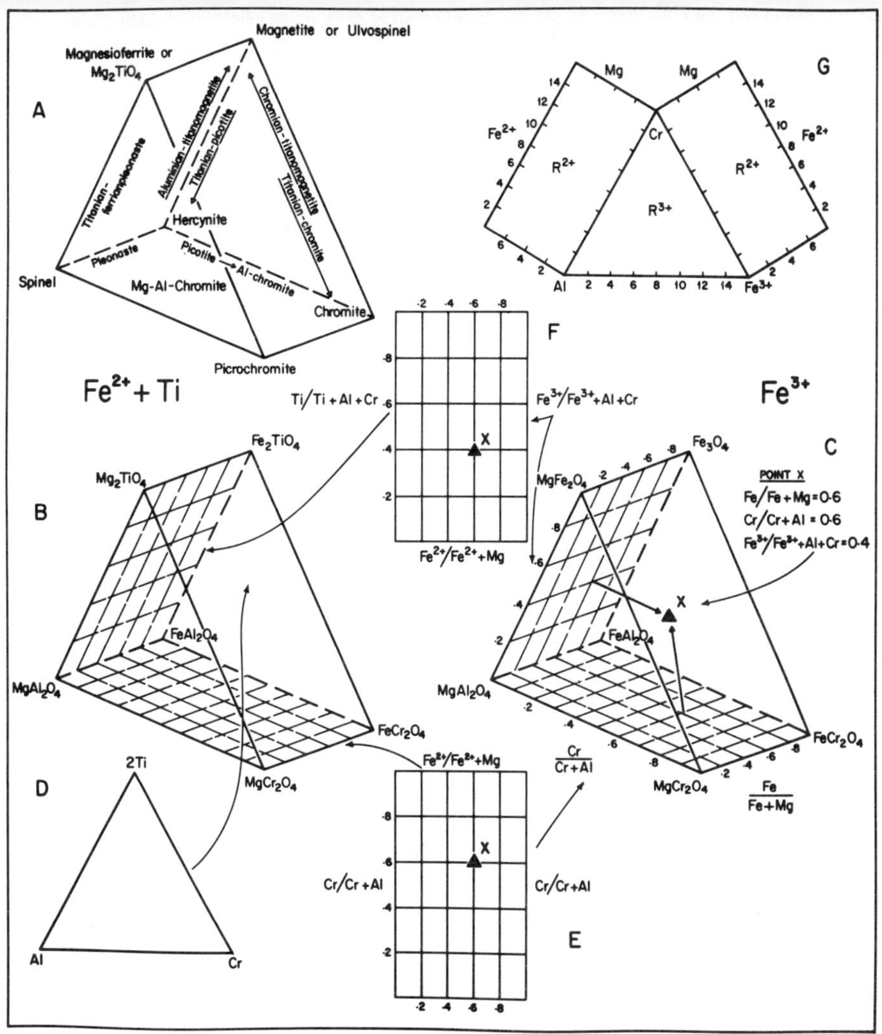

Figure Hg-24. Nomenclature and plotting procedures for compositions in the multicomponent spinel prism. The bases of the prisms are defined by normal spinels and the apecies by inverse spinels. Inverse spinels are divided into Fe^{2+}+Ti (b) and Fe^{3+} (c). The ternary on the left of each of these prisms has Mg as the common element, the ternary to the right has Fe as the common element. For the divalent prism Ti, Al and Cr are the variables. Because Fe^{2+} and Mg are common factors for each of the respective ternaries, the ratios shown in (e) define variations along the prism base. A useful projection is shown in (d), but the most widely used projection is that shown in (f) with the respective ratios required for projection above the prism base. These prisms are known as modified Johnson spinel prisms (Stevens, 1944; Irvine, 1965; Jackson, 1969; Haggerty, 1972) which differ slightly from the sliced and folded back prism shown in (g) which is known as the Pavlov projection (Pavlov, 1949; Hor, 1975; Smith and Dawson, 1975). In this Ti-free projection an analysis is first plotted in the ternary using the trivalent components, and a second plot of the same analysis is transferred to the rectangle on the left if Al > Fe^{3+}, or to the rectangle on the right if Fe^{3+} > Al; this second plot is specific to the divalent ions. The scale values represent the number of cations in tetrahedral and octahedral coordination for the spinel formula based on 32 oxygens and 24 cations.

Figure Hg-25

Mineral Morphology

(a) Dark gray chromian-spinel core mantled by titanomagnetite which is partially oxidi along the peripheral margins to titanomaghemite (white). 0.09 mm.

(b) Euhedral primary pseudobrookite crystal containing a cylindrical silicate glass in clusion with an attached crystal of chromian-titanomagnetite. 0.09 mm.

(c) Skeletal ilmenite with glass and finely crystalline inclusions. 0.09 mm.

(d) An ilmenite crystal with glass inclusions and with renewed second generation growt of T-shaped crystallites along the basal plane. 0.09 mm.

(e) - (h) Skeletal growth morphologies of titanomagnetite illustrative of a progressiv trend towards euhedral morphology. This trend is classified as the *cruciform* type 150 μm.

(i) This titanomagnetite crystal is also of the *cruciform* type with the distinction th growth, parallel to {111} spinel planes, takes place along the entire length of th primary cross-arms (*cf.* with e-h). 0.11 mm.

(j) Titanomagnetite crystal of the *multiple* cross-arm type. 80 μm.

(k) - (m) Titanomagnetite crystals of the *complex* type contain orthogonal and non-orth gonal multiple cross-arms. Growth patterns are evenly or haphazardly initiated at terminations, and along primary, secondary or tertiary cross-arms. k = 0.13 mm; l = 0.12 mm; m = 750 μm.

NOTE: The scale given after each caption or sets of captions is equal to the width of the photomicrograph in micrometers, or mm. *All plates are in reflected light oil immersion* (Figs. Hg-25-36).

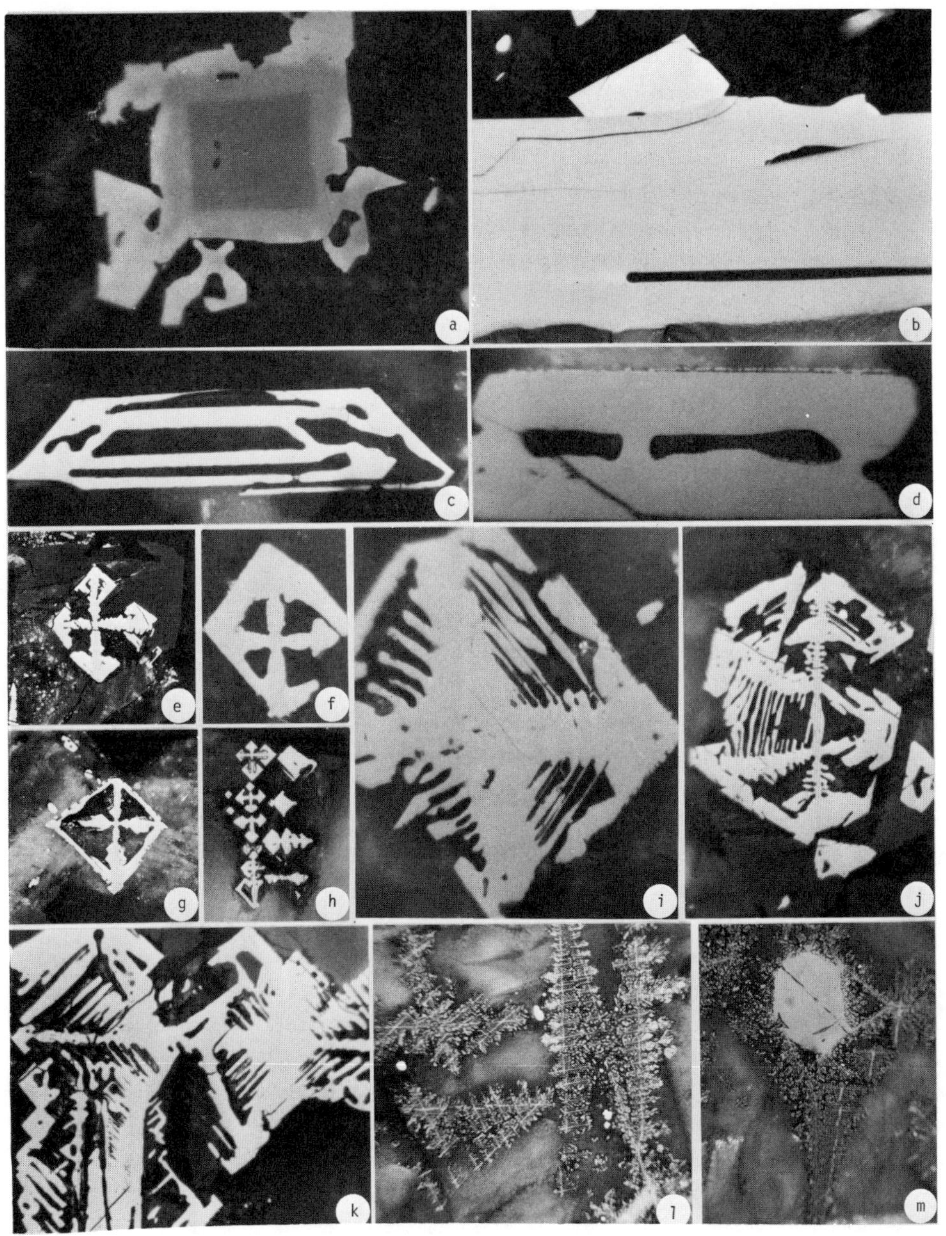

Figure Hg-25

Subsolidus reactions

Oxidation or reduction "exsolution" results from processes of subsolidus reaction. Textures produced by these mechanisms are akin to the crystallographically oriented intergrowths formed by exsolution as defined above. These textures do not form, however, by cooling a one-phase solid solution into a P-T range where a solvus is stable.

Oxidation "exsolution" is restricted to the formation of Ilm-Hem$_{ss}$ along {111} cubic spinel planes by oxidation of Usp-Mt$_{ss}$. Reduction "exsolution" is limited to Ilm-Hem$_{ss}$ with the formation of Usp-Mt$_{ss}$ along {0001} rhombohedral planes.

The term pseudomorphic oxidation is used to describe assemblages which result from the decomposition of titanomagnetite-ferrian ilmenite intergrowths at high T°C and f_{O_2}.

Metailmenite and metatitanomagnetite are used in the broadest sense to denote the detection of oxidation without the specific designation or identification of phase assemblages for members of the Ilm-Hem$_{ss}$ series and for members of the Usp-Mt$_{ss}$ series, respectively.

Maghemitization is restricted to the oxidation of Usp-Mt$_{ss}$ at relatively low T°C and results in titanomaghemite (Fig. Hg-23b).

Oxide assemblages and textures

Two-dimensional mineral morphology

The crystal habits of chromites, pseudobrookites, ilmenites and titanomagnetites are summarized in Figure Hg-25a-m and other examples are illustrated throughout this section in Figures Hg-26-36. These minerals are the major primary oxides in igneous rocks and the order in which they are listed is the typical paragenetic crystallization sequence. All minerals exhibit euhedral characteristics with chromites (Fig. Hg-25a) and titanomagnetite (Fig. Hg-25e-m) displaying cubic or modified octahedral forms, and with ilmenites (Fig. Hg-25c-d) and psuedobrookites (Fig. Hg-25b) in plate-like and lath-shaped forms typical of sections through crystals with rhombohedral and orthorhombic symmetry, respectively. The most prominently developed skeletal forms are seen in ilmenites and in titanomagnetites from lavas which have undergone rapid chilling. For ilmenites the direction of most rapi growth is normal to the c-axis, whereas secondary nucleation is more typically along {000] basal planes (Fig. Hg-25d). Crystal growth patterns for titanomagnetites are varied but these may be classified into the following common forms: (i) the *cruciform* type consists of a simple set of cross-arms at right angles which correspond to the crystallographic axes; growth is initiated at the extremes of these arms and arrow-heads develop which continue to grow until all sets coalesce (Fig. Hg-25e-h). An alternative variation to this pattern is illustrated in Figure Hg-25i with crystallization extending along the entire length of each of the primary cross-arms along directions parallel to {111} spinel planes; (2) the *multiple cross-arm* type is shown in Figure Hg-25j and in this type the cross-arms are neither orthogonal nor is there a preferred pattern of growth along either the cross-arms or at the extremities of the cross-arms; and (3) the *complex* types are illustrated in Figures Hg-25k-m and these are characterized by dendritic arrays of crysta lites which are attached to a central stem from which cross-arm growth is initialized alon the entire length of the stem at fairly regular intervals. This type is commonly observe on titanomagnetite xenocrysts as shown in Figure Hg-25m. The three types, cruciform, multiple cross-arm, and complex, are rarely observed in the same lava flow and although the explanation for these varied growth patterns is probably related to cooling rates and to the onset of co-crystallizing silicates the systematics remain to be established.

Chromian spinel$_{ss}$

Representative compositions of chromian spinels from a variety of geological settings are listed in Table Hg-20, and examples of the textures and assemblages commonly observed are illustrated in Figures Hg-26 and Hg-27; other examples which form exsolution bodies in pircoilmenite and in titanian-chromites are shown in Figures Hg-29 and Hg-33, respectively, and are discussed below.

Chromian spinel$_{ss}$, as defined here, is a term used in the broadest sense to denote compositions which are Cr-rich but which show extensive solid solubility among members on the base of the spinel prism (Fig. Hg-24), and for limited solid solubility with members above the prism base. These compositions are typical of mafic and ultramafic suites and in many cases chromian spinels may be identified optically because of extraordinarily complex zoning. This feature is particularly prevalent in basaltic suites and in kimberlites. For basalts (Figs. Hg-26a-f) the cores of crystals in the groundmass are most commonly enriched in Cr_2O_3, Al_2O_3, and MgO whereas the mantles are enriched in FeO, Fe_2O_3, and TiO_2; these distributions are apparent in the electron microprobe element distribution x-ray scanning images shown in Figures Hg-26c-f for a chromian spinel core mantled by titanomagnetite of the type comparable to the assemblage shown in Figures Hg-25a and Hg-26a. These mantles result by the reaction of early formed chromite with an Fe-Ti rich liquid and contrast with those chromites which are included in either olivine or pyroxene where the silicate-enclosed crystals commonly exhibit very thin reaction mantles or no mantles at all, as illustrated in Figure Hg-26b. For kimberlites, the basaltic trend of early Cr, Mg, Al and later Fe + Ti is also observed (Fig. Hg-26g-h); but the paragenetically later mantling phase can also become the cores for nucleation and growth of picroilmenite as shown in Figure Hg-26i. In rare instances for groundmass spinels, but commonly in garnet kelyphitic rims, the cores show a preferred enrichment of Mg, Al, and Fe^{3+}, with intermediate zones which are enriched in Cr and Fe^{2+}, and with outermost zones which are enriched. These complex distributions are illustrated in Figure Hg-26j for a groundmass crystal, and in Figures Hg-26k and l for two crystals in kelyphite.

Figure Hg-27 illustrates a variety of textures between chromites and silicates, and chromites and titanomagnetites. Silicates, glass and sulfide inclusions are shown in Figure Hg-27a; the effects of magmatic corrosion and resorption are in Figures Hg-27b-c; mantling by titanomagnetite in Figures Hg-27a-c; coprecipitating growth of chromite + olivine in Figure Hg-27; the oxidation at high temperatures (>600°C) of chromite + titanomagnetite in basalts (Figs. Hg-27e-f); and the alteration associated with serpentinization in partially decomposed harzburgite in Figures Hg-27g-h. Additional discussions of the relationships of these overgrowths and of the oxidation trends are to be found in Table Hg-19 and in Chapter 4. Neither the resorption characteristics, nor the high-temperature oxidation assemblages of chromian spinels have received attention in the literature which is in contrast to the reaction of chromite to titanomagnetite (*e.g.*, Evans and Moore, 1968; Beeson, 1976) and of the alteration of chromite to ferritchromite and magnetite (*e.g.*, Onyeagocha, 1974; Engin and Aucott, 1971; Springer, 1974; Ulmer, 1974; Hamlyn, 1975; Bliss and Maclean, 1975).

Figure Hg-26

Chromian Spinel$_{ss}$

(a) Asymmetrically mantled, chromian-spinel cluster partially enclosed in olivine. The mantles are titanomagnetite and the intermediate zones are chromian-titanomagnetites The adjacent discrete crystal of titanomagnetite, although incomplete, would be classified as the *cruciform* type. 0.16 mm.

(b) Euhedral spinel core mantled by titanomagnetite. Mt$_{ss}$ are also present along the crack and in patches associated with inferred cracks below the polished surface. 0.16 mm.

(c-f) X-ray scanning images obtained by electron microprobe illustrating the distributi of major elements for a chrome spinel-titanomagnetite core-mantle relationship. Th intensity of the spots are approximately proportional to concentration. The elemen displayed are Fe (in c), Ti (in d), Cr (in e) and Al (in f); the core is thus virtu ally Ti-free (d) and the mantle virtually Cr (e) and Al-free; both the mantle and t core contain Fe (c). 0.11 mm.

(g-h) Multiple zoning in chromian-spinels from kimberlites with chromite cores and magnetite mantles. g = 0.15 mm; h = 0.08 mm.

(i) Euhedral magnetite core, epitaxially overgrown by picroilmenite. 0.09 mm.

(j) Oscillatory zoning in spinel where the white areas are Mg, Al and Fe^{3+} enriched, t darker areas are Cr+Fe^{2+} enriched, and the mantle is Ti enriched.

(k-l) Compositionally similar to (j), but these spinels are present in garnet kelyphit rims associated with Ti-phlogopite. k = 0.09 mm; l = 0.07 mm.

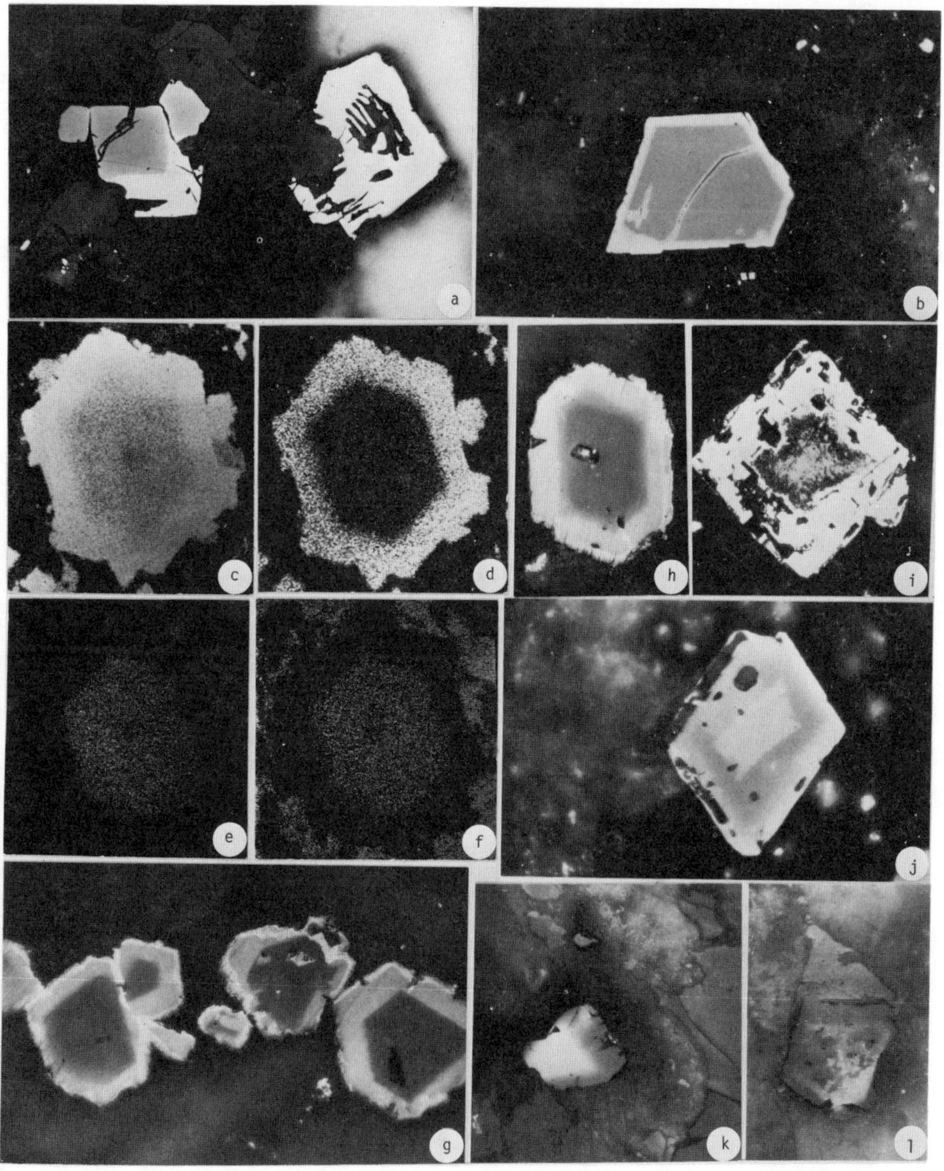

Figure Hg-26

Figure Hg-27

Chromian Spinel$_{ss}$ Reactions

(a) Glass inclusions in a phenocrystic chromian spinel mantled by titanomagnetite; an island of chromite in the glass suggests magmatic corrosion. The lighter mantle is titanomagnetite. 0.14 mm.

(b) Coarse web-shaped chromian spinel with partially crystalline, and dominantly glass inclusions. Note that the extent of the titanomagnetite mantle is a function of th size of chromite and that the internal cores have a similar morphology to the outer edges. 0.16 mm.

(c) A symplectic internal core of chromian spinel mantled by successive barriers of later chromite and an outer margin of titanomagnetite; most of the cuniform segmen in the core are also mantled and the smallest areas contain the largest mantles of Mt_{ss}. 0.15 mm.

(d) Euhedral chromian spinel core mantled by subgraphic chromite + glass + olivine. T overall outer-morhpology of the symplectite is broadly similar to that of the core 0.16 mm.

(e) Irregular glassy inclusions in chromite mantled by titanomagnetite which has under gone oxidation "exsolution" and subsequent partial decomposition to R + Hem_{ss}. 0.16 mm.

(f) The dark central core is chromian spinel, the outer assemblage is oxidized titano-magnetite, and the white attached areas were Ilm_{ss} but are now R + Hem_{ss}. The lig oriented {111} trellis lamellae were also originally Ilm_{ss} but are now R + Hem_{ss}; residual host Mt_{ss} are still apparent, and these contain dark, oriented pleonaste rods. 0.15 mm.

(g-h) These two examples illustrate the features typical of "ferritchromit" which res during the serpentinization of chromite in ultramafic rocks. The outer mantles a Fe^{3+}-rich chromian magnetites, and the exchange chemistry is complex and varied (see Table Hg-21). 0.15 mm.

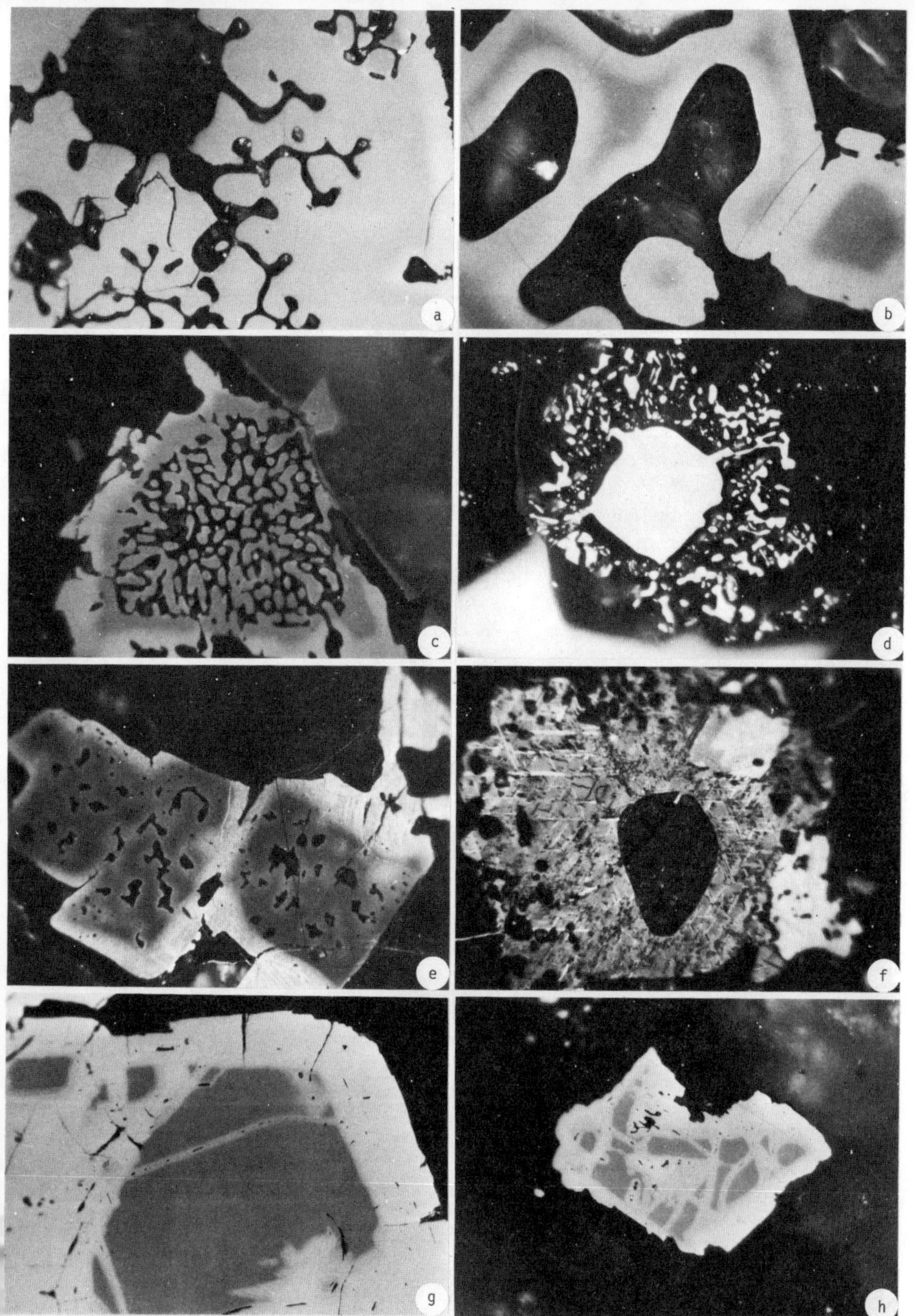

Figure Hg-27

Figure Hg-28

Ilmemite-Hematite$_{ss}$

(a-b) The light-gray hosts are titanohematite and the oriented darker-gray lenses are ferrian-ilmenite. In (a) individual grains show sharp terminations and large con centrations of ilmenite at the grain boundaries. In (b) the grains are rounded, and these preferred concentrations are not as evident. Each of the larger ilmenit lenses is surrounded by a depletion zone, and the regions between the large lenses are occupied by finer lenses which are assumed to have formed at T°C lower than those of the larger bodies. This distribution is the synneusis texture. 0.15 mm.

(c-d) The cores of these Hem$_{ss}$ grains contain similar distributions in Ilm$_{ss}$ lenses to those shown in (a-b), but here the development of thick mantles of ilmenite has re sulted from very extensive migration during exsolution. These mantles contain a second generation of exsolved Hem$_{ss}$, and for all sets which are in optical continuity, the plane of exsolution is {0001}. 0.15 mm.

(e-f) Low and high magnifications, respectively, of Hem$_{ss}$ cores with exsolved lenses of Ilm$_{ss}$ and with mantles of Ilm$_{ss}$. In addition to these lenses are coarse lamellae of pleonaste which share the same plane of exsolution as those of the ilmenite. The outer margins result in a symplectic intergrowth with Mt$_{ss}$. At high magnification it is evident that each of the pleonaste lamella is surrounded by Ilm$_{ss}$, suggesting that the spinel predated Ilm$_{ss}$ exsolution. Whether these spinels are the result of exsolution or an exsolution-like process is not clearly understood. e = 0.15 mm; f = 400 µm.

(g-h) The reverse relationships are illustrated here for ilmenite hosts and Hem$_{ss}$ exsolution. The plane of exsolution is still {0001} but the notable difference her is that the lenses are extremely fine grained and that depletion zones, or zones non-exsolution, are concentrated along the grain boundaries. 0.15 mm.

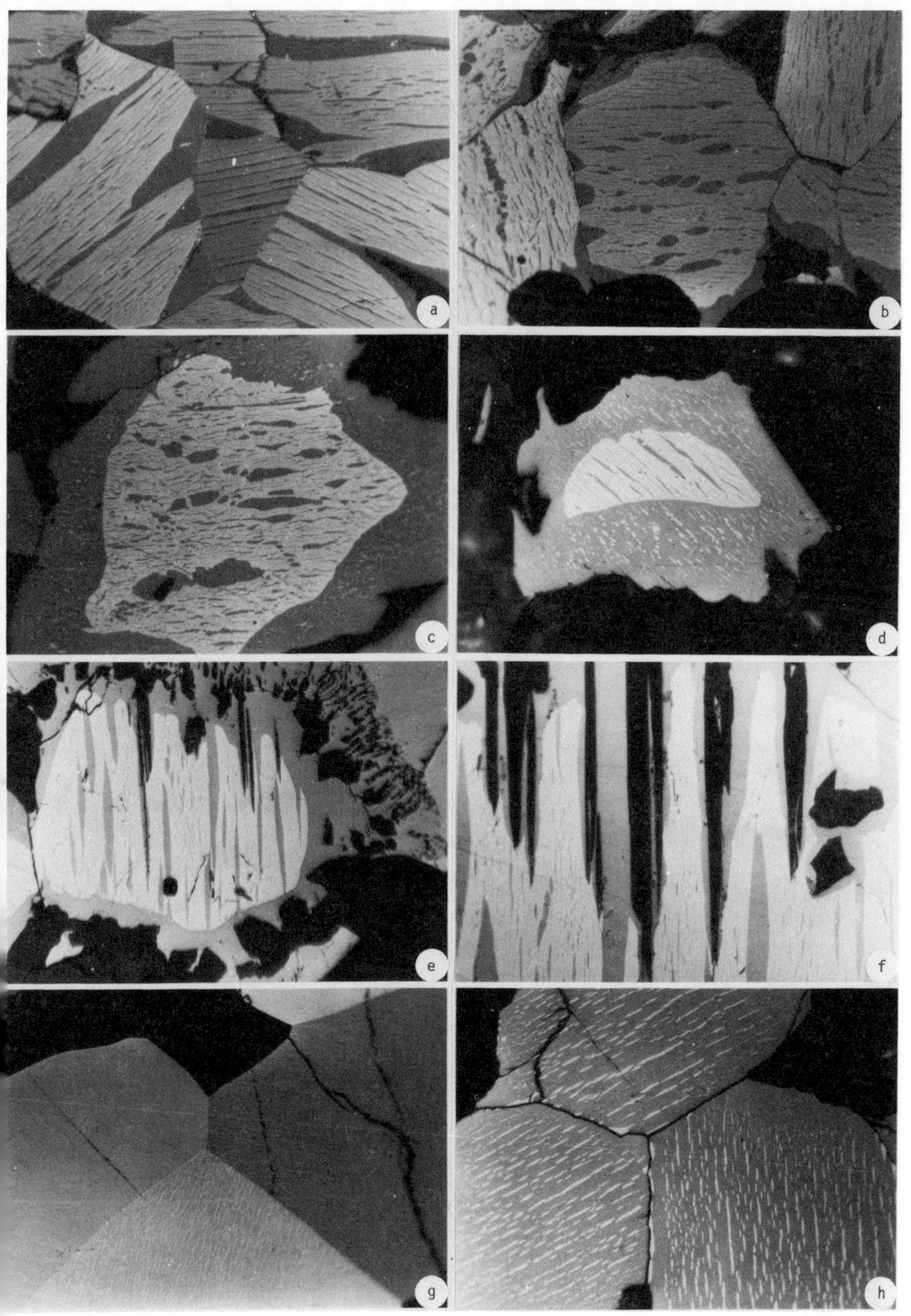

Figure Hg-28

Figure Hg-29

"Exsolution" in $Ilmenite_{ss}$, $Hematite_{ss}$, and Rutile

(a-b) Discontinuous oriented rods of titanian chromites in kimberlitic picroilmenites. Both grains show a varied distribution of lamellae with respect to Ilm_{ss} grain boundaries but the rods are uniform in size. The plane of exsolution is assumed to be parallel to {0001}. 0.15 mm.

(c-d) Although the oriented lamellae in these picroilmenites have distinctly different optical contrasts with respect to each other and with respect to their hosts, the lamellae in both grains are Mg-Al-titanomagnetites; the lighter lamellae are high in Fe^{3+}, and the darker lamellae are enriched in Mg and Al. The bleached zones in (c) are the result of partial oxidation and the surrounding groundmass crystals a perovskite. Peripheral rutile is present in grain (d) and a small amount is also present in grain (c). 0.15 mm.

(e-h) The host in these grains is titanohematite; the darker gray lenses are Ilm_{ss}; t z-shaped lighter gray lamellae are rutile in the typical *blitz* texture. One ruti lamella is present in (e), along which abundant Ilm_{ss} have exsolved; no Ilm is present in (g). The ilmenite exsolved plane is {0001} and the rutile exsolved planes are assumed to be $\{01\bar{1}1\}$ and $\{01\bar{1}2\}$; the former results by exsolution *sens stricto*, whereas the latter is more likely the result of an "exsolution"-like process related to oxidation. 0.08 mm.

(i) A rutile phenocryst containing sigmoidal lenses of Ilm_{ss} and mantled by Ilm_{ss}, Mt and perovskite. 0.15 mm.

(j) A picroilmenite crystal with a core of rutile; the rutile contains irregular incl sions of Ilm_{ss} rather than the oriented arrays typical of the association. The m ginal and groundmass grains are perovskite. 0.15 mm.

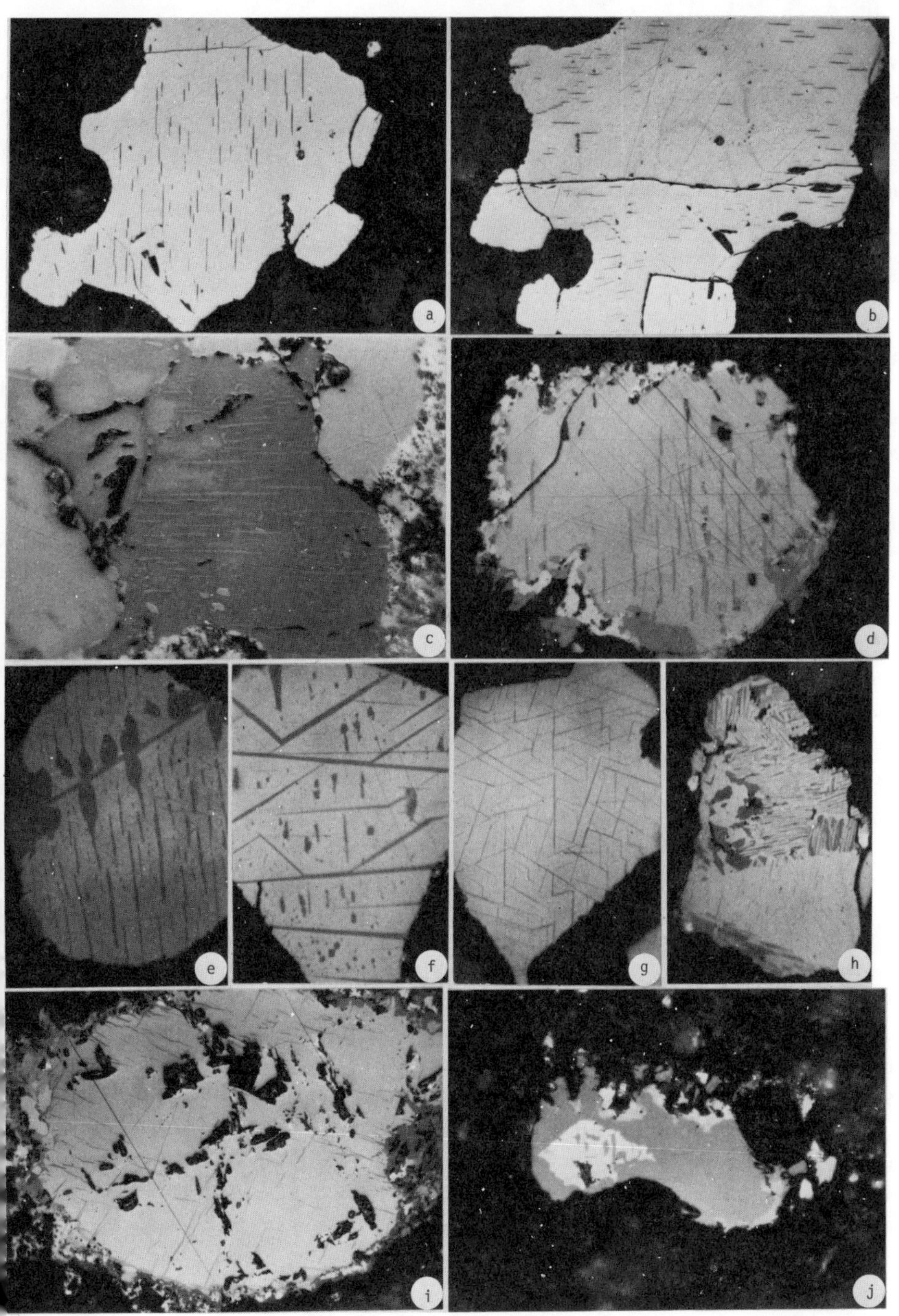

Figure Hg-29

Ilmenite-hematite$_{ss}$

A review of the textural assemblages associated with members of the Ilm-Hem$_{ss}$ series are illustrated in Figures Hg-28-31. Exsolution characteristics are illustrated in Figure Hg-28, and the experimental phase relationships are discussed in Chapter 2 for the series Fe_2O_3-$FeTiO_3$. Exsolution of Hem$_{ss}$ from Ilm$_{ss}$, and of Ilm$_{ss}$ from Hem$_{ss}$ is restricted to deep-seated intrusions and are particularly characteristic of anorthosite associations and other basic suites, but are also present in granitic suites. The plane of exsolution is parallel to the {0001} rhombohedral direction, and exsolution takes place in accord with decomposition as a consequence of slow cooling and solvi-intersection. The growth and distribution of large and finer lamellae results in a synneusis texture, *i.e.*, with finer lenses of the solute (exsolving phase) distributed between thicker lenses in the solvent (host-dissolving medium). Under conditions of extreme and prolonged cooling, diffusion of the solute migrates towards crystal boundaries and atoll textures result. Equally common, although not within the resolution of these photomicrographs, are second and in some cases tertiary exsolution bodies within the primary lamellae; these relationships suggest a process of exsolution which yields successive sets of compositional pairs of Ilm$_{ss}$ and of Hem$_{ss}$ which conform respectively to conjugate sets of lamellae whose compositions are controlled by the slopes and the limiting boundaries defined by the immiscible solvus region. Textural interpretations of the series are considered by Edwards (1965) and by Ramdohr (1969), and the magnetic self-reversal properties of the series are discus[sed] by Carmichael (1961,1962). Solvus relationships were established by Carmichael (1961) on natural material, and the revised experimental solvus by Lindsley (1973) is discussed in Chapter 2. A detailed treatment of exsolution mechanisms in the series has been published by Kretchsmar and McNutt (1971) in which they suggest that exsolution of compositi[ons] with >Ilm$_{85}$ is continuous, whereas with initial compositions which are more enriched in Hem$_{ss}$, exsolution proceeds discontinuously. This interpretation is consistent with the textures illustrated in Figure Hg-28 inasmuch as there is a greater preponderance of discontinuous or second generation lenses in grains where hematite solid solution members fo[rm] the host and Ilm$_{ss}$ members the exsolved solute; for the reverse relationship of Ilm$_{ss}$ hos[ts] the exsolved Hem$_{ss}$ solute is a relatively evenly-distributed single generation constituen[t] (compare for example, Figs. Hg-28g and Hg-28h with Figs. Hg-28b and Hg-28f).

Other examples of previously assumed exsolution are shown in Figure Hg-29, and in these crystals the exsolving solutes are spinels from picroilmenite and of rutile from titanohematite. These examples are typical of Ilm$_{ss}$ in kimberlites, and of Hem$_{ss}$ in more acid suites. The exsolved phases have an exsolution appearance with respect to crystallographic control, and with respect to the morphology of crystal lamellae. However, for both the Ilm$_{ss}$ and the Hem$_{ss}$ the solvent and the solute have differing crystal symmetries and neither the host nor the exsolved phase form members of either a continuous or a discontinuous solid solution series. The exsolved constituents of the kimberlitic picroilmenites are titanian-chromites (Fig. Hg-29a-b), and Mg-Al-titanomagnetites (Fig. Hg-29b-d); these phases are oriented along {0001} rhombohedral planes which is consisten[t] with the parallel planes of exsolution exhibited in Ilm-Hem$_{ss}$ (*i.e.*, exsolution *sensu stricto*). In the case of Hem$_{ss}$ (Fig. Hg-29e-h), two of the crystals shown contain Ilm$_{ss}$ lenses in addition to rutile. The grain in Figure Hg-29e has a single thick lamella and many finer lamellae which are clearly earlier than the exsolution of Ilm$_{ss}$. In Figure Hg-29f the paragenetic relationships are not definitive but here rutile is in far reater abundance. Figures Hg-29g and h show extreme examples of the development of rutile n Hem_{ss} with well-defined crystallographic control along {01$\bar{1}$1} rhombohedral planes re-ulting in Ramdohr's (1969) "blitz" texture. The abundance of rutile in these two latter nstances is far in excess of the limits for the solid solubility of TiO_2 in members along he join Fe_2O_3-$FeTiO_3$, in the absence of Pb_{ss} members. Therefore, these assemblages are ost likely the result of an exsolution-like process which is related to oxidation in much he same sense that Ilm_{ss} "exsolve" by oxidation exsolution from Usp-Mt_{ss} members.

The reverse relationship of Ilm_{ss} lamellae in rutile hosts is illustrated in Figures g-29i and j and in two groundmass grains from kimberlites. For both grains the perplexing roblem of whether true exsolution or an exsolution-like process is responsible for this abric is compounded by the fact that rutile and Ilm_{ss} are at times present in approximately ual concentrations; this bears a much closer relationship to the expected modal concen-ations which are likely to develop if decomposition of a Pb_{ss} member, in the system $_2TiO_5$-$FeTi_2O_5$-$MgTi_2O_5$, is invoked as the precursor to such bimodal assemblages, and since ch compositions have been recognized (*i.e.*, compositions approaching those of lunar ar-lcolites (FeMg)Ti_2O_5 (Haggerty, 1975), the likelihood of decomposition rather than ex-lution is favored. For the cases of minor Ilm_{ss} in rutile the probability of unmixing om a system with limited solid-solubility may still be possible from an initially high mperature and homogeneous one-phase mineral. The mechanisms remain unresolved.

Reactions involving ilmenite-hematite$_{ss}$

In contrast to the uncertainties aired above, the effects of magmatic and metasomatic actions and the resulting assemblages which form from early and late stage deuteric idation and reduction are reasonably well defined. Examples of the assemblages and the xtures derived from each of these modifying processes are illustrated in Figures Hg-30 l Hg-31.

The first process to be considered is magmatic metasomatism which falls into at least following three distinct categories characterized by the major newly developing Ti-iched constituent: (1) the formation of sphene ($CaTiSiO_5$); (2) the formation of perov-te ($CaTiO_3$); and (3) the formation of aenigmatite ($Na_2Fe_5TiSi_6O_{20}$). Each of these ctions is illustrated in sets of photomicrographs in Figure Hg-30. Sphenitization relatively common in regional low-grade metamorphic terrains, but in many underformed tectonically undisturbed deep-seated intrusives and in recent hypabyssal suites, the elopment of sphene as a reaction product of Ilm_{ss} or of titanomagnetite (discussed er) is pervasive, suggesting that it constitutes a product of autometasomatism by ction of residual liquids with early formed Fe-Ti oxides. The formation of primary secondary perovskite is restricted to undersaturated suites, and the most spectacular nples are in melilite basalts, kimberlites, and in carbonatites. For the latter two es, Ilm_{ss} are enriched in $MgTiO_3$ and decomposition results also in the formation of and Ti-rich spinels some of which are magnesian-titanomagnetites in composition while rs show variable enrichment in Cr_2O_3, Al_2O_3 and FeO. In the cases of sphene and vskite reactions, the additive metasomatic ions are Ca + Si, and Ca, respectively. perovskite reactions Fe from the ilmenite is accounted for in the formation of a el, and in low abundance levels of Fe in $CaTiO_3$. However, for sphene reactions the ss Fe is restricted to the very limited formation of Hem_{ss}; Fe is most commonly nt in analyses of $CaTiSiO_5$, and the formation of associated late-stage amphiboles phyllosilicates are usually regarded as the potential mineral sinks that develop

Figure Hg-30

Ilmenite$_{ss}$ Reactions

(a-b) Sphene ($CaTiSiO_5$) replacement of Ilm_{ss} are illustrated in both grains; rutile is an associated phase in (a), and Hem_{ss} lamellae are exsolved from ilmenite in (b). The variation in color which is particularly evident in (a) results from variations in crystal orientation and reflection pleochroism. 0.15 mm.

(c-f) The progressive decomposition of picroilmenite illustrated in this series results initially in the marginal formation of Mg-rich Mt_{ss} + perovskite; the spinels are dark gray, and perovskite ($CaTiO_3$) is white. With more intense decomposition (f) Hem_{ss} + R are associated phases. (c, d & f) = 0.15 mm; (e) = 750 μm.

(g-h) The grains illustrate the replacement of Ilm_{ss} by aenigmatite ($Na_2Fe_5TiSi_6O_{20}$) which appears as the dark gray constituent. Grain (g) is a discrete ilmenite crystal, whereas grain (h) has {111} oriented Ilm_{ss} lamellae in Mt_{ss} and an *external composite* ilmenite. Note that it is only the Ilm_{ss} which is selectively replaced by aenigmatite (also known as cossyrite) and that the Mt_{ss} remains unaffected. 0.15 mm.

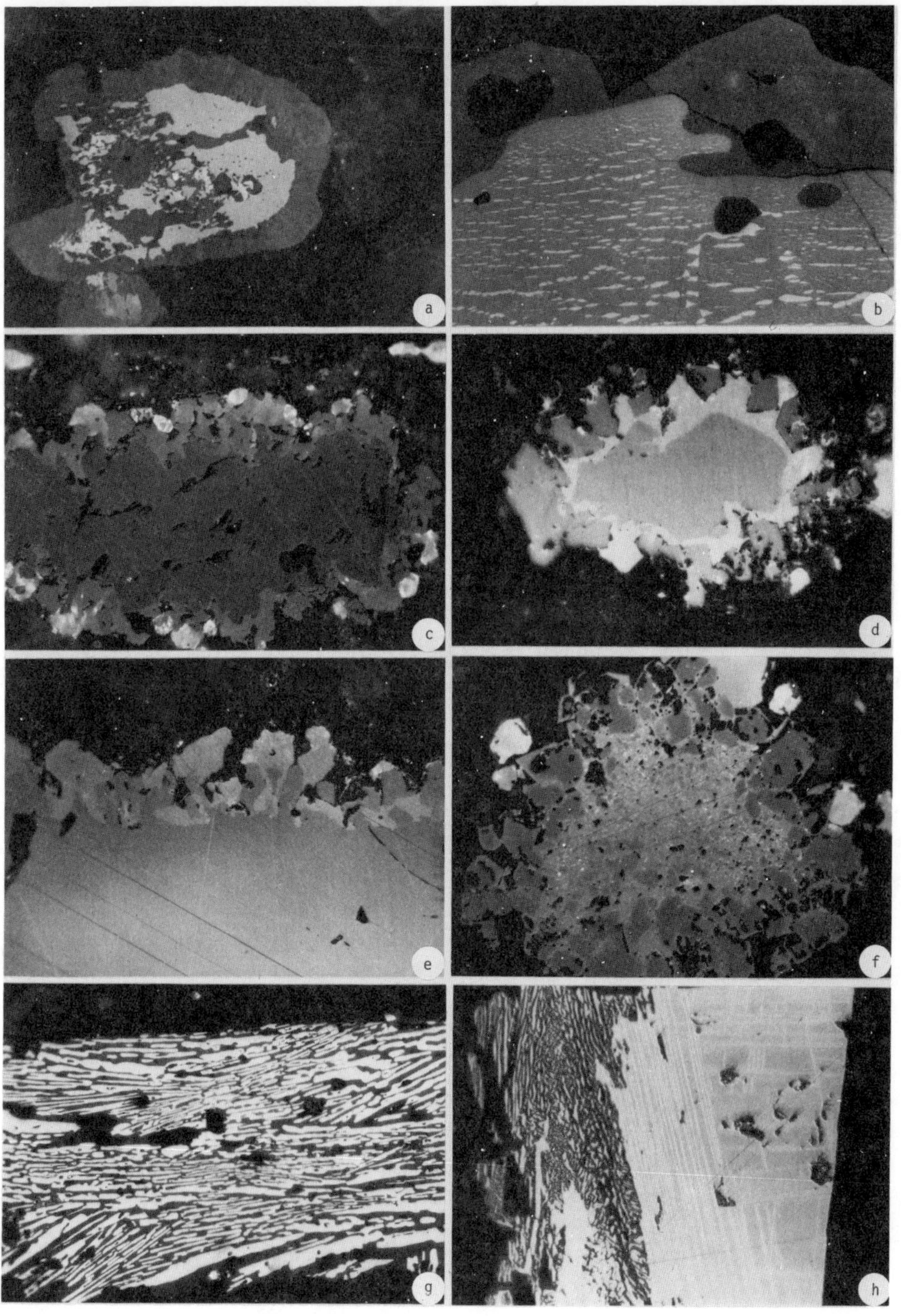

Figure Hg-30

during metasomatism which are capable of accommodating fairly sizeable proportions of Fe. In many cases, however, the Fe-exchange process is not evident in the metasomatized halos adjacent to individual crystals, so that the Fe is either redistributed among other Fe-bearing minerals, which may be possible at elevated temperatures, or alternatively removed from the system in aqueous solutions if the reactions take place at correspondingly lower temperatures.

Aenigmatite reactions differ from those of sphene and perovskite insofar as the problem of Fe reconstitution is accounted for, and in the sense that the metasomitizing ions are Na and Si rather than Ca + Si or Ca. Aenigmatite replacement of ilmenite and of the ilmenite constituents of oxidized titanomagnetites are observed in pegmatoid zones in thick differentiated basalts, and these zones in common with the primary crystallization of aenigmatite are typical of peralkaline host rock compositions. Examples of aenigmatite replacement are illustrated in Figure Hg-30g-h and in the case of the titanomagnetite grain (Fig. Hg-30h), it is clear that aenigmatite replacement post-dates the oxidation exsolution of Ilm_{ss} from the titanomagnetite. This is consistent with experimental data for aenigmatite which show that synthetic compositions have a maximum stability below 900°C (Lindsley, 1971), and with the textural evidence that aenigmatite is a late-stage interstitial liquid reaction product.

Reduction "exsolution" and the oxidation of Ilm_{ss} are illustrated in Figure Hg-31. Buddington *et al.* (1963) and Buddington and Lindsley (1964) have discussed the formation of Mt_{ss} from $Ilm\text{-}Hem_{ss}$ by subsolidus reduction. These authors have noted the occurrences of subsolidus reduction of Ilm_{ss} in hornblende granite gneisses from the Adirondacks and in paragneisses of the Franklin, New Jersey area. $Magnetite_{ss}$ develop as lenses or lamellae along planes parallel to the basal planes of the ilmenite; the texture has been simulated experimentally (Buddington and Lindsley, 1964), and the only known wide-spread occurrences in igneous rocks are in trachybasalts from Teneriffe (Haggerty *et al.*, 1966) and in andesites from Peru (Haggerty and Wilson, unpublished). Examples of the latter are illustrated in Figure 31a-b, and in the three ilmenite grains shown, the reduction exsolution lamellae of Mt_{ss} are partially oxidized to titanomaghemite. An explanatio for the trachybasalts cannot be given, but the widespread occurrence of these intergrowths in the Peru andesites suggests that this assemblage may have resulted during deuteric cooling; an alternative mechanism could also be invoked which would relate reduction to sulfide hydrothermal activity.

Apart from these examples of subsolidus reduction, $spinel_{ss}$ are also present in kimberlitic picroilmenites as discussed above with reference to Figure Hg-29a-d, and a second example from the deeply weathered pipes of west Africa are shown in Figure Hg-31c-d. Thes examples are for ilmenite xenocrysts and contrast with the groundmass ilmenites discussed previously. Surface weathering is highly selective in replacing the Mt_{ss}, with a tendency to leave the more resistant Mg-rich ilmenites virtually untouched. The overall similarity between these kimberlitic ilmenites and those in andesites would tend to favor a subsolidu reduction mechanism, and the reducing agent is perhaps more clearly defined in the former and is related to $CO:CO_2$ equilibria. However, the possibility still exists that the formation of Mt_{ss} is not related to subsolidus reduction but to a process of low-temperatu chemical weathering. It is relevant to note: (1) that oxidation exsolution lamellae of Ilm_{ss} in Mt_{ss}, that coexist with Ilm_{ss} having reduction lamellae of Mt_{ss}, are moderately common in the Peru andesite suite but extremely rare in kimberlites; and (2) that there are no visibly apparent reduction modifications to coexisting $spinel_{ss}$.

The high-temperature oxidation of Ilm_{ss} specific to basalts is discussed in Chapter 4. For Ilm_{ss} in other igneous rocks the progressive sequences of oxidation are less well established, and in rock types other than those of the extrusive basic suite, the prevalent decomposition assemblage is Hem_{ss} + TiO_2. Examples in which the TiO_2 polymorph is rutile (based on x-ray data) are illustrated in Figure Hg-31e-g for three grains from a gabbroic anorthosite; additional examples in which the TiO_2 polymorph is anatase (based on x-ray data) are illustrated in Figure Hg-31h-j for a progressive sequence of oxidation in a kimberlitic ilmenite megacryst. Adjacent to the ilmenite the assemblage is finely textured, and is anatase + Hem_{ss} (h); in an intermediate zone the Hem_{ss} is progressively removed and coarse crystals of anatase result (i); at the outermost grain boundary the Hem_{ss} is largely removed and a coarse cavernous network of anatase is the end product. This example is most typical of kimberlitic Ilm_{ss} and is most likely the result of low-temperature magmatic fluidal interaction and later supergene dissolution.

Ulvöspinel-magnetite$_{ss}$

The consolute point for the solvus of the series Fe_2TiO_4-Fe_3O_4 is at approximately 600°C and compositions along this join are sensitive to T°C and f_{O_2} conditions in magmas which co-crystallize members of the Fe_2O_3-$FeTiO_3$ solid solution series as discussed in Chapter 2. The relatively low temperature of the critical point yields compositions which may be quenched from above 600°C in most extrusives and in some high-level hypabyssal suites; hence, compositions which span the entire series are theoretically possible although the extremely low f_{O_2} values required for the formation of stoichiometric Fe_2TiO_4 dictates that it is relatively rare as a discrete mineral. Under conditions of slow cooling, however, for members within the low temperature immiscible region, unmixing or phase exsolution into Usp_{ss}-rich and Mt_{ss}-rich solid solution members form when the solvus is intersected. This results in exsolution and the plane of exsolution nucleation is parallel to {100} spinel planes. Textures are described as woven, cloth, parquet or lit-par-lit. Examples of Mt_{ss} exsolving Usp_{ss} and of the reverse relationship are illustrated in Figure Hg-32. These photomicrographs are most typical of those examples described in the literature by Ramdohr (1953,1969), Vincent and Phillips (1954), Nickel (1958), Vincent (1960), Vaasjoki and Heikkinen (1962), and Tsvetkov (1965). These examples, and from the reports by many other investigators, *e.g.*, Anderson (1966), Morse and Stoiber (preprint), there is a clearly defined relationship between exsolution of members along the join and deep-seated basic intrusives of the gabbroic and anorthositic suites. Hypabyssal occurrences have been described by Vaasjoki and Heikkinen (1962) and Jensen (1966) but in general even dykes appear to quench in members with intermediate compositions and exsolution is inhibited.

The examples illustrated in Figure Hg-32 are from the Kiglapait intrusion, Labrador, and from the Cumberland stock, Rhode Island. In samples from both localities Ilm_{ss} are associated oxidation "exsolution" constituent and the relative parageneses between exsolution of Usp_{ss}, or of Mt_{ss}, and of Usp_{ss} oxidation to Ilm_{ss} are variable. For example, in some samples Ilm_{ss} predate Usp_{ss} exsolution, and oxidation must therefore have taken place above 600°C; in other instances the formation of Usp_{ss} predates that of the Ilm_{ss}, and oxidation was hence below 600°C. Exsolution produces two constituents which are both more highly enriched in Usp_{ss} and therefore FeO, and a second phase which is more highly enriched in Fe_3O_4 and therefore in Fe^{3+}. The former is more susceptible to oxidation and the latter consequently less susceptible to oxidation than the original

Figure Hg-31

Ilmenite$_{ss}$ Reactions

(a-b) The hosts in these grains are Ilm_{ss}, and the oriented lamellae along {0001} planes are Mt_{ss} which show partial to complete decomposition to titanomaghemite. Note that the cores abound with lamellae whereas the margins are lamellar-free. The lamellae result by subsolidus reduction and subsequent low T°C oxidation. 0.15 mm.

(c-d) Deformed Mt_{ss} lamellae in picroilmenite which show partial oxidation to Hem_{ss} (c) and dissolution of these lamellae as shown in (d). The darker mantles which are free of lamellae are Mg-titanomagnetites. 0.15 mm.

(e) The dark gray host is ilmenite, the white ellipsoidal bodies are Hem_{ss}, and the lighter gray sigmoidal phase is rutile. Note that R alone is absent in the ilmenite so that although the texture is suggestive of exsolution, the reaction $2FeTiO_3 + 1/2O_2 = Fe_2O_3 + 2TiO_2$ is more likely. 0.09 mm.

(f) A core of Hem_{ss} with oriented Ilm_{ss}; the mantle is ilmenite with fine exsolution bodies of Hem_{ss}. Ilmenite within the core as well as ilmenite in the same outer rim, show decomposition to an intimate intergrowth of Hem_{ss} + R. The resulting assemblage is recognized optically by yellow to red internal reflections which are characteristic of rutile, and by the distinction that the Hem_{ss} are whiter than either the exsolving Hem_{ss} or the exsolved Hem_{ss}. Because the decomposed hematite is associated with rutile, it is Fe^{3+}-rich and Ti-poor, in contrast to the exsolution-associated Hem_{ss} which are titanohematites. Apart from color differences the former also have a higher reflectivity. 0.13 mm.

(g) A more advanced stage of $Ilm_{ss} \rightarrow R + Hem_{ss}$ than that illustrated in (h). 750 μm.

(h-j) This series illustrates the decomposition of Ilm_{ss} to anatase (TiO_2) + Hem_{ss} with the preferred dissolution of Fe_2O_3 and the development of a porous network of euhedral TiO_2 crystals. A portion of the unaltered ilmenite is shown in the upper left hand corner of (h). 0.1 mm.

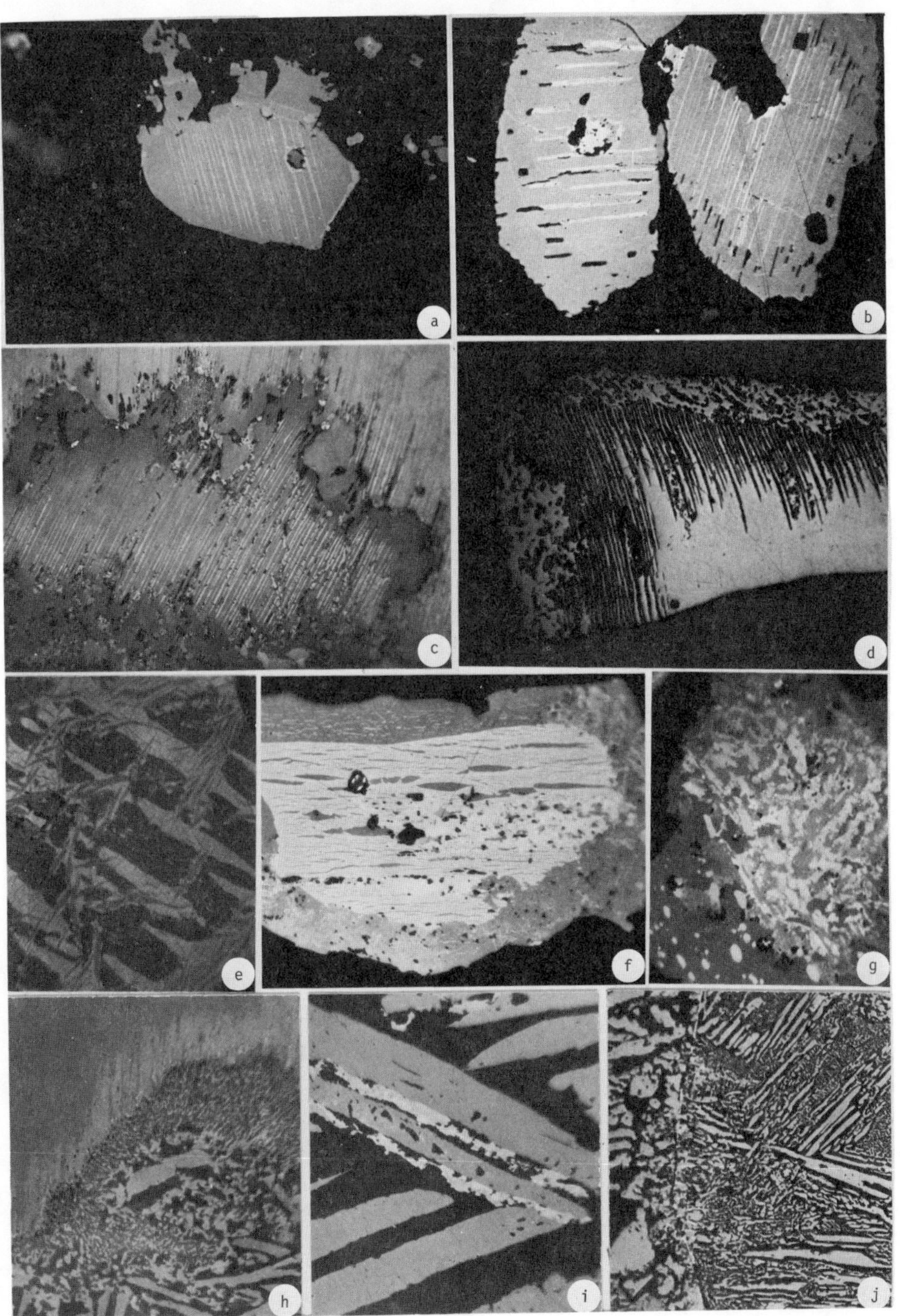

Figure Hg-31

Figure Hg-32

Ulvöspinel-Magnetite$_{ss}$

(a) A parquet-textured intergrowth of exsolved Mt_{ss} (white) and Usp_{ss} (gray in the photo-micrograph but brown in reflected light oil immersion) from an initial composition which must have been close to $Usp_{50}Mt_{50}$. 0.15 mm.

(b) *Blitz*-textured Usp_{ss} in Mt_{ss} with an attached grain of pyrrhotite. 0.15 mm.

(c) A transitional series from *lit-par-lit* Usp_{ss} to lamellar Usp_{ss}. The planes of exsolution are parallel to {100} spinel planes. The assemblage is weakly anisotropic and Ilm_{ss} are inferred, although these cannot be identified as discrete oxidation constituents. 0.15 mm.

(d) The black rods are pleonaste and the host is exsolved Usp_{ss} from Mt_{ss}; the pleonaste results from a higher temperature solvi intersection than the Usp-Mt_{ss} assemblage. Note that Usp_{ss} are absent in zones adjacent to pleonaste. 0.15 mm.

(e) The area to the right of this grain contains abundant lamellae of Ilm_{ss} whereas the area to the left contains almost equal proportions of Usp_{ss} + Mt_{ss} in a cloth-like texture. This is indicative of extremely steep oxidation gradients and although ilmenite cannot be observed within the Usp_{ss}-rich portion, weak optical anisotropy is apparent and incipient Ilm_{ss} are hence inferred. 0.15 mm.

(f-g) Blocky Mt_{ss} (white) with finely dispersed original Usp_{ss}. These areas are dark gray and display very distinct optical anisotropy; Ilm_{ss} are once again inferred. The term *protoilmenite* (Ramdohr, 1963) has been employed to describe these optically unresolvable areas. These grains show a distinct marginal texture which is apparent also in (a) and in (e). In addition to *protoilmenite*, grain (g) also has a sandwich lath of Ilm_{ss}. 0.15 mm.

(h) A thick sandwich lath of Ilm_{ss} divides this grain into: a relatively unoxidized and *protoilmenite*-rich zone (left); and a highly oxidized zone with Ilm_{ss} trellis lamellae and minor Usp_{ss} exsolution. 0.15 mm.

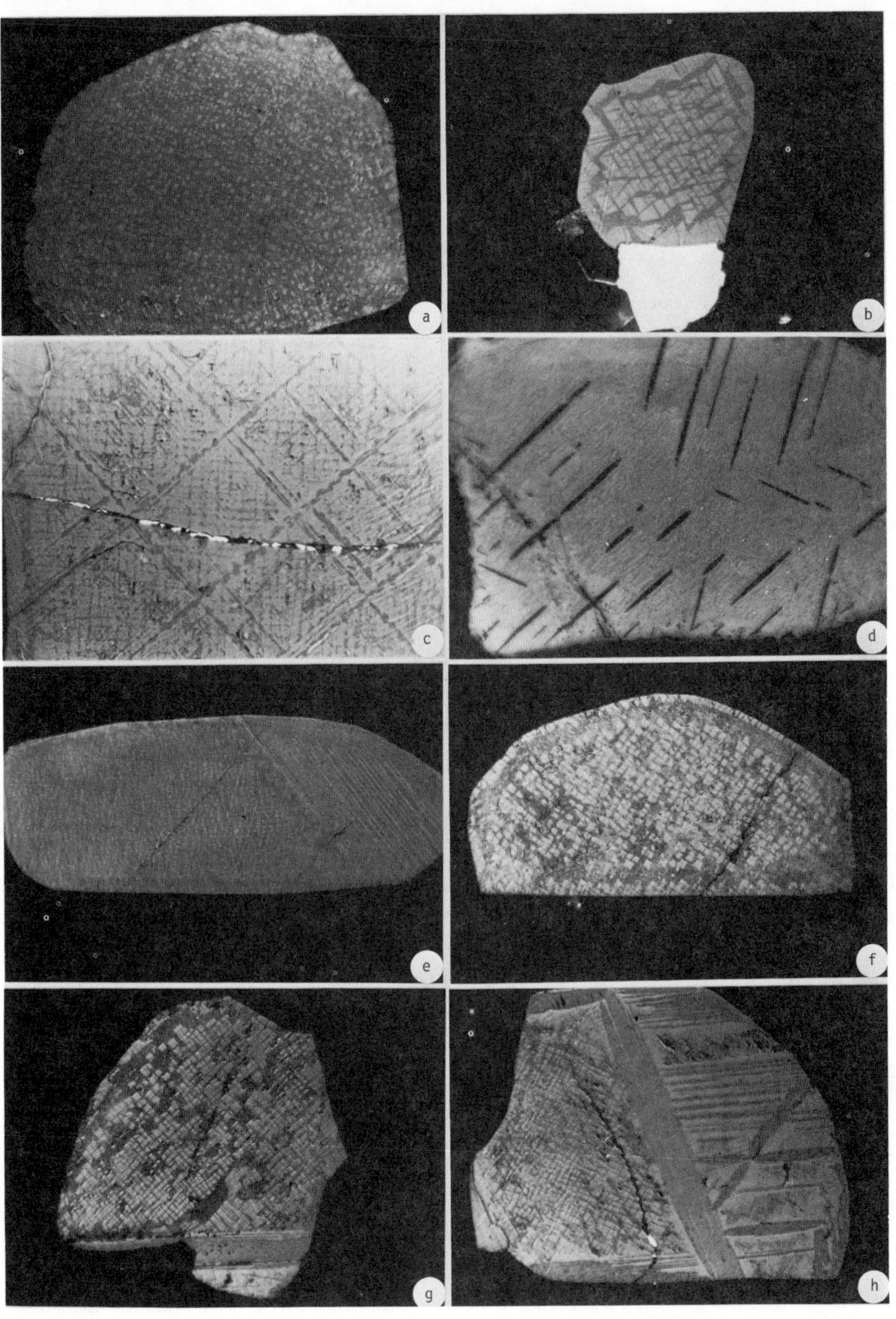

Figure Hg-32

one-phase supersolvus solid solution member; therefore, it is most common that selective oxidation of the Usp_{ss} takes place and that the Mt_{ss} remains unaffected. Each of the >600° and <600°C, Ilm_{ss}-Usp_{ss} relationships are illustrated in Figure Hg-31 with the most pronounced example shown in Figure Hg-31h. Associated with the exsolution of Usp_{ss} from $Usp\text{-}Mt_{ss}$ is the exsolution of Mg-Al-spinel which is commonly of pleonaste composition ($MgAl_2O_4$-$FeAl_2O_4$). This spinel is characteristically earlier than the Fe-Ti oxide exsolution, and this reflects a higher temperature solvus intersection than that of Fe_3O_4-Fe_2TiO_4 (Fig. Hg-31d); other examples are discussed in the following section.

Titanomagnetite-pleonaste$_{ss}$

The experimental relationships between magnetite and hercynite have been determined by Turnock and Eugster (1962), between ulvöspinel and magnetite by Vincent *et al.* (1957), and between ulvöspinel and hercynite by Muan *et al.* (1972). The magnesium-bearing analog system has been partially established, and from the experimental and analytical information available and from the widespread occurrences of pleonaste exsolution lamellae in $Mt\text{-}Usp_{ss}$ members, it is safe to assume that mutual solid solubility is extensive among members for reasonably large proportions of the system $MgAl_2O_4$-$FeAl_2O_4$-Fe_2TiO_4. However, Muan *et al.* (1972) have established that the series $FeAl_2O_4$-Fe_2TiO_4 and $MgAl_2O_4$-Mg_2TiO_4 are both interrupted by regions of immiscibility which span approximately 75 percent of the series at 1000°C. In the presence of titanium, therefore, it is only those compositions which approach end-member proportions that are likely to remain as single phases without exsolution. For Usp_{ss}-rich members, the maximum solubility of $FeAl_2O_4$ in Fe_2TiO_4 that can be attained at 1000°C is approximately 20 mole % hercynite.

Natural occurrences and textural relationships of exsolved pleonaste in titanomagnetite are illustrated in Figure Hg-33a-d. As noted above, the exsolution of Usp_{ss}, or Mt_{ss} from $Mt\text{-}Usp_{ss}$ in general postdates that of pleonaste exsolution which takes place along spinel planes which are coincident with those of the titanates and is parallel to {100}. With respect to oxidation "exsolution" ilmenite, however, pleonaste exsolution may either pre- or postdate the oxidation event, and in many instances there is abundant evidence which supports simultaneous exsolution and oxidation; an example is illustrated in Figure Hg-33d which shows both abundant pleonaste along the margins and within Ilm_{ss} lamallae, and in addition to these, there are pleonaste lamellae which are offset by Ilm_{ss} lamellae. For cases where oxidation precedes exsolution, the partitioning of Mg and Al are preferentially towards Ilm_{ss} and Mt_{ss}, respectively. However, the decidedly higher partitioning coefficient is of Al into the spinel, and in the presence of the less strongly partitioned Mg, the Mt_{ss} host rapidly approaches supersaturation levels and pleonaste solid solutions exsolve. This type of exsolution can be most clearly demonstrated in basalts and is discussed in Chapter 4. The process of exsolution is, in summary, a process related to oxidation and may be viewed as an *imposed* exsolution induced by saturation of the host specifically in Al_2O_3 but also in MgO. From this discussion it should now be obvious that spinels which are so often observed to be closely associated with Ilm_{ss} lamellae (Fig. Hg-33d) cannot have resulted by exsolution from the ilmenite because of Al_2O_3 deficiencies. Detailed studies, however, have not been undertaken to establish whether spinels in Ilm_{ss} differ in composition from spinels in Mt_{ss}, and new data may be quite enlightening which will either strengthen, or diminish, our present notions of partitioning coefficients of minor elements between the cubic and rhombohedral constituents.

Magnetite-chromite-hercynite$_{ss}$

The joins $FeCr_2O_4$-Fe_2TiO_4 and $MgCr_2O_4$-Mg_2TiO_4 show complete solid solubility down to at least 1000°C (Muan *et al.*, 1972). With the addition of Al_2O_3 as $FeAl_2O_4$ and as $MgAl_2O_4$, as noted above, assymmetrical solvi are intersected for the joins $FeAl_2O_4$-Fe_2TiO_4 and for Mg_2TiO_4-$MgAl_2O_4$. For the system Fe_2TiO_4-Fe_3O_4-$FeAl_2O_3$ the immiscible region along the join Usp-Her is maintained, and the solvi are approximately similar for the case of magnetite substituting for chromite at 1385°C (Muan, pers. comm., 1975). The system for which Fe_2TiO_4 is replaced by Fe_3O_4 in the ternary Fe_3O_4-$FeAl_2O_4$-$FeCr_2O_4$ has not been determined experimentally; however, there is now evidence in natural occurring spinels in this system which suggest that the solvi relationships are quite similar. Two examples are illustrated in Figure Hg-33e-f, and these are respectively from an olivine pyroxenite at the Giant Nickel Mine, Hope, B. C. (Muir and Naldrett, 1973) and from ultramafic rocks in the contact aureole of the Nain anorthosite complex, Labrador (Berg, 1976). In both examples the assemblages are a chromian-hercynite (or more correctly Cr-Mg-hercynite) intimately associated with an aluminous chromian-magnetite. The hercynitic constituent is generally homogeneous except for occasional coarse lamellae of ilmenite (Fig. Hg-33e), and this contrasts with the magnetite which is typically heterogeneous and contains a delicate cloth-like fabric reminiscent of Usp_{ss} exsolution. This exsolved phase has not been identified and cannot be ulvöspinel because of the uniformly low TiO_2 contents of these chromian-magnetites.

Two additional examples of chromite-magnetite associations are illustrated in Figure Hg-33 (g & h). These intergrowths are from partially serpentinized harzburgites in the Josephine Creek, Oregon, peridotite body, and in these examples the chromite is dominant and the magnetite subordinate. Figure Hg-33h is of particular interest because an alloy of metallic Ni-Fe (Ni_3Fe) is an associated mineral and because these oxide assemblages show complete "unmixing" of Mt_{ss} to the grain boundaries of Chr_{ss}.

Reactions involving magnetite-ulvöspinel$_{ss}$

Modifications to member of the Mt-Usp_{ss} series fall into at least the following six categories: (1) oxidation "exsolution"; (2) high-temperature pseudomorphic oxidation; (3) sphenitization; (4) maghemitization; (5) martitization; and (6) oxyhydration.

Oxidation "exsolution" results in the formation of Ilm_{ss} lamellae along {111} spinel planes from initial Usp_{ss}-rich compositions (Fig. Hg-34a-b). Pseudomorphic oxidation describes the decomposition of Mt_{ss} + Ilm_{ss} to R + Hem_{ss} + Pb_{ss} (Fig. Hg-34c-d). These processes of oxidation take place at high temperatures (above 600°C in most cases); the oxide systematics for assemblages in basalts is discussed more completely in Chapter 4. Oxidation "exsolution" is widespread in basic suites and in rocks of intermediate composition; the assemblage is occasionally present in more acid suites and appears to be equally abundant in extrusives as well as in hypabyssal and plutonic settings. Assemblages which are characteristic of the more intensely oxidized pseudomorphic oxidation associations are rare in suites other than those of basic extrusives, and this is a product of the variations in TiO_2 contents, the abundance of volatiles, and the degree to which volatile interactions take place.

Sphenitization as reviewed previously is a metasomatic process where the additive components are Ca + Si and the depleted exchanged component is Fe. Examples of both the selective replacement of oxidation "exsolved" Ilm_{ss} and of Mt_{ss} hosts are illustrated

Figure Hg-33

Titanomagnetite-Spinel$_{ss}$ Relationships

(a-d) The coarse, very coarse and very fine black rods in these grains are pleonastic spinels along the join $MgAl_2O_4$-$FeAl_2O_4$. The plane of exsolution, in common with that of ulvöspinel, is along {100} spinel planes. Pleonaste exsolution predates that of Usp exsolution (a) and in some cases exhibits several generations of exsolution (b & d). In relation to coexisting oxidation-"exsolved" Ilm$_{ss}$, pleonaste exsolution is either earlier than the oxidation of titanomagnetite (c), or contemporaneous with oxidation (d). Ilm$_{ss}$ lamellae are either lighter or darker gray in the photomicrographs depending on the orientation. Some pleonaste rods are offset by Ilm$_{ss}$ (d), whereas in other regions (d also) it appears that oxidation and nucleation of Ilm$_{ss}$ has occurred along the discontinuities created by high-temperature exsolution. In these cases the exsolved spinels are fragmented and are totally enclosed in ilmenite (thick sandwich lath in grain d). 0.15 mm.

(e-f) These are two examples of coexisting Cr-Mg-hercynite ($FeAl_2O_4$, black) and Al-Cr-magnetite. The cloth-like fabric in the lighter Mt$_{ss}$ areas, although similar in appearance to Usp-Mt$_{ss}$ exsolution, has not been identified and is not Usp because of analytically determined low Ti contents. An Ilm$_{ss}$ lath is present in (e) and the bright fleck in the center of (f) is pyrrhotite. 0.15 mm.

(g-h) The relationships of chromite and of magnetite differ only in the degree of phase separation. Although the series is known to be complete at high temperatures and that coexistence is dependent on f_{O_2} (see Fig. Hg-39), these assemblages are present in highly reduced rocks (serpentinites) which probably formed at $T < 500°C$. Exsolution is inferred but the assemblage may have originated by an entirely different process, *e.g.*, recrystallization at high pressure or by oxidation. Mt$_{ss}$ appear white in (g) and light gray in (h); the white areas in (h) are Ni_3Fe. 0.15 mm.

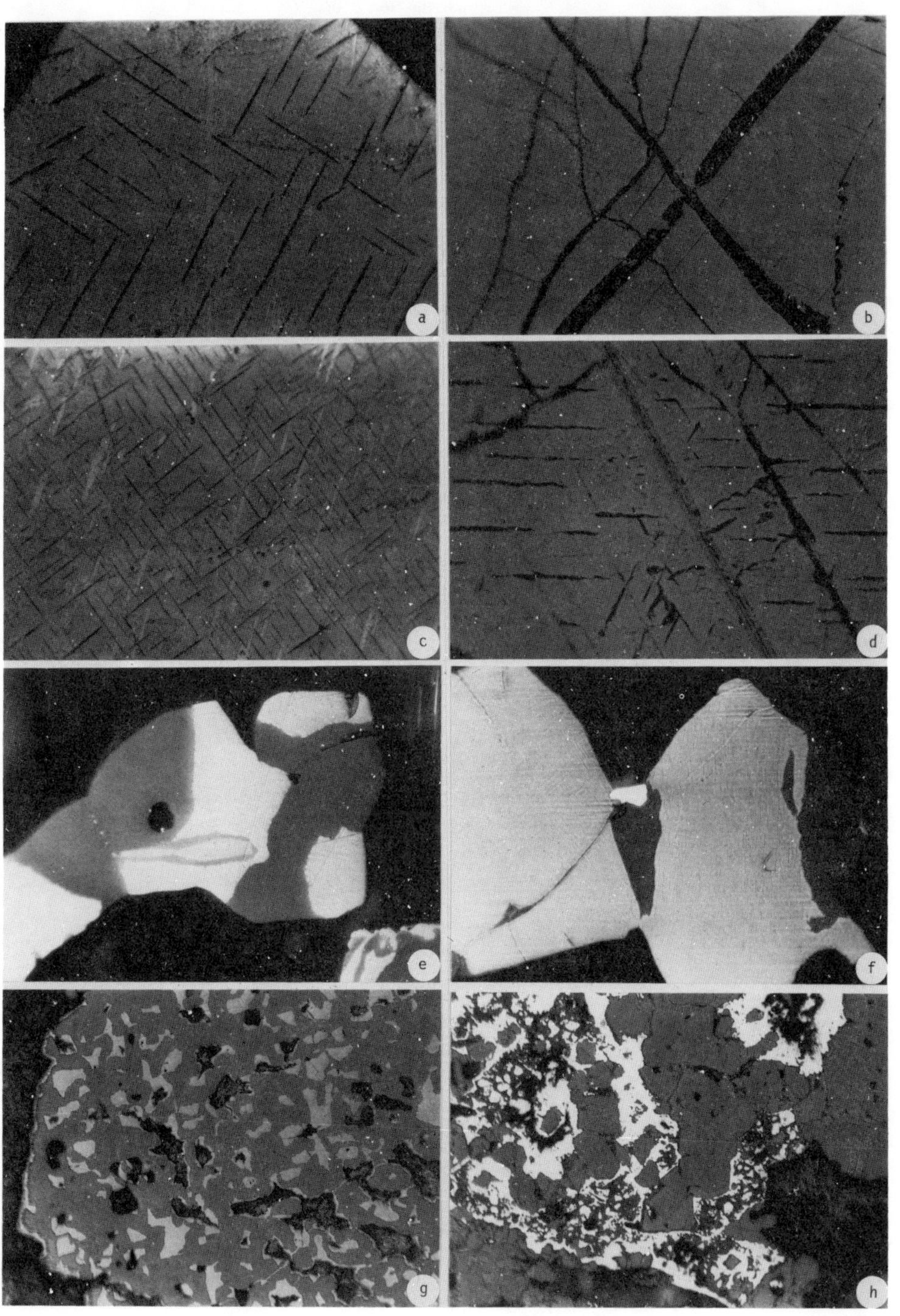

Figure Hg-33

Figure Hg-34

Ulvöspinel-Magnetite Reactions

(a) Dark exsolution rods of pleonaste and associated oxidation-"exsolution" lamellae of Ilm_{ss} with finely textured Usp_{ss} in the Mt_{ss} host. 0.15 mm.

(b) An unusual development of *internal composite* Ilm_{ss} grading progressively into *sandwich laths* across a discontinuity in titanomagnetite. 0.15 mm.

(c-d) These crystals of original titanomagnetite are now highly oxidized (at T > 600°C) and the constituent phases are gray Pb_{ss} lamellae along relic {111} planes, which were originally Ilm_{ss}, with the associated white phase which is Hem_{ss}. Grain (c) contains some unreacted Mt_{ss} with abundant rods of pleonaste. Note that where the magnetite is oxidized (SSW in c) that the resulting assemblage is hematite and that beads of *undigested* spinel persist because of the limited solid solubility of Mg (but also Al) into the Ti-poor Fe_2O_3 structure. 0.15 mm. (See Chapter 4 for additional and more complete coverage.)

(e-h) Sphene (either light gray or dark gray) replacement of titanomagnetite-ilmenite assemblages which have resulted by oxidation-"exsolution." Grain (e) is an unusual example because of the preferred replacement of the Mt_{ss} constituent. Grain (f) is most typical because replacement by $CaTiSiO_5$ is restricted to the Ti-rich lamellae which were originally Ilm_{ss}. Grains (g) and (h) show the partial to complete replacement of Mt-Ilm; in (g) the Mt is totally replaced and the thick Ilm lamellae are partially replaced; in (f) a sphene pseudomorph has resulted with ghost-like lamellae of original Ilm_{ss} still apparent along {111} planes (a result of chemical destruction but maintained structural integrity). The contrast in optical properties of these sphenes is related to the degree of crystallinity; sphene after Ilm is coarsely crystalline (*cf*. Fig. Hg-30a-b), whereas sphene after Mt is extremely fine grained. 0.15 mm.

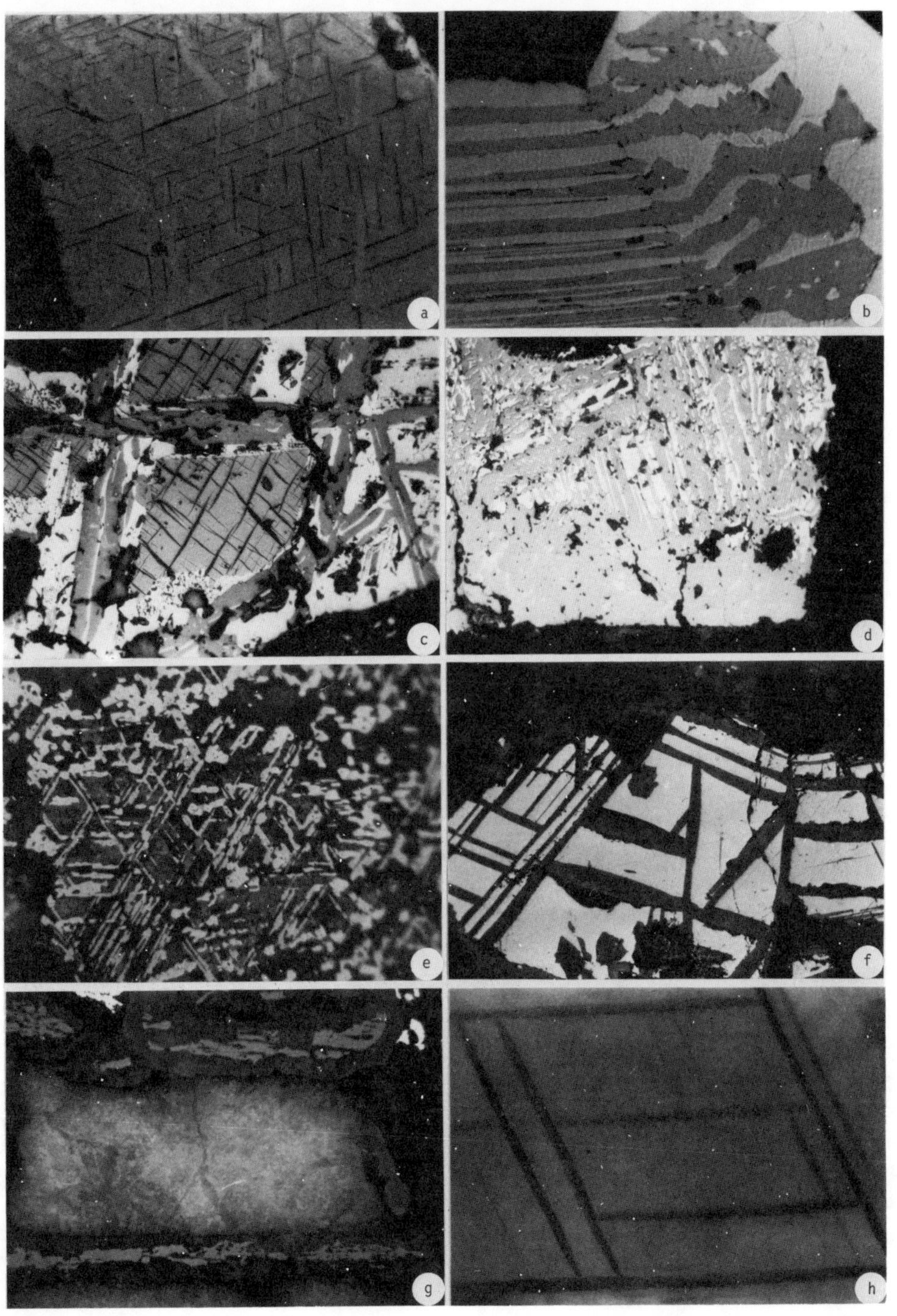

Figure Hg-34

respectively in Figures Hg-34e and Hg-34f. Two additional and more advanced examples are illustrated in Figures Hg-34g-h and in these there are distinctive optical properties for each of sphene replacing Mt_{ss} and of sphene replacing Ilm_{ss}. For Ilm_{ss} replacement the sphene is coarsely polycrystalline and compact, in contrast to sphene replacing host titanomagnetite which is microcrystalline and porous; the former yields a well-defined x-ray powder pattern whereas the latter yields a diffuse pattern. The compositions, as determined by electron microprobe for both types, show that FeO is rarely present in concentrations which exceed 0.5 wt. % and that the two types are similar in every other respect. The example shown in Figure Hg-34e is from an altered basalt from Jamaica, and the remaining examples are from the mafic body of the Kalgoolie Gold Mine, Australia. Similar examples are typically present in both deformed and undeformed metasomatized mafic complexes, and it is considered by some authors (*e.g.*, Desborough, 1963; Davidson and Wyllie, 1968) that the extensive removal and remobilization of Fe^{2+} + Fe^{3+} may result in the formation of large iron-ore deposits of the Cornwall type. The temporal relationships of sphenitization are difficult to determine but in some cases the process may result from autometasomatism with late stage fluids, or from autometasomatism associated with synkinematic intrusions. Other examples of sphenitization can be more clearly related to post-supergene oxidation or to post low-temperature deuteric oxidation. For these associations, sphene paramorphs the curvilinear planes of pre-existing maghemite as shown in Figure Hg-35a-b. This form of replacement, which appears as the murky gray to dark gray phase in Figure Hg-35b has not been fully characterized, and diverse opinions range from R + Hem_{ss}, to sphene, and to amorphous Fe-Ti oxide. More recent studies by electron microprobe and coupled x-ray analyses show that sphene is present in many instances, but in an equal number of cases neither the detection of Ca nor the characteristic peaks for rutile or for sphene have been found in Debye-Scherrer powder patterns. In all cases, however, both Fe and Ti are typically present, but data yield low analytical totals and a complex submicroscopic assemblage is hence inferred.

Maghemitization (γFe_2O_3) and martitization (αFe_2O_3) have been discussed rather extensively in the literature (*e.g.*, Basta, 1959; Akimoto and Kushiro, 1960; Katsura and Kushiro, 1961; Ramdohr, 1969). Examples of each of these oxidation products of Mt_{ss} are illustrated respectively in Figures Hg-35c-d and Hg-35e-f. Maghemite is a cationic deficient spinel, is hence cubic and optically isotropic, and is commonly either white or bluish-white in reflected light oil immersion. Oxidation planes are typically ill defined, and the characteristic fabric as illustrated in Figure Hg-35c is along curved, conchoidal directions which are usually accompanied by fine hairline cracks; these cracks are considered to develop as a result of the reduction in volume from Mt_{ss} to $maghemite_{ss}$ and are perhaps also the consequence of octahedral cationic deficiencies in γFe_2O_3. With increasing distance from these cracks there are gradational color changes from white immediately adjacent to the cracks, to tan at contacts with unoxidized titanomagnetite. These variations represent a continuous sequence of oxidation solid solution from titanomaghemite to titanomagnetite, and members of the series are therefore somewhat akin to the omission solid solution series of kenotetrahedral and keno-octahedral magnetites described by Kullerud *et al.* (1969). This topic is treated by Lindsley in Chapter 1. A second example of γFe_2O_3 is shown in Figure Hg-35d in association with the high-temperature oxidation "exsolution" of Ilm_{ss} lamellae along {111} spinel planes. In this example wide variations in the intensity of maghemitization, which are similar to those shown on a finer scale in Figure Hg-35c, is well demonstrated. Division of this large titanomagnetite crystal into subunits of smaller areas results in a patch-quilt appearance with each subarea showing a higher or lower degree of oxidation intensity; the white areas are at the most advanced stage, and the tan areas (gray in the photomicrograph), or areas of intermediate color, are at low intensity levels, or are at intermediate levels in the transition from titanomagnetite to titanomaghemite. Oxidation to maghemite is typically associated with more titaniferous Mt_{ss} components, whereas αFe_2O_3 (hematite) is more typical of $magnetite_{ss}$ (*i.e.*, Ti-deficient) compositions. The inversion of γFe_2O_3 to αFe_2O_3 takes place between 250° and 550°C and is dependent on grain size and a number of other factors as discussed in Chapter 1.

An additional distinction, apart from the characteristic association of γFe_2O_3 and titanomagnetite, is that hematite oxidation of Mt_{ss} takes place preferentially along {111} spinel planes (Fig. Hg-35e), and this fabric bears a close resemblance to the orientation and form of Ilm_{ss} which develop by oxidation exsolution (see, *e.g.*, Fig. Hg-34a) at elevated temperatures from Usp_{ss}-rich starting compositions. As oxidation becomes more intense, the {111} primary fabric is disrupted (Fig. Hg-35f) and pseudomorphs of hematite after magnetite result. The end product is colloquially referred to as *martite* and the oxidation process as *martitization*. Note that in contradistinction to the high-temperature oxidation of titanomagnetite, no additional phases such as Pb_{ss} or R form (see Chapter 4). The optical distinctions between hematite and maghemite are a higher reflectivity constant for hematite, marked pleochroism and optical anisotropy; maghemite is optically isotropic or very, very weakly anisotropic and has a reflectivity value which is close to that of titanomagnetite but substantially less than that of hematite.

Oxyhydration is the final process to be considered, and in ideal cases the hydrated products of magnetite or of hematite is goethite, $\alpha FeO.OH$ (Fig. Hg-35g); oxyhydration of maghemite (γFe_2O_3) results in lepidocrocite ($\gamma FeO.OH$); and akaganeite ($\beta FeO.OH$) which is relatively rare appears to be more closely associated with FeCl decomposition in meteorites rather than with oxide hydration. Hydration oxidation of titanomagnetite or of titanohematite is uncommon in basic igneous rocks and appears to be most commonly associated with the less titaniferous varieties in more silica-rich suites, and as a byproduct of sulfide decomposition.

The processes of maghemitization and oxyhydration are low-temperature events. These may result from the final stages of alteration associated with deuteric cooling, but the more likely associations are in supergene weathering and in burial metamorphism. Hematite derived from low titanium magnetites, however, can be either a high-temperature product or alternatively it may form under conditions which are similar to those of maghemite formation.

Oxides derived from silicates

The replacement of ilmenite by aegnigmatite and by sphene is reviewed in Figure Hg-30 and in Figures Hg-30 and Hg-34, respectively. The reverse relationships of silicates yielding oxides by exsolution, partial decomposition through processes of magmatic interaction, or of oxidation are illustrated in Figure Hg-36.

Exsolution of Mt_{ss} and in some instances Ilm_{ss} is relatively common in plagioclase and in pyroxene from plutonic suites. Particularly striking examples are in larvikites and in labradorite-bearing anorthosites where oxide exsolution is in part responsible for the blue irridescent *schiller* effect. The planes of exsolution are typically parallel to

Figure Hg-35

Magnetite and Titanomagnetite Reactions

(a-b) The gray and white areas are sphene but have also been described as amorphous Fe-Ti oxide, and this is so particularly for grain (b). A comparison of the curvilinear control on the distribution of these secondary products with textures typical of maghemite replacement (c) suggests that sphene postdates the formation of γFe_2O_3. Ilm is absent in grain (a) but is present in (b). Small residual areas of Mt_{ss} are still present in (b) but the grain is largely oxidized to titanomaghemite. 0.15 mm.

(c) This is the most typical development of titanomaghemite (white) after titanomagnetite, *i.e.*, along curved conchoidal fractures which are inferred to result from changes in structural volume induced by the inversion of an inverse spinel to a cationic deficient spinel. 0.15 mm.

(d) Variable compositions and a spectra of intermediate phase inversions are apparent in this partially oxidized titanomagnetite grain. The lamellae are Ilm_{ss} and the variations in color among segments which are bounded by these lamellae are representative of stages in the transition from titanomagnetite to titanomaghemite. Note that the volume-change cracks are not necessarily associated with the most intensely maghemitized areas, and that the crystal interior is more highly oxidized than the grain boundaries. 0.15 mm.

(e-f) Hematite is the white phase in both grains, and oxidation of these Ti-free crystals along {111} spinel planes is in stark contrast to the lack of crystallographic control so typical of maghemitization (b & c). These distinctive textural habits and the anisotropy of Hem (maghemite is cubic and hence isotropic) are the optically distinguishing properties for identification. 0.15 mm.

(g-h) The dark gray regions in these photomicrographs are goethite; in (g) it develops as a botryoidal (or reniform = kidney shaped) overgrowth on Mt but some areas of the Mt also exhibit partial oxyhydration; in (h) hematite is an associated phase, and this assemblage is typical of vesicle infillings in volcanic suites. 0.15 mm.

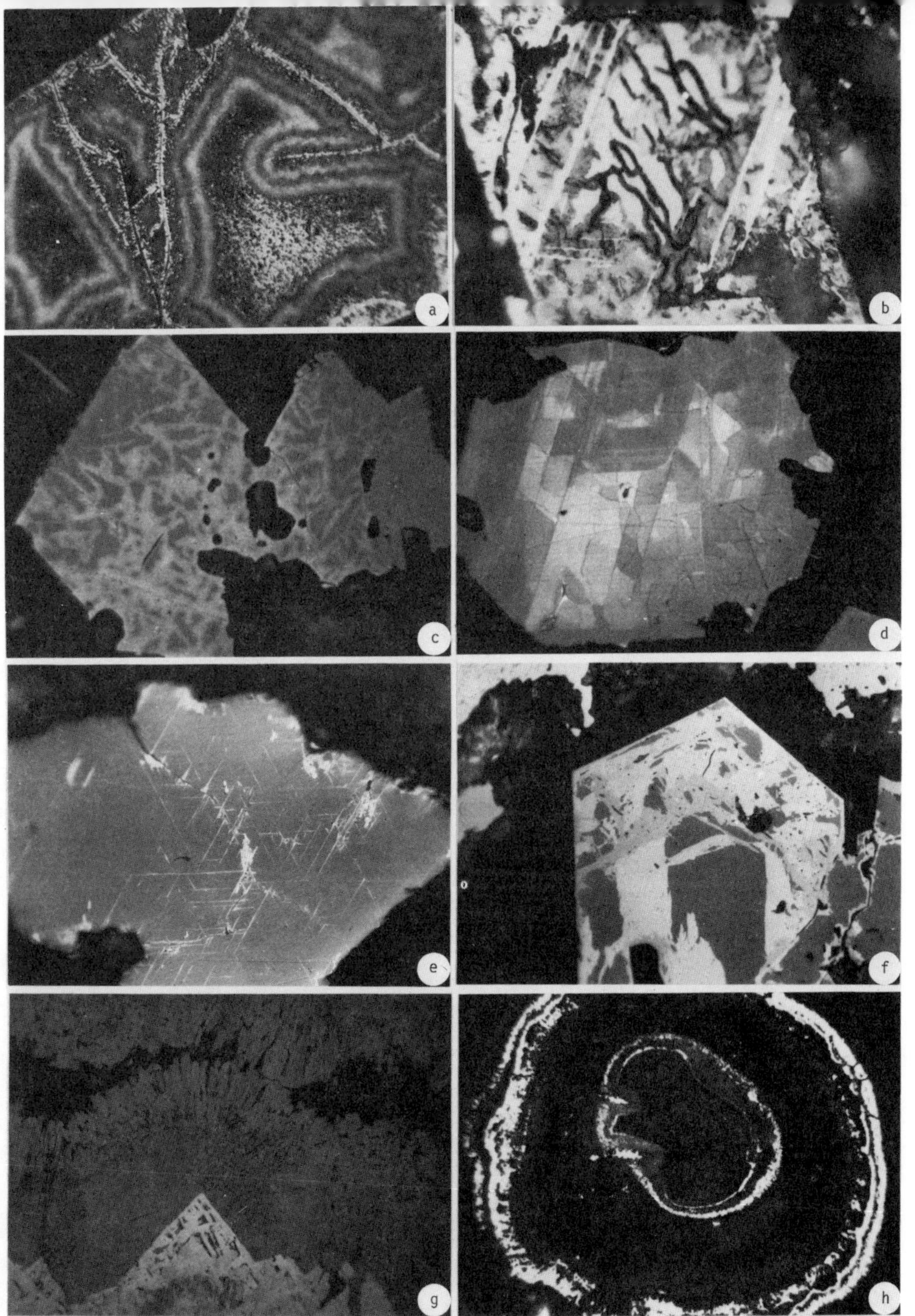

Figure Hg-35

Figure Hg-36

Oxides Derived from Silicates

(a) Usp-Mt_{ss} exsolved as rods and discontinuous blebs in plagioclase. The large adjacent oxides are Ilm_{ss}. 0.15 mm.

(b) Symplectic arrays of Al-rich spinels in clinopyroxene. 0.15 mm.

(c) Magnetite derived from biotite and concentrated along cleavage planes. 0.15 mm.

(d) Magnetite associated with the peripheral decomposition of amphibole. 0.15 mm.

(e) Coarse symplectic magnetite in a highly oxidized crystal of olivine. Oxidation occurred at $T > 600°C$ based on the presence of associated Pb_{ss}. 450 μm.

(f) The outer margin of this highly oxidized olivine crystal ($T > 600°C$) has a diffus[illegible] mantle of Hem (white) + magnesioferrite (gray). The highly reflective zone adjace[illegible] to this mantle is predominantly Hem; the core of the olivine is dark and diffuse [illegible] at high magnifications, symplectic magnetite can be resolved which is similar to that shown in grain (e). 0.15 mm.

(g-h) These olivine crystals are typical of those which are generally described as having undergone *iddingsitization*. Because *iddingsite* is not mono-mineralic, as is clearly apparent from the photomicrographs, the name has only descriptive connotations. The identifiable white phase is goethite and the associated darker phases (note cleavages in grain g) are smectite-related constituents. This assemblage parallels maghemitization of titanomagnetite; ilmenite remains unaffected as shown in Figure Hg-35c. 0.15 mm.

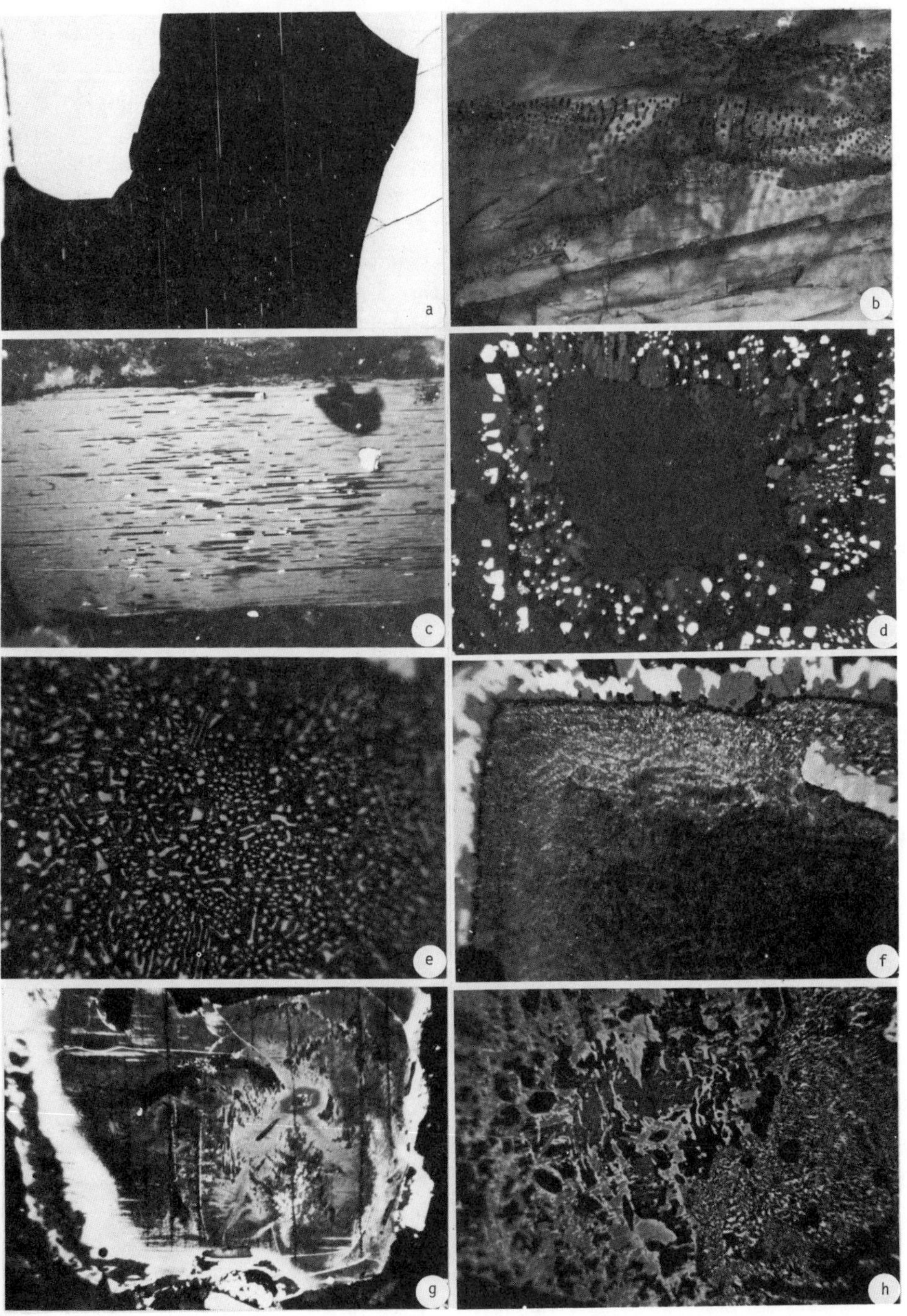

Figure Hg-36

twin and cleavage planes in the host silicate, and the oxides are generally lamellar or plate-like in form as illustrated in Figure Hg-36a. These finely textured oxides are an important constituent in magnetic studies, and several investigators (*e.g.*, Hargraves and Young, 1969; Evans and McElhinny, 1969) have demonstrated that the major stable component of remanence resides in these microscopic arrays rather than in the larger discrete crystals. More aluminous oxides in the form of green or brown spinels are also relatively common in pyroxenes (Fig. Hg-36b) but appear to be rare in plagioclase (Whitney and McLelland, 1973) in cases other than those which involve the reaction of olivine + plagioclase = aluminous pyroxene + spinel. These latter reactions are typically the result of metamorphism although convincing arguments have been made, for example by Griffin (1971) and by Griffin and Heier (1973), which suggest that a pressure increase during cooling is a probable mechanism to trigger the reaction at subsolidus temperatures. The products of the reaction result in a symplectic intergrowth and the assemblage, which may also be accompanied by garnet, develops as a corona around olivine or at the grain junctions between plagioclase and olivine.

Figures Hg-36c and Hg-36d illustrate the partial decomposition of biotite and of amphibole to magnetite; these magnetites are typically very close to stoichiometric Fe_3O_4, but minor element concentrations of 1-2 wt. % ($MgO + Al_2O_3 + TiO_2$) may also be present. Hydration of these oxides is also very common and this leads to the formation of goethite.

The most widespread form of silicate decomposition is the oxidation of olivine, which at high temperatures results in magnetite + hematite or in hematite + magnesioferrite (Fig. Hg-36e-f). At lower temperatures the typical breakdown product is *iddingsite* which has a major constituent of goethite intimately associated with smectite (Fig. Hg-36g-h). *Iddingsite* formation may develop during the volatile saturated stages of deuteric cooling but is more clearly demonstrated to result as a consequence of supergene weathering processes. The distinction between high and low temperatures for these olivine byproducts is also reflected in the oxidation assemblages of the discrete oxides which show that the magnetite-hematite-magnesioferrite suite is accompanied by rutile + hematite, or by pseudobrookite + hematite, whereas *iddingsite* develops more typically in parallel with the oxidation inversion of titanomagnetite to titanomaghemite (Baker and Haggerty, 1967; Haggerty and Baker, 1967). For all of the above cases, the earlier comments related to magnetic stability and remanence apply here also.

PRIMARY OXIDE DISTRIBUTIONS

Introduction

Paragenetic sequences, compositions and model abundances of primary opaque oxides in igneous rocks depend for the most part on the initial bulk chemistry of the host rock, on the depth of emplacement, and on the prevailing oxygen fugacity of the crystallizing magma.

As a general principle because FeO, TiO_2 and Cr_2O_3 contents increase with decreasing SiO_2, basic rocks tend to contain larger concentrations of oxides than either intermediate suites or acid end members. To a first approximation this distribution is controlled largely by the increases in FeO contents which result with increasingly more mafic compositions. Titanium variations, on the other hand, in the range from acid to basic suites determine both the distribution and composition of mineral solid solutions within the system FeO-Fe_2O_3-TiO_2. Members of each of the solid solution series Fe_2TiO_4-Fe_3O_4

Jsp-Mt$_{ss}$), $FeTiO_3$-Fe_2O_3 (Ilm-Hem$_{ss}$), $FeTi_2O_5$-Fe_2TiO_5 (FPb-Pb$_{ss}$) vary in TiO_2 content and ı the ratio of Fe^{2+}:Fe^{3+}. Therefore, oxide distributions and compositions depend on ıitial titanium abundances and on f_{O_2}. These coupled parameters lead to Mt-rich$_{ss}$ and ematite-rich Ilm$_{ss}$ in acid and intermediate rocks, and to Usp-rich$_{ss}$ and Ilm-rich$_{ss}$ in ısic and ultrabasic suites. Because both Fe^{2+}:Fe^{3+} and Fe:Ti ratios are dependent on emperature and f_{O_2}, for which coequilibrated oxide pairs yield unique solutions, the ıitial temperatures of crystallization and the cooling paths followed within the sub-lidus are the limiting factors chiefly responsible for compositional variations between tanomagnetites and ilmenites. This distribution implies, therefore, that suites of ıitially lower temperature origins, *e.g.*, acid extrusives, result in Mt-rich solid solu-ons + Ilm$_{ss}$, but by the same token suites which reequilibrate over long periods of time sult in products of closely comparable compositions. Members of the Ti-enriched FPb-Pb$_{ss}$ ries are restricted to more basic rock types and are more abundant as secondary oxidation oducts, of Usp-Mt$_{ss}$ and Ilm-Hem$_{ss}$, than as primary precipitates. The TiO_2 polymorphs utile, anatase, brookite) are also commonly of secondary origin but minor primary con-ntrations are occasionally present in granites, syenites and diorites. Chromium abun-nces vary from low to very low in acid and intermediate rocks, increase with more mafic aracter, and peak in the ultrabasic. Chrome-bearing members of Mt-Usp$_{ss}$ are restricted most exclusively to the basalt suite, whereas more complex solid solutions of Al_2O_3 and O in chromite ($FeCr_2O_4$) are common and abundant in picrites, peridotites, dunites and kimberlites.

The major controls on the distribution of oxides as a function of depth dependence e the effects of crystal settling processes and the likely but controversial differences at may exist in f_{O_2} levels among rocks of equivalent chemistry which have evolved as tholiths, minor intrusions, extrusives, or explosive tephra. Crystal fractionation oxides is widespread in basic intrusives, less common in rocks of intermediate compo-tion, and rare in acid intrusives. This generalized distribution also applies to suites truded at hypabyssal levels and for volcanic equivalents, with rare exceptions, is sent in all but exceptionally thick basic lava flows. In the absence of crystal set-ing, the oxides remain a minor component regardless of the mode of emplacement but sume radically different textural and paragenetic relationships with the silicates. ese textures may vary from equigranular in intrusives, to inequigranular in dikes and lls, and to interstitial in extrusives. With respect to the effects of oxidation ntrol on oxide distribution with depth, the problem is one of open and closed systems l on the retention capacities of the host rock to volatile loss during magmatic em-acement. Volatile retention at depth in the closed system model will lead to higher dation and greater proportions of Fe^{3+}, and with the higher probability of an approach equilibrium for the partitioning of iron species and of titanium between oxides and icates. In a generalized sense the extrusive situation is one of rapid volatile loss l of partial disequilibrium among oxides and silicates. In hypabyssal regions, con-ions may vary between these two extremes although it is probably closer to that of plutonic environment.

The final and perhaps the most fundamental parameter in controlling the distribution, ndance and compositions of opaque oxides in igneous rocks is the level of initial gen fugacity and the f_{O_2} path which is followed with cooling. At magmatic temperatures is a major controlling factor in elemental partitioning among the oxides, and in mental partitioning between the oxides and silicates. For high values of f_{O_2}, oxides will crystallize in preference to Fe-rich silicates whereas at relatively lower values of f_{O_2} the available iron is competitively partitioned between these crystallizing phases. The former results in the *Bowen* trend and the latter in the *Fenner* trend. In the case of titanium and for coexisting oxides and silicates, titanium will preferentially enter oxide solid solutions at low values of f_{O_2}, and hence the precipitation of Fe_2TiO_4-, $FeTiO_3$-, and $FeTi_2O_5$-rich solid solutions are favored. With progressively increasing f_{O_2}, which is the generalized progression from basic to acid suites, the ratio of Fe^{3+}:Fe^{2+} also increases, and this results in the stabilization of Mt_{ss} and of Hem-rich Ilm_{ss}. There is also a strong interdependence between silica activity and f_{O_2} which dominates in highly undersaturated rocks. In these suites TiO_2 is depleted in Usp_{ss} and in Ilm_{ss} to the extent that members of the Ilm_{ss} series are commonly absent. Either sphene ($CaTiSiO_5$) or perovskite ($CaTiO_3$) may precipitate, with the former developing at relatively high silica activities and high oxygen fugacities, and the latter at correspondingly lower values of a_{SiO_2} and f_{O_2}.

The degree of volatile retention as a function of depth is clearly of crucial significance in defining f_{O_2} levels and is important also in evaluating the relative contributions of internal versus external buffering, that is, whether the environment controls the trend or crystallization, or whether crystallization itself has the overriding effect in controlling the nature of the environment. The problem is complex with many variable and unknown parameters, but some insight is to be gained in the context of oxide distributions and compositions as a function or rock type, as discussed below, and in the context of T°C and f_{O_2} of coequilibrated oxide pairs as discussed in a later section entitled *T and f_{O_2} Variations in Igneous Rocks*.

Chromian spinel distributions

Spinel, chromian-spinel and related spinel species are restricted almost exclusively to mafic and ultramafic suites, and as early precipitates are enriched in Cr, Al and Mg and depleted in Fe + Ti. Complex zoning is widespread, and this reflects an interaction of early formed crystals with the liquid as temperature decreases. These reactions result in Fe + Ti enrichment with specific increases in Fe^{3+}. The onset of olivine and pyroxene crystallization dominate the partitioning of Mg, whereas Al is accounted for by the precipitation of plagioclase. Late-stage reaction mantles (Figs. Hg-25 and Hg-26) and discretely crystallizing spinels are thus most commonly depleted in the more refractory elements (Mg and Al), but it is important to emphasize that before the onset of crystallization of olivine or of pyroxene that a substantial proportion of the available Cr in the magma is locked up in early chromite.

To illustrate the complexity of spinel zonal trends, the most extreme examples, which are those from kimberlites, are shown in detail in Figure Hg-37, and these provide a contrast for the more general but varied trends exhibited in other petrologic suites which are summarized in Figure Hg-38 from data given in the accompanying tables. The important aspects to note initially are the general variations in trends along the bases of the multicomponent spinel prisms. These trends are: (1) early enrichment of Cr and later enrichment in Al--the kimberlite trend; (2) early enrichment of Mg and later enrichment in Fe^{2+}--magmatic ore deposit trend; and (3) early enrichment of Al and later enrichment in Cr--the xenolith and peridotite trend. Starting and terminal points within individual suites differ from one complex to another, as do the subtleties in trend

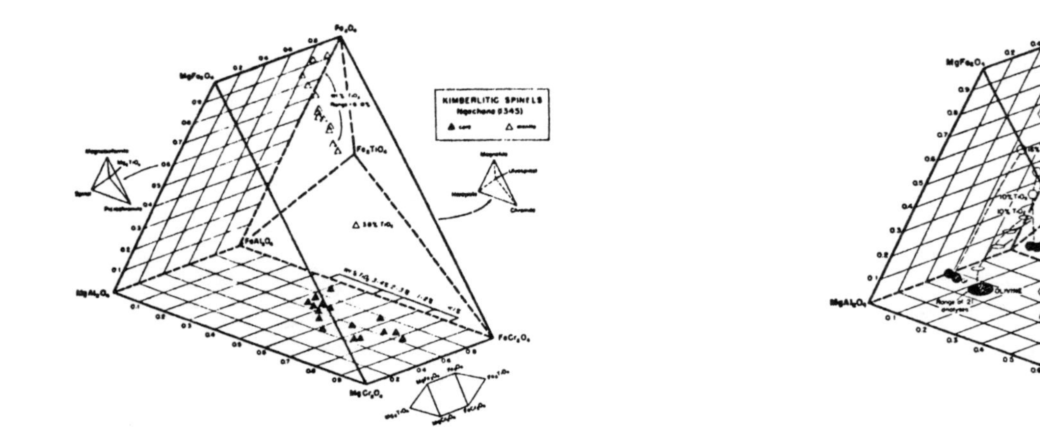

Figure Hg-37a. A modified multicomponent spinel prism consisting of a front triangular Mg-face that includes an extension to Mg_2TiO_4 (inset), and a rear triangular Fe^{2+}-face that includes Fe_2TiO_4. The Ti-phase extensions are attached tetrahedra. These tetrahedra are intended for diagrammatic representation only, and the analyses have not been subjected to a second-order projection to account for the relative proportions of $(FeMg)_2TiO_4$. Ranges of wt. % TiO_2 instead are listed. Note that the base of the prism is Ti-free, and that the lowermost inset represents a projection from the front rectangular face. The data points are for spinels from Nqechane showing simple core-mantle relationships.

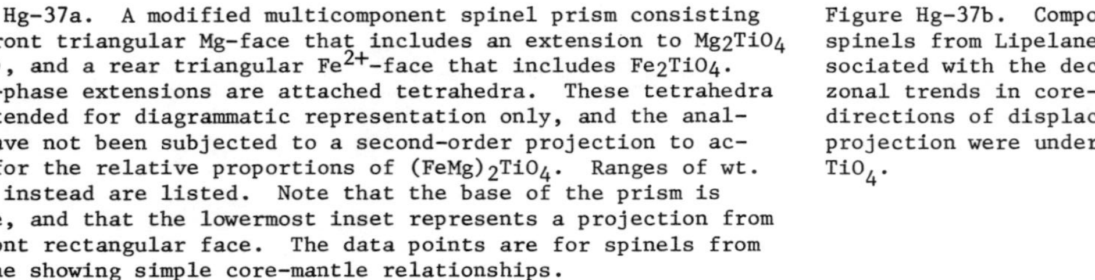

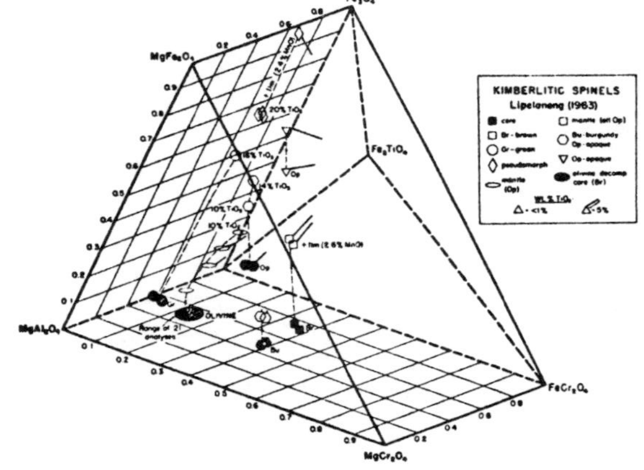

Figure Hg-37b. Compositional data for a variety of optically distinct spinels from Lipelaneng. The large shaded ellipse is for spinel associated with the decomposition of olivine. The dashed lines show zonal trends in core-mantle relationships. The solid lines show the directions of displacement within the spinel prism if a second-order projection were undertaken to account for the presence of $(FeMg)_2TiO_4$.

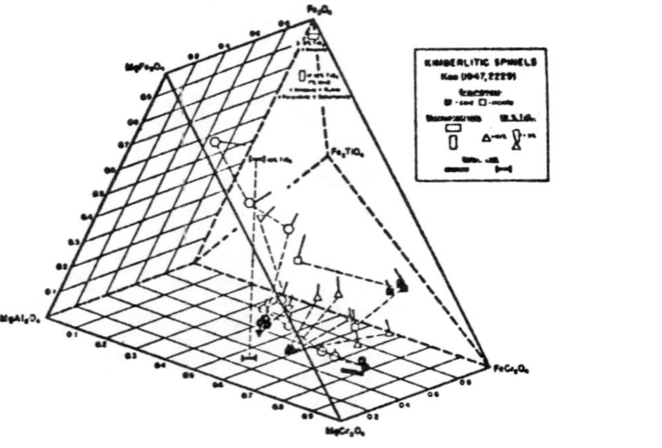

Figure Hg-37c. Complex oscillatory zoning in spinels from Kao. Dashed lines join core-mantle overgrowths, and the solid lines represent relative concentrations of wt. % TiO_2.

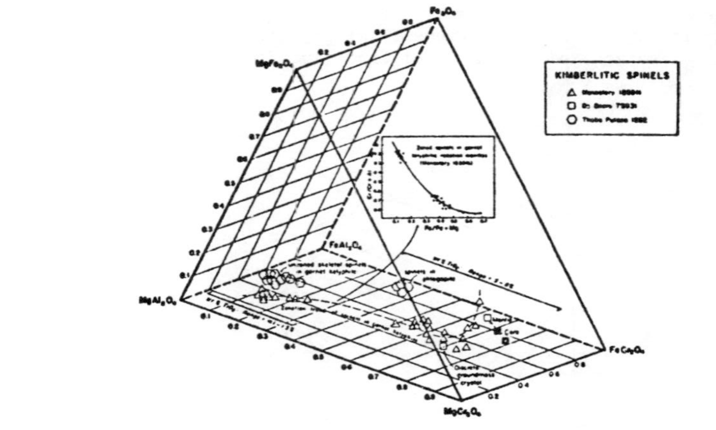

Figure Hg-37d. Compositional data for spinels in garnet kelyphitic rims. The inset is a projection onto the base of the spinel prism.

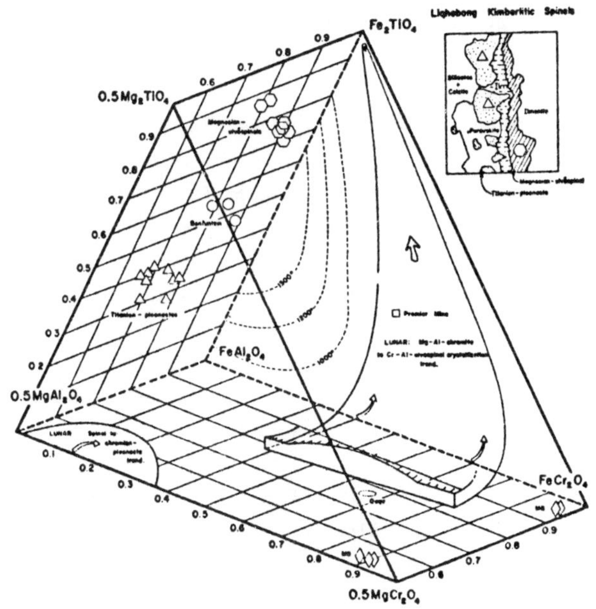

Figure Hg-37e. Multicomponent spinel prism showing the compositions of spinels in the reaction mantle as illustrated in the inset. These spinels are labelled magnesian-ulvöspinel and titanian-pleonaste but to account for the presence of Fe^{3+} the correct terminology for each are magnesioferrite-ulvöspinel and titanian-ferrianpleonaste, respectively. Spinels labelled Benfontein are from Dawson and Hawthorne (1973); Qaqa (= Letseng-la-terae) are from Nixon *et al.* (1963); those marked MB are chromites in diamond inclusions reported by Meyer and Boyd (1972), and Premier Mine are titian-chromites exsolved from ilmenite (Danchin and D'Orey, 1972). The solvi at 1000°-1300°C for the Fe-bearing ternary are from Muan *et al.* (1972). The trends for the lunar spinels are predominantly from Haggerty (1972), but include a large number of unpublished analyses.

directions; but in a general sense these distinguishing trends are those most typically observed.

Migrating upwards from the base of the spinel prism towards Ti + Fe^{3+} we note that there is a preferred directional sweep towards the Fe-ternary defined by chromite, hercynite, and magnetite-ulvöspinel, and away from the Mg-ternary defined by spinel, picrochromite and Mg_2TiO_4--magnesioferrite. The sole exception to this preferred trend is exhibited by kimberlites which continue to show relatively high MgO contents with increasingly higher concentrations of Fe^{3+} + Ti. In contrast, the xenolith and peridotite suites exhibit little or no enrichment in Ti or Fe^{3+}, and these spinels are essentially confined to the spinel prism base. These differences are of interest because of their respective high-pressure affinities, but it should be noted that kimberlitic xenolith suites and xenoliths in basalts are broadly within the field defined by Alpine-type peridotites and that the kimberlitic trend above the prism base is specifically that associated with groundmass spinels which undoubtedly did not form under the same P-T conditions as those of the accompanying xenoliths. From a slightly different viewpoint, the contrasts in Cr/Al variations ought to indicate variations in both P and T if one considers the distribution of Al among plagioclase, spinel and garnet in lherzolite suites. There is now abundant experimental and field evidence to show that the partitioni of Al among these three minerals in indeed pressure dependent, with spinel-bearing lherzolites forming an intermediate regime between the low-pressure plagioclase suites and the high-pressure garnet suites. Cr-bearing spinels are indeed more characteristic of low-pressure suites, such as stratiform deposits, but spinel tends to be replaced in the extremely high-pressure suites by Cr-bearing diopside + Cr-bearing garnet. The apparent and unresolved anomaly in this otherwise orderly progression is that spinels containing among the highest chromite concentrations yet recognized are those included in diamond.

Turning now to the remaining suites illustrated in Figure Hg-38 which are for the three basalt trends, we note that trends 1 and 2 (Fig. Hg-38b&c) share the common factor of Fe^{3+} enrichment but are distinguished also on the basis of Ti which is significant in trend 1 but reduced in trend 2, and on the basis of Al which is high in trend 2 but low in trend 1. For the third basalt trend (Fig. Hg-38c), Ti and Fe^{3+} are present in only minor proportions although the range in Cr and Al is approximately equivalent to that of trend 3. Although the data are sparse for trends 2 and 3, the marked differences among these trends can be related to their respective petrologic settings. Trend 1 appears to be most typical of subaerially extruded tholeiitic suites as characterized by spinels in Hawaiian lavas (*e.g.*, Evans and Moore, 1968; Beeson, 1976); trend 2 is characterized by spinels associated with island-arc volcanism (*e.g.*, Arculus, 1974); whereas trend 3 is that of spinels in mid-oceanic ridge basalts (*e.g.*, Sigurdsson and Schilling, 1976; Ayuso *et al.*, 1976; Dick, 1976). The latter are of particular interest because of their similarity in spinel mineral chemistry to the xenolith and peridotite suites and because the variation within the cylindrical volume shown includes the range from picritic basalts with low Cr/Cr+Al ratios, to olivine tholeiitic basalts with intermediate R^{3+} ratios, to alkali olivine basalts with high Cr/Cr+Al ratios. These variations among rock types appear, therefore, to be related to differences in the initial bulk chemistry of the magma and to paragenetic relationships of coprecipitating Mg, Al and Fe-bearing silicates; reaction trends above and beyond the limits of the base of the prism are related to the nature of the residual liquids as crystallization proceed but are also clearly related to the extent to which such reactions are possible as is

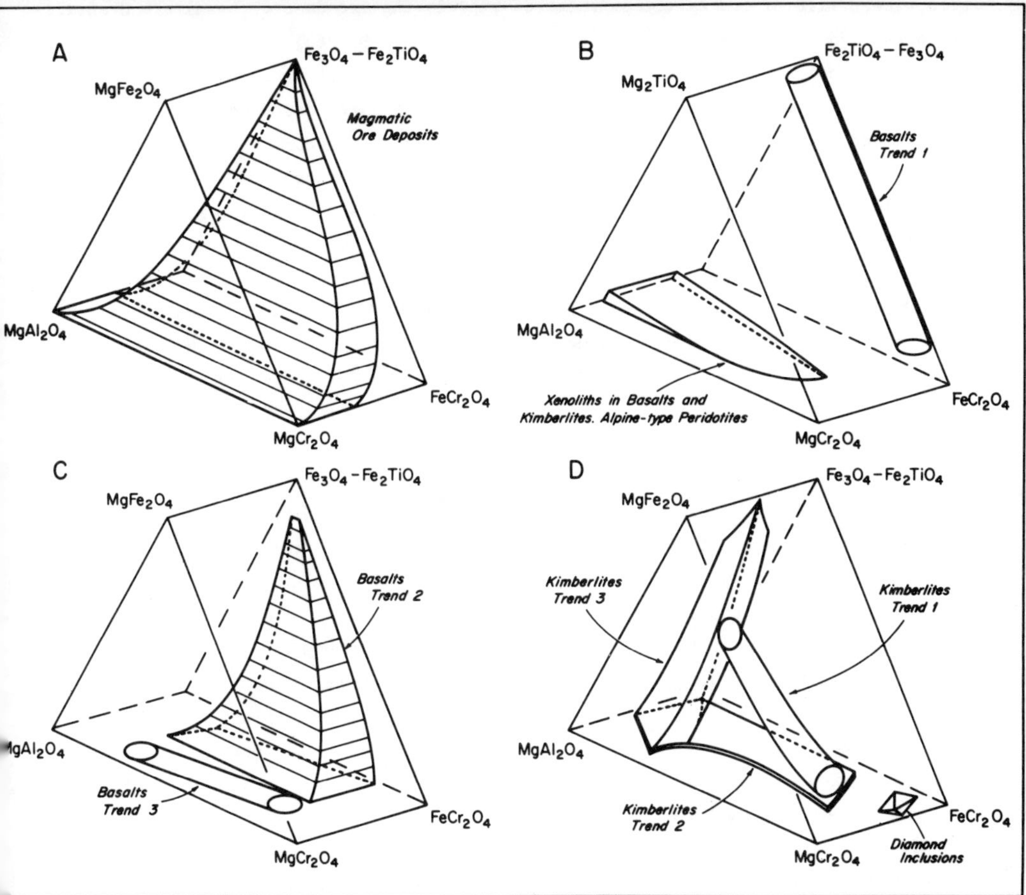

Figure Hg-38. Spinel distributions for a variety of rock types and geological settings. Endmembers for each prism base are constant, but the apices differ (refer to Fig. Hg-24a for nomenclature) and are dependent on Fe^{2+}, Fe^{3+}, and Ti. Where Fe^{3+} is dominant the series is given as Fe_3O_4-Fe_2TiO_4; where Fe^{2+} and Ti are dominant, the endmembers are reversed. Basalt trends 1, 2 and 3 are respectively for subaerially extruded olivine basalts, island-arc volcanism, and deep-sea basalts. Kimberlite trends 2 and 3 are shown in greater detail in Figure Hg-37, and trend 1 is from Mitchell and Clarke (1967). These volumes were compiled from the data given in the accompanying tables at the end of this chapter.

perhaps best illustrated by the examples of rapidly chilled deepsea basalts and their subsequent confinement to the prism base (Trend 3, Fig. Hg-38c).

An evaluation of the underlying controlling factors for these contrasting distributions is limited in part by the availability of experimental data and in part by the nature of field and laboratory studies. Spinel mineral chemistry as a function of P and T is not known and hence a coherent interpretation of natural occurrences of widely variable petrologic settings cannot be made aside from the comments made above with respect to the pressure-controlling distributions of Al. At 1 bar, however, experimental data which do exist provide important clues to the interpretation of low-pressure natural systems, and the three examples summarized in Figure Hg-39 and Figure Hg-40 are instructive from the point of view of providing information with respect to T and f_{O_2}. In Figure Hg-39 data by Ulmer (1969) and by Muan *et al.* (1972) are given for a range of f_{O_2} at constant temperature (1300°C) and for a range in temperature and variable f_{O_2} in spinel multicomponent systems which differ only in the apex constituents; these are magnesioferrite and magnetite for Ulmer (1969), and Mg_2TiO_4 and ulvöspinel for Muan *et al.* (1972). The Ti-free system (Ulmer) shows a progressive destabilization of magnesioferrite and subsequently also of magnetite with decreasing f_{O_2}. The series spinel-picrochromite is stable for the entire range in f_{O_2}, and a particularly significant point to note is that chromite ($FeCr_2O_4$) is unstable between $f_{O_2} = 10^{-0.21}$ to 10^{-9} atm at 1300°C. The data by Muan *et al.* (1972) for the Fe^{3+}-free but Ti-bearing system show that there are extensive immiscibility gaps between the magnesian and iron titanate join (Mg_2TiO_4-Fe_2TiO_4) and the pleonaste series ($MgAl_2O_4$-$FeAl_2O_4$) which project towards picrochromite and chromite (Fig. Hg-39f). The ternary solvi for each of these respective planes, although shown as continuous regions of immiscibility, are in part conjectural and are perhaps more complex for intermediate ratios of Fe/Mg.

The relevancy of these data to natural systems should be pivoted on the stable volume as outlined in Figure Hg-39c. At 10^{-7} atms both a plausible distribution for the range in magmatic ore deposits and a trend which encompasses the diverse distributions which are evidently also exhibited by basaltic suites is satisfied. At high values of f_{O_2} (*e.g.*, 10^{-5} and 10^{-2} atms, Figs. Hg-39b&C), the extensions towards magnesioferrite are in excess, whereas at lower f_{O_2} values (10^{-9} atm, Fig. Hg-39e) compositions are restricted to the prism base, and both magnetite and magnesioferrite are unstable, noting that although hercynite is stable, chromite remains unstable. Within the limitations of these broadly defined terms one should note also that terrestrial igneous system encompass both a wide range in Fe^{2+}/Fe^{3+}, and in $Fe^{2+} + Fe^{3+}/Ti$ ratios, so that neither of the two idealized systems provide for a unique interpretation of T or of f_{O_2} of chromi spinel formation. These data do, however, suggest that f_{O_2} are probably moderate and based on coexisting and coequilibrated $Usp_{ss} + Ilm_{ss}$, an envelope which is close to the FMQ buffer curve seems reasonable.

Although the volumes illustrated in Figure Hg-39 are shown as continuous trends, there are in reality a paucity of data for regions of intermediate composition between the normal spinels on the base of the prism and the inverse spinels at the respective spinel apices. The photomicrographs illustrated in Figure Hg-23 and Figure Hg-24 show, furthermore, that the contacts between chromite cores and titanomagnetite mantles are relatively sharp, although in many instances these are also gradational. Irvine (1965, 1967) has suggested that these discontinuities are most likely the result of a peritectic involving the precipitation of pyroxene in the interval between chromite crystallization

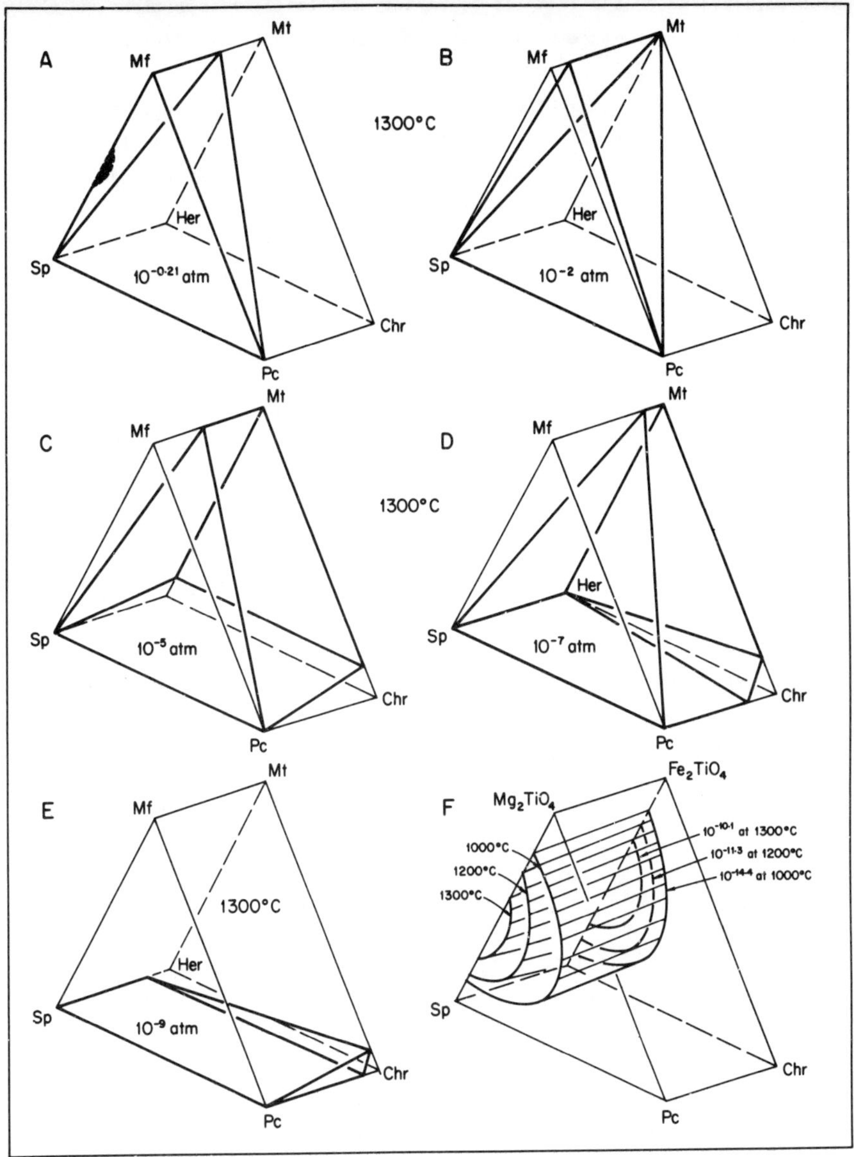

Figure Hg-39. Spinel stability as a function of f_{O_2} at 1300°C (Ulmer, 1969); stable volumes are bounded by the heavy outlines in (a-e). The hatched area along Sp-Mf is a narrow region of immiscibility. In (f) the apex components differ and are Mg_2TiO_4 and Fe_2TiO_4. Values shown adjacent to T°C are for f_{O_2}. The experimental data are strictly for the Mg and Fe ternaries (Muan *et al.*, 1972) and the solvi corridors between these two faces are inferred, based in part on the distributions of natural occurrences but also for purposes of clarity.

and titanomagnetite crystallization. Experimental data on the compositions of spinels in basalts crystallized over a range of T and f_{O_2} by Hill and Roeder (1974) have confirmed Irvine's suggestion and as shown in Figure Hg-40, provide a number of additionally interesting trends for the behavior of major elements at fixed T and variable f_{O_2}, and at fixed f_{O_2} for a range in temperature. For the phase regions illustrated in Figure Hg-40 and for a path of crystallization along A-B, starting at an initial temperature of ∿1250°C and f_{O_2} ∿$10^{-7.5}$ atms and proceeding parallel to the synthetic buffer curves, the regions intersected below the liquidus are as follows: (1) Ol + Sp + Liq; (2) Pl + Px + Ol + Liq; and (3) Pl + Px + Ol + Sp + Liq. The intervening spinel-absent field is at approximately 10^{-9} atms and 1125-1150°C. For values of $f_{O_2} > 10^{-8.5}$ atms spinel crystallization is continuous, demonstrating therefore that in basalts at least interruptions in trend-continuities from the base of the spinel prisms to the apices are a function of f_{O_2}, a feature which is shown also by Ulmer's study as illustrated in Figure Hg-39. Turning briefly to the compositional trends shown in Figure Hg-40b&c, we note firstly that Al_2O_3, Cr_2O_3 and MgO increase with increasing temperature, whereas FeO, Fe_2O_3 and TiO_2 decrease under experimental conditions of constant f_{O_2} (10^{-7} atms). At constant T (1200°C) and variable f_{O_2}, the oxides Cr_2O_3, Al_2O_3 and FeO increase with decreasing f_{O_2}. Slight increases are also shown by TiO_2, but MgO and Fe_2O_3 decrease. High f_{O_2} values are therefore conducive to the precipitation of magnesioferrite (equivalent to the *Bowen* trend); relatively lower f_{O_2} values favor the crystallization of aluminian-chromites. These trends also demonstrate that high temperatures and constant f_{O_2} (10^{-7} atms) tend to favor Mg-Al-chromites, whereas at lower temperatures (1150°C) the stable spinels are Mg-Al- and Cr-bearing titanomagnetites. These experimental trends closely parallel those of natural systems.

Pseudobrookite distributions

Members of the psuedobrookite series (Fe_2TiO_5-$FeTi_2O_5$-$MgTi_2O_5$) assume a number of distinct modes of occurrence in igneous rocks: (1) as primary precipitates in rapidly quenched, glass-spattered, pahoehoe flow tops (Anderson and Wright, 1972); (2) in open cavities in rhyolites associated with bixbyite + cassiterite (Lufkin, 1976); (3) as skeletal crystals in association with ulvöspinel in plagioclase in rapidly quenched dikes (Rice *et al.*, 1971); (4) in lamproitic dikes in association with sanidine, richterite, phlogopite, olivine and diopside (Velde, 1975); (5) in ultramafic nodules (Cameron, 1974) and in groundmass kimberlites in association with rutile + picroilmenite (Haggerty, 1975) (6) as oxidation products of ulvöspinel and ilmenite solid solutions (*e.g.*, Ottoman and Frenzel, 1965; Traut, 1975; and as described in Chapter 4); and (7) as reduction products of primary oxides in meteorite-impacted basalts (El Goresy and Chao, 1976).

Within the binary Mg-free system, the end member $FeTi_2O_5$ is stable at high temperatures and low values of f_{O_2}, and there is a progressive decrease in the lower thermal stability limit towards Fe_2TiO_5 which exists only at relatively high values of f_{O_2}. These relationships are illustrated in Figure Hg-13, p. Hg-57, Chapter 4, and the data show, in addition, that the compositions of coexisting titanomagnetites and ilmenites vary sympathetically with the compositions of Pb_{ss}; members enriched in $FeTi_2O_5$ coexist with Usp-rich and Ilm-rich solid solution members, whereas Fe_2TiO_5-enriched members coexist with Hem-rich and Mt-rich solid solutions. For the Mg-bearing system the join $MgTi_2O_5$-$FeTi_2O_5$ has been examined by Lindsley *et al.* (1974), and their data show firstly that

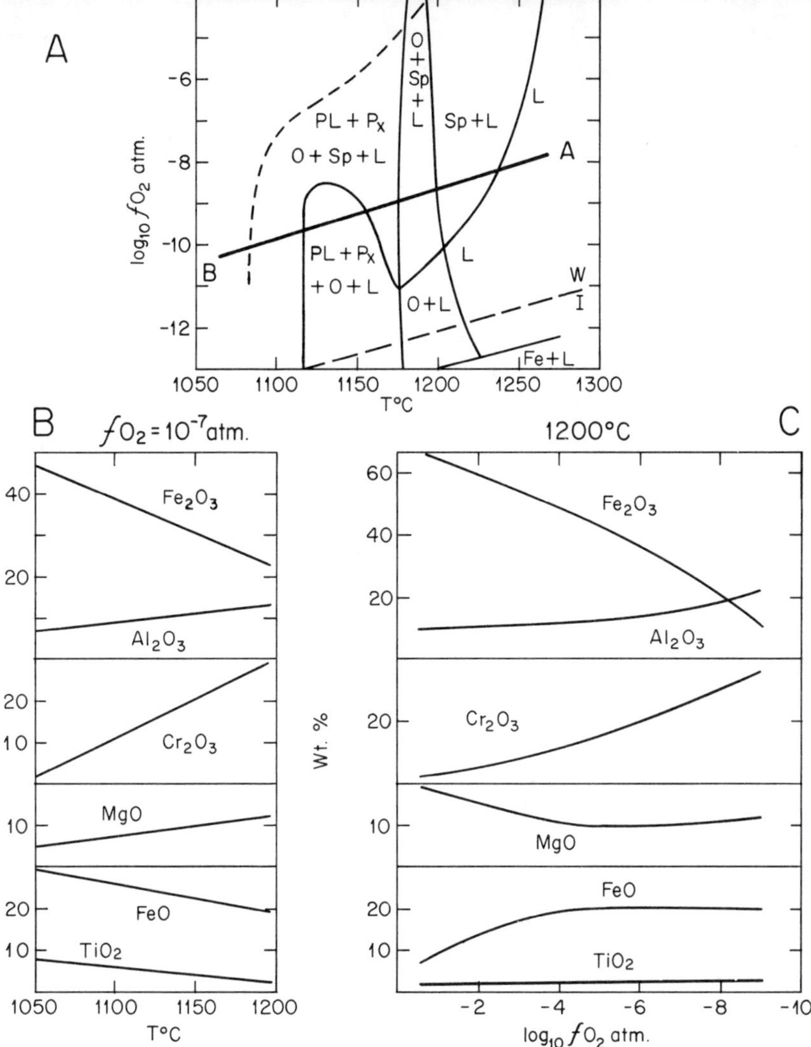

Figure Hg-40. These figures are redrawn from Hill and Roeder (1974) and are for spinel relationships in basalts recrystallized from glasses from above the liquidus. Pl-plagioclase; Sp-spinel; Px-pyroxene; O-olivine; L-liquid; and Fe-iron; the dashed line marked WI is the wustite-iron buffer curve; the curved dashed line is the bulk solidus. The cooling path A-B is discussed in the text and is relevant to terrestrial basalts; paths below the WI buffer are relevant to lunar basalts. Spinel compositions at fixed f_{O_2} and at fixed T are shown in (b) and (c), respectively.

$MgTi_2O_5$ has a lower thermal stability limit of 700°C and that this temperature increases as a function of pressure. The end members Fe_2TiO_5 and $FeTi_2O_5$ have lower thermal stability limits of 585°C and 1140°C, respectively (Haggerty and Lindsley, 1970), so that the addition of both Mg and Fe^{3+} tends to decrease the lower thermal stability limit of $FeTi_2O_5$ members, whereas with the addition of Mg the stability is increased with pressure to a maximum of ∿20 kb.

Compositional data on members of the series Pb-FPb-Kar show that the end members are not realized in natural systems and that characteristically, compositions fall within the three-component system rather than within sets of each of the binaries. This general statement applies to terrestrial rocks only, and it does include lunar armalcolites, $FeMgTi_2O_5$, of intermediate composition between $FeTi_2O_5$ and $MgTi_2O_5$ which has a lower thermal stability limit of 1110°C (Lindsley *et al.*, 1974). Compositions which are intermediate between Fe_2TiO_5 and $MgTi_2O_5$ (kennedyite) have been described from a sill in the Mateka Hills, Rhodesia (von Knorring and Cox, 1961), and in this example (Haggerty, 1976) as well as in those listed above, variable to substantial proportions of Fe^{2+} and Fe^{3+} are present. In addition, the range in MgO values is equally variable so that the stability limits inferred for each of the variety of petrologic settings is at best difficult and for the most part impossible. The single common factor among all settings, however, and it is a factor which includes lunar basalts, is that relatively rapid chillin appears to be a necessary prerequisite for the preservation of primary precipitates. In the absence of rapid chilling, ternary Pb_{ss} react with the residual liquid and magnesian-rich ilmenites result; if the reaction is incomplete, mantles of ilmenite, which are similar in appearance to the relationships for zoned spinels (Fig. Hg-26), develop. These mantles become saturated in Ti (but also Cr + Al), and spinels + rutile form in oriented intergrowths.

The distribution of primary Pb_{ss} and secondary Pb solid solutions in igneous rocks appears to indicate that although primary precipitates are present in a variety of settings, their occurrences are relatively rare; secondary Pb_{ss}, on the other hand, are extremely common, but these are restricted almost exclusively to subaerially-extruded basaltic suites as high-temperature oxidation products of Ilm_{ss} and of Usp_{ss}-Mt_{ss}, as discussed in Chapter 4.

Ilmenite distributions

Ilmenite, in common with magnetite and chromite, varies from one extreme by being present as a minor constituent of most igneous rocks to the other extreme where concentrations in ore deposits are of economic proportions. The systematic variations in Ilm-Hem_{ss} which coexist and coequilibrate with Usp-Mt_{ss} are now well established and have received widespread application as a geothermometer and oxygen geobarometer (Buddington and Lindsley, 1964). The consolute point for the series Ilm-Hem is at approximately 800°C, exsolution is widespread and is typically present in deep-seated intrusions for the entire range from acid to basic suites. Ilmenite is most commonly present as a primary magmatic precipitate but is almost equally common as an oxidation byproduct of Usp-Mt_{ss}; decomposition leads to oxidation-"exsolution" trellis lamellae, and the plane of exsolution is parallel to {111} spinel planes.

Examples of the generalized distributions of the rhombohedral phases in the quaternary $FeTiO_3$-Fe_2O_3-$MgTiO_3$-$MnTiO_3$ is illustrated in Figure Hg-41a for a variety of rock types. These distributions show that basic suites for the most part are confined to the join $FeTiO_3$-Fe_2O_3; that acid suites, granitic pegmatites and carbonatites lie in the ternary Ilm-Hem-Pyh; and that hematite-rich solid solutions are restricted to acid intrusive suites and to anorthosite suites, where Hem_{ss} are exsolved, and these coexist with ilmenite-rich solid solutions.

These marked compositional variations among ilmenites can be broadly correlated with the bulk chemistry of the rock, and the best known example, because it has been employed as a prospecting guide, is the case for kimberlites where ilmenites form a distinctive population with high MgO contents. Equally common, although not as widely recognized, is that acid intrusive and extrusive suites can also be characteristically earmarked because of their high MnO contents. The range and distribution of these two oxides, however, does not appear to be progressive between the two extremes. It seems rather that there are characteristic peaks for these oxides which is consistent at least for MgO which does reach a maximum in ultramafic suites, but is inconsistent for MnO which does not peak in acid suites; the average bulk rock ranges for ultramafic suites is 1500-1600 ppm Mn and for granitic suites is 400-600 ppm Mn. Although apparently anomalous, these differences are resolved in the context of widely differing modal proportions and in the context of partitioning coefficients for Mn between ilmenite and the two distinct assemblages of associated minerals.

Temperature is also a related factor as illustrated in Figure Hg-41 which shows that the MnO contents of ilmenite decrease with increasing temperature. This relationship has been treated for example by Buddington and Lindsley (1964), Lipman (1971), Neumann (1974), Czamanska and Mihálik (1972) and by Tsusue (1973). Within the context of these studies attention has also been given to the partitioning of Mn between coexisting titanomagnetite and ilmenite, and there is uniform agreement that Mn is strongly partitioned towards the rhombohedral constituent regardless of whether the ilmenite is primary or results by oxidation "exsolution." The characteristics of this partitioning as a function of rock type are shown in Figure Hg-41, and although a more detailed discussion is given in the notes to Tables Hg-7 and Hg-12, the results by Neumann (1974) and by Lipman (1971) bear repeating. Neuman has shown, in addition to the generalized trends noted above, that: (a) Mn_{Ilm} increases with increasing $(Mn^{2+}/Fe^{2+})_{magma}$ and is more pronounced when crystallization takes place at low T°C; (b) $Mn^{3+}_{Ilm}/Mn^{3+}_{Mt}$ decreases with increasing f_{O_2}; (c) ilmenites in peralkaline suites contain the highest concentrations of MnO (7-30 wt %), monzonites range between 2-5 wt % MnO, and diorites contain ∿2 wt % MnO; and (d) although there are large intra-grain variations, little or no zoning in MnO is detected within individual crystals. In Lipman's study, there is a clear correlation between SiO_2, T°C, MnO, and Al_2O_3 contents in a suite of samples ranging from rhyolites to latites. These data show that: (a) low T°C (∿650°) and high SiO_2 (78 wt %) correlates linearly with high T°C (∿900°C) and low SiO_2 (66 wt %); and (b) that MnO in ilmenite and Al_2O_3 in magnetite decrease and increase respectively as T increases.

Crystallochemical considerations by Czamanske and Mihálik (1972), and by Neumann (1974) on the preferred distributions of Mn^{2+} and Mn^{3+} between coexisting cubic and rhombohedral constituents argue convincingly in favor of Mn occupation in the ilmenite structure based on the following considerations and in the context of $FeTiO_3$-$MnTiO_3$ and Fe_3O_4-Mn_3O_4-$MnFe_2O_4$ solid solubility: (1) Hausmannite is tetragonal but inverts

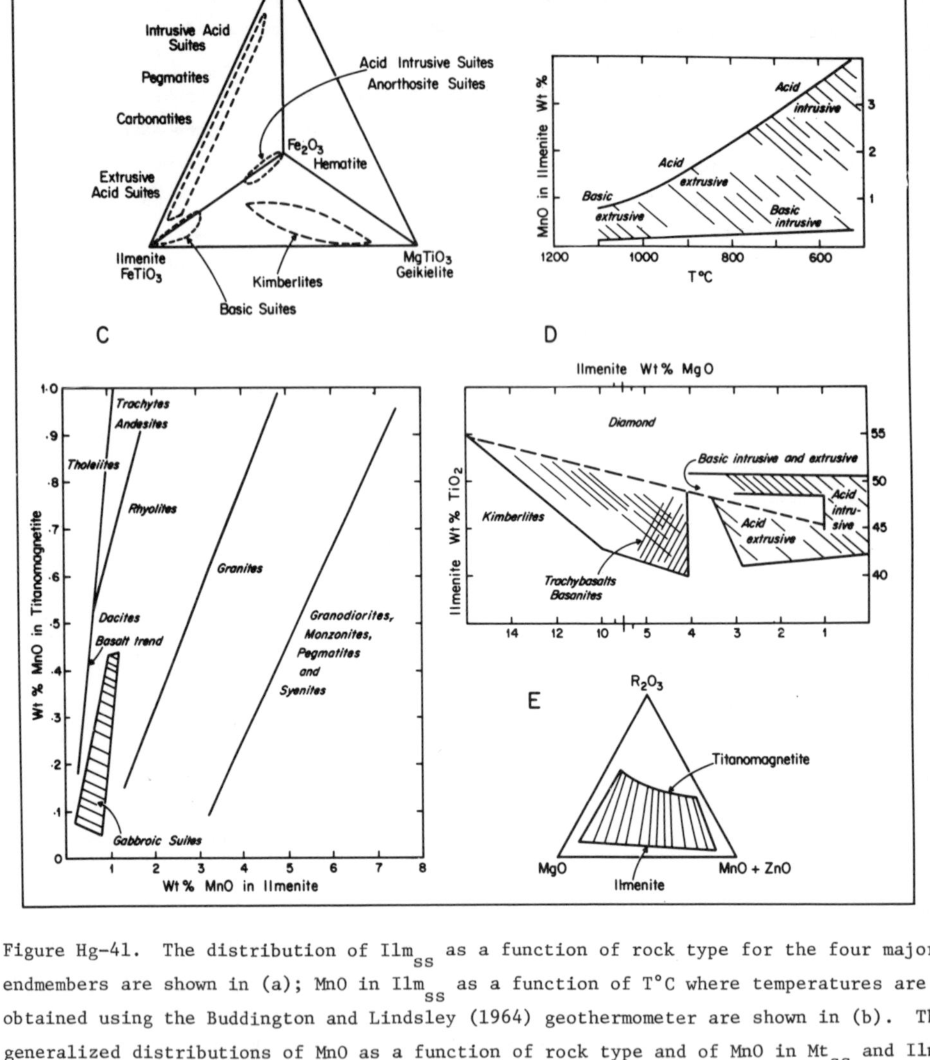

Figure Hg-41. The distribution of Ilm_{ss} as a function of rock type for the four major endmembers are shown in (a); MnO in Ilm_{ss} as a function of T°C where temperatures are obtained using the Buddington and Lindsley (1964) geothermometer are shown in (b). The generalized distributions of MnO as a function of rock type and of MnO in Mt_{ss} and Ilm_{ss} are illustrated in (c). TiO_2 *vs.* MgO and their concentrations among rock types for Ilm_{ss} are shown in (d); the dashed line in this figure is defined as the *kimberlite constraint* curve and the range for Ilm_{ss} intergrown with diamonds is equivalent to the span of the label; inclusions in diamond (not marked) are close to stoichiometric $FeTiO_3$. In (e), the distribution of minor elements in coexisting Mt_{ss} and Ilm_{ss} for a variety of rock types demonstrates the preferred partitioning of divalent ions in Ilm_{ss}, and of trivalent ions, R^{3+} (Cr_2O_3, Al_2O_3, V_2O_3) in Mt_{ss}.

to the spinel structure at high temperatures; (2) both $FeTiO_3$ and $MnTiO_3$ are rhombohedral and have only octahedral cation sites; (3) the ionic radii for octahedral coordination are Fe^{2+} = 0.86 Å; Fe^{3+} = 0.73 Å; Mn^{2+} = 0.91 Å; Mn^{3+} = 0.58-0.59 Å; and Ti^{4+} = 0.69 Å; in tetrahedral coordination Fe^{3+} = 0.57 Å; Mn^{2+} = 0.75 Å; and Mn^{3+} = 0.57 Å. For low titanium magnetites, therefore, with Fe^{3+} distributed both between tetrahedral and octahedral sites, tetrahedral substitution in spinel appears reasonable for Mn^{3+}. For titanomagnetites, however, with Fe^{2+} distributed between octahedral and tetrahedral sites, Mn^{2+} substitution may be possible in octahedral coordination but is less likely in tetrahedral coordination. Mn^{2+} and Fe^{3+} have approximately equivalent tetrahedral site preference energies which differ substantially from those of Fe^{2+} and Mn^{3+} which have preferred octahedral site preference energies, and this is so particularly for Mn^{2+} which is higher than Fe^{2+} (Navrotsky and Kleppa, 1967); Mn^{2+} in the spinel structure is hence most probably confined to the tetrahedral site. In summary, the inferences are that at very high f_{O_2}'s and at high temperatures, Mn^{3+} can be stabilized in the spinel structure, whereas at lower values of f_{O_2} and with the dominance of Mn^{2+}, the rhombohedral structure is decidedly preferred.

The other major component of Ilm_{ss} in igneous rocks is geikielite ($MgTiO_3$), and the generalized distributions among rock types are illustrated in Figure Hg-41. The distribution of MgO and of other minor components between Ilm_{ss} and titanomagnetite are shown in Figure Hg-41; specific examples of ilmenite compositions in kimberlites are given in Figure Hg-42. Magnesian concentrations in Ilm_{ss} can once again be broadly related to magma chemistry, with acid intrusive suites containing the lowest concentrations and the more mafic suites the highest concentrations (Table Hg-20). It is of interest to note, however, that ilmenite is characteristically absent in most ultrabasic suites with the exception of kimberlites where it is present in the groundmass, in eutectic-like (symplectic) intergrowths with pyroxene, and in some nodule associations (Table Hg-18). Ilmenite is also commonly absent in feldspathoidal suites where the Ti-rich constituent is either perovskite or sphene.

It had long been considered that the high MgO contents in ilmenites, so characteristic of kimberlites, was probably a pressure-related property; that would seem now not to be the case, and although experimental data are sparse, the composition of the primary magma is probably the most influential factor on mineral chemistry. The distribution of Mg among the oxides and between the oxides and the silicates is clearly also related to paragenetic sequences of crystallization, temperature of formation, and on rates and paths of equilibration. In previous discussions an emphasis was placed on the early enrichment, and hence depletion, of the more refractory elements (Mg, Al, Cr) in chromian spinels which are followed by the later precipitation of titanomagnetites. It was noted also that the onset of olivine, pyroxene, and plagioclase crystallization could account in large part for continuous decreases in Mg and Al in the oxides. The normal sequence of oxide crystallization is chromite followed by ilmenite followed by titanomagnetite. The interval between chromite and ilmenite crystallization is most commonly filled in basic suites by the precipitation of olivine, so that both the abundance of Mg and the rate at which olivine crystallizes are constraints on the amount of geikielite which can form; the abundance of Ti is also clearly of importance. An additional factor is the preferred partitioning of

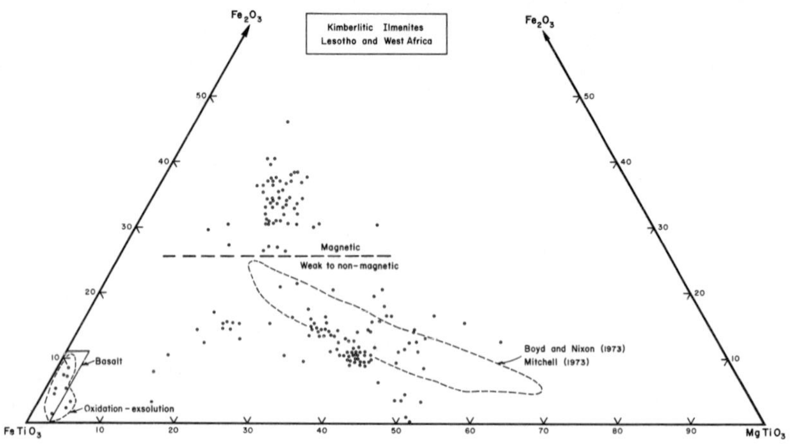

Figure Hg-42a. Electron-microprobe data for ternary solid solution kimberlitic ilmenite

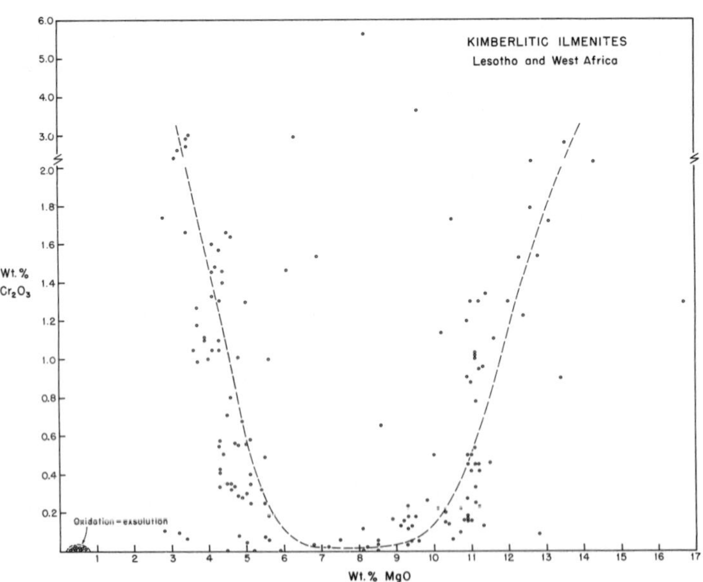

Figure Hg-42b. Cr_2O_3 versus MgO for kimberlitic ilmenites. The parabolic curve drawn through the data maxima is estimated, and does not represent a computed best fit.

Mg towards Ilm_{ss} rather than towards titanomagnetite (Fig. Hg-41e); this partitioning is effective at magmatic temperatures and is particularly apparent in the subsolidus as observed in Ilm_{ss} which develop by oxidation "exsolution" from Usp-rich solid solutions.

Diamond inclusion studies, kimberlitic autolith* series, and basaltic phenocryst studies provide insights to the behavior of Mg in Ilm_{ss}. Ilmenite inclusions in diamonds (Meyer and Svisero, 1975) have been shown to contain very low MgO contents, in common with ilmenites in autolith nucleii (Ferguson *et al.*, 1975; Danchin *et al.*, 1975). Phenocrystal ilmenites in kimberlites are commonly reversely zoned, with higher MgO contents at the margins and lower contents in the cores (Haggerty, 1975), and this trend is observed also in ilmenite phenocrysts in basalts (Anderson and Wright, 1972). In contrast, ilmenites intergrown with diamonds have moderately high MgO contents (Sobolev, 1972) in common with the encasements on autolith nucleii, the compositions of groundmass ilmenites, and the compositions of ilmenites in pyroxene sympletic intergrowths. The significance of these examples would tend to suggest, therefore, that either the crystallization of olivine was drastically different between early and late ilmenite or that pressure was an important parameter. This factor has been addressed in part by Mitchell (1973) and by Stormer (1972). Stormer has shown, based on free energy data, that for a number of spinel-yielding and ilmenite-yielding reactions, that Mg is favored in the rhombohedral structure in the presence of Ti. In addition, the volume change associated with the reaction Fo + Ilm = Geik + Fa is positive, and therefore, high pressures will tend to reduce the geikielite component of ilmenite solid solutions. In summary, the comparative rarity of very high MgO ilmenites in basic suites is probably related to the early crystallization of olivine (Stormer, 1972); the range in high MgO contents in kimberlitic assemblages is most likely the result of variable olivine crystallization and probably pressure, although both the initially high MgO contents of these rocks and the nature of late stage, but Ti-enriched liquids are undoubtedly also contributory factors.

For the most part, minor elements such as ZnO, NiO, ZrO, V_2O_3, Al_2O_3 and Cr_2O_3 are relatively low (Fig. Hg-41e), but once again the exceptions are kimberlitic ilmenites as illustrated in Figure Hg-42b for which an unexplained but parabolic relationship exists between MgO and Cr_2O_3. Although this variation was previously considered to be related to pressure (Haggerty, 1975), it now seems more probable that spinel crystallization, the redistribution of Cr among garnets and clinopyroxenes coupled with the inability to distinguish among true groundmass, phenocrystal, xenocrystal, and fragmented modular ilmenite, are all factors which are likely to have an overriding influence on the parabolic relationship.

Igneous suites containing Ilm_{ss} of exceptionally high MgO or MnO contents and suites in which Ilm_{ss} are absent prevent the determination of T°C and f_{O_2} for coequilibrated oxide pairs (Buddington and Lindsley, 1964). Indirect methods of determination and the effects of a_{SiO_2} in ilmenite-absent suites are discussed in a subsequent section.

---

*An ovoid inclusion containing a nucleus of a mineral or rock fragment encased in fine-grained equigranular kimberlite. Autoliths are present as discrete, coherent bodies in much the same way that nodules and xenocrysts are present.

Magnetite-ulvöspinel distributions

The voluminous literature on ore deposit petrology and on paleomagnetic, rock magnetic, and T-f_{O_2} determinations is an excellent measure of the ubiquitous distribution of the series Usp-Mt_{ss} in all igneous rock types. Modal concentrations and compositions vary as noted in previous sections, and T, f_{O_2} and Ti contents are the major parameter-defining variables on composition; silica activity is a related variable, and its influence also determines Ti distributions. Ulvöspinel-rich components of the series are sensitive to subsolidus oxidation, and byproducts of Ilm_{ss} result. Because the series is sensitive also to T and f_{O_2}, both initial magmatic temperatures and the extent to which equilibration is possible defines whether Usp-rich or Mt-rich members will result. The former are stable at low values of f_{O_2}, and the latter at higher values. Slow cooling in deep-seated intrusives results in exsolution, but because the consolute point is so low (∿600°C), compositions along the entire series are not only possible but widespread.

The series has an inverse spinel structure, and extensive solid-solubility towards members of the normal spinel series are also encountered (Fig. Hg-38). Earlier discussions have emphasized the role of partitioning of minor elements, the effects of paragenesis, and the role of magma chemistries on the compositions of coexisting oxides and silicates. These reviews are pertinent here also, but it is perhaps most instructive to examine in greater detail the distribution of minor elements both as a function of rock type and in relationship to coexisting Ilm_{ss}. The data which are given in Table Hg-20 are not surprising in view of preceding comments, and in summary the variations which should be noted are as follows: (1) Mg_{Ilm} and Mn_{Ilm} are greater than Mg_{Mt} and Mn_{Mt}; (2) Al_{Mt} and Cr_{Mt} are greater than Al_{Ilm} and Cr_{Ilm}; (3) Cr_{Mt} increases with decreasing SiO_2; (4) Al_{Mt} is episodic, is uniformly low in acid intrusives but peaks in rhyodacitic suites, in trachybasalts, trachyandesites and again in troctolitic and noritic suites; and (5) V_{Mt} and Ni_{Mt} are greater than V_{Ilm} and Ni_{Ilm}, and these peak respectively in basaltic and in the more mafic suites.

Oxidites

Magmatic ore deposits as reviewed in Tables Hg-13, 14 and 19 are divisible into the following categories based on petrologic settings: (1) segregations associated with large layered intrusive bodies; the oxides are most commonly chromite, but magnetite and ilmenite are also present; (2) segregations association with anorthosite suites; the oxides are typically ilmenite and titanomagnetite; (3) segregations associated with Alpine-type peridotites; the oxide is invariably chromite; magnetite is present but ilmenite is rare; (4) magmatic segregations associated with apatite, where the oxide is typically titanomagnetite; and (5) oxide segregations associated with carbonatites, where the most abundant oxides are magnetite, rutile, and brookite in association with multiple and complex oxides of vanadates and niobates. Oxide associations not considered are those typically related to granitic suites, where the oxides are stannates, tungstates, and tantalates.

The origins of these deposits are for the most part controversial with the possible exception of magmatic sedimentation in rhythmically-layered basic bodies. A more general discussion is beyond the scope of this chapter, but a number of specific points will be briefly treated with a bearing on future studies. These are: (1) the anorthosite associations; (2) the apatite associations; and (3) stratiform and Alpine-type associations.

Anderson and Morin (1970) have suggested that significant differences exist between andesine anorthosites and labradorite anorthosites; that the former are demonstrably intrusive and are dominated by titanomagnetite and Mt-Usp exsolution; that the latter result by partial fusion of continental crustal material and that these are dominated by $Ilm\text{-}Hem_{ss}$ and by mutual exsolution. The observation and conclusions are provocative and worthy of more intensive investigation.

Although the association of titanomagnetite and apatite has long been recognized in anorthosite, gabbroic and in carbonatitic suites, little new experimental data has been forthcoming in elucidating the association since the original study by Philpotts (1967) which demonstrated that the liquidus of magnetite could be depressed from ∿1600°C to more reasonably accepted magmatic temperatures (∿1200°C) by phosphate fluxing. A renewed interest in this problem seems timely in view of the current debate surrounding the origin of what had previously been considered the type magmatic example, Kiruna (Freitsch, 1973; Parák, 1975).

Studies of the variations between, and contrasts within, stratiform and Alpine-type (podiform) chromite deposits have continued for decades and will undoubtedly continue for future generations. The distinctions in chemistry and in settings show that the former are related to cratonic regions with relatively small variations in mineral chemistry; the latter on the other hand are characteristically associated with island-arc volcanism, or active and mobile orogenic belts of Paleozoic or younger age. These deposits are characterized by a highly variable chemistry, and this is so particularly for Al_2O_3 and Cr_2O_3. Earlier comments related to the pressure dependence of Al among plagioclase, spinel and garnet lherzolites showed that spinels are typically absent in the highest pressure regimes (garnet) and that chromite replaces spinel in the lower pressure regimes (plagioclase). The Al contents of spinels ought, therefore, to be pressure dependent, an observation which is qualitatively apparent in a comparison of spinel-bearing inclusions in basalts with those in kimberlites. More significantly perhaps are the following points: (1) spinel lherzolites occupy the experimental field at magmatic temperatures between 5 and 10 kb which is equivalent approximately to 25-50 km; the oceanic geotherm intersects the spinel-garnet transition at magmatic temperatures; and (3) recent experimental and theoretical calculations show that the Al contents of enstatites are not pressure dependent. Barometric estimates, therefore, cannot be determined for this critical and intermediate region in the mantle, a region of some importance to basaltic but also to andesitic evolution. Since the enstatites are not pressure sensitive but the spinels at the two extremes of plagioclase and garnet are, the proposal and possibility for an Al or an Al/Cr high-pressure barometer seems promising and reasonable. If such variations are indeed P dependent, the ranges in Al/Cr for Alpine-type peridotites may reflect the enormous and expected pressure gradients associated with orogenic events.

Introduction

The experimental data for which T and f_{O_2} values can be determined for coequilibrated oxide pairs (Buddington and Lindsley, 1964) is discussed in Chapter 2. This section is a compilation of these determinations from the literature inclusive of and since 1964 as detailed in Tables Hg-7-22. A total of 434 determinations are plotted in the accompanying diagrams, and the Guinness citation goes to I.S.E. Carmichael; one published datum point (Annual Report C.I.W. Yearbook 69, 1971) is due to the originator D.H. Lindsley, as a jest in part perhaps to test the experimental data in natural systems. The results show conclusively that the data are not *quid pro quo** but *Q.E.D.*

Extrusive suites

The data for extrusive suites (Fig. Hg-43a-c) show the following notable points: (1) 98% of the values for acid extrusive suites plot on or above the FMQ buffer curve (Fig. Hg-43a); (2) the intermediate suites have equilibrated at higher temperatures and the range in f_{O_2} is larger than that of the acid extrusive suites (Fig. Hg-43b); (3) the basic extrusive suites form a tight envelope along the FMQ buffer, and at lower temperatures are between the FMQ and WM buffer curves (Fig. Hg-43c), a trend which is depicted by a few data points in the intermediate suite (Fig. Hg-43b); values for deep-sea basalts are comparable to those of subaerially extruded suites which have quenched at high temperatures; and (4) subsolidus equilibration for all suites closely parallels the FMQ buffer curve.

Intrusive suites

Values for T and f_{O_2} for intrusive suites are divided into acid-intermediate suites (Fig. Hg-44a) and layered series (Fig. Hg-44b-c). The most notable observation when these data are compared with the extrusive suites is that the range in T and in f_{O_2} are substantially lower for the more slowly cooled bodies. This is a reasonable expectation, but what is perhaps more surprising is that the trend of equilibration is precisely parallel to that of the FMQ buffer curve. Within these suites it is of interest to note that the distinctions between the anorthositic (Fig. Hg-44b) and the gabbroic suites (Fig. Hg-44c) are not significantly different, except for the somewhat higher temperatures of equilibration for the latter. The trends for the Labrieville anorthosite are, however, distinctly different as shown by A-A' and B-B' (Fig. Hg-44b). These trends correspond to the andesine-anorthosites discussed earlier (Anderson and Morin, 1970) which are characterized by $Ilm\text{-}Hem_{ss}$, and T and f_{O_2} are inferred. The labradorite-anorthosites which contain $Usp\text{-}Mt_{ss}$ are within the range for both the anorthositic suites and the gabbroic suites.

The acid-intermediate intrusive suites exhibit a somewhat greater scatter about the FMQ buffer, but this range is diminished if the suites are separated into rock type and localities; the granitic and pegmatitic suites lie on the oxidizing side of the buffer curve; syenites lie on or close to the curve; shonkinites below the curve; and monzonites both above and below the curve.

*Blunder made by putting one thing for another.

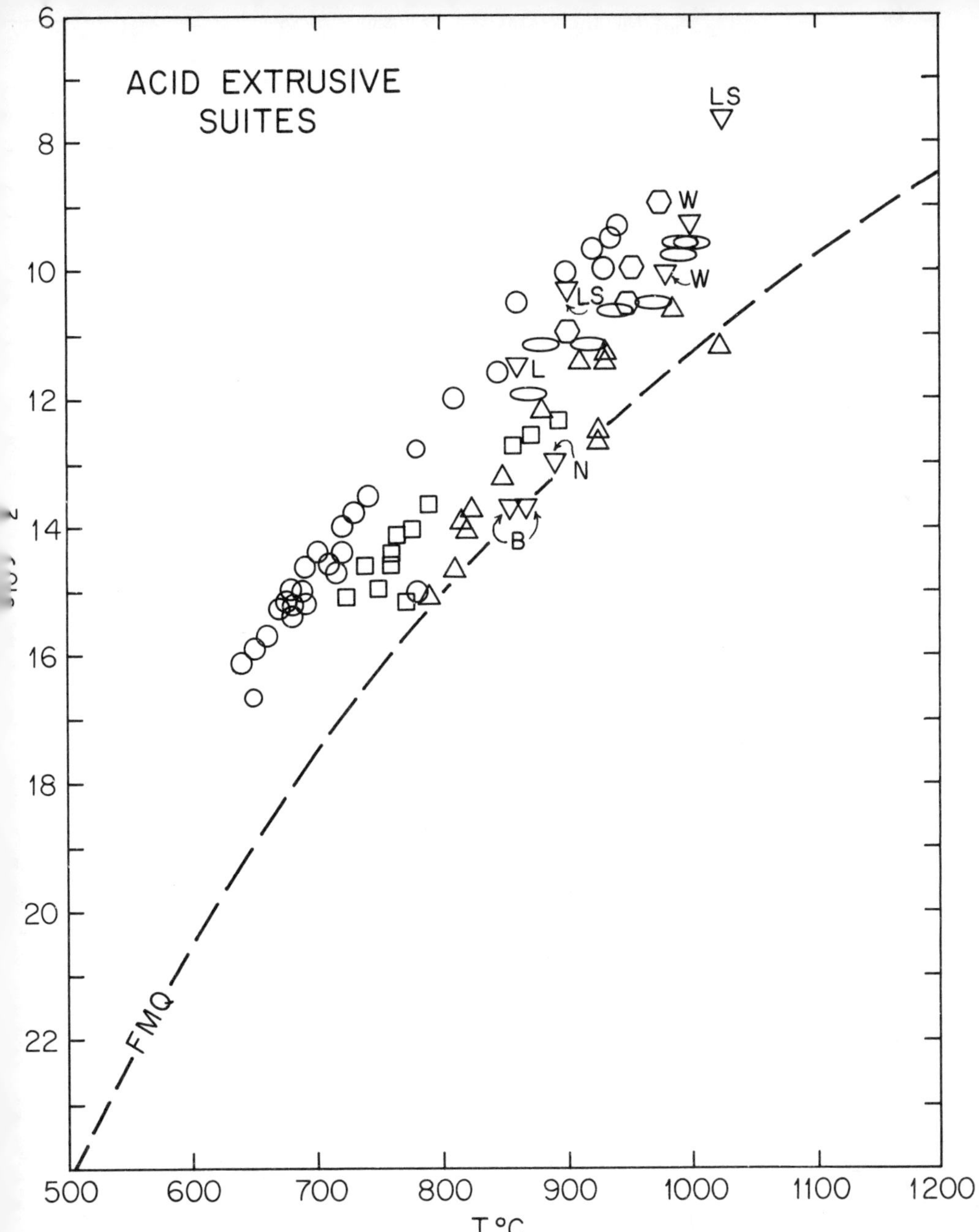

Figure Hg-43a. Symbols refer to the following sources: triangles--Carmichael (1967b); circles--Lipman (1971); hexagons--Buddington and Lindsley (1964); squares--Ewart *et al.* (1971); ellipses--Heming and Carmichael (1973). B refers to Bauer *et al.* (1973); L to Lowder (1970); LS to Lerbekmo and Smith (1973); W to Wilkinson (1971); and N to Nicholls (1971). FMQ is the fayalite-magnetite-quartz buffer curve.

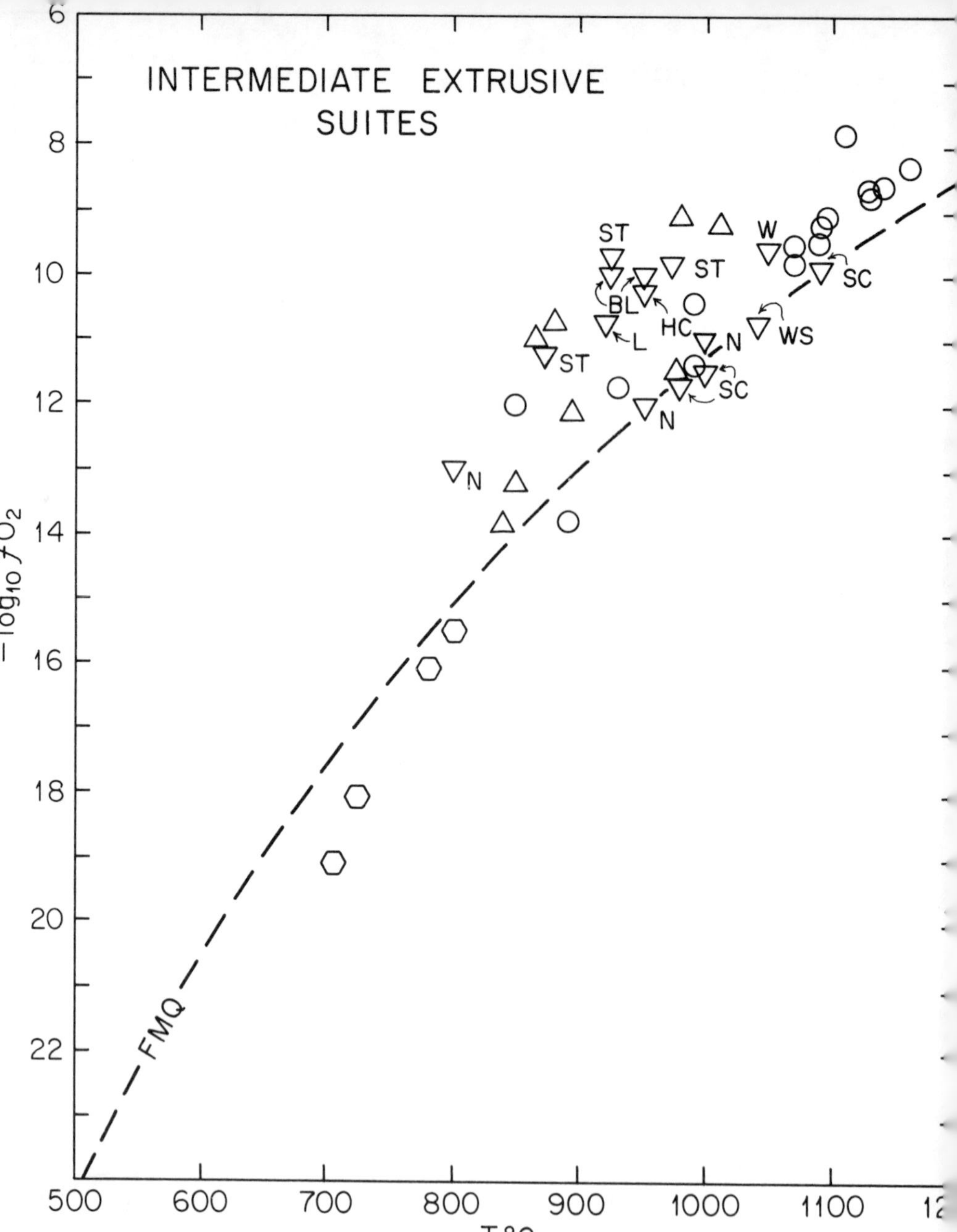

Figure Hg-43b. Symbols refer to the following sources: circles--Anderson (1968); hexagons--Ridley and Baker (1973); the labelled triangles are: ST--Stormer (1972); BL--Buddington and Lindsley (1964); N--Nicholls (1971); W--Wilkinson (1971); SC--Smith and Carmichael (1969); WS--Whitney and Stormer (1976); L--Lowder (1970); HC--Heming and Carmichael (1973). FMQ is the fayalite-quartz-magnetite buffer curve.

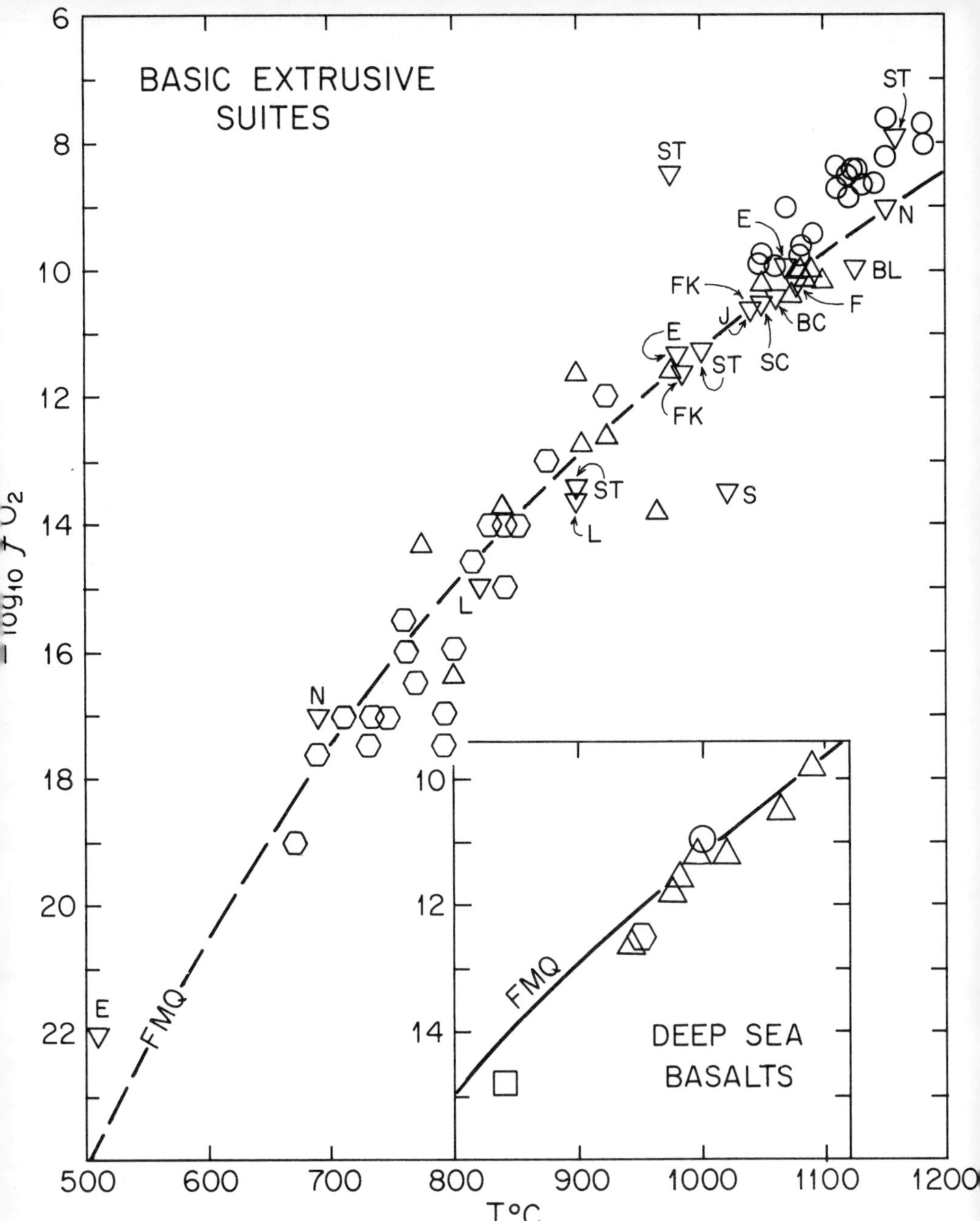

Figure Hg-43c. Symbols refer to the following sources: apex-up triangles--Carmichael (1967a); circles--Anderson and Wright (1972); hexagons--Gidskehang and Davison (preprint 1975). Apex-down triangles: S--Steinthorsson (1972); J--Jakobsson *et al.* (1973); E--Evans and Moore (1968); BC--Brown and Carmichael (1971); N--Nicholls (1971); L--Lindsley and Haggerty (1971); ST--Stormer (1972); F--Fodor (1975); FK--Fodor *et al.* (1972). Deep-sea basalts: triangles--Mazzullo and Bence (1976); circles--Prinz *et al.* (1976); squares--Bence and LaBorde (1976); hexagons--Ridley *et al.* (1974). FMQ is the fayalite-magnetite-quartz buffer curve.

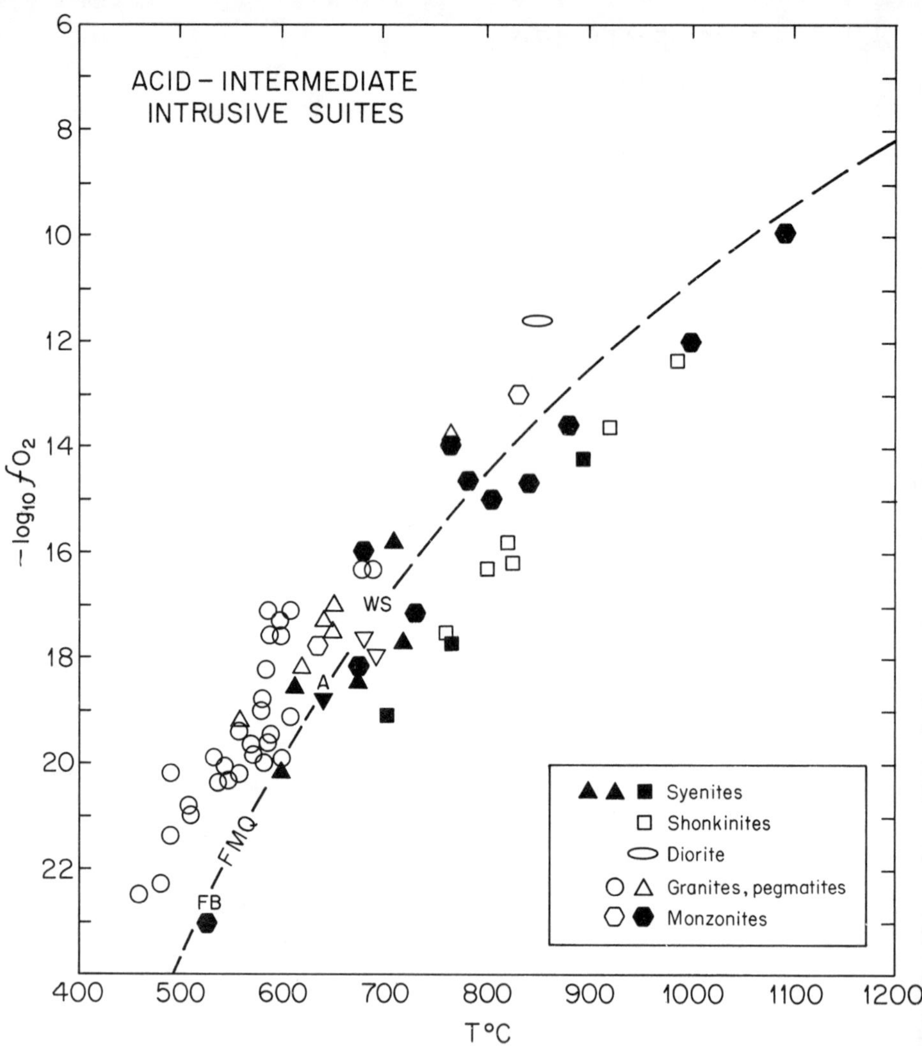

Figure Hg-44a. Symbols refer to the following sources: triangles--Buddington and Lindsley (1964); circles--Lipman (1971); hexagons--Neumann (1974,1976); ellipse--Philpotts (1967); WS--Whitney and Stormer (1976); FB--Frisch and Bridgewater (1976); A--Anderson (1968). FMQ is the fayalite-magnetite-quartz buffer curve corrected to 5 Kb.

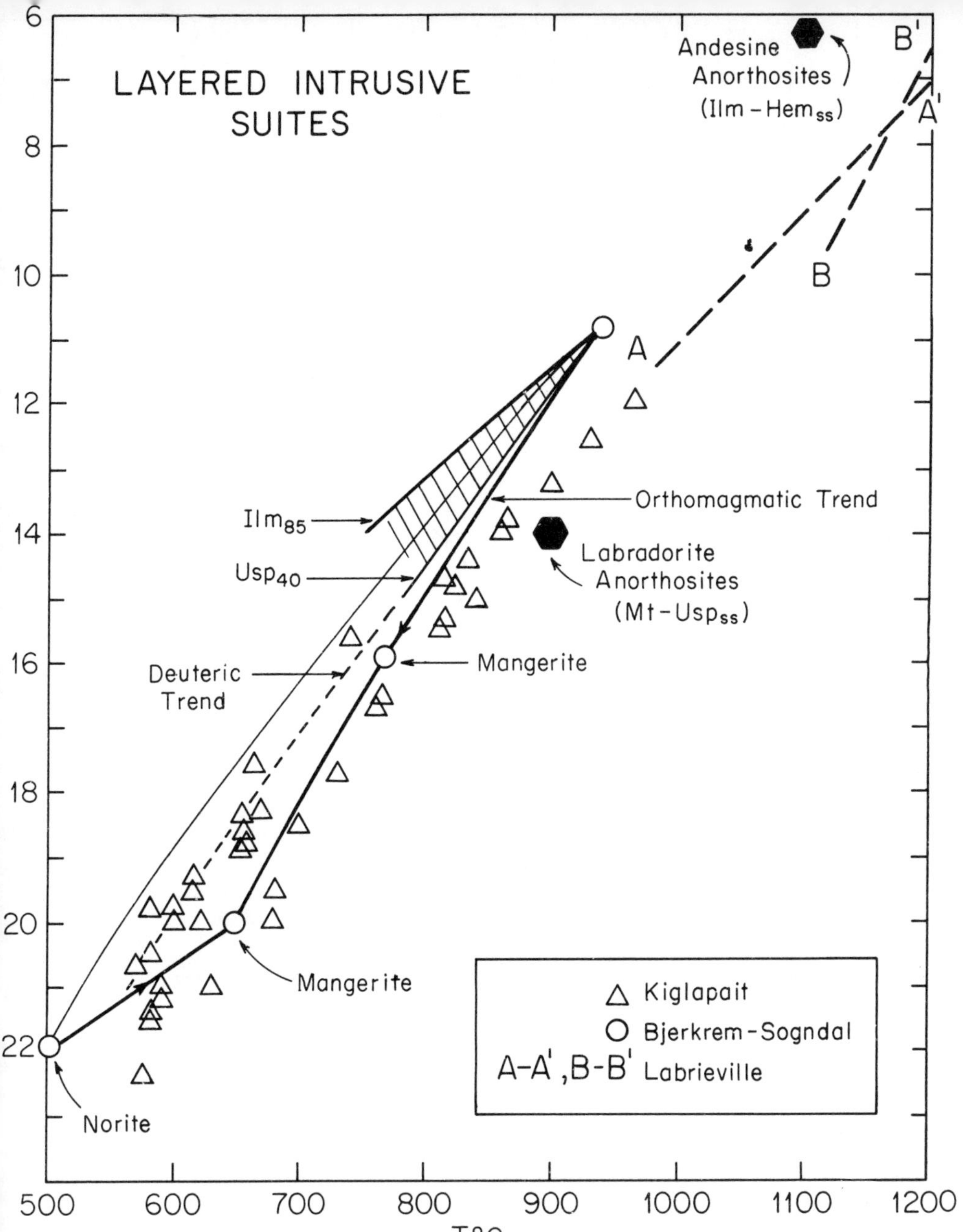

Figure Hg-44b. The Kiglapait suite are from Morse and Stoiber (preprint 1970). For the Bjerkrem-Sogndal suite (Duchesne, 1972), the envelope bounded by norite, mangarite and Usp_{85}-Usp_{40} (which refers to compositions from the curves by Buddington and Lindsley (1964)) encloses the maximum range in T-f_{O_2}; the orthomagmatic trend defines the lower limit, and the deuteric trend is the inferred average trend based on oxidation'"exsolution." The Labrieville trends (A-A',B-B') are from Anderson (1966), and the filled hexagons are from Anderson and Morin (1971).

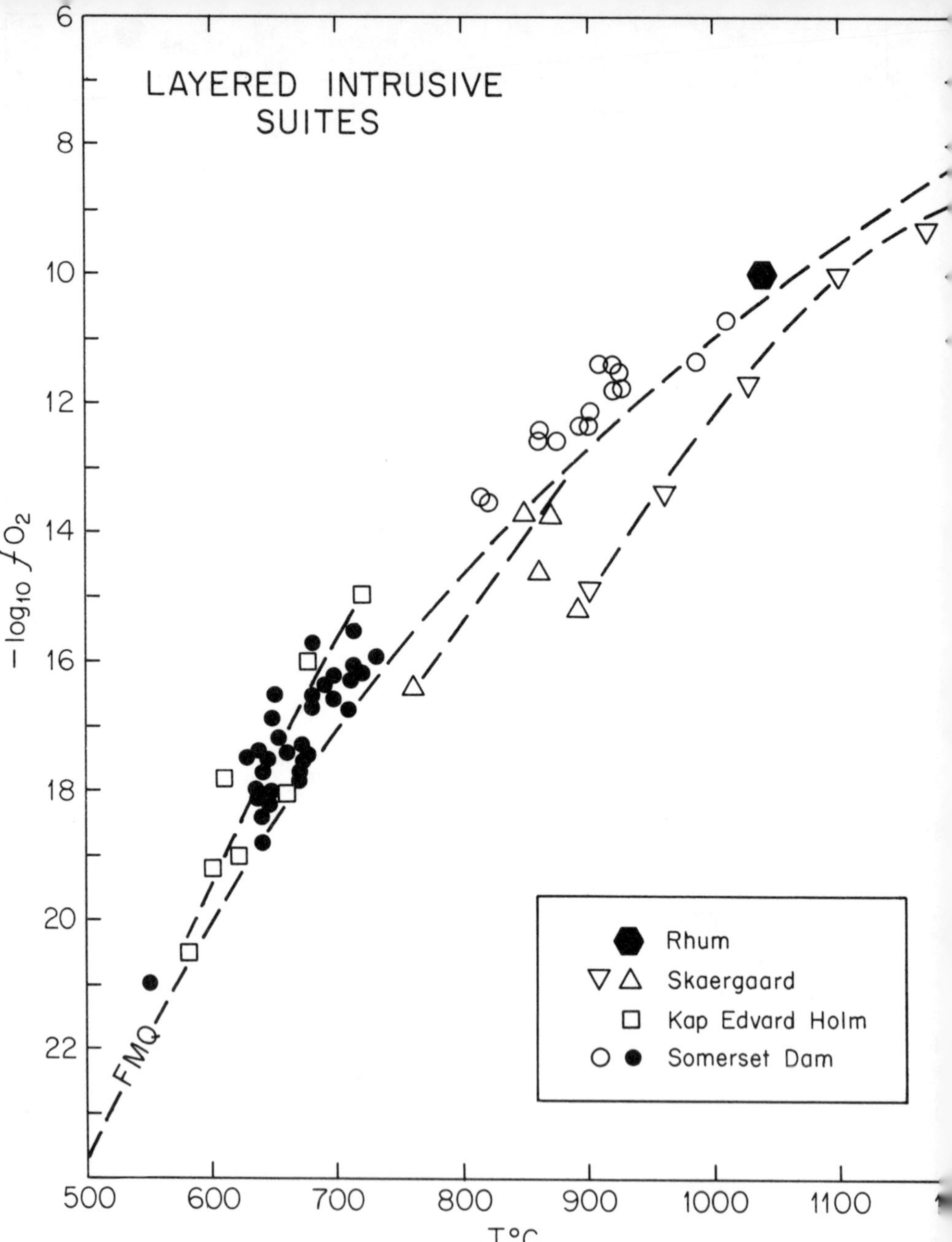

Figure Hg-44c. Rhum (Donaldson, 1975); Skaergaard, apex-up (Buddington and Lindsley, 1964); apex-down, values based on a_{SiO_2} (Williams, 1971); Kap Edvard Holm (Elsdon, 197 Somerset Dam (Mathison, 1975). FMQ is the fayalite-magnetite-quartz buffer curve corrected to 5 Kb.

Magmatic ore deposits

Here once again, using the FMQ buffer as the reference curve and separating the data according to rock type and to location, Figure Hg-45 shows that the anorthosite associations are both above and below the curve; the apatite associations fall on or above the buffer curve; and the gabbroic suites are on or below the curve. In general there is a far greater dispersion of the data here than in any of the other suites considered, and this is due in part to the variety of associations but is due also to fairly sizable analytical errors induced by minor element components. Notwithstanding this limitation it should be noted that approximately 80% of the data conform to parallel equilibration trends which closely match that of the FMQ buffer curve. Within the context of the spread in data, it is of interest to note that the ore-rich bands of layered intrusive sequences are always slightly more oxidized than the oxides in closely associated or contact host rocks.

Data summary

The seven sets of data discussed above are summarized in Figure Hg-46, and these are divided into extrusive and intrusive suites. The conclusion of this compilation follows: (1) That at high, but particularly at intermediate temperatures, acid extrusive suites are more highly oxidized than either of the intermediate suites or the basic suites; (2) by sliding the acid extrusive suite on the FMQ buffer to lower temperatures a perfect match with the granitic and pegmatitic suites results (compare Fig. Hg-43a with Fig. Hg-44a); (3) at high temperatures the intermediate suite is comparable to the acid suite but extends to more reducing conditions with respect to the FMQ buffer curve; (4) the basic suite, on average, is more reduced than either of the acid or intermediate suites; (5) the phonolite and trachyte field is calculated, based on a_{SiO_2}, to be even more reducing than that of the basalt field (Nash *et al.*, 1969); (6) the high-temperature acid-intermediate intrusive envelope is comparable to the intermediate extrusive suite, and if the latter is slid down an FMQ equilibration path to lower temperatures a close match results; (7) the basic extrusive suites if permitted to equilibrate would be superimposed on the intrusive suites of comparable composition; (8) although the range for rocks associated with magmatic ore deposits extends to values approaching the MH buffer curve, a large proportion of the data fall within the field defined by other intrusive suites, with the notable difference of increasingly greater departure from parallelism with progressive equilibration.

The overall conclusion from these data suggest that if *all* intrusive suites are compared with *all* extrusive suites between 800 and 1000°C that the differences in f_{O_2} are minimal; distinctions among intrusive suites are also minimal, but basic suites tend to cluster more tightly as a group than do either intermediate or acid suites. Granites and pegmatites are more highly oxidized with respect to FMQ than either gabbros or anorthosite suites. The distinction among extrusive suites is that rhyolites, pumices and ignimbrites are more highly oxidized with respect to FMQ than basalts. Although calculated, the evidence for phonolites and trachytes suggest that these rocks equilibrated, on the average, below the FMQ buffer curve. Thus the problem of open (extrusive) and closed (intrusive) systems remains unresolved; however, the notion that acid rocks tend to be more highly oxidized than basic rocks appears to be endorsed.

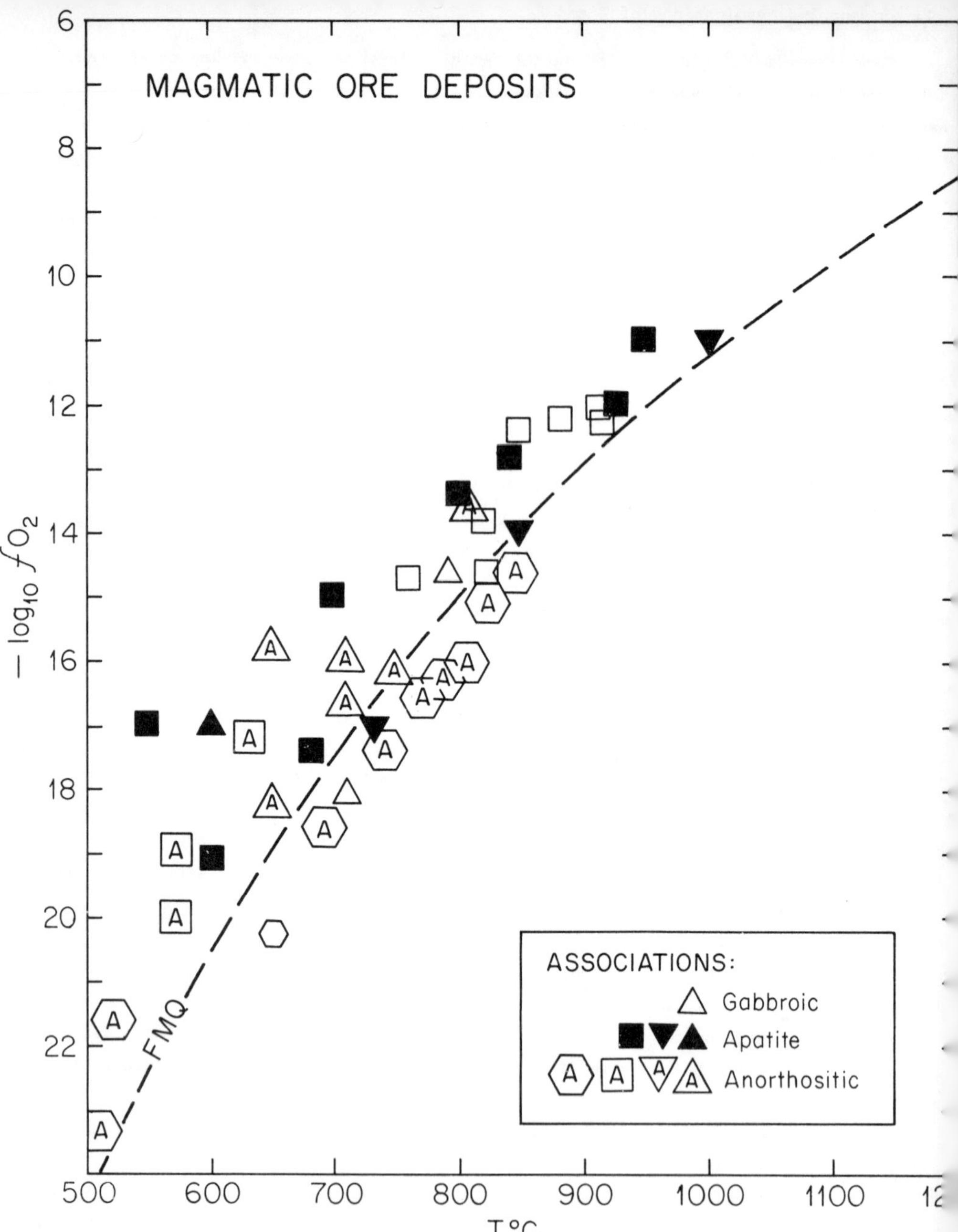

Figure Hg-45. Symbols refer to the following sources: apex-up triangles--Buddington and Lindsley (1964); apex-down triangles--Philpotts (1967); squares--Lister (1967); hexagons--Anderson (1968). FMQ is the fayalite-magnetite-quartz buffer curve.

Experimental determinations of T and f_{O_2}

The results of electrochemical, f_{O_2} reequilibration at bulk rock liquidus temperatures, crystallization of basalts under conditions of known f_{O_2}, and the composition of gasses from active volcanic vents are shown in Figure Hg-47 and are summarized in detail in Table Hg-21.

The ZrO_2 electrolytic cell technique (Sato, 1970, 1971, 1972) has provided data on samples from the Oka carbonatite complex, the Bushveld and Stillwater complexes, a kimberlite from the Premier mine, and *in-situ* determinations of f_{O_2} from actively cooling Hawaiian lava lakes. The liquidus equilibration experiments are on basalts and andesites; values for the andesites are at 1200°C, and these fall between the ranges determined for basalts (Fudali, 1965). The remaining suites are basic and ultrabasic, and a comparison of these data with the relevant coequilibrated oxide data discussed above (Figs. Hg-43c, Hg-44b, Hg-44c, and Hg-46) show that the range is comparable to the basalt and to the gabbro-anorthosite equilibration trends with respect to the FMQ buffer. The experimental envelope for comparison (Fig. Hg-47) is the one defined by the Makaopuhi lava lake (drill hole #9) which is above the FMQ curve and the H_2O/H_2 and CO_2/CO curves for Surtsey which lie on the reducing side of the FMQ buffer curve. It is this gaseous envelope which closely conforms to the upper and lower limits of the values determined for the oxides. There are several encouraging (E) and surprising (S) results: (1) The range for the Picture Gorge basalt is an oxide determination path obtained from crystallized products under known f_{O_2} (very E); (2) the kimberlite path and the carbonatite value lie within the range of the gas curves (E but very S); (3) diamond oxidation is very close to the H_2O/H_2 curve but is below the CO_2/CO (S); (4) both the Stillwater and Bushveld values fall below the FMQ buffer, which is approximately between the gas curves and are in stark contrast to the oxide values determined for other layered complexes (very S). The full impact of E and S can only be gained by reference to Table Hg-21; however, in summary there seems little doubt that the oxide geothermometer really works and that in concert with the electrochemical technique the perplexing problem and the diversity of opinion with regard to open and closed systems may finally be resolved.

Silica activity

This discussion is concerned specifically with ilmenite-absent suites. The role of silica activity (a_{SiO_2}) and of f_{O_2} in relation to the distribution of Ti among oxides and silicates was first noted by Verhoogen (1962) who showed, based on thermodynamic considerations, that low f_{O_2} favors the entry of Ti into oxides, but that at high temperatures and low a_{SiO_2} Ti is more likely to enter pyroxenes. He showed in addition that the assemblage magnetite + perovskite or magnetite + sphene is most likely to occur at low temperatures and high f_{O_2}, and that these assemblages are more stable than diposide + ilmenite. TiO_2 has a larger affinity for CaO than either FeO or MgO and none at all for SiO_2, so that under conditions of low a_{SiO_2} and high temperatures sphene ($CaTiSiO_5$) will decompose to SiO_2 + perovskite ($CaTiO_3$).

The instability of ilmenite, therefore, is of interest in attempts to apply the Buddington and Lindsley thermobarometer, and this problem has been addressed more completely by Carmichael and coworkers (Carmichael and Nicholls, 1967; Carmichael, 1967c; Nicholls *et al.*, 1971; Smith and Carmichael, 1968; Carmichael *et al.*, 1970; Nash *et al.*,

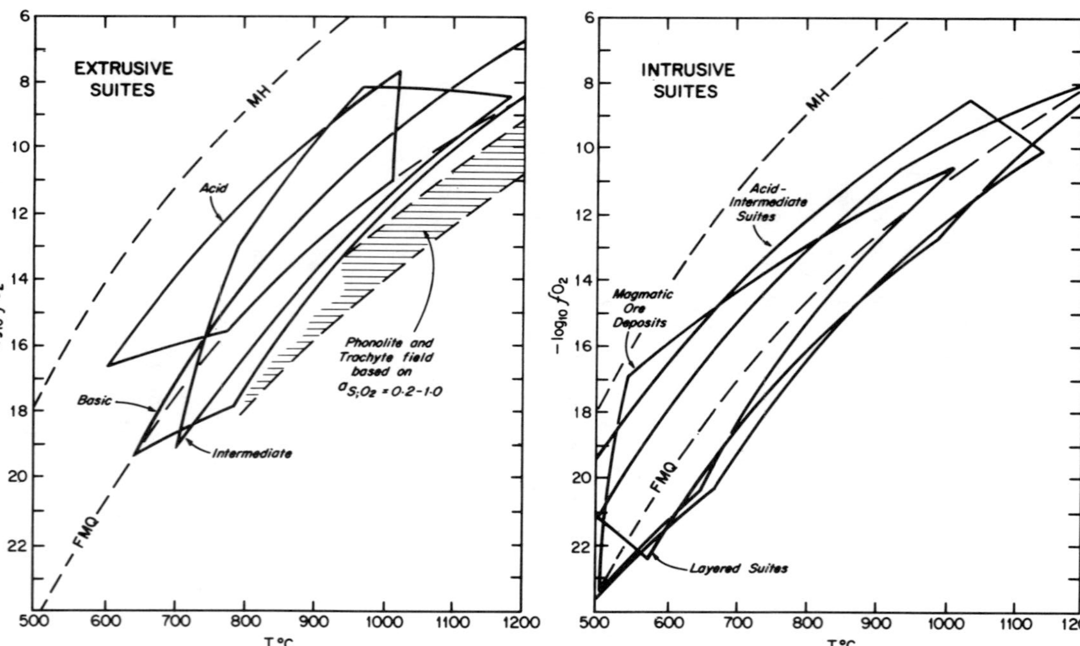

Figure Hg-46. The envelopes illustrated are from data given in Figures Hg-43-46; the shaded envelope is from Nash *et al.* (1969). FMQ is the fayalite-magnetite-quartz buffer curve; MH is the magnetite-hematite buffer curve.

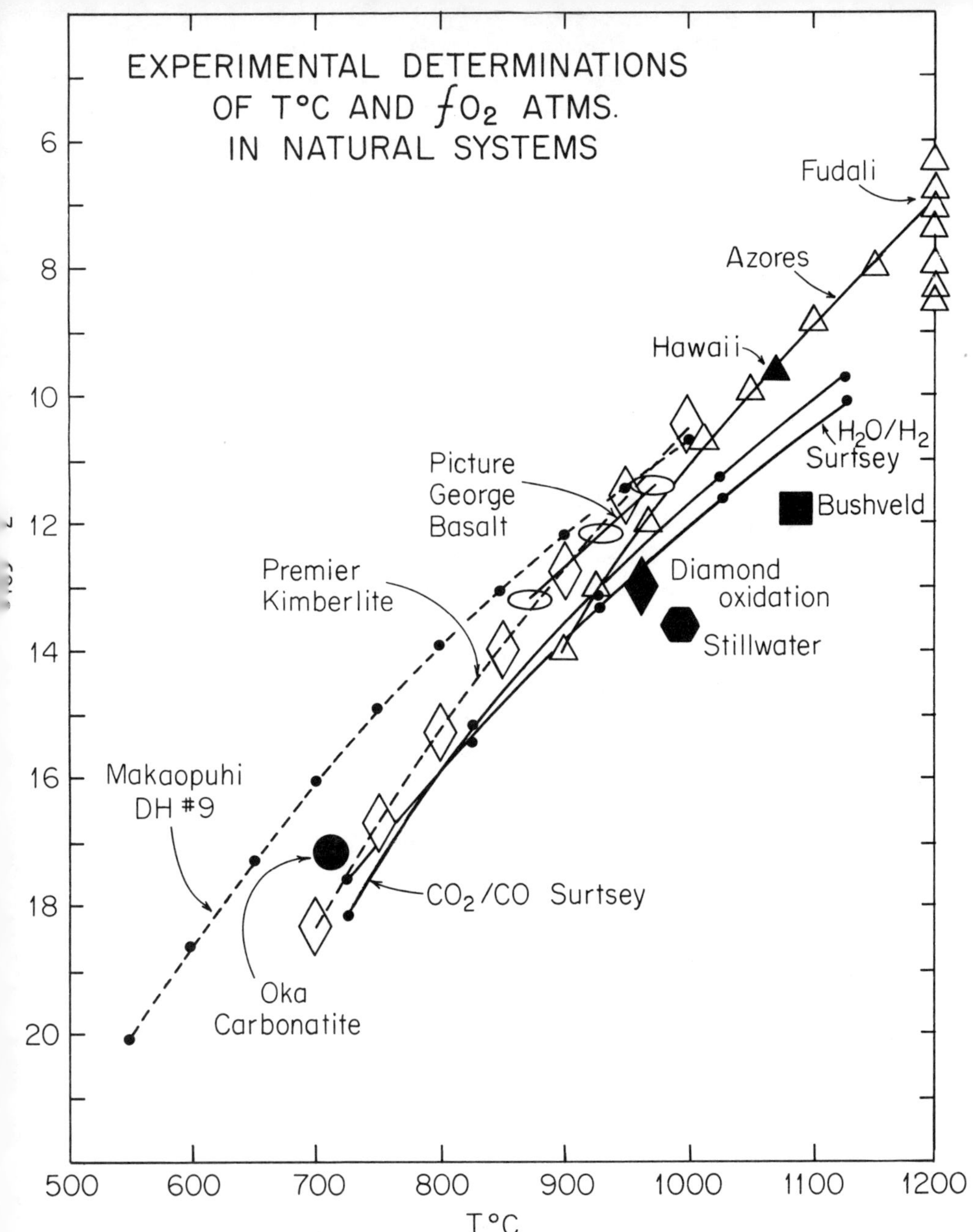

Figure Hg-47. The Makaopuhi curve (drill hole #9) and the filled triangle marked Hawaii are *in-situ* determinations and are from Sato and Wright (1966) and Peck and Wright (1966), respectively. Surtsey gas curves are from Steinthorsson (1972) and are computed based on gas analysis. The 1200°C values are from Fudali (1965); the Azores and Picture Gorge trends are from Duke (1974) and Tuthill Helz (1973), respectively. The kimberlite trend and the value for diamond oxidation are from Ulmer *et al.* (1976). The remaining complexes are as follows: Oka (Friel and Ulmer, 1974); Stillwater (Sato, 1972); Bushveld (Flynn *et al.*, 1972). See Table Hg-21.

1969; Smith, 1970). In these detailed studies the relationship of sphene and or perovskite with coexisting magnetite are evaluated in terms of a_{SiO_2} and f_{O_2}. The associations are most commonly present in highly undersaturated suites (*e.g.*, nephelinites and leucitites), and distinctive assemblages result which are olivine + Mt_{ss} + perovskite and Mt_{ss} + sphene. It is inferred that the former assemblage is very close to the FMQ buffer curve, whereas the latter requires relatively high oxygen fugacities. Low a_{SiO_2} will favor the formation of perovskite, and higher a_{SiO_2} the formation of sphene. The formation of either of these phases is at the expense of potentially crystallizing oxides with the net result that Mt_{ss} are depleted in Ti, and the partitioning is so effective that ilmenite is absent.

Three idealized curves showing the relationships of two perovskite-yielding reactions and a sphene-yielding reaction are illustrated in Figure Hg-48 (Carmichael and Nicholls, 1967). The perovskite 1 reaction is for olivine + Usp + O_2 = Mt + perovskite + enstatite; the perovskite 2 reaction at lower values of f_{O_2} is for åkermanite + Usp + O_2 = perovskite + Mt + diopside; the sphene reaction which falls between the two perovskite curves is for diopside + Usp + O_2 = Mt + sphene + enstatite. The FMQ buffer curve lies between the perovskite 2 and the sphene curve, and the perovskite 2 curve is encompassed by the computed trachyte and phonolite field for limits of a_{SiO_2} between 0.2 and 1.0 (Nash *et al.*, 1969). The perovskite-sphene silica buffered curve is shown in Figure Hg-49 (Carmichael *et al.*, 1970), and this diagram is contoured as a function of f_{O_2} (Mithcell, 1973) to emphasize the dependencies not only on a_{SiO_2} for sphene or perovskite formation but also the related effects of oxygen fugacity (Smith, 1970; Carmichael *et al.*, 1970). Although the perovskite 2 curve is above that of the sphene curve in Figure Hg-48, note that the reaction is consistent with respect to the decomposition of olivine (*i.e.*, olivine + perovskite + Mt_{ss} is close to the FMQ buffer curve).

In summary the demise of ilmenite and the depletion of Ti in Mt_{ss} can be related to a_{SiO_2} and f_{O_2}. If a_{SiO_2} is considered alone, Mt_{ss} + perovskite will form in preference to Mt_{ss} + sphene at low values of a_{SiO_2} ($CaSiTiO_5 = CaTiO_3 + SiO_2$). If f_{O_2} is considered in addition, high f_{O_2} and high a_{SiO_2} will favor sphene; lower f_{O_2} and lower a_{SiO_2} will favor perovskite (Fig. Hg-49), *i.e.*, for any given f_{O_2} sphene or perovskite are dependent on a_{SiO_2}.

Sphene or perovskite are typical of the undersaturated suites and are rarely observed to coexist (Smith, 1970). Sphene has not been identified in kimberlites, and Mitchell (1973) has presented a convincing case based on a_{SiO_2} for olivine that the value of T and f_{O_2} for the crystallization of groundmass kimberlite is at approximatel 600°C and 10^{-20} atms., in which perovskite abounds. Although he correctly considered that the f_{O_2} was somewhat lower than that of basalts, his calculation for these intrinsic values falls precisely on the extrapolated curve obtained by Ulmer *et al.* (1976) for the Premier mine kimberlite (Fig. Hg-47).

Apart from the absence of ilmenite in applying the thermobarometer, deleterious minor elements which frequently reach major proportions (Table Hg-20) prevent determination because of the limiting effects of these elements on the pure synthetic syste The delinquents are MnO and MgO in ilmenite, and one example is shown in Figure Hg-48 where the application of a_{SiO_2} has successfully provided an estimate of the range of T and of f_{O_2} for a suite ranging from monzonites to granodiorites to granites (Czamanske

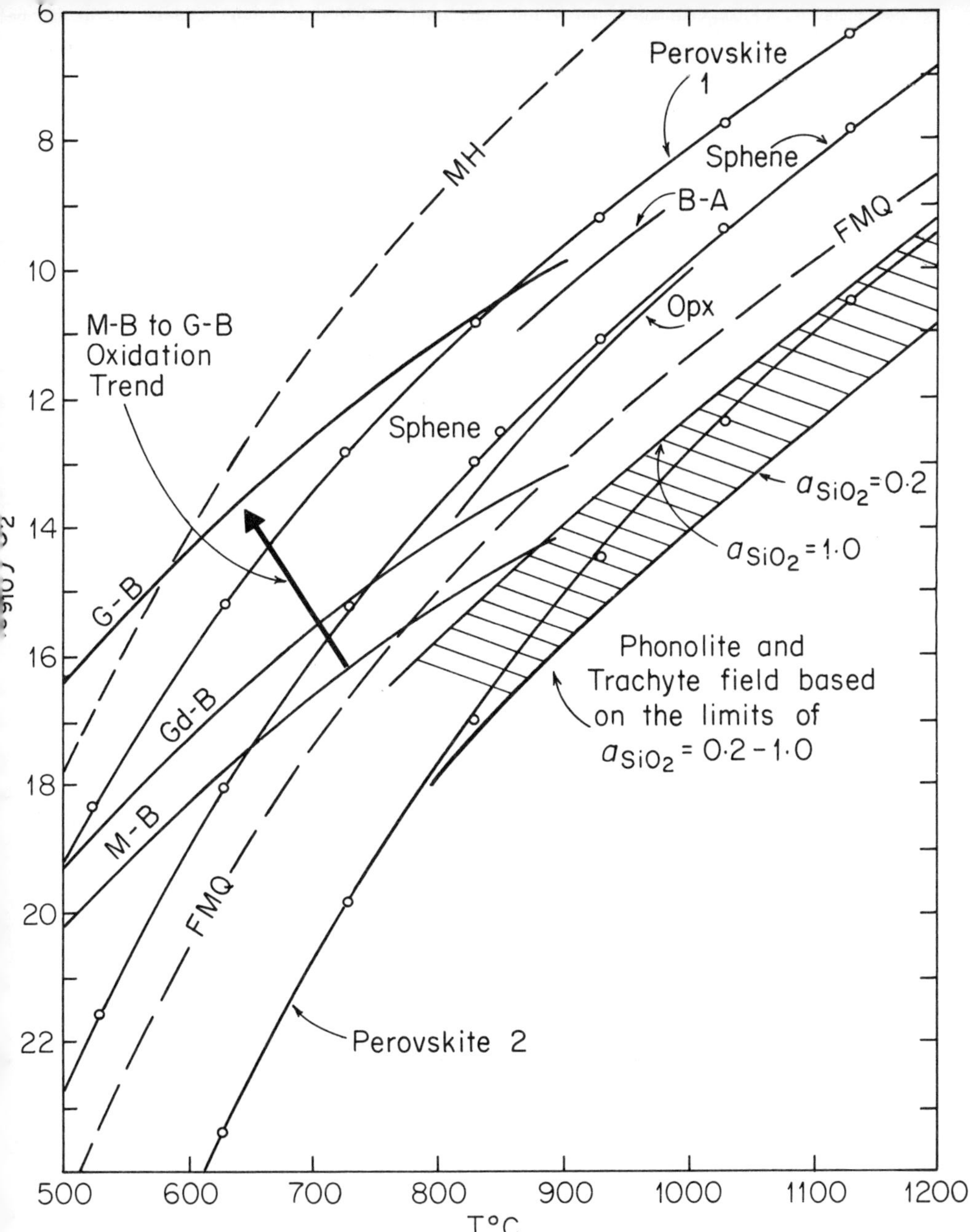

Figure Hg-48. The curves are from the following sources: shaded envelope--Nash *et al.* (1969); perovskite 1 & 2 and sphene--Carmichael and Nicholls (1967); Opx--orthopyroxene and B-A--biotite-amphibole (Carmichael, 1967a); the M-B to G-B oxidation trends are for monzonitic biotites and granitic biotites, respectively; Gd-B is granodioritic biotite (Czamanske and Wones, 1973). MH and FMQ are the experimental buffer curves, and for this and all previous figures the magnetite-hematite curve and the fayalite-magnetite-quartz buffers are from Eugster and Wones (1962) and Wones and Gilbert (1969), respectively.

and Mihálik, 1972; Czamanske and Wones, 1973). The curves illustrated refer specifical to the trends calculated for biotite, with the monzonitic biotite having the lowest f_{O_2} and the granitic biotites the highest f_{O_2}. This study is of particular interest with respect to the data for intrusive suites (Fig. Hg-44a), and in the context of an early study by Carmichael (1967a) who showed that for a suite of salic rocks (rhyolitic obsidians to pitchstones and dacites) that a clearly defined distinction could be made among these rocks on the basis of whether olivine, orthopyroxene, or biotite and amphibole were present. The olivine-bearing samples fall on the FMQ buffer curve; the Opx suite fall above this curve (Fig. Hg-48), and the biotite and amphibole dacitic suite (marked B-A on Fig. Hg-48) are at even higher levels of f_{O_2} than the Opx curve. The Opx curve is almost coincident with the sphene curve discussed above, and closely approximates that of the granodioritic biotite curve; the high f_{O_2} granitic biotite curve, which is close to the MH buffer at low temperatures, intersects the perovskite curve and is above the dacitic biotite-amphibole curve. However, if the dacitic curve is extrapolated to lower temperatures, a reasonably close estimate results which shows that the computed trends for the intrusive suites are well within the range of chemical equivalent extrusive suites. These results are a clear endorsement to an earlier conclusion which suggested that in broad terms, acid suites equilibrate under more highly oxidizing conditions than those of basic suites, whether extrusive or intrusive.

The computed phonolite and trachyte field (Nash *et al.*, 1969) is more highly reducing with respect to the FMQ buffer and with respect to basalts (Fig. Hg-48). The upper limit is defined approximately by the silica buffer of perovskite-sphene as illustrated in Figure Hg-49b (Carmichael *et al.*, 1970); the lower limit is not known although it is doubtlessly above the larnite-wollastonite curve (Fig. Hg-48b). These limits define a_{SiO_2} for the nephelinites; between sphene and forsterite are the alkali olivine basalts; above this field are the tholeiitic suites, and beyond that the quartz-bearing suites (Carmichael *et al.*, 1970).

From the above discussion it should finally be concluded that a_{SiO_2} and f_{O_2} play their respective and important roles in defining not only the compositions of the opaque mineral oxides, but determining also whether both Usp-Mt_{ss} and Ilm-Hem_{ss} can coexist. The overall and more general conclusion is that many more new studies, systematic in intent and detailed in context, are required to refine the trends and distributions of oxides as outlined in this chapter.

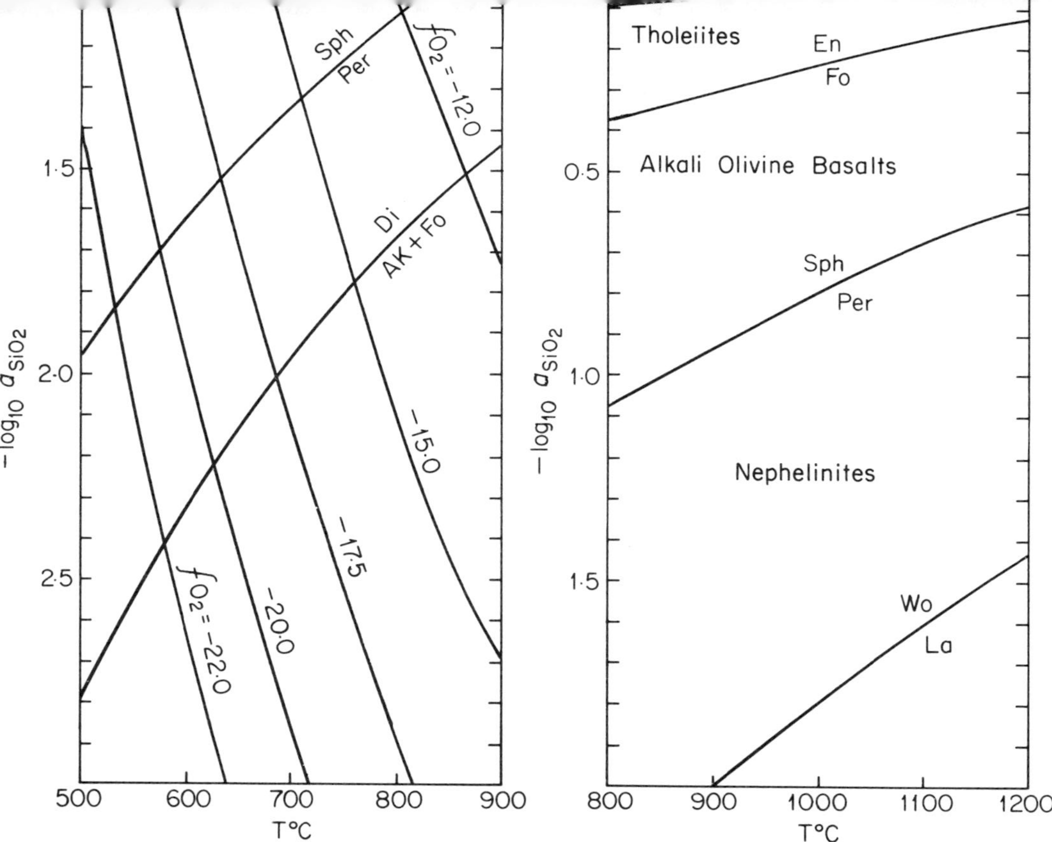

Figure Hg-49. (a) and (b) are from Mitchell (1973) and Carmichael *et al.* (1970), respectively. The silica buffer curves are for perovskite-sphene; åkermanite-forsterite-diopside; forsterite-enstatite; and larnite-wollastonite.

TABLES OF MINERALOGICAL, PETROLOGICAL, AND CHEMICAL PROPERTIES OF OPAQUE MINERAL OXIDES IN IGNEOUS ROCKS

Tabulation of Opaque Mineral Oxides in Igneous Rocks

Table Guide

The general format for Tables Hg-7-12 allow for ready access to the literature with respect to volcanic suites, hypabyssal suites, and acid-intermediate intrusive suites, as a function of the reported mineralogy. Suites have on occasion been grouped or subdivided as deemed necessary to maintain uniformity within the study reported or to allow for comparisons with suites from other localities; therefore the division between basic and acid is to be viewed only in broad terms and a specific example is that the under-saturated suites may appear either in the intermediate Table or alternatively in the basic Table. The mineralogical constituents are divided into primary oxides (ulvöspinel-magnetite$_{ss}$; ilmenite-hematite$_{ss}$; and chromian spinel$_{ss}$) and secondary or oxidation assemblages which are designated as follows: Ilm$_{ss}$ refers to Ilm-Hem$_{ss}$ formed by oxidation "exsolution"; Pb$_{ss}$ are members of the pseudobrookite series; Hem$_{ss}$ refers specifically to hematite-rich solid solutions of low Ti contents derived from low Ti-magnetites or by exsolution from

Ilm-Hem_{ss}; R is rutile, but also anatase or brookite; γ is γFe_2O_3, maghemite or titanomaghemite; OH refers to goethite. The second set of data relate to T°C and f_{O_2} for coequilibrated oxide pairs as determined from the curves defined by Buddington and Lindsley (1964). Within the T°C column are values also for Curie temperatures where reported, and where the emphasis of the study is magnetic (Mag) rather than petrologic (Pet).

In Tables Hg-13-19 the format differs inasmuch as short notes are provided to somewhat greater detail for the more mafic plutonic suites, and for the xenolith and magmatic ore deposit occurrences. The choice for layered sequences was refined on the basis of the availability of data dealing specifically with the oxide mineral group in the context of their petrologic setting.

Compositional data on the oxides for a variety of suites is, for the most part given in Table Hg-20 to allow for rapid and direct comparisons; however, other data are also disseminated throughout the mafic, ultramafic, and ore deposit suites where variations among rock types are recognized in particular complexes, or where the compositions of the oxides have been used to distinguish among rocks of otherwise identical mineralogies.

Table Hg-21 is restricted to T°C and f_{O_2} determinations of natural systems and these include the range from lava lake drilling programs to laboratory determinations by a variety of techniques.

The final table (Hg-22) is a reference guide to the techniques which are most commonly employed in the characterization of the oxide minerals and it includes a list of recent review articles and standard texts. No evaluation or description of the techniques is provided but it is of interest to note that within the context of the review in Tables Hg-7 to Hg-21 that not a single sample has yet received the attention of all techniques and that at the very most, no more than three techniques have been employed. There is clearly a need for at least one complete characterization which could become the *oxide normalizing parameter* for future studies.

Acid Extrusive Suites

Notes: (1) Electron microprobe analysis. (2) Coexisting Fe-Mg silicates are iron-rich olivine, or Opx, or biotite and amphibole; the olivine group fall on or straddle the FMQ buffer curve; the Opx group fall above the FMQ curve; and above this is the biotite + horneblende assemblage which is dacitic in composition and is calculated to have crystallized at f_{H_2O} -400 bars. (3) Data for the pitchstone reported here is for one sample of a suite of rocks from the Thingmuli volcano, which includes icelandites, basaltic andesites and olivine tholeiites. For the most part crystallization is close to, slightly above, or slightly below that of the FMQ buffer curve. (4) The range for rhyolites is separated here from data which also includes a study of pumices (see appropriate Table). A comparison of the two suites shows that the rhyolites are lower temperature (T = 735-780°C) extrusives when compared with the explosive pumice suite, where T = 860-890°C. The rhyolites contain amphibole phenocrysts + Opx + biotite whereas the pumices contain only hypersthene and augite. Calculated f_{H_2O} at 735°C and 745°C = 1100 and 1300 bars respectively for the rhyolites. (5) The T-f_{O_2} datum point falls precisely on the FMQ buffer curve and is consistent with data reported by Carmichael (1967a,b) which show that associated rock types, although differing in T and f_{O_2} have oxide equilibration characteristics which are similar to the cooling paths defined by the slope of the FMQ buffer curve. The dacite reported in this study is consistent also, inasmuch as it falls above the NNO buffer curve but at a lower value than those reported by Carmichael above. (6) The rhyolites contain amphibole + biotite and plot above the NNO buffer curve. (7) The oxides which are associated with Pb_{ss} + Hem_{ss}, are bixbyite + cassiterite. The entire assemblage is present in cavities in miarolitic, lithophysal rhyolite and T > 500°C is considered necessary for crystallization, probably from a vapor phase. (8) T-f_{O_2} data for the rhyolites form part of a detailed study of pumice (see appropriate Table), and Lipman's demonstration that T°C correlates with SiO_2 (higher silica with lower T°C) in the pumices is borne out also by two suites of rhyolites; in addition, these rhyolites have similarly high MnO contents in the ilmenites (3-4.5 wt% MnO) to those in the pumices but T = 700-780°C is relatively low; the correlation patterns are similar and H_2O is not regarded as being a significant factor in the differing modes of eruption. A third suite of rhyolites (810-860°C) are considered to have had a low magmatic water content and these plot off the T-SiO_2 trend. Post-caldera lavas have Fe-Ti phenocrysts with very large compositional variations and these are considered to have formed by extensive magma mixing as a result of caldera collapse.

Table Hg-7

ACID EXTRUSIVE SUITES

Rock Type	Locality	Primary Mineralogy			Oxidation Mineralogy						T°C	log10 fO_2	Emphasis		Notes	References
		$Mt\text{-}Usp_{ss}$	$Ilm\text{-}Hem_{ss}$	Chr_{ss}	Ilm_{ss}	Pb_{ss}	Hem_{ss}	R	γ	OH			Pet	Mag		
Obsidian	Iceland	x	x								925	-12.5	x		1,2	Carmichael (1967b)
Obsidian	California	x	x								930	-11.4	x		1,2	Carmichael (1967b)
Obsidian	California	x	x								825	-13.7	x		1,2	Carmichael (1967b)
Porphyritic Obsidian	California	x	x								880	-12.2	x		1,2	Carmichael (1967b)
Rhyolitic Obsidians	California	x	x								815-985	-13.9 to -10.6	x		1,2	Carmichael (1967b)
Pitchstone	Arran	x	x								900-925	-12.9 to -12.2	x		1,2	Carmichael (1967b)
Porphyritic Pitchstone	Scotland	x	x								910	-11.4	x		1,2	Carmichael (1967b)
Porphyritic Pitchstone	Iceland	x	x								925	-12.6	x		1,3	Carmichael (1967a)
Rhyolites	California	x	x								790-810	-15.1 to -14.7	x		1,2	Carmichael (1967b)
Rhyolites	New Zealand	x	x								735-780	-15.2 to -13.8	x		1,4	Ewart et al. (1971)
Ryodacites	Hawaii	x	x								855-865	-11.7	x		1	Bauer et al. (1973)
Ryodacite	Greece	x	x								890	-13.0	x		1,5	Nicholls (1971)
Pantellerite	Pantelleria	x	x								1025	-11.2	x		1,2	Carmichael (1967b)
Rhyolite	New Britain	x	x								860	-11.5	x		1,6	Lowder (1970)
Rhyolite	New Mexico					x	x						x		1,7	Lufkin (1976)
Rhyolite	India	x	x								500-600			x		Klootwijk (1975)
Rhomb-porphyry	Norway	x	x		x		x	x			570-670			x		Storetvedt and Petersen (1970)
Rhyolites	Nevada	x	x								700-780	-14.5 to -13.0	x		1,8	Lipman (1971)
Rhyolites	Nevada	x	x								810-860	-12.0 to -10.5	x		1,8	Lipman (1971)

Pumice, Ash, Ignimbrites

NOTES: (1) Wet chemical analysis. (2) Electron microprobe analysis. (3) This pumice falls on Carmichael's Opx curve and is close to the NNO buffer curve (see note 4, Acid Extrusive Table). (4) This pumice study also includes a suite of rhyolites. The rhyolites are amphibole and biotite-bearing and T = 860-890°C. (5) The two T-f_{O_2} values are for samples from a bi-lobate rhyodacitic tephra which have identical mineralogies, but based on the differences in Ilm_{ss} contents ($Ilm_{73.5}$ $Hem_{26.5}$ vs Ilm_{51} Hem_{41}) and coexisting Mt_{ss} (Mt_{84} vs Mt_{82}) two distinct events are postulated. (6) Identification and stratigraphic mapping of tephra based on the thermomagnetic properties of Fe-Ti oxides, or "ferromagnetic-tephrochronology". (7) T = blocking temperature. (8) The entries for Lipman are for two major tuff units (a) Paintbrush Tuff which comprises the Topopals Spring Member, the Yucca Mt. member, and the Tira Canyon member; and (b) Timber Mt. Tuff which comprises the Rainer Mesa member and the Ammonia Tanks member. For four of the five members rhyolitic + qtz latite units show that the rhyolitic tuffs (i.e. higher SiO_2) have crystallization temperatures which are lower than those of the associated qtz latites, and the relationship between inferred T°C and SiO_2 is very nearly linear (low T and high SiO_2, 78 wt%; high T and low SiO_2, 66%). MnO (1-8 wt% in Ilm_{ss}; 1-4 wt% in Mt_{ss}) and Al_2O_3 (0.2-1.5 wt% in Mt_{ss}) decrease and increase respectively as T increases (Buddington and Lindsley, 1964). Their respective relationships with SiO_2 therefore are that: Al_2O_3 in Mt_{ss} increases as SiO_2 decreases; and MnO in Ilm_{ss} increases as SiO_2 increases. The range of T-f_{O_2} for the pumice suite falls along a curve which is closely parallel to the synthetic buffer curves and is between the Ni-NiO curve and the MnO-Mn_3O_4 curve. Values for Mt_{ss} with oxidation exsolution Ilm_{ss} yield T = 650°C, and $10^{-16.7}$ atms; and T = 780°, and $10^{-12.8}$ atms. In a number of rhyolitic tuffs the uppermost portions of the unit contain Mt_{ss} + sphene rather than the assemblage Mt_{ss} + Ilm_{ss}. This change in assemblage [(which is accompanied also by Cpx: ($3CaTiSiO_5 + Fe_3O_4 = 3FeTiO_3 + 3CaSiO_3 + 1/2\ O_2$)]is indicative of more highly oxidizing conditions and suggests that the upper units of each cogenetic ash-flow sequence represents a more highly oxidized magma than those of the initial eruptions. Inferred water pressures for T = 625-725°C range from 500 to 1200 bars respectively. The oxide data suggest that the differentiated upper rhyolitic parts (high SiO_2 and low modal phenocrysts) of the magma chamber (T = 700° ± 50°C) were very nearly saturated in H_2O; the lower portions of the chamber are represented by the qtz latite (low SiO_2, high modal phenocrysts) suite (T = 900°C) and these crystallized at low values of P_{H_2O}.

(9) This study is an important extension of the above (Lipman, 1971) inasmuch as it is directed towards an oxygen isotope comparison of T°C for the rhyolite and qtz latite tuff suites. δO^{18} values (per mil) in Mt_{ss} range between 2.0-3.4 (qtz latites) and 0.3-2.5 (rhyolites); feldspar + Mt_{ss} yield the respective ranges in temperature quoted in the Table. The upper limits of

Table Hg-7 (2)

PUMICE, ASH, IGNIMBRITES

Rock Type	Locality	Primary Mineralogy			Oxidation Mineralogy						T°C	log10 fO_2	Emphasis		Notes	References
		$Mt\text{-}Usp_{ss}$	$Ilm\text{-}Hem_{ss}$	Chr_{ss}	Ilm_{ss}	Pb_{ss}	Hem_{ss}	R	γ	OH			Pet	Mag		
Dacitic Pumices	Japan	x	x								900-975	-11.5 to - 9.0	x		1	Buddington and Lindsley (1964)
Pumice	California	x	x								820	-13.9	x		2,3	Carmichael (1967b)
Pumice	New Zealand	x	x								860-890	-12.2 to -12.8	x		2,4	Ewart et al. (1971)
Tephra	Alaska	x	x								900-1025	-10.2 to - 7.7	x		2,5	Lerbekmo and Smith (1972)
Pumice	New Zealand	x	x		x		x						x	x	1,2	Ewart (1967)
Pumice	Japan	x									240-555			x	6	Momose et al. (1968)
Pumice	Japan	x									450			x	6	Momose and Kobayashi (1972)
Pumice	Japan	x												x	6	Kobayashi and Momose (1969)
Ignimbrites	Ireland	x	x		x	x	x	x			400-500			x	7	Deutsch and Somayajulu (1970)
Welded Tuffs	Iceland	x	x		x		x		x	x			x	x	2	Haggerty (unpublished)
Ignimbrites	Peru	x	x				x		x	x			x	x	2	Wilson and Haggerty (unpublished)
Rhyolitic tuff Qtz latite tuff	Nevada, Topopah Spring	x	x		x						720-715 845-900	-14.4 to -14.7 -11.6 to -10.2	x		2,8	Lipman (1971)
Rhyolitic tuff	Nevada Yucca Mt.	x	x								640-690	-16.1 to -14.6	x		2,8	Lipman (1971)
Rhyolitic tuff Int. Rhy-Qtz latite	Nevada Tiva Canyon	x	x								660 720-730	-15.7 -14.0 to -13.8	x		2,8	Lipman (1971)
Rhyolite, Qtz latite	Nevada Renier Mesa	x	x								710-680 920-935	-14.5 to -15.4 - 9.7 to - 9.5	x		2,8	Lipman (1971)

730° (rhyolites) and 130°C (qtz latites) correspond closely to those obtained for coequilibrated oxide pairs; a significant observation is that Mt_{ss} become consistently richer in O^{18} as the unit becomes more mafic, a trend which is not exhibited by coexisting feldspars; the latter is considered to have reequilibrated in the subsolidus as a result of the introduction of meteoric water to the crystallizing magma in the interval between rhyolite eruption and qtz latite eruption. (10) Values for T-f_{O_2} are from Table 5, p. 9 in Heming and Carmichael (1973) but the absolute range in T is 1035-835°C $\approx 10^{-9}$ and $10^{-11.2}$ respectively. These values fall above the FMQ buffer curve, above the Opx curve and below the biotite-amphibole curve as defined by Carmichael (1976b)--see note 4 Acid Extrusive Table. Utilizing T from co-equilibrated oxide pairs and the mole fraction of FeS in coexisting pyrrhotites, $f_{S_2} = 10^{-2}$ to 10^{-4} atms. Calculations suggest that Opx + Mt_{ss} in the rhyolitic pumice equilibrated at P = 2.2 to 2.6 Kb. (11) Mt_{ss} compositions (Al_2O_3 = 2.11-2.21; MgO = 1.28-1.46; MnO = 0.51-0.52; V_2O_3 0.49-0.59) and coexisting plag + Opx + Cpx from two widely separated areas suggest a common submarine source. (12) The papers by Whitten and by Ramdohr are of interest with regard to the fumarolic precipation of very pure magnetite and associated sulfides. (13) The T and f_{O_2} range are for hornblende-hypersthene dacites. An EMPA for titanomaghemite is also provided and is of interest because it contains 2.39 wt% Al_2O_3 which is consistent with the values for unoxidized Mt_{ss}; for MgO (0.18 wt%) however the value is below the range for unoxidized equivalents (0.2 to 1.54 wt%).

Table Hg-7 (3)

PUMICE, ASH, IGNIMBRITES

Rock Type	Locality	Primary Mineralogy			Oxidation Mineralogy						T^oC	log 10 fO_2	Emphasis		Notes	Reference
		$Mt\text{-}Usp_{ss}$	$Ilm\text{-}Hem_{ss}$	Chr_{ss}	Ilm_{ss}	Pb_{ss}	Hem_{ss}	R	γ	OH			Pet	Mag		
Rhyolite Qtz latite	Nevada Ammonia Tanks	x	x		x						670- 680 930- 940	-15.3 to -15.0 -10.0 to - 9.3	x		2,8	Lipman (1971)
Rhyolites Qtz latites	Nevada	x	x								560- 730 650- 930		x		2,9	Lipman and Friedman (1975)
Rhyolitic pumice	New Guinea	x	x								880 995	-11.2 - 9.6	x		2,10	Heming and Carmichael (1973)
Dacitic pumices	New Guinea	x	x								870- 990	-12.0 to - 9.6	x		2,10	Heming and Carmichael (1973)
Andesitic pumice	New Guinea	x	x								955	-10.2	x		2,10	Heming and Carmichael (1973)
Dacitic pumices	Tonga, Australia	x											x		11	Bryan (1971)
Tuffisites	Ireland	x									585		x		12	Whitten (1959)
Fumaroles		x											x		12	Ramdohr (1962)
Dacitic ignimbrites	Australia	x	x						x		980 1000	-10.1 - 9.3	x		13	Wilkinson (1971)

Intermediate Extrusives

NOTES: (1) Wet chemical analysis. (2) Electron microprobe analysis. (3) The andesites fall between the Opx and biotite-amphibole curves as defined by Carmichael (1967b). (4) The andesites plot on the FMQ buffer curve and conform with the Thingmuli trend as defined by Carmichael (1967b) (5) Neither of the trachybasalts contain discrete Ilm_{ss}; the rhombohedral phase is associated with Mt_{ss} or is enclosed in augite. Interesting and unusual zonal trends are observed for the oxides in the trachybasalts; Ilm_{ss} are zoned from cores with Hem_{27} to margins of Hem_{15} when in contact with magnetite; in the same manner Mt_{ss} have interior compositions of Usp_{19} and Ilm_{ss}-contact compositions of Usp_{65}. Thus both phases become progressively more reduced towards their respective common contact; although Ilm_{ss} and Mt_{ss} are intimately intergrown in typical sandwich and composite textures, the more highly reduced nature of the conterminous regions suggests that the oxides crystallized under progressively more reducing conditions and that the assemblage does not result by subsolidus oxidation. The MgO contents of Ilm_{ss} in the trachybasalts are high and range from 1.9 to 6.7 wt%; Mt_{ss} have a range from 1.3-4.2 wt% MgO and Al_2O_3 varies from 0.5-4.5 wt%. For the trachyandesites the respective ranges are: Ilm_{ss}, MgO = 1.5-2.6 wt%; Mt_{ss}, MgO = 0.9-2.2 wt% and Al_2O_3 wt%. The T-f_{O_2} relationships for the trachybasalts differ by only a small factor from those of the Hawaiian sub-alkaline basalts in spite of substantial differences in Fe_2O_3/FeO ratios (0.14-0.24 vs. 0.09) Anderson considers that the higher values observed in the Tristan suite may be compensated for by higher basicity to maintain comparable T-f_{O_2} trends. (6) FHIC = ferrohypersthene, iron-cordierite. The range in T-f_{O_2} corresponds to a trend which parallels the FMQ buffer curve and falls on the more reduced side of the curve which is consistent with the coexisting mineralogy--tridymite + Fe-Opx + Mt_{ss}; titanomagnetite contains 5.35-4.78 wt% Al_2O_3, and 0.36-1.41 wt% MgO. Ilmenites contain MgO + Al_2O_3 + MnO ≃ 2 wt%. (7) The ranges in T and f_{O_2} were obtained indirectly because only titanomagnetite compositions were analytically possible. These data are based on $\underline{a}_{SiO_2}$ for the compositions of coexisting Ol + Mt_{ss}; the ranges are respectively for $\underline{a}_{SiO_2}$ = 1.0 and for $\underline{a}_{SiO_2}$ = 0.2 which is derived from nepheline + alkali feldspar (because qtz is absent the upper value is also the upper limit). The resulting regime for T-f_{O_2} is an envelope which parallels the FMQ buffer curve and is on the reduced side of the curve. (8) The datum point for this porphyritic pitchstone falls between the trends of amphibole-biotite (dacitic trend) and Opx as defined by Carmichael (1967b). (9) This dacite value falls slightly below the biotite-amphibole curve as defined by Carmichael (1967b). (10) Trace element data on Mt_{ss} and Ilm_{ss} show that the ilmenites are richer in Mg than coexisting Mt_{ss} which are enriched in Al, V, Cr, Ni, Co, and Fe. In the normal course of fractional crystallization, Cr and Ni, and to a lesser extent V and Co, tend to decrease in Mt_{ss} in successive crystal fractions. Enrichment of Cr

Table Hg-8

INTERMEDIATE EXTRUSIVES

Rock Type	Locality	Primary Mineralogy			Oxidation Mineralogy						T°C	log10 fO_2	Emphasis		Notes	References
		$Mt-Usp_{ss}$	$Ilm-Hem_{ss}$	Chr_{ss}	Ilm_{ss}	Pb_{ss}	Hem_{ss}	R	γ	OH			Pet	Mag		
Hornblende-mica andesite	Japan	x	x								925	-10.0	x		1	Buddington and Lindsley (1964)
Hornblende-andesite	Idaho	x	x								950	-10.0	x		1	Buddington and Lindsley (1964)
Andesite	New Mexico	x	x								925	- 9.8	x		2,3	Stormer (1972)
Andesite	Greece	x	x								950-1000	-11.0 to -12.0	x		2,4	Nicholls (1971)
Trachybasalt	Tristan da Cunha	x	x		x						990-1140	-10.4 to - 8.6	x		2,5	Anderson (1968)
Trachybasalt	Tristan da Cunha	x	x		x						850-1070	- 8.6 to -12.0	x		2,5	Anderson (1968)
Trachyandesite	Tristan da Cunha	x	x		x						1110	- 7.8	x		2,5	Anderson (1968)
Trachyte	Tristan da Cunha	x	x		x						930	-11.7	x		2,5	Anderson (1968)
FHIC	St. Helena	x	x								705-800	-15.5 to -19.1	x		2,6	Ridley and Baker (1973)
Phonolites	Kenya	x									1200-900	-12.0 to -16.0	x		2,7	Nash et al. (1969)
Trachytes	Kenya	x									1200-900	-10.0 to -14.0	x		2,7	Nash et al. (1969)
Trachytic-Pitchstone	France	x	x								1010	- 9.2	x		2,8	Carmichael (1967b)
Dacite	California (Medicine Lake)	x	x								895	-12.1	x		2,8	Carmichael (1967b)
Dacite	California (Lassen)	x	x								865-980	-10.9 to - 9.1	x		2,8	Carmichael (1967b)

correlates in specific Mt_{ss} when co-existing Ilm_{ss} are Cr-poor; Wilkinson concludes that a relatively small number of nucleating spinel centers are capable of incorporating extraordinarily high concentrations of Cr, thus rapidly depleting the magma prior to the onset of Ilm_{ss} or Fe-Mg silicate crystallization. The Cr and Co contents in Ilm_{ss} decrease with increasing differention. (11) This paper discusses the role of Al in titanomagnetite phenocrysts and compositions with as much as 11.52 wt% Al_2O_3 are reported with MgO = 1.31-4.52. The range in compositions are: (a) Spinel (max) = $Sp_{23.7}Usp_{33.4}Mt_{42.9}$; (b) Usp (max) = $Sp_{10.0}Usp_{55.4}Mt_{34.6}$; and (c) Mt (max) = $Sp_{22.6}Usp_{30.7}Mt_{46.7}$. Aoki concludes that the relationship of the ternary solvus for the system Fe_2TiO_4-Fe_3O_4-(FeMg) Al_2O_4 is an extension from Mt-Usp towards (FeMg)Al_2O_4 to a maximum value of approximately 25 mole% spinel + hercynite. (12) A particularly common intergrowth in these andesites are Ilm_{ss} with reduction exsolution lamellae of Mt_{ss} along {0001} planes; the Mt_{ss} areas exhibit partial to complete oxidation to titanomaghemite. (13) The locations of samples are from Mt. Pisgah, Amboy Crater, Cima Dome. The single value for T and f_{O_2} and the range in values are for microphenocrysts and for roundmass oxides respectively; these values lie on the reducing side of the FMQ buffer. Of particular interest are the high MgO (1.3-5.0 wt%) and Al_2O_3 (1.1-3.9 wt%) contents of Mt_{ss} microphenocrysts, and the observation that titanomagnetite preceeds the cyrstallization of Ilm_{ss} which is so typical of tholeiitic magmas. (14) A comparison of T°C from the oxides with values determined from coexisting feldspars show good agreement: 1040°C for the oxides and 1000°C for the feldspars. The qtz trachyites are associated with epizonal granites which yield T = 650-700°C and $f_{O_2} = 10^{-18}$ bars. (15) In this study of maghemitization, the titanomaghemite constituent is shown to increase in TiO_2 (from 26.2 to 31.7 wt%), Al_2O_3 (from 4.5 to 5.2 wt%) and MnO (from 1.0 to 3.5 wt%) with respect to unoxidized titanomagnetite, which is enriched in FeO (61.8 vs 51.1) and MgO (6.4 vs 5.0). The authors conclude that the widely held concept of cationic deficiency in γFe_2O_3 is probably erroneous since mass migration of minor elements also takes place; hence vacancies can be satisfied by ions such as Mn and perhaps also Al. (Note: (1) the initial compositions of these titanomagnetites are extraordinary for andesites; (2) there are no comments with regard to primary zoning). (16) Compositions of Mt_{ss} obtained from unit cell determinations yield Usp_{45} and Usp_{50}. (17) A comparative study of the irreversible changes in Curie temperatures as a function of heating in air vs vacuum shows that heating in vacuum depresses the Curie point and decreases saturation magnetization; these changes, based on x-ray and microscopic observations suggest decomposition of titanomaghemite to Ilm-Hem_{ss}.

Table Hg-8 (2)

INTERMEDIATE EXTRUSIVES

Rock Type	Locality	Primary Mineralogy			Oxidation Mineralogy						T°C	log10 fO_2	Emphasis		Notes	References
		$Mt\text{-}Usp_{ss}$	$Ilm\text{-}Hem_{ss}$	Chr_{ss}	Ilm_{ss}	Pb_{ss}	Hem_{ss}	R	γ	OH			Pet	Mag		
Dacite	Greece	x	x								800	-13.0	x		2,9	Nicholls (1971)
Dacite	New Mexico	x	x								875	-11.2	x		2,3	Stormer (1972)
Dacite	New South Wales	x	x								1050	- 9.6	x		2,10	Wilkinson (1971)
Shoshonite	Indonesia	x	x										x		2	Joplin et al. (1972)
Bytownite-Andesite and Dacites	Tonga	x											x		2	Bryan et al. (1972)
Trachyandesites	Japan	x		x	x				x				x		1,11	Aoki (1966)
Trachybasalts, Trachyandesites Trachytes	St. Helena	x	x										x			Baker (1969)
Trachytes Andesites	Northwest Territories	x	x		x		x							x		Park et al. (1973)
Andesites	Andes, Peru	x	x		x		x	x	x	x			x	x	2,12	Wilson and Haggerty (unpublished)
Trachybasalts	California	x	x	x							1090 995- 980	- 9.9 -11.4 to -11.7	x		2,13	Smith and Carmichael (1969)
Qtz Trachyte	New Hampshire	x	x								1040	-10.7	x		2,14	Whitney and Stormer (1976)
Dacites	New Britain	x	x								920 930	-10.2 -10.5	x			Lowder (1970)
Leucite	Italy	x	x						x				x			Cundari (1975)
Andesites	France	x							x				x	x	2,15	Prévot et al. (1968)
Trachyandesites	New Zealand	x	x						x		~580		x	x	16	Wright (1967)
Andesites Dacites	Japan	x	x						x		380- 510			x	17	Ozima and Larson (1967)
Andesites	Bulgaria	x	x		x		x		x		600			x		Vollstädt et al. (1967)

Basic Extrusives

Notes: (1) Wet chemical analysis. (2) Electron microprobe analysis. (3) The latter part of this paper is an important contribution from the standpoint of providing a synopsis of T-f_{O2} data for salic and basaltic rocks and with respect to thermodynamic calculations limiting the fields of Pb_{ss} formation as a function of T and f_{O2}; a host of other reactions pertinent to the formation of perovskite and sphene are also included. The range in T and f_{O2} given here are summarized in greater detail in the Acid-Intermediate Extrusive Table. The distribution and compositions of Mt-Usp_{ss} suggest that basalts having Curie temperatures in excess of 200°C have suffered subsequent oxidation. (Note: since deuteric oxidation results in the formation of relatively pure magnetites (i.e. with Tc ≈ 585°C) the term "subsequent" simply implies post-liquidus; the criteria of coexisting Pb_{ss} is unequivocal evidence of deuteric oxidation as firmly stated by the authors, and as discussed in Chapter 5). (4) These data represent one aspect of a detailed study of the Thingmuli volcano. For the suite of samples the inferred T and f_{O_2} relationship is one that closely parallels the FMQ buffer curve; MgO in the ilmenites and V_2O_3 in the magnetites show sharp decreases as a function of bulk rock Fe^2 + Fe^{3+} + Mn/Fe^{2+} + Fe^{3+} + Mn + Mg ratios; for the same ratio ZnO increases in Mt_{ss} and MnO increases in Ilm_{ss}. The picritic basalt contains Mg-Al-chromite with MgO = 13.5 wt% and Al_2O_3 = 24.3 wt%. (5) T and f_{O_2} data from Surtsey in relationship to those from Carmichael (1965a, b), and with respect to the FMQ buffer curve, fall on the reduced side of the curve but coincide with the f_{O_2} values obtained for gasses emanating from the active volcano when considered as CO_2/CO and H_2O/H_2 mixtures. Coexisting Mg-Al-chromites contain MgO = 15.74 wt% and Al_2O_3 = 39.76 wt%. Lavas and coequilibrated oxide pairs yield T = 1,000°C and $f_{O_2} = 10^{-11.5}$ which closely approximates that of the host lava. One xenolith contains hornblende + pargasite + kaersutite and it is inferred that this assemblage resulted by reaction with the lava at P<8Kb and at a gas pressure > 1Kb. (7) The respective values at 980°C and 1070°C are for a diabase vein segregation and for the tholeiite host lava which were collected at 96' and 7' from the top of the ancient Makaopuhi lava lake. The range in Mt_{ss} compositions is considerable and one sample (62') yields T = 500°C, $f_{O_2} = 10^{-22}$ atms. Chromian spinels exhibit an equally wide range in compositions and the significant observations are as follows: (a) spinels exhibit a continuous variation from Chr_{60} to Usp_{80} to Mt_{70}; (b) for the most part Cr correlates with Al, but a pre-eruption trend shows the reverse relationship with Al_2O_3∿18 wt%; (c) the Usp component decreases as a function of depth and the most Cr-rich chromites are in the upper chill zone; (d) the spinels are typically unzoned and are included in olivine; and (3) it is the view that the transition from chromite to ulvöspinel, or of Chr_{ss} to Mt_{ss}, results from mass migration and inward "diffusion" of Fe^{2+}, Fe^{3+}, and Ti, and mass outward "diffusion" of Cr, Al and Mg which is evidently most effective at depths of 168-190' in the lava lake. (8) The range in magnetite compositions from the 1955 Kilauea eruption show cores with Usp_{42} to Usp_{60}; margins with Usp_{49} to Usp_{68}; and areas adjacent to

Table Hg-9

BASIC EXTRUSIVES

Rock Type	Locality	Primary Mineralogy			Oxidation Mineralogy						T°C	log 10 fO_2	Emphasis		Notes	Reference
		Mt-Usp_{ss}	Ilm-Hem_{ss}	Chr_{ss}	Ilm_{ss}	Pb_{ss}	Hem_{ss}	R	γ	OH			Pet	Mag		
Basalt	Idaho	x	x								1125	-10.0	x		1	Buddington and Lindsley (1964)
Basaltic-type Rocks	Iceland and Cascades	x	x								1090-980	-10.0 to -12.0	x		2,3	Carmichael and Nicholls (1967)
Olivine-tholeiites	Iceland	x	x		x						775-1100	-14.3 to -10.1	x		2,4	Carmichael (1967a)
Iron-rich Tholeiites	Iceland	x	x								980-1090	-11.6 to -10.0	x		2,4	Carmichael (1967a)
Basaltic-andesites	Iceland	x	x		x						840-975	-13.7 to -11.6	x		2,4	Carmichael (1967a)
Icelandites	Iceland	x	x		x						905-965	-12.7 to -11.8	x		2,4	Carmichael (1967a)
Pitchstone	Iceland	x	x								925	-12.6	x		2,4	Carmichael (1967a)
Picrite Basalt	Iceland	x	x	x							1050	-10.2	x		2,4	Carmichael (1967a)
Basalt	Surtsey	x	x	x							1020	-11.5	x		2,5	Steinthorsson (1972)
Mugearite-Hawaiite	Iceland	x	x		x						1030-1050	-10.5	x		2,6	Jakobsson et al. (1973)
Alkali Olivine Basalt	Hawaii	x	x	x	x			x			980-1070	-11.4 to -10.0	x		2,7,13	Evans and Moore (1968)
Basaltic	Hawaii	x	x	x	x						1050-1180	-10.2 to - 8.1	x		2,8	Anderson and Wright (1972)
Olivine Basalt	Kenya	x	x	x							1075	-10.3	x		2	Brown and Carmichael (1971)
Alkali Olivine Basalt	Kenya	x	x	x							1060	-10.4	x		2	Brown and Carmichael (1971)
High Alumina Basalt	Greece	x	x		x						690-1150	-17.0 to - 9.0	x		2,9	Nicholls (1971)
Alkali Olivine Basalt	New Mexico	x	x								900	-13.5	x		2,10	Stormer (1972)
Basalt	New Mexico	x	x								1150-1010	- 8.0 to -11.0			2,10	Stormer (1972)

ilmenite with Usp_{54} to Usp_{66}, representing progressive outward reduction with crystallization. Ilmenites show core compositions of Hem_{15-22} and margins with Hem_{9-21} (i.e. also reducing); lamellar ilmenites range from Hem_{17} to Hem_{48}. Coexisting primary pseudobrookites have cores = $Fpsb_{47-65}$ and margins = Fsb_{43-66} (this is a narrow range but is also indicative of reduction at the margins); Pb_{ss} are mantled by Mg-ilmenites. Similar trends are exhibited for suites of samples from the Kamakaia lava, and for lavas from Yellow Cone. An interpretation for the paragenesis of the 1955 eruption suggests that (a) initial (i.e. oxide core compositions) crystallization took place at T = 1120°C, $f_{O_2} = 10^{-8.1}$; (b) during eruption T and f_{O_2} ranged from 1180°C and $10^{-8.1}$ to 1050°C and $10^{-10.2}$; (c) crystallization of Ilm_{ss} and Mt_{ss} in an oxidizing, fractionated, hybrid magma; (d) heating and reduction induced by magma-mixing and precipitation of Pb_{ss}, followed by reaction with the liquid to yield Ilm_{ss}. This sequence accounts for the reduced trend in zoned crystals and is blessed by the fact that marginal increases in Cr_2O_3 and in MgO are also observed. An estimate of the nature of these magmas is as follows: Magma (a) T = 1120°C, $f_{O_2} = 10^{-8.1}$, Fe_2O_3 = 2.0 wt%, K_2O = 1.2 wt%; magma (b) T = 1200°C, $f_{O_2} = 10^{-8}$, Fe_2O_3 = 1.5 wt%, K_2O = 0.5 wt%. By assuming magma (a) = 43% + magma (b) = 57%, magma (c) is derived with T = 1166°C, $f_{O_2} = 10^{-8}$, Fe_2O_3 = 1.7 wt%, K_2O = 0.8 wt%, which is virtually identical to one of the erupted flows (T = 1180°C, $f_{O_2} = 10^{-8.1}$, Fe_2O_3 = 1.8 wt%, K_2O = 0.8 wt%). An added source for the increase in heat is considered to be by gas loss, primarily SO_2. An overall evaluation of differences in f_{O_2} paths for different suites of rocks suggests that differences in redox state may be due simply to differential gas loss prior to eruption by effervescence. (9) The lower temperature reported is for Mt_{ss} + oxidation exsolution Ilm_{ss}; this value is coincident with the NNO buffer whereas the unoxidized sample lies on the FMQ buffer. (10) This study also includes andesites and dacites. A significant feature is that Ilm_{ss} in basanites contain up to a max of 40 mole% Geik. Curves derived for the distribution of Mg in olivine vs. Mg in Ilm_{ss} as a function of T°C show a good correlation at 727°C. Stormer suggests that high P will reduce the geikielite component and that the paucity of high Mg ilmenites in basalts is related to the early precipitation of olivine. (11) The two values for T°C are lower limits which lie on the reducing side of the FMQ buffer. Based on $\underline{a}_{SiO_2}$ and the composition of coexisting olivine with T = 900°C an upper estimate of $f_{O_2} = 10^{-12.6}$ atm (i.e. above FMQ). The pegmatoid segregation is characterized by the presence of aenigmatite replacing Ilm_{ss}, and experimental data show that it is not stable much above the FMQ buffer curve so that the range in f_{O_2} at 900°C is approximately that of the FMQ buffer curve. (12) This is the pioneering paper which was among the first to show a correlation between oxidation and reversed magnetic polarity. (13) These studies consider the variations in oxide and magnetic properties within traverses across single lavas--see Chapter 5 for additional examples.

Table Hg-9 (2)

BASIC EXTRUSIVES

Rock Type	Locality	Primary Mineralogy			Oxidation Mineralogy						T°C	log 10 fO_2	Emphasis		Notes	Reference
		$Mt\text{-}Usp_{ss}$	$Ilm\text{-}Hem_{ss}$	Chr_{ss}	Ilm_{ss}	Pb_{ss}	Hem_{ss}	R	Y	OH			Pet	Mag		
Basanite	New Mexico	x	x								975	- 8.5	x		2,10	Stormer (1972)
Basalt	Oregon	x	x		x						820	-15.0	x		2,11	Lindsley and Haggerty (1970)
Basaltic Pegmatoid	Oregon	x	x		x						900	-13.7	x		2,11	Lindsley and Haggerty (1970)
Tholeiitic Dike Segregation	Hawaii	x	x								980	-11.4	x		2,7,13	Evans and Moore (1968)
Basalt	Mull	x	x		x									x	12	Ade-Hall and Wilson (1963)
Basalt	Iceland, Hawaii, India, Mull, Ireland	x	x		x						100-600			x	2	Ade-Hall et al. (1965) Ade-Hall (1964a)
Basalts	Mull	x	x		x			x						x	12	Ade-Hall (1964b)
Basalts	Skye	x	x		x	x	x	x					x	x	13	Ade-Hall et al. (1968a)
Basalts	Skye	x	x		x	x	x	x			405-610			x	13	Ade-Hall et al. (1968b)
Basalts	Skye	x	x		x	x	x	x						x	13	Ade-Hall et al. (1968c)
Basalts	Iceland, Mull, Skye, Scotland	x	x		x									x		Ade-Hall (1969)
Basalts	Canary Islands	x	x		x		x	x						x		Ade-Hall and Watkins (1970)
Basalts	Canada	x	x		x		x	x					x	x	14	Ade-Hall et al. (1971)
Basalt	France	x		x									x		2,15	Babkine et al. (1965)
Basalt	The Earth	x	x		x				x				x	x		Basta (1960)
Basalt	India	x	x		x								x			Bose (1965)
Basalt + + +	The Earth	x	x		x								x	x	1,16	Buddington et al. (1955)
Basalt	Nova Scotia	x	x		x		x		x		300-585			x		Carmichael and Palmer (1968)
Basalt	Michigan	x	x		x		x						x			Cornwall (1950)
Basalt	Basutoland	x	x		x		x	x	x				x			Cox and Hornung (1966);Haggerty(u.p
Basalt	Summary	x	x		x				x						2,17	Creer (1971)
Basalts	Argentina, Turkey, Azores, Canary Islands, Japan, Bulgaria, Aden Pacific	x	x		x				x		x			x	2,17	Creer and Ibbetson (1970)

(14) Detailed correlations and characteristic assemblages are related to thermomagnetic curves for variation in the intensity of deuteric oxidation, and for variations as a function of regional hydrothermal alteration. (15) Core-mantle relationships are discussed for Mg-Al-chromite zoned by titano-magnetite; a comparison is made between xenolithic peridotite spinels which are present in the basalt with basaltic chromites and these show that the former are enriched in Al_2O_3 (49 vs. 35-47 wt%), and MgO (24 vs. 11-18 wt%), and depleted in Cr_2O_3 (15 vs. 17-27 wt%) and FeO (3 vs. 10-21); a definitive pattern for Fe_2O_3 is not apparent. (16) This is the foundation paper for the bulk of the analyses which supplemented and resulted in the oxide-benchmark publication by Buddington and Lindsley (1964). (17) These papers provide a detailed discussion of the geophysical interpretation of remanent magnetization in oxidized basalts. (18) Chromites in quenched basalts and pumices from the Kilauea Iki crater exhibit minimal zoning but show a gradual progression of decreasing Cr_2O_3, MgO and Al_2O_3 at the expense of increasing FeO, Fe_2O_3, TiO_2 and V_2O_3, a trend which is comparable to that shown for chromites in the ancient Makaopuhi lava lake (see note 7). These latter chromites are poorer in Cr_2O_3 (38 vs. 43 wt%) and this is reflected in bulk rock Cr_2O_3 contents which range from 0.14 - 0.20 for Kilauea Iki, and from .056 - .059 for Makaopuhi. The ranges in Al_2O_3, MgO and TiO_2 are approximately equal. Eruption T = 1200-1225°C, but attempts to utilize coexisting Ol + Chr yield 2000-2300°C. (19) These papers consider the magnetic and oxide variations in drill cores from the Alai and Makaopuhi lava lakes, Hawaii. The oxide data are discussed in Chapter 5. (20) This, in common with Creer (1971)--see note 17, is an excellent review of correlative parameters between magnetic property data and oxide mineralogy; cooling histories and their relationship to T and f_{O_2} are also considered. (21) Compositional data for Pb_{ss} are reported from 29 localities; this early report is still the most comprehensive study in the literature and was the first to draw attention to the high MgO contents typical of the series, and the fact that preferred distributions may be present in lavas exhibiting differential high temperature oxidation profiles. See chapter 5. (22) For three generations of primary titanomagnetite systematic variations are observed which show progressive increases in TiO_2 and MnO contents, and progressive decreases in Al_2O_3 = 9.5 wt%; MgO = 5.1 and Cr_2O_3 = 1.8 wt%. The systematic increase in TiO_2 from $Usp_{31.9}$ to $Usp_{78.7}$ is considered to result from a decrease in f_{O_2} with crystallization. (23) Columnar basalt study. (24) On the suitability of basalts of variable oxidation states for magnetic field intensity studies; these data show that only the most highly oxidized samples from a single lava study (lava HS--Chapter 5) obey the necessary criteria for reliable paleointensity determinations. (25) Although it had long been recognized that the particle sizes of Mt_{ss} are related to magnetic stability, this paper was the first to point out that particle size reduction (and hence increased stability) may be accomplished by oxidation exsolution, i.e. large crystals being reduced to a large number of finer sub-crystals separated by {111} Ilm_{ss} lamellae. (26) A detailed electron microprobe study of Ti-Al-chromian-ulvöspinels show that crystals are zoned from cores (high Cr) to

Table Hg-9 (3)

BASIC EXTRUSIVES

Rock Type	Locality	Primary Mineralogy			Oxidation Mineralogy						T°C	log 10 fO_2	Emphasis		Notes	Reference
		$Mt\text{-}Usp_{ss}$	$Ilm\text{-}Hem_{ss}$	Chr_{ss}	Ilm_{ss}	Pb_{ss}	Hem_{ss}	R	γ	OH			Pet	Mag		
Basalts	Hawaii			x									x		2,18	Evans and Wright (1972)
Basalts	New Mexico	x	x								1070	-10.3	x		2	Fodor (1975)
Basalts	Hawaii	x	x								{1040-	{-10.5 to	x		2	Fodor et al.(1972)
Basalts	Germany	x	x		x	x	x	x	x	x	{ 980	{-11.5	x			Frenzel (1953)
Basalts	S.W.Africa	x	x		x	x	x	x	x					x	2	Gidskehaug and Davison (Pre-Print)
Basalts	Hawaii	x	x		x	x	x	x						x	13,19	Gromme et al. (1969)
Basalts	Iceland, Hawaii	x	x	x	x	x	x	x					x		2,13,19	Haggerty (1971)
Basalts	General	x	x		x	x	x	x						x	20	Hargraves and Petersen (1971)
Basalts	Iceland	x	x		x		x	x			150-580			x		Hargraves and Ade-Hall (1975)
Basalts	Nevada, Oregon	x	x		x	x	x	x						x		Larson et al. (1971)
Basalts	Ireland	x	x		x						100-300			x	13	Lawley and Ade-Hall (1971)
Basalts	Scotland	x	x		x	x	x	x					x		13	MacDonald (1967)
Basalts	Japan	x									200			x		Nishida and Sasajima (1974)
Basalts	Germany					x							x		2,21	Ottemann and Frenzel (1965)
Basalts	France	x							x		140-250		x	x	2,22	Prévot and Mergoil (1973)
Basalts	India	x												x		Radhakrishnamurty et al. (1972)
Basalts	India	x	x						x					x	23	Radhakrishnamurty et al. (1971)
Basalts	Skye, Wales	x	x											x		Radhakrishnamurty et al. (1973)
Basalts	Germany	x	x		x	x	x	x	x				x			Ramdohr (1952)
Basalts	Germany	x							x					x		Schult (1968)
Basalts	Iceland	x	x		x	x	x	x						x	13,24	Smith (1967a)
Basalts	Oregon, Iceland	x	x		x	x	x	x			575-650			x	2	Smith (1967b)
Basalts	Iceland, Scotland, Oregon	x	x											x	2	Smith (1967c)
Basalts	General	x	x		x									x	25	Strangway et al. (1968)
Basalts	Scotland	x	x		x		x	x					x			Stumpfl (1961)
Basalts	Idaho	x		x									x		2,26	Thompson (1973)
Basalts	Skye, Mull	x												x	27	Watt et al. (1973)

margins (low Cr) and exhibit a progressive sequence from chromite to titanomagnetite as a function of crystallization: chromites in olivine have the highest overall Cr contents (22.9 wt%); intermediate Cr contents are found in crystals protruding from olivines (19 wt%); and the lowest Cr contents are found in the matrix glass (17 wt% Cr_2O_3). Similarly decreasing trends are present for MgO, Al_2O_3 and Fe_2O_3; increasing trends in MnO, FeO and TiO_2 are in keeping with trends shown for spinels in the Hawaiian lava lakes (Evans and Moore, 1968--note 7; Anderson and Wright, 1972--note 8; Evans and Wright, 1972--note 18) and are in accord with experimental data (Hill and Roeder, 1974) for crystallization in the range 1000-1150°C and $f_{O_2} \sim 10^{-8}$ atms. (27) A Mössbauer study of optically homogeneous titanomagnetites in basalts show that samples having Curie temperatures in the range 150-300°C yield Fe^{3+}/Fe^{2+} ratios between 0.8 to 0.9, whereas samples with Curie temperatures in the range 550-600°C yield Fe^{3+}/Fe^{2+} = 1.1 to 2.0. Note: Increased T_c with increasing oxidation is consistent with experimental data but is inconsistent for homogeneous Mt_{ss} or for compositions typical of basaltic titanomagnetite. (28) Mg-Al-chromites exhibit core to mantle zoning from Usp_3 to Usp_{60}; Cr, Al and Mg decrease progressively while Fe + Ti gradually increase towards the margins, a distribution which is consistent with decreasing T and with partial reaction of Chr_{ss} with liquid. (29) A statistically significant data set shows that Al-rich chromites form a continuous series with titanomagnetite. Crystals enclosed in augite, which are in turn enclosed in olivine, contain zoned spinels which have Cr-free cores (51.50 wt% Al_2O_3) and Cr + Fe^{3+}-rich margins; a progressive increase in TiO_2 and MnO are observed with a concomitant decrease in MgO from core to margin. Spinels within the mantling olivine are Al-Cr-titanomagnetites and Al-chromites; titanomagnetites are closely associated but are in the groundmass. The observed trend therefore is: (1) ferrian pleonaste, chromian ferrian pleonaste, aluminian chromite, Al-chromian titanomagnetite, and titanomagnetite. The composite data show that these spinels cover the entire range of compositions exhibited by spinels in Alpine type ultramafics and spinels present in layered intrusions. (30) Spinel compositions in glassy basalts and in dolerites from an ophiolite sequence exhibit a similar trend of Cr_2O_3 enrichment from core to margin as those discussed above (Arculus, 1974); the zonation is less striking and the trends for Mg, Al, Fe and Ti are similar. (31) A comparison of chromites in basalts from the Bridget Cove volcanic complex, with chromites in closely related ultramafic suites, are in remarkable agreement exhibiting moderate Cr_2O_3 (42-46 wt%), high MgO (1-14 wt%) and low Al_2O_3 (7-11 wt%) contents. The magnetite phenocrysts have Al_2O_3 = 0.5-8.5; TiO_2 = 0.5-7.5; Cr_2O_3 < 0.1-10; and MgO = 4-5 wt%. (32) A comprehensive analytical data bank suggests extensive to complete solid solution between Usp and Chr. (33) Al-titanomagnetites have reported ranges from 6-10 wt%. Al_2O_3 and Cr_2O_3 are 1.56 max; TiO_2 = 15.12-21.31; and MgO = 1.56-8.92 wt%. Al and Mg may be slightly overestimated because of silicate contamination.

Table Hg-9 (4)

BASIC EXTRUSIVES

Rock Type	Locality	Primary Mineralogy			Oxidation Mineralogy						T^oC	log 10 fO_2	Emphasis		Notes	Reference
		$Mt\text{-}Usp_{ss}$	$Ilm\text{-}Hem_{ss}$	Chr_{ss}	Ilm_{ss}	Pb_{ss}	Hem_{ss}	R	γ	OH			Pet	Mag		
Basalts	Iceland	x	x		x	x	x	x	x					x	13	Watkins and Haggerty (1965)
Basalts	Iceland	x	x		x	x	x	x	x					x	13	Watkins and Haggerty (1967)
Basalts	Iceland	x	x		x	x	x	x	x					x	13	Watkins and Haggerty (1968)
Basalts	France	x	x		x	x	x				556			x		Whitney et al. (1971)
Basalts	General	x	x		x	x	x							x		Wilson and Haggerty (1966)
Basalts	Oregon	x	x		x	x	x							x		Wilson and Watkins (1967)
Basalts	Iceland	x	x		x	x	x	x	x						13	Wilson et al. (1968)
Basalts	New Zealand	x	x	x	x				x				x	x		Wright (1967)
Basalts	New Zealand	x	x		x		x				400-520		x	x		Wright and Lovering (1965)
Basalts	California	x	x								1050	-10.6	x			Smith and Carmichael (1968)
Basalts	S.W. Africa	x	x								670-920	-19 to -12.0		x		Gidskehaug et al. (1975)
Basalts	Ireland	x	x											x	23	Symons (1967)
Ankaramite	Crozet	x		x	x								x		28	Gunn et al. (1970)
Alkali basalts	Granada			x									x		29	Arculus (1974)
Basalts	Taiwan			x									x		30	Liou (1974)
Basalts	Alaska	x		x									x		31	Irvine (1973)
Basalts	Hawaii	x	x	x	x								x		32	Beeson (1976)
Alkali basalts	Japan	x		x										x	1,33	Sasajima et al. (1975)

Table Hg-10

DEEP SEA BASALTS

Locality	Rock Type	Primary Mineralogy				Oxidation Mineralogy			T°C	log 10 fO_2	Emphasis		Notes	Reference
		$Mt-Usp_{ss}$	$Ilm-Hem_{ss}$	Chr_{ss}	Sulfides	Ilm_{ss}	γ	OH			Pet	Mag		
DRILL CORE STUDIES														
Mohole test, Guadalupe	basalt	x	x		Po				355			x		Cox and Doell (1962)
MAR Legs 2 and 3	basalt			x							x			Frey et al.(1974)
North Pacific, Legs 5,6 and9	basalt	x	x				x		100-320					Lowrie et al. (1973)
Philippine Sea, Caroline ridge, Leg 6	basalt	x	x	x					950	-12.5	x			Ridley et al. (1974)
Atlantic. Leg 11	basalt	x	x	x	x						x			Ayuso et al. (1976)
Caribbean, Leg 15	basalt	x			x						x			Bence et al. (1975)
EastPacific rise, Leg 16	basalt	x	x						215			x		Heinrichs (1973)
Central Pacific, Leg 17	basalt	x		x	x						x			Myers et al. (1975)
Indian Ocean, Leg 25	basalt	x							~500			x		Wolejszo et al. (1973)
Leg 26	basalt	x	x	x	Po, Py	x	x	x	198-564		x			Ade-Hall (1974a,b)
Pacific Ocean, Leg 33	basalt	x	x	x	Py, Cpy	x	x	x	280-340 310-370 550-580			x		Cockerham and Hall (1976)
Nazca Plate, Leg 34	basalt	x	x	x	x				1090-940	-9.8 to -12.6	x			Mazzullo and Bence (1976)
Nazca Plate, Leg 34	basalt	x	x	x	Py, Po, Cpy	x	x	x	124-490			x		Ade-Hall et al. (1976a,b)
Nazca Plate, Leg 34	basalt	x							125-165			x		Lowrie and Kent (1976)
Nazca Plate, Leg 34	basalt	x	x				x		125-380					Gromme and Mankinen (1976)
Nazca Plate, Leg 34	basalt	x							145-420			x		Deutsch and Pätzold (1976)

Table Hg-10 (2)

DEEP SEA BASALTS

Locality	Rock Type	Primary Mineralogy				Oxidation Mineralogy			$T^{o}C$	log 10 fO_2	Emphasis		Notes	Reference
		$Mt\text{-}Usp_{ss}$	$Ilm\text{-}Hem_{ss}$	Chr_{ss}	Sulfides	Ilm_{ss}	γ	OH			Pet	Mag		
DRILL CORE STUDIES														
Nazca Plate, Leg 34	basalt	x	x						840	-14.7	x			Bunch and LaBorde (1976)
MAR Leg 37	lherzolite, gabbro	x	x	x	x						x			Hodges and Papike (1976)
MAR. Portable bottom drill	basalt	x	x	x	Py, Po, Cpy		x	x	200-550			x		Brooke et al. (1970)
MAR. Portable bottom drill	basalt	x	x									x		Ade-Hall et al. (1973)
PILLOW BASALT STUDIES														
Reykjanes ridge	basalt	x	x						150-190			x		De Boer et al. (1969); De Boer (1975)
Juan de Fuca ridge	basalt	x	x				x		175-235			x		Marshall and Cox (1972)
Central Indian ridge and Juan de Fuca ridge	basalt	x	x				x		200-300			x		Marshall and Cox (1971a)
Vancouver Island	basalt	x	x		Cpy	x					x		Hem_{ss}	Surdam (1968)
South Pacific	basalt	x	x		x				250-400			x		Watkins and Paster (1971)
Scotia Sea	basalt	x		x	Py				250-400					Watkins et al. (1970)
Iceland	basalt	x									x			Yagi (1964)
DREDGE HAUL STUDIES														
MAR	basalt	x										x		Matthews (1961)
MAR and Pacific	basalt	x	x			x	x					x		Ade-Hall (1964)
MAR	basalt	x					x					x		Vogt and Ostenso (1966)
MAR	basalts, greenstones	x										x	sphene	Luyendyk and Melson (1967)
MAR 45°N	basalt, gabbro	x										x		Opdyke and Hekinian (1967)

Table Hg-10 (3)

DEEP SEA BASALTS

Locality	Rock Type	Primary Mineralogy				Oxidation Mineralogy			T°C	log 10 fO_2	Emphasis		Notes	Reference
		$Mt\text{-}Usp_{ss}$	$Ilm\text{-}Hem_{ss}$	Chr_{ss}	Sulfides	Ilm_{ss}	γ	OH			Pet	Mag		
DREDGE HAUL STUDIES														
MAR 45°N	basalt, gabbro, peridotite	x	x	x	Po,Py, Cpy	x	x	x				x		Haggerty (1970); Haggerty and Irving (1970); Irving et al. (1970a); Irving (1970)
MAR 45°N	basalt	x	x				x		165-525			x		Irving et al. (1970b); Park and Irving (1970)
MAR 45°N	basalt	x	x	x	Py,Po	x	x		200-540			x	$R+Pb_{ss}$	Carmichael (1970)
MAR 45°N	basalt	x	x				x		100-340			x		Schaeffer and Schwarz (1970)
MAR 45°N	basalt	x										x		Evans and Wayman (1972)
MAR. Famous	basalt			x							x			Dick (1976)
MAR 30-40°N	basalt			x							x			Sigurdsson and Schilling (1976)
MAR 0°-10°N	lherzolites, norite, tholeiite teschenite	x	x	x					1000	-11	x		T°C-fO_2 for tholeiite	Prinz et al. (1976)
MAR 22°-52°N Puerto Rico Trench Caryn seamount	basalt	x	x			x	x		200-550			x		Wazilewski (1968)
North Atlantic, Caribbean	basalt, metabasalt, gabbro, peridotite	x	x	x	x	x	x	x				x		Fox and Opdyke (1973); Haggerty unpublished
MAR 29 and 45°N, 22°S, Leg 11, Red Sea, S.W. Pacific	basalt	x		x							x			Bryan (1972)
Galapagos rift Costa Rica rift	basalt	x	x					x	140-165 200-215 240-260			x		Anderson et al. (1975)
Cobb seamount	basalt	x	x			x			540			x		Merrill and Burns (1972)

Table Hg-10 (4)

DEEP SEA BASALTS

Locality	Rock Type	Primary Mineralogy				Oxidation Mineralogy			T^oC	log 10 fO_2	Emphasis		Notes	Reference
		$Mt\text{-}Usp_{ss}$	$Ilm\text{-}Hem_{ss}$	Chr_{ss}	Sulfides	Ilm_{ss}	γ	OH			Pet	Mag		
DREDGE HAUL STUDIES														
Pacific Ocean, Mohole test site	basalt	x							250			x		Ozima and Ozima (1967)
Pacific Ocean, Atlantic	basalt, andesite	x					x		300-550			x		Ozima and Oxima (1971)
Pacific Ocean	basalt	x	x			x	x		180-660			x		Joshima (1973)
Macquarie Is.	basalt, gabbro	x	x			x	x					x	Pb_{ss}+ Hem_{ss} +R	Butler and Banerjee (1973)
Macquarie ridge	basalt, harzburgite gabbro, troctolite	x	x		x						x	x	Hem_{ss}	Gunn and Watkins (1971)
OXIDATION MECHANISMS $Usp_{ss} \rightarrow Fe_2O_3$; self-reversal processes; deep-sea and sub-aerial basalt comparisons; mineralogical dependencies in the context of oceanic magnetic quiet zones.														Haggerty (1970); Haggerty and Irving (1970); Irving et al. (1970a); Irving (1970). Banerjee (1971); Kono (1971); Marshall and Cox (1971b); Ozima and Larson (1970); Prévot and Grommé (1975); Readman and O'Reilly (1972)

Hypabyssal Suites

NOTES: (1) This paper discusses and shows a correlation between normal and reversed directions of magnetizations and Fe-Ti oxide oxidation. (2) Replacement of Fe-Ti oxides by sphene suggests that Fe is remobilized and may lead to large concentrations which result in Cornwall-type deposits. (3) Data are also presented for T-f_{O_2} determinations of recrystallized qtz diorites and a gabbroic-anorthosite; the qtz diorites yield T = 550°C and $f_{O_2} = 10^{-19}$ atms; the gabbroic anorthosite yields T = 670-760, $f_{O_2} = 10^{-19.5}$ to $10^{-15.5}$ atms. (4) This study is detailed, in oxide characterization and in the effects of deuteric oxidation and weathering in profiles across two dikes. (5) This composite dike consists of a central acid core flanked on either side by dolerite margins. The Fe-Ti oxides are concentrated largely in the basic margins and exhibit maximum oxidation variations of Ilm_{ss} in Mt_{ss}, and of γFe_2O_3. (6) Magnetite (3.18 wt% TiO_2) is present along veinlets in joint systems which suggests leaching by hydrothermal activity. (7) Skeletal crystallization of Pb_{ss} is reported coexisting with Usp_{ss} enclosed in plagioclase; this assemblage is thought to be in disequilibrium and is considered to have developed as a result of rapid quenching. (8) Kennedyite (Fe_2TiO_5-$MgTi_2O_5$) is described for the first time with MgO = 6.45 wt%, Al_2O_3 = 2.15 wt% and FeO = 2.0 wt%. Additional details are to be found in Haggerty (1976). (9) Mt_{ss} compositions show MgO = 0.79-2.33; Al_2O_3 = 2.31-3.55; Cr_2O_3 = 0.4 wt%; TiO_2 = 23.05-29.66 wt%. (10) Titanomagnetites and ilmenites are zoned from core to margin as follows: Usp_{34}-Usp_{41} and Ilm_{86}-Ilm_{96} respectively. As the Fe-Mg silicates become Fe-enriched in the differentiation sequence ilmenite compositions become enriched in MnO and exhibit decreases in Hem_{ss}, and decreases in MgO contents, a trend which is shown also by zoning within single crystals. Useful criteria are provided to distinguish primary Ilm_{ss} from oxidation exsolution Ilm_{ss}. In traverses approaching Ilm_{ss} lamellae convincing evidence is provided for progressive increases in Cr_2O_3 contents in Mt_{ss} and in considerable enrichment in MgO and MnO in the Ilm_{ss} constituent. These criteria are substantiated by the observation also that primary Ilm_{ss} have higher Hem_{ss} and consistently lower MnO + MgO concentrations than those formed by oxidation exsolution.

Table Hg-11

HYPABYSSAL SUITES

Rock Type	Locality	Primary Mineralogy			Oxidation Mineralogy						T°C	log 10 fO_2	Emphasis		Notes	Reference
		$Mt\text{-}Usp_{ss}$	$Ilm\text{-}Hem_{ss}$	Chr_{ss}	Ilm_{ss}	Pb_{ss}	Hem_{ss}	R	γ	OH			Pet	Mag		
DIKES																
Dolerites	Mull, Skye	x	x		x			x						x		Ade-Hall et al. (1971)
Dolerites	Mull	x	x		x			x						x	1	Ade-Hall and Wilson (1969)
Dolerites	Mull	x	x		x			x			450			x		Ade-Hall et al. (1972)
Diabases	Pennsyl-vania	x	x		x		x						x		2	Davidson and Wyllie (1968)
Diabases	Finland	x	x		x		x				~925	-12.5	x		3	Haapala and Ojanperä (1972)
Dolerites	Denmark	x	x		x		x	x	x		870,950	-14.0,-12.5	x		4	Jensen (1966)
Diabases	Wyoming	x	x					x	x		~550			x		Spall (1971)
Dolerites	Scotland	x	x								580			x		Unan (1970)
Diabases	Sweden	x	x		x			x					x			Uytenbogaardt (1953)
Diabases	Finland, Greenland	x	x		x								x			Vaasjoki and Puustinen (1966)
Dolerites	Iceland	x	x		x				x				x	x		Watkins and Haggerty (1968)
Composite	Iceland	x	x		x				x				x	x	5	Haggerty and Watkins (1966)
Microdiorites	Peru	x	x		x				x				x	x		Wilson and Haggerty (unpublished)
Dolerites	MacQuarie Is.	x	x		x						580			x		Banerjee et al. (1974)
Diabase	New Jersey	x													6	Puffer and Peters (1974)
Camptonite	New Hampshire	x													7	Rice et al. (1971)
Trachyandesitic	Jamaica	x	x		x				x		150-575					Watkins and Cambray (1971)
SILLS																
Diabases	Pennsyl-vania	x	x		x				x					x		Beck (1966)
Diabase	New Jersey	x	x								550-600			x		Hargraves and Young (1969)
Diabase	Arizona	x	x		x						580-590			x		Helsley and Spall (1972)
Dolerite	S. Rhodesia	x	x	x									x		8	Von Knorring and Cox (1961)
Teschenite	Australia	x	x						x				x		9	Wilkinson (1957)
Theralite-Tinguaite	Australia	x							x				x		9	Wilkinson (1965)
Olivine Diabase	Arizona	x	x		x										10	Smith (1970)

Acid-Intermediate Intrusive Suites

Notes: (1) Wet chemical analyses; (2) fayalite - ferrohedenbergite - mesoperthite granite; (3) hornblende (+ biotite) mesoperthite granites; (4) biotite mesoperthite granites; (5) EMPA Manganoan-ilmenite (Hem_6 Ilm_{66} Pyh_{28}) = 12.7 wt% MnO in qtz-biotite-diorite; Mn-Ilm (Ilm_{72} Pyh_{28}) = 12.7 wt% MnO in adamellite; (6) Mt_{ss} in plag, in hornblende and in biotite. (7) EMPA. Mn-Ilm: granite, MnO = 29.8-30.1 wt%; granodiorite, MnO = 10.8-19.5 wt%; monzonite, MnO = 2.9-4.1%. Values for MgO are uniformly low, max. = 0.17 wt%. Coexisting magnetites for the three rock types contain max. MnO contents of 0.75, 0.54, and 0.18 respectively; TiO_2 + MgO + Al_2O_3 + Cr_2O_3 + V_2O_5 is typically <1 wt%. Average values for co-existing Mt + Ilm are: granites, $Mt_{98.6}$ + $Ilm_{29.9}$ $Hem_{4.5}$ $Pyh_{63.0}$; granodiorites, $Mt_{98.9}$ + $Ilm_{60.3}$ $Hem_{6.1}$ $Pyh_{32.4}$; monzonite, $Mt_{98.3}$ + Ilm_{85} $Hem_{6.2}$ $Pyh_{7.3}$. Estimates of f_{O_2} at T = 700°C and P_{H_2O} = 1000 bars suggest an upper limit defined by the MH buffer (granites), and a lower limit defined by the NNO buffer (monzonites), with the granodiorites being intermediate between these extremes and closer to the NNO buffer. (8) Spatial distribution of primary and secondary magnetite; Qtz. monzonite, primary Mt = 65%; oxidation Mt = 25%; alteration Mt = 10% by vol.; monzonite, primary = 45%; oxidation = 30%; alteration = 25% by vol. (9) EMPA. Mn-Ilm.MnO = 11.1-15.2; Mn_2O_3 = 3.3 wt% based on $L_{\alpha 1.2}/L_{\beta 1}$. (10) EMPA. Mt_{ss} show a restricted range in compositions with TiO_2 = <1-3.5 wt% but with an extremely wide range in the Ilm_{ss} compositions with TiO_2 = 2-48 wt%; MnO contents of Ilm_{ss} = <1-6.7 wt%. Contrasting assemblages are: (a) Mt-Usp_{ss} + Ilm-Hem_{ss}; and (b) Mt_{ss} + Hem_{ss}. The three ranges in T and f_{O_2} given are for the first of the two assemblages for pegmatites from Vermilion, Minn., Pegmatite Points, Co., and Kingman, Ariz. This suite yields f_{H_2O} = 10^2 to 10^3 bars and confining pressures are estimated to be 5-10 Kb. For the Mt + Hem pegmatitic suite T = 600-700°C based on (a) electrolytic cell determination for a garnet-bearing but Hem_{ss}-free assemblage from the Benson mine which yields T = 600-670°C, f_{O_2} = 10^{-17} to 10^{-21} bars (Palmer 1967); and (b) on coequilibrated oxide pairs in gneisses from Benson which yield T = 600-650°C and f_{O_2} = 10^{-17} to 10^{-19} bars (Buddington and Lindsley, 1964). By using the range in T = 600-700°C and the MH buffer, the range in f_{O_2} = 10^{-12} to 10^{-15} bars for Mt + Hem pegmatites. (11) EMPA. These data form part of the more extensive study discussed above for pegmatites from Pegmatite Points, Co. (12) The ranges in T and f_{O_2} are for the Rosalie granite, the Indian Creek granite, and the Mt. Evans granodiorite respectively. The first of these is considered to have been emplaced as a high T°C, rather dry magma; the remaining two have undergone recrystallization and reequilibration at low T°C, and in a very dry environment. Respective f_{H_2O} for the three intrusions are 3600-3900 bars; 85-180 bars; and 25-240 bars. (13) These values represent values for a range of T and f_{O_2} between 600-750°C and f_{O_2} = 10^{-16} to 10^{-21} bars. The feldspar geothermometer yields T = 650-700°C for the granite, and T = 710-1065°C for the syenite; f_{O_2} determinations at 690°C (syenite) and 680°C (granite) yield independent values of $10^{-17.8}$ and $10^{-18.1}$ bars respectively, based on the assemblage Qtz + Ol + Mt. An associated qtz-trachyte lava yields T = 1040°C and f_{O_2} = $10^{-10.7}$ bars based on Usp_{ss} + Ilm_{ss}, and T = 810-1095°C based on

Table Hg-12

ACID-INTERMEDIATE INTRUSIVE SUITES

Rock Type	Locality	Primary Mineralogy			Oxidation Mineralogy						T°C	log 10 fO_2	Emphasis		Notes	Reference
		$Mt\text{-}Usp_{ss}$	$Ilm\text{-}Hem_{ss}$	Chr_{ss}	Ilm_{ss}	Pb_{ss}	Hem_{ss}	R	γ	OH			Pet	Mag		
Granites	Adirondacks	x	x								765	-13.8	x		1,2	Buddington and Lindsley (1964)
Granites	Adirondacks	x	x								650	-17.5	x		1,3	Buddington and Lindsley (1964)
Granites	Adirondacks	x	x								620-640	-18.2 to 17.0			1,4	Buddington and Lindsley (1964)
Granite	Ireland		x										x		5	Elsdon (1975)
Granites	Oklahoma		x				x				>550			x		Spall (1968)
Granite	Oklahoma	x	x		x		x			x	~650			x		Spall and Noltimier (1972)
Granite	Texas	x	x				x				~640			x		Spall (1972)
Granodiorite	California	x	x		x						529-572			x	6	Wu et al. (1974)
Granites, Monzonites, Diorites	Wyoming	x	x				x				565			x		Lidiak (1974)
Granites, Monzonites, Granodiorites	Norway	x	x		x		x	x	x				x		7	Czamanske and Mihalik (1972); Czamanske and Wones (1973)
Monzonite, Quartz-Monzonite	Nevada	x											x		8	Sanderson (1974)
Felsite	Virginia	x	x						x	x	535-575			x		Lovlie and Opdyke (1974)
Adamellite	California		x					x					x		9	Snetsinger (1969)
Adamellites, Monzonites	Australia	x	x		x						520-560			x		Robertson (1963)
Granite-Pegmatite	Adirondacks	x	x								560	-19.2			1	Buddington and Lindsley (1964)
Granite-Pegmatites	North American Locations	x	x								560-590 490-580 590-610	-20.2 to -19.6 -20.2 to -19.0 -17.6 to -17.1			10	Puffer (1975)
Granitic Pegmatites	Colorado	x	x		x						460-535	-22.5 to -20.0	x		11	Puffer (1972)

feldspars. f_{H_2O} = 300-400 bars (syenite), and up to 800 bars for the granite. Low initial H_2O accounts for high initial temperatures of crystallization. Significant concentrations of MnO and Nb_2O_3 are present in all ilmenites: trachyte (MnO = 4.4-9.8, Nb_2O_3 = 0.7-2.0); syenite (MnO = 1.6-1.9, Nb_2O_3 = 0.2-2.0); granite (MnO = 1.7-2.9; Nb_2O_3 = 1.4-4.0 wt%). (14) Monzonites and plagifoyaites, from the southern part of Oslo rift, fall on the reducing side of the FMQ buffer. (15) Olivine diorite and monzonites from North of Oslo also fall on the reducing side of the FMQ buffer. (16) Monzonites and Qtz. monzonites from the central rift system are close to or fall above the Ni-NiO buffer curve and there is evidence to suggest that in cooling from 780-740°C, f_{O_2} changed from $10^{-14.6}$ to $10^{-16.3}$ atm. based on oxidation exsolution. (14-16) EMPA. Neumann's paper details a comparison of several geothermometers and makes use also of coexisting aegerine + arfvedsonite to estimate f_{O_2} in alkali granites which show a range in T°C = 700-780°C and $f_{O_2} = 10^{-18.5} \pm 10^{-1.3}$ atm. (i.e. a range between the WM and FMQ buffer curves); nepheline syenites in the suite contain sphene + (Mn-Ilm) + (Mn-Mt) and are estimated to have crystallized well above the FMQ buffer curve at T = 855-925°C. (17) Compositional data for T - f_{O_2} determinations given in Neumann (1976) above are contained in this publication which is a detailed evaluation of the distribution of Mn among Ilm_{ss}, Mt_{ss} and bulk rock. These data show: (a) that Ilm_{ss} are enriched in MnO relative to Mt_{ss}; (b) that there are large intra-grain variations but little or no zoning within single crystals; (c) that MnO is strongly partitioned towards Ilm_{ss} which form by oxidation exsolution from $Mt\text{-}Usp_{ss}$; (d) that Mn^{2+}_{Ilm} increases with increasing $(Mn^{2+}/Fe^{2+})_{Magma}$ and is more pronounced when crystallization takes place at low T°C; and (e) that $Mn^{3+}_{Ilm}/Mn^{3+}_{Mt}$ decreases with increasing f_{O_2}. For the suites of samples discussed in notes 15-17 the peralkaline suites contain the highest concentrations of MnO = 7.7-30.2 wt% ($Ilm_{79.6}$ $Hem_{3.9}Pyh_{16.5}$ - $Ilm_{34.7}$ $Pyh_{65.3}$); the monzonites range between MnO = 2-5 wt%; the diorites contain ~2 wt% and the plagifoyaites 5-7 wt% MnO. For Ilm_{ss} in all suites MgO + Al_2O_3 <1 wt%. Unoxidized Mt_{ss} contain a max. of 4.96 wt% MnO (nepheline syenite); oxidized Mt_{ss} with Ilm_{ss} lamellae contain a max. of 2.9 wt% MnO and for the majority of cases Fe_2O_3 increases by a factor of 2 relative to unoxidized grains in the same sample. (18) Compositional data for two pyrophanites yield $Ilm_{23.08}$ $Hem_{2.41}$ $Pyh_{74.51}$ and $Ilm_{10.01}$ $Hem_{3.74}$ $Pyh_{86.25}$. (19) EMPA. $Mn\text{-}Ilm_{ss}$ show a good correlation between MnO and FeO and the data fall into two groups: one group is high in MnO = 13.1-17.2 wt%, and the second group has a lower range, MnO = 4.51-6.97 wt%. Individual grains are zoned from core to margin in both groups with Mn/Fe increasing towards the grain boundaries. K_D of Mn^{2+} and Fe^{2+} between Ilm_{ss} and the granitic magma = 5.5. (20) EMPA. Ilmenites contain 2.55-2.96 wt% MnO, and Mt_{ss} have MnO - 0.09-0.32 wt%; Al_2O_3 + MgO ≃ 1 wt% in each primary constituent. Estimates of T°C for orthopyroxenes and for biotite, in concert with values obtained for oxide pairs, suggest a T-f_{O_2} cooling curve equivalent to the FMQ buffer curve. For T = 500°C and $f_{O_2} = 10^{-24.2}$ the value for f_{H_2O} = 24 bars. (21) Oxide-rich diorite in anorthosite, Sault-aux-Cochons complex. (22) Anderson's paper is a broad and detailed study of the Labrieville

Table Hg-12 (2)

ACID-INTERMEDIATE INTRUSIVE SUITES

Rock Type	Locality	Primary Mineralogy			Oxidation Mineralogy						T^oC	log 10 fO_2	Emphasis		Notes	Reference
		$Mt\text{-}Usp_{ss}$	$Ilm\text{-}Hem_{ss}$	Chr_{ss}	Ilm_{ss}	Pb_{ss}	Hem_{ss}	R	γ	OH			Pet	Mag		
Granite, Granodiorites	Colorado	x	x		x						680-685 505-570 480-585	-10.3 to -16.3 -20.8 to -19.1 -22.3 to -18.2	x		12	Puffer (1972)
Granite, Qtz syenite	New Hampshire	x	x								~680 ~690	-17.7 -18.0	x		13	Whitney and Stormer (1976)
Monzonites	Norway	x	x		x						1000 805 730 880	-12.0 -15.0 -17.1 -13.6	x		14	Neumann (1976)
Plagifoyaite	Norway	x	x		x						850	-14.4	x		14	Neumann (1976)
Olivine Diorite	Norway	x	x		x						840	-14.7	x		15	Neumann (1976)
Monzonites	Norway	x	x		x						1090 675	-10.0 -18.2	x		15	Neumann (1976)
Monzonites	Norway	x	x		x						780 680 765	-14.6 -16.3 -14.0	x		16	Neumann (1976)
Qtz Monzonites	Norway	x	x		x						830 640	-13.0 -17.8	x		16	Neumann (1976)
Olivine Monzodiorites to Granites	Norway	x	x		x								x		17	Neumann (1974)
Nepheline Syenite Pegmatite	Norway		x										x		18	Neumann (1964)
Granite	Japan		x										x		19	Tsusue (1973)
Qtz Monzonite Mafic Granite	Greenland	x	x								525	-23	x		20	Frisch and Bridgewater (1976)
Diorite	Quebec	x	x								615	-18.5	x		1,21	Buddington and Lindsley (1964)
Augite Syenite	Quebec	x	x								720	-15.7	x		1	Buddington and Lindsley (1964)
Syenite	Quebec	x	x								640	-18.8	x		22	Anderson (1968)
Syenitic Pegmatite	N. J. Highlands	x	x								600	-20.1	x		1	Buddington and Lindsley (1964)

anorthosite complex in which he also reports T = 620°C and $f_{O_2} = 10^{-19.5}$ for an oxide-rich gabbro. In the sequence from anorthosite (early), to gabbroic anorthosite, to Fe-Ti oxide rich gabbro, to syenite (latest): Mt_{ss} become impoverished in V_2O_3/Fe_2O_3 by a factor of 0.5; and Ilm_{ss} become enriched in MnO/FeO by a factor of 4.3. These changes are associated with a progressive decrease in Hem_{ss} contents which are: Hem_{34} (anorthosite) to Hem_6 (syenites and gabbros). f_{H_2O} at 880°C = 2000 bars for the gabbro. (23) EMPA. The ranges in T and f_{O_2} are for initial and final conditions of crystallization. For the interval of T = 700-985°C the estimate of f_{H_2O} = 310 bars. (24) The value of T and f_{O_2} represents one of several values obtained from a study of the association of Mt_{ss} + apatite (see Tables on Magmatic Ore Deposits). (25) Textural discussion of the distribution of exsolved Usp_{ss} and of the oxidation to Ilm_{ss}. (26) Magnetite is present in two forms: (a) with exsolved Usp_{ss} + Ilm_{ss}; and (b) Ti-poor Mt_{ss} free of exsolution. Sphene is a selective replacement product and the authors consider that Fe-mobilization may lead to Cornwall type magnetite deposits. (27) Hornblende-biotite-qtz diorite contains a reversely magnetized Hem_{ss} with exsolved Ilm_{ss}; and a normally magnetized Mt. The Hem-Ilm_{ss} association is in the range Ilm_{5-10}; self-reversal although non-reproducible in the laboratory is considered to involve a negative charge interaction between a Ti-ordered constituent and a Ti-disordered constituent (i.e. within the Ilm-Hem exsolution array). (28) Syenites are within an alkaline complex (2111 my) and because these rocks have high blocking temperatures (500-650°C) the directions of magnetization are those at the time of emplacement. (29) The oxide assemblage is similar to that reported by Merrill and Gromme (1969) - note 27. The two rhombohedral constituents are Ilm_{90} and Ilm_{40} respectively in exsolution association; intrusion T = 800°C.

Table Hg-12 (3)

ACID-INTERMEDIATE INTRUSIVE SUITES

Rock Type	Locality	Primary Mineralogy			Oxidation Mineralogy						$T^{o}C$	log 10 fO_2	Emphasis		Notes	Reference
		$Mt\text{-}Usp_{ss}$	$Ilm\text{-}Hem_{ss}$	Chr_{ss}	Ilm_{ss}	Pb_{ss}	Hem_{ss}	R	γ	OH			Pet	Mag		
Syenites	Montana	x	x								890-700 <765	-14.2 to -19.1 <-17.8	x		23	Nash and Wilkinson (1970)
Quartz-Syenites	N.J. Highlands Adirondacks	x	x								675 710	-18.4 -15.8			1	Buddington and Lindsley (1964)
Shonkinites	Montana	x	x								985-800 985-820 985-825 920-760	-12.3 to -16.3 -12.3 to -15.8 -12.3 to -16.2 -13.6 to -17.6	x		23	Nash and Wilkinson (1970)
Diorite	Quebec	x	x								850	-12	x		24	Philpotts (1967)
Syenites	India	x	x		x								x		25	Bose (1965)
Granophyres	Pennsylvania	x	x		x								x		26	Davidson and Wyllie (1968)
Diorite	California	x	x					x			595-640; 585			x	27	Merrill and Gromme (1969)
Syenite	N.W. Territories	x	x				x				500-650			x	28	Irving and McGlynn (1976)
Granite	Oklahoma	x	x		x		x				578-600			x		Vincenz et al. (1975)
Granite	Switzerland	x	x				x	x			560-600			x	29	Heller (1971)

Table Hg-13

BASIC INTRUSIVE SUITES

LAYERED COMPLEXES

BUSHVELD

Assemblages	Review	References
Chromite + Magnetite	General geological setting; chromite seams in critical zone ($\pm$ 3500' norite interlayered with pyroxenite and anorthosite); magnetite seams ($\pm$ 3400') in central and lower sections of upper zone (ferrodiorites, gabbros, troctolites, anorthosites).	Willemse (1969); Coertze (1970; De Villiers (1970); Groeneveld (1970)
Chromite	Detailed characterization of chromite deposits in critical zone.	Cameron (1963); Cameron and Desborough (1969); Cameron (1970); Cameron and Emerson (1959)
Chromite	Liquid immiscibility consideration. Postcumulus and subsolidus equilibration of chromite and coexisting silicates.	McDonald (1959); Cameron (1975); Cameron (1969)
Titanian chromites	Critical zone and Merensky reef; exsolution of ilmenite.	Frankel (1942); Cameron and Glover (1973); Legg (1969)
Magnetite-ulvöspinel-hercynite-ilmenite; Vanadiferous magnetite	Monomineralic adcumulate magnetite closely associated with anorthosite and pyroxenites. Magnetite in seams and pegmatoidal plugs distinguished on V_2O_5 content. Antipathetic relationship between V_2O_5 and TiO_2: lowermost seam (V_2O_5=2%; TiO_2=14%); uppermost seam (V_2O_5=0.3%; TiO_2=18-20%). Extensive exsolution of $Usp\text{-}Mt_{ss}$.	Molyneux (1970); Willemse (1969)
Olivine-apatite magnetites	Correlated with main zone; apatite is normally restricted to upper zone. c.f. Table Hg-19 Mt-Ap associations.	Grobler and Whitfield (1970)
Titaniferous magnetite + ilmenite	Upper zone. Cumulus in upper sections, pegmatitic at lower levels.	Molyneux (1972)
Chromite + anorthite + bronzite	Zr-cell determination of T^oC and fO_2 for the F-unit (first appearance of cumulus plagioclase). $T = 1085 \pm 20^oC$; $fO_2 = 10^{-11.85\pm0.4}$ atms.	Flynn et al. (1972)

Table Hg-13 (2)

BASIC INSTRUSIVE SUITES

SKAERGAARD

Assemblages	Review	References
Fe-Ti oxide rich horizons	General geological setting; Fe-Ti oxide segregations in upper sections of hypersthene olivine gabbro horizon of layered series.	Wager and Deer (1939) Wager and Brown (1968)
Ulvöspinel-magnetite-hercynite-ilmenite	Mineral petrography, chemistry and modes of oxides in hypersthene olivine gabbro; middle gabbro (olivine free); hortonolite ferrogabbro; ferro-hortonolite ferrogabbro; and fayalite ferrogabbro. Modes for Mt and Ilm vary systematically: from Fe-Ti rich band (36.3 and 13.8% vol. respectively) in hyp-ol-gabbro, to 1.2 and 4.7% vol. respectively in fayalite ferrogabbro. V_2O_3 in Usp-Mt_{ss} is constant between hyp-ol-gabbro and middle gabbro (85% solidified) and decreases to zero at 95% solidified in hort-ferrogabbro. Al_2O_3 (0.17 wt%) in Ilm-Hem_{ss} in hyp-ol-gabbro (69% solidified) decreases to zero in middle gabbro.	Vincent and Phillips (1954)
Ulvöspinel-magnetite-hercynite-ilmenite	Mineral chemistry; petrography; postulated Usp-Mt_{ss} solvi and miscibility gap; magnetic property measurements.	Vincent (1960); Chevellier et al. (1954); Vincent et al. (1957); Wright (1959); Wright (1969)
Usp-Mt_{ss} + Ilm-Hem_{ss}	T°C / fO_2: Fayalite ferrogabbro 890 -13.2 Ferrohortonolite ferrogabbro 760 -16.4 Hortonolite ferrogabbro 860 -14.6 Middle gabbro 850 -13.7 Hypersthene olivine gabbro 870 -13.7	Buddington and Lindsley (1964)
Ilm_{ss}-Opx	Myrmekitic intergrowth of Ilm_{ss} ($Ilm_{52.7}Geik_{40.6}Hem_{6.7}$) - Opx($En_{68.0}Fs_{29.8}Wo_{2.2}$); considered to have formed by subsolidus oxidation of Usp-Mt_{ss} inclusion in Opx.	Haselton and Nash (1975)

Table Hg-13 (3)

BASIC INTRUSIVE SUITES

STILLWATER

Assemblages	Review	References
Chromites	General geological setting; differentiation sequence of upper quartz monzonite, intermediate banded zone (gabbro + anorthosite) and lower ultramafic zone divided into an upper bronzitite zone and a lowermost peridotite zone (principally bronzitite, harzburgite and chromite). Division of ultramafic chromitite layer into 15 cyclic units.	Jones et al. (1960); Hess (1960); Jackson (1961); Jackson (1963)
$Chromite_{ss}$	Liquid immiscibility considerations; chemical petrography and variations in Fe:Mg ratios of coexisting olivine + chromite. Mg and Fe^{2+} in olivine and Cr, Al and Fe^{3+} in chromites are related to modal abundances of both minerals. Thermodynamic evaluation and suggestion for olivine-chromite geothermometer. Footwall to hangingwall temperature variations: 1300-900°C respectively. Chromite in marginal facies contains higher Fe^{3+} contents than $FeCr_2O_4$ towards the interior of intrusion.	Jackson (1966); Jackson (1969); Jackson (1970)
Chromite-ferritchromit	$Chromite_{ss}$ oxidation - genetically related to chlorite formation. See Table Hg-20 for summary of chromite → ferritchromit trends.	Beeson and Jackson (1969)
Chromite + olivine	Zr-cell determination of T^{o}C and fO_2 for the G-Zone. Data yield an intersection at T = 993°C and fO_2 = -13.6 atms. for Ol + Chr.	Sato (1972)

Table Hg-13 (4)

BASIC INTRUSIVE SUITES

KIGLAPAIT, LABRADOR

Assemblage	Review	References
Fe-Ti Oxides	General geological setting: intrusion is elliptical in plan and funnel-shaped in section with a total exposed thickness of 7800m; division into a Lower Border Zone (700m) and an Upper Border Zone (400m) with a 6700m layered series sandwiched between these two contact zones. The layered series is divided into a Lower Zone of cumulate troctolites, and an Upper Zone which is divided into 6 subunits, according to the compositions and nature of the cumulus phases (plagioclase, augite, olivine). Demarcation of the Lower Zone-Upper Zone contact is the onset of cumulus augite at the 84% solidified level. Cumulus oxides appear at the 88.6% level and in layers between 92-94% solidified. The main ore band (0.5m thick) in association with apatite is at the 93.5% solidified level above which oxide abundances fall off rapidly and essentially to zero at the 100% solidified level. Lower Zone oxides, and oxides in the Upper Zone up to 88.6% solidified, are Usp_{15-20} and Ilm_{4-6}. At 88.6% solidified Usp_{55-65} is present with extensive exsolution of $Mt-Usp_{ss}$. Maximum TiO_2 contents of titanomagnetites are reached at the 93% solidified level with Usp_{66} and $Ilm_{94.6}$. Oxides within the Upper Border Zone range from Usp_{19}-Usp_{55}, and Ilm_{98}-Ilm_{96}. T^oC and fO_2 values fall into two groups: silicate-rich rocks $\leq 740^oC$ and $10^{-15.8}$ bars; and oxide-rich layers $\leq 815^oC$ and $10^{-15.3}$ bars. Highest values are in silicate-free bands; $810-965^oC$ and $10^{-15.5}$ to $10^{-11.9}$ bars. Low group is close to the FMQ buffer curve and the high group falls between the FMQ and WM buffers.	Morse (1969); Morse and Stoiber (preprint)

Layered series Upper Zone

Cumulus	% Solidified	Zone	T^oC	log 10 fO_2
augite	85.8	UZa	580-655	-21.5 to -18.7
	86.0	UZa	590-615	-21.0 to -19.5
titanomagnetite	92.0	UZb	815-832	-14.7 to -14.4
	93.4	UZb	680-815	-19.5 to -15.3
	93.5	UZb	810-965	-15.5 to -11.9
apatite	94.1	UZc	680-840	-20.0 to -15.0
antiperthite	97.0	UZd	615-740	-19.3 to -15.8
pyrrhotite	98.25	UZe	600	-20
mesoperthite	99.94	UZf	575-590	-22.4 to -21.2
	99.98	UZf	570-580	-20.7 to -20.5
	Upper Border Zone		620-865	-20.0 to -13.8

Table Hg-13 (5)

BASIC INTRUSIVE SUITES

MUSKOX

Assemblages	Review	References
Magnetite-ilmenite-chromite	General geological setting; strongly differentiated layered series from uppermost granophyres and mafic granophyres, to granophyre-rich gabbro, gabbro and olivine gabbro, to feldspathic peridotites, peridotites, and dunites; an intermediate marginal zone ranging from peridotite to bronzite gabbro; to the lowermost zone of feeder dikes (picrites to bronzite gabbros). Disseminated chromite throughout olivine-rich units (1-3%); two chromite seams near top of main ultramafic section of layered series in units between cryptic bands of orthopyroxenite and peridotite. Mt + Ilm in marginal zone bronzite gabbro and in feeder dike; lower Mt-rich zone and an upper Ilm + Mt horizon are in sequence in the upper granophyre-bearing gabbro.	Smith and Kapp (1963); Findley and Smith (1965); Irvine and Smith (1967); Irvine and Smith (1969)
Chromite	Crystallization sequences; phase petrography, mineral chemistry and thermodynamics of chromite formation and of chromite-silicate crystallization interactions.	Irvine (1965); Irvine (1966); Irvine (1975)

GREAT DIKE

Assemblage	Review	References
Chromite	General geological setting; layered sequence, cryptic variation and differentiation; theories of in-situ crystal fractionation vs. multiple intrusions. Three major units: (1) Gabbroic; (2) pyroxenitic, picrite and harzburgite, chromite seam 1, harzburgite, chromite seam 2; (3) pyroxenitic suite. Compositional variations in chromites, petrography (see Table Hg-19).	Lightfoot (1940); worst (1960); Worst (1964); Hess (1950); Bichan (1969); Hughes (1970)

Table Hg-13 (6)

BASIC INTRUSIVE SUITES

KAP EDVARD HOLM, EAST GREENLAND

Assemblage	Review	References
Fe-Ti Oxides	General geological setting; 7500m layered sequence divided into a Lower layered series (plagioclase, augite, olivine cumulates) which has a thickness of 3800m and an Upper layered series. The distinction between these series is compositional rather than in the nature of the cumulates. The Upper series is further divided into a cumulate sequence of plagioclase, augite, olivine and a capping of plagioclase-augite cumulates (1200m). Evidence for magma replenishment; profusion of xenoliths of layered cumulates in Upper layered series. Cumulus and intercumulus oxides are more abundant in the Upper layered series. Subsolidus oxidation of $Usp\text{-}Mt_{ss}$; oxidation of Ilm_{ss} to rutile, and reduction of Ilm_{ss} to Usp_{ss}. Higher levels of the intrusion equilibrated at lower T^oC with the upward diffusion of water along a vertical temperature gradient.	Abbott (1962); Deer and Abbott (1964); Wager and Brown (1968); Elsdon (1972)

	Strat. Ht. (meters)	Assemblage	T^oC	log 10 fO_2
	6950	plag-aug-oxide-orthocumulate	600	-19.2
	6750	do	620	-19.0
Upper	6700	do	* 660	-18.0
	6650	do	610	-17.8
	5100	do	* 580	-20.5
	3150	do	680	-16.0
Lower	2950	do	700	-16.5
	2950	gabbro pegmatite	* 720	-15.0

* Oxidized samples.

Table Hg-13 (7)

BASIC INTRUSIVE SUITES

SOMERSET DAM, S.E. QUEENSLAND

Assemblage	Review	References
Fe-Ti Oxides	General geological setting; Triassic subvolcanic magma chamber; layered sequence consists of 20 saucer-shaped conformable layers, 500m in thickness. Layering in repetitive sequences: troctolite, olivine gabbro, ferrigabbro, and leucogabbro. Relative proportions of $Mt\text{-}Usp_{ss}$ peak in ferrigabbro zones, whereas $Ilm\text{-}Hem_{ss}$ peak in leucogabbroic horizons. Minor elements Al_2O_3, Cr_2O_3, V_2O_3, NiO and CoO decrease systematically from the base to the top of each zone (troctolite, olivine gabbro, ferrigabbro and leucogabbro) in both Usp_{ss} and Ilm_{ss}. Fe_2O_3 and MgO show positive correlations with stratigraphic height and peak in the leucogabbro zones. Crystallization and subsolidus equilibration is parallel to and approximately coincident with the NNO buffer.	Mathison (1967); Mathison (1975)

Zone	Rock Type	T°C	log 10 fO_2	Zone	Rock Type	T°C	log 10 fO_2
1	Olivine gabbro	660	-16.8	4	Troctolite	640-680	-18.0 to -16.7
	Ferrigabbro	695	-16.7		Olivine gabbro	655-660	-17.3 to -17.4
	Leucogabbro	550	-21.0		Ferrigabbro	670-685	-17.8 to -17.4
2	Troctolite	670-715	-17.3 to -15.5		Leucogabbro	640-675	-18.4 to -17.5
	Olivine Gabbro	715-730	-16.1 to -15.9	5	Troctolite	640	-18.0
	Ferrigabbro	695-710	-16.5 to -16.3		Ferrigabbro	720	-16.2
	Leucogabbro	640	-17.4		Leucogabbro	650	-18.1
3	Troctolite	650-670	-16.5 to -15.8	6	Troctolite	640	-18.8
	Olivine gabbro	650	-16.9		Olivine gabbro	710	-16.7
	Leucogabbro	630-645	-17.5		Ferrigabbro	700	-16.3
					Leucogabbro	645	-18.2

Table Hg-13 (8)

BASIC INTRUSIVE SUITES

RHUM

Assemblages	Review	References
Chrome spinel$_{ss}$	Layered sequence (7000 ft) characterized by four cumulus minerals: olivine, plagioclase, clinopyroxene, chrome spinel. Intercumulus liquid is broadly basaltic in composition. Cumulus chrome spinel is present at the junctions of a felspar cumulate forming the top of layer 11, and at the base of an olivine cumulate in overlying layer 12.	Brown (1956); Wager and Brown (1968)
Chromite-Picotite$_{ss}$	Systematic variations in spinel compositions for layer 11: Al_2O_3 variation is parabolic; Cr_2O_3 decreases in spinel cumulate; and MgO shows a slight increase. Interpretation is that cumulus chromites have reacted with ol + plag + interstitial liquid to form aluminous-rich picotites.	Henderson and Suddaby (1971)
Chrome spinels	Further comparison of layers 11 and 12 with additional horizon between units 7 and 8 (chrome spinel between upper olivine cumulate and lower plagioclase cumulate). Al-trend as above but newly defined Fe trend is shown to result by cumulus chromite + liquid only.	Henderson (1975)
Chrome spinels + Usp$_{ss}$ + Ilm$_{ss}$	Harrisite from Rhum pluton. Chrome spinels increase in Cr + Al at the expense of Fe^{3+} suggesting a higher temperature reaction of early formed chromites (anomalous) Ilm$_{ss}$ show high but variable MgO contents (0.38-7.69 wt%). Coexisting oxides in harrisite give T = 1040°C; $fO_2 = 10^{-10}$ atms.	Donaldson (1975)

Table Hg-13 (9)

BASIC INTRUSIVE SUITES

SUDBURY, ONTARIO

Assemblage	Review	References
Fe-Ti Oxides	General geological setting of nickel irruptive; gravity differentiate. South Range section (1200') consists of quartz-rich norite, norite, upper gabbro and capped by micropegmatite. The North Range section (7000') is based by mafic norite, followed by felsic norite and oxide-rich gabbro, and capped by micropegmatite. Mt_{ss} in Usp_{ss} and the Fe_2O_3 contents of Ilm_{ss} increase systematically from micropegmatite to mafic norite horizon in the North Range section. For the South Range section the lowermost quartz-rich norite also contains high Fe_3O_4 contents in Usp_{ss} and higher Fe_2O_3 contents in Ilm_{ss} than the overlying norite. Extensive exsolution and oxidation exsolution, hence T^oC and fO_2 values are estimates.	Naldrett et al. (1970); Gasparrini and Naldrett (1972)

South Range	T^oC	log 10 fO_2	North Range	T^oC	log 10 fO_2
Quartz-rich norite	600	-21	Mafic norite	775-825	-12.8 to -13.0
			felsic norite	850-890	-13.0 to -11.6
			oxide-rich gabbro	850	-14.2
			micropegmatite	850-1025	-14.5 to -11.0

Table Hg-13 (10)

ANORTHOSITE SUITES

Locality	Review	T°C	log 10 fO_2	References
U.S., Canada, Norway, India	Proposed distinctions between labradorite anorthosites and andesine anorthosites. Andesine types are demonstrably intrusive and the suggestion that these types result by partial fusion of labradorite types which are primordial continental crust. Distribution of oxides and assemblages. Labradorite anorthosites: Massive oxide deposits. Dominant $Usp\text{-}Mt_{ss}$ in the range of basalts and layered basic intrusions; exsolution of Mt_{ss} and Usp_{ss}; Ilm_{ss} restricted with rare exceptions to granule exsolution. Lower fO_2 values in comparison to andesine anorthosites; estimated value at 900°C = 10^{-14} atms. Andesine anorthosites: Massive oxide deposits. Dominant $Ilm\text{-}Hem_{ss}$; extensive exsolution of Ilm_{ss} and Hem_{ss}. Range Hem_{40} to Hem_{20}. With fractionation, Ilm_{ss} decrease from Hem_{35} to Hem_5 in gabbroic and syenitic facies. Higher fO_2 ($\approx 10^{-5}$) at 1100°C in comparison to basalts and to labradorite types.			Anderson and Morin (1970)
Ontario	Whitestone andesine anorthosite. Oxides predominantly in border facies. 80-90 modal % is $Ilm\text{-}Hem_{ss}$ (Ilm_{90} to Ilm_{50}); exsolution of compositions $>Ilm_{85}$ is continuous, whereas more $Hem\text{-}rich_{ss}$ exsolve discontinuously. Average compositions of exsolved Ilm_{ss} and Hem_{ss} lamellae are Ilm_{95} and Hem_{72} respectively. $Mt\text{-}Usp_{ss}$ = 5-22 modal %; composition = $Usp_{3.5}$.	1200	-5	Kretchsmar and McNutt (1971)
Quebec	Labrieville andesine anorthosite massif. Early to late lithologic sequence is anorthosite, gabbroic anorthosite, oxide-rich gabbro, syenite. Exsolution of Hem_{ss} and Ilm_{ss} from $Ilm\text{-}Hem_{ss}$; rare $Mt\text{-}Usp_{ss}$ reduction exsolution from $Ilm\text{-}Hem_{ss}$; oxidation exsolution of Ilm_{ss} from $Usp\text{-}Mt_{ss}$; oxide exsolution in plagioclase. T°C and fO_2 determinations not possible because assemblages fall into highly oxidized regions of thermo-barometric curves. Possible crystallization paths give approximate values in the range of 1400-975°C and 10^{-5} to 10^{-12} atms.			Anderson (1966)
	Gabbroic facies	620	-19.5	
	Syenetic facies	640	-18.8	

Table Hg-13 (11)

ANORTHOSITE SUITES

Locality	Review		T^oC	log 10 fO_2	References
Norway					
Bjerkrem-Sognal massif. Differentiated sequence with anorthosites at the base of lopolithic structure followed by leuconorites, norites, monzonorites, mangerites and quartz-mangerites. Rythmic layering in leuconorite horizon is divided into 5 units. Coexisting oxides; homogeneous Usp-Mt_{ss} and Ilm-Hem_{ss}; exsolution of Hem_{ss} in Ilm_{ss} in grains when Hem_{ss} exceeds 7-9 mole %; exsolution of Usp_{ss} and Zn-Al-pleonaste in Usp-Mt_{ss}; oxidation exsolution of Usp-Mt_{ss} to form trellis intergrowths of Ilm-Hem_{ss}. On a large scale Hem_{ss} decrease from base to top, whereas the TiO_2 contents in Usp_{ss} increase. The base is characterized by Ilm-Hem_{ss} (Hem_{16-20}) alone and by exsolution; the top is characterized by homogeneous Ilm_{ss} (Hem_2). Magnetites are homogeneous at the top (<2% TiO_2) and Usp_{ss} exsolution is restricted to the base of the intrusion. V_2O_3 correlates positively with Fe_2O_3 in Mt_{ss} and in Ilm_{ss}; MnO also correlates positively with FeO in Ilm_{ss} and with TiO_2 in Mt_{ss}.					Duchesne (1972)
	Layered Leuconoritic Horizon				
Rhythm 1	pyroxene-ilmenite-anorthosite	$Ilm_{16.6}$-$Ilm_{19.0}$			
	leuconorite	$Ilm_{15.5}$-$Ilm_{19.6}$			
Rhythm 2	ilmenite-anorthosite	$Ilm_{15.5}$			
	leuconorite	$Ilm_{14.5}$			
Rhythm 3	pyroxene-ilmenite-anorthosite	$Ilm_{15.5}$			
	leuconorite		550	-21	
	norite		550	-21	
	melanocratic norite		550	-21	
	leuconorite		500-550	-22	
	norite		500-550	-22	
	norite		550-600	-20 to -21	
Rhythm 4	pyroxene-ilmenite-anorthosite	$Hem_{19.0}$			
	olivine leuconorite		550-600	-20	
	leuconorite	$Hem_{16.7}$			
	norite and leuconorite		500	-21	
	banded leuconorite		550-600	-21	
	pyroxenite		600-660	-21 to -19	
	norite		600	-20	

Table Hg-13 (12)

ANORTHOSITE SUITES

Locality	Review	T^oC	log 10 fO_2	References
Rhythm 5	leuconorite	500-550	-22 to -21	
	norite	620-660	-20.2 to -19	
	Monzonoritic and Mangeritic Suites			
	monzonorites	630-660	-19.2 to -20.0	
	peridotite	670	-19.2	
	mangeromonzonorites	600-650	-22.0 to -19.7	
	olivine mangerite	610-640	-21.5 to -20.5	
	olivine quartz mangerite	650-670	-19.5 to -20.0	
South Africa Limpopo belt.	Anorthositic suite contains chromite, chromian magnetite and ilmenite.			Horr et al. (1975)
	Magnetic Studies: Allard Lake, Quebec. Exsolution in Ilm-Hem_{ss}. Self-reversal and bipolarity in single specimens.			Carmichael (19610; Carmichael (1962); Carmichael (1964); Hargraves (1959); Hargraves (1962); Hargraves and Burt (1967)
	Michikamau, Labrador. Evidence for single domain magnetite in exsolved oxides from plagioclase.			Murthy et al. (1971)
	Jimberlana, Australia. Single domain magnetite in oxidized Mt-Usp_{ss} in oxidation exsolved crystals in norite.			McClay (1974)

Table Hg-14

MAFIC AND ULTRAMAFIC SUITES

Locality / Review	References
Dun Mt, Red Mt., and Redhills, New Zealand Layering of variable thicknesses between dunite and herzburgite. Spinels are picotites which exhibit minor compositional variations among dunites, harzburgites, massive chromitites, and olivine chromitites. $Al_2O_3 \approx 21$ wt%; Cr_2O_3 = 38-45 wt% and MgO = 11-15 wt%. Challis notes that the Cr contents of chromites in nodules are lower than those in intrusive peridotites. The New Zealand ultramafic plutons are considered to be derived by gravitational differentiation from a tholeiitic magma.	Challis (1965)
Red Mt. - Del Puerto ultramafic mass, California This body is composed of partially serpentinized harzburgite, dunite and clinopyroxenite. Olivine compositions vary between Fo_{84}-Fo_{91} for dunite and harzburgite whereas Ol in the clinopyroxenite is Fo_{75}. Opx is constant at En_{90} but shows variable Al_2O_3 contents (0.7-2.8 wt%). Chromites in the dunites and harzburgites exhibit highly variable Al_2O_3 and Cr_2O_3 contents (Al_2O_3 = 7.4-36.2 wt%; Cr_2O_3 = 27.9-56.3 wt%) whereas FeO, MgO and Fe_2O_3 are relatively constant. Comparison of rock and primary mineral compositions suggests that serpentinization occurred with minimal changes in chemistry thereby implying an appreciable volume increase.	Himmelberg and Coleman (1968)
Vulcan Peak, S. W. Oregon This locality forms a part of the Josephine ultramafic complex and is composed of partially serpentinized foliated harzburgite with 10% dunite in concordant and discordant layers. Chromian spinel concentrations which are restricted to the dunite have the highest $Mg/Mg+Fe^{2+}$ ratios and coexist with the most magnesian-rich olivines.	Himmelberg and Loney (1973)

Oxide	Harzburgite		Dunite		Chromitite	
Cr_2O_3	47.8	44.3	53.8	35.2	46.4	42.3
Al_2O_3	21.9	25.7	13.3	27.9	22.4	28.2
Fe_2O_3	0.6	0.4	3.7	6.7	2.7	1.2
FeO	16.4	15.5	17.6	17.5	13.4	12.1
MgO	12.0	13.1	10.0	11.9	14.1	15.7
MnO	0.26	0.22	0.35	0.27	0.21	0.17
TiO_2	0.05	0.08	0.48	0.15	0.11	0.07
Total	99.01	99.30	99.23	99.62	99.32	99.74

Chromites are compositionally heterogeneous but Ol-Chr pairs yield T = 915-1365°C and an average of 1150°C. The peridotite may have originated as a refractory residue during partial fusion or alternatively by crystallization of an ultramafic or picritic magma.

Table Hg-14 (2)

MAFIC AND ULTRAMAFIC SUITES

Locality / Review	References
Mt. Albert, Gaspe, Quebec, Canada A semi-circular intrusion (6x3 miles) composed predominantly of serpentinized dunite and peridotite with associated harzburgite and pyroxenites. Chromites vary in Cr_2O_3 contents from 35.1 - 58.4 wt%; Al_2O_3 = 10.6 - 31.3 wt%; MgO = 3.4 - 15.9 wt%; FeO = 12.7 - 28.3 wt%; Fe_2O_3 = 0 - 8.1 wt%; TiO_2 < 1 wt%; and NiO < 0.5 wt%. The intrusion is zoned with respect to chromite compositions: higher Cr and Fe chromites occur near the margins of the intrusion, and higher Al_2O_3 and MgO chromites towards the interior of the complex. The marginal facies suggest T < 900°C and the central-most region T = 940-1100°C; P varies between 10 and 30 Kb. The zonal variation is related broadly to the degree of serpentinization and deformation, and marginal Opx and Ol are more magnesian; this distribution results from high fO_2 and P_{H2O}, $Fe^{2+} \rightarrow Fe^{3+}$ and the formation of high Fe-Cr spinels and high Mg silicates.	MacGregor and Smith (1963) MacGregor and Basu (1976)
General; Duke Island and Union Bay, Alaska; Tulameen, British Columbia Chromian spinel as a petrogenetic indicator: a theoretical development of expressions which permit contouring of the multicomponent spinel prism according to (a) spinel compositions which may exist with Ol or Opx having specific MgO/FeO ratios; (b) the activity ratios of Cr/Al, Cr/Fe^{3+} and Al/Fe^{3+} for spinel solid solutions forming under a variety of geological conditions; and (c) fO_2 estimates for spinels coexisting with Ol + Opx. Application of the above theoretical treatment to podiform and stratiform chromites, to mafic inclusion suites in extrusives, and to ultramafic intrusive bodies. Comparative compilation shows that chromite, in general, forms simultaneously with Ol and that their crystallization is terminated by a peritectic which yield Opx. Mg/Fe ratios between spinels and mafic silicates correlate and appear to be T°C sensitive (see below, Loney et al. , 1971). Chromitites in stratiform intrusions exhibit a range of fO_2 conditions which contrast with Alpine-type podiform chromites which have a narrow and low fO_2 range. Data for massive veins and pods from Duke Island, Union Bay and Tulameen have high Fe_2O_3 contents (20.8-32.4 wt%) and widely variable ranges in Al_2O_3 (6.9-15.8 wt%), and Cr_2O_3(25.2-43.9 wt%); MgO is restricted and falls between 7.1-10.0 wt%. The $Fe^{3+}/(Fe^{3+}+Cr+Al)$ ratio for this chromite suite is approximately a factor of 2 larger than most other Alpine-types and although fO_2 is inferred, extremely low a_{SiO_2} is favored.	Irvine (1965 and 1967)
S. W. Oregon Peridotites contain the high pressure assemblage forsterite, aluminous enstatite and diopside, and spinel. Sp compositions from three localities have virtually identical Cr/Cr+Al ratios (0.137-0.189) with $Al_2O_3 \simeq$ 52 wt% and $Cr_2O_3 \simeq$ 12.5-17.5 wt%. For a fourth locality Al_2O_3 = 36.54 wt%; Cr_2O_3 = 27.85 wt% and Cr/Cr+Al = 0.338. This latter spinel is representative of recrystallization at lower pressures, in the plagioclase lherzolite field, in contrast to the high Al_2O_3 spinels which are indicative of crystallization in the spinel lherzolite field. Estimates of T for recrystallization are in the range 1100-1200°C over a range of P=19 to 5 Kb.	Medaris (1972)

Table Hg-14 (3)

MAFIC AND ULTRAMAFIC SUITES

Locality / Review	References
Lizard Area, Cornwall, U.K. The Lizard peridotite has a coarse grained primary core, within a cataclastic finely foliated and recrystallized marginal zone: the primary assemblage is Fo_{89}, Al-enstatite, Al-diopside and chromian-spinel (Al_2O_3 = 47.58; Cr_2O_3 = 17.99; Fe_2O_3 = 5.07; FeO = 10.13; MgO = 19.18); the recrystallized assemblage yields Ol + En + Di (pyroxene with normal Al_2O_3 content), plagioclase and chromite. The differences in the primary and recrystallized assemblages (i.e. Al-pyroxenes + spinel → Ol + plagioclase) are considered to be due to initial differences in P_{TOTAL} at the time of intrusion. $(MgAl_2O_4 + FeCr_2O_4) + CaMgSi_2O_6 + 2MgSiO_3 = CaAl_2Si_2O_8 + 2Mg_2SiO_4 + FeCr_2O_4$ Sp_{ss} Di En An Fo Chr Peridotite core → Marginal shell	Green (1964)
Connemara, Ireland The Dawros peridotite is a layered ultramafic complex which contains spinels of two contrasting compositions. Primary spinels are orbicularly textured and resemble podiform types in Alpine peridotites, with the following composition: Cr_2O_3 = 46.40; Al_2O_3 = 13.10; Fe_2O_3 = 10.0; FeO = 20.60; MgO = 8.40; MnO = 0.31; TiO_2 = 0.53; and V_2O_5 = 0.25 wt%. Spinels considered to be of secondary, recrystallization origin are highly aluminous with the following composition: Cr_2O_3 = 9.30; Al_2O_3 = 51.30; Fe_2O_3 = 6.83; FeO = 18.87; MgO = 14.0. The primary Cr-rich spinel occurs principally with cumulate olivine + Cpx + Opx, whereas the recrystallized spinel is present in association with either Ol + Cpx or with bronzite (En_{72}; Al_2O_3 = 6.02 wt%). This distribution differs markedly from those reported by Green (1964) - see above.	Rothstein (1972)
Burro Mountain, California, U.S.A. An Alpine type peridotite of which approximately 60% is foliated harzburgite, and the remainder is dunite which is present as dikes, sills and irregular pods in the harzburgite. Harzburgite: Ol = Fo_{91}; Opx = En_{90}; Cpx = $Ca_{47.0}Mg_{50}Fe_{3.0}$ (Al_2O_3 = 1.3-3.0 wt%); Chr = $Cr/(Cr+Al+Fe^{3+})$ = 0.37-0.55. Dunite: Ol = Fo_{91}-$Fo_{92.7}$; Chr = $Cr/(Cr+Al+Fe^{3+})$ = 0.30-0.75. Chrome-spinel compositions from the harzburgite are distinct from those in the dunite; although the dunite (in sills, dikes and pods) is litholigically similar, compositions vary as follows and are divisible into three types: Type 1 are pods; Type 2 are dikes and sills; and Type 3 are sills.	Loney et al. (1971)

Table Hg-14 (4)

Loney et al. (1971) cont.

Oxide	Harzburgite		Dunites Type 1		Dunites Type 2		Dunites Type 3		Ortho-pyroxenite
Cr_2O_3	*32.7 -	45.7	*50.1 -	58.0	*44.7 -	47.6	*33.9 -	26.9	49.7
Al_2O_3	37.4	24.8	20.6	10.4	19.1	23.9	35.6	41.2	17.6
Fe_2O_3	0.1	0.0	0.6	3.6	6.2	0.5	0.6	2.3	2.7
FeO	13.8	16.6	15.2	17.0	16.2	12.9	12.1	11.3	16.9
MgO	15.6	12.3	12.8	10.5	11.9	14.7	16.2	17.4	11.4
MnO	0.14	0.19	0.19	0.30	-	0.17	0.14	0.14	-
TiO_2	0.08	0.12	0.16	0.20	-	0.17	0.14	0.15	-
Total	99.8	99.7	99.6	100.0	98.1	99.9	98.7	99.4	98.3
$Cr/\Sigma R^{3+}$	0.37	0.55	0.62	0.75	0.56	0.57	0.39	0.30	0.63

*Choice of analyses based on min. and max. values for Cr_2O_3.

The differences among spinel compositions are considered to be due to differences in the initial bulk composition of the rock; differences in texture (viz. subhedral vs anhedral) are probably due to differences in crystallization, with Cr-rich spinels crystallizing early and Al-rich spinels crystallizing later with respect to the mafic silicates. Utilization of the exchange reaction between olivine and chromian spinel (Irvine, 1965; Jackson, 1969):
$Fe^{2+}Si_{0.5}O_2 + Mg(Cr_\alpha Al_\beta Fe^{3+}_\delta)_2O_4 = MgSi_{0.5}O_2 + Fe^{2+}(Cr_\alpha Al_\beta Fe^{3+}_\delta)_2O_4$ where $\alpha+\beta+\delta=1=R^{3+}$ cats; and the thermodynamic equilibrium distribution coefficient ($\bar{K}$) for the reaction between olivine and spinel

$$K = \frac{X^{Ol}_{Mg} X^{Chr}_{Fe^{2+}}}{X^{O}_{Fe^{2+}} + X^{Chr}_{Mg}}$$

, where X^{Ol}_{Mg} and $X^{Ol}_{Fe^{2+}}$ are mole fractions of the end members in Ol, and X^{Chr}_{Mg} and $X^{Chr}_{Fe^{2+}}$ are the fractions of divalent cations in spinel; T is calculated from $K_{\mathbf{reaction}} = \exp(-\Delta G^o_{reaction}/RT)$. For the Burro Mt. complex: spinel + Ol pairs yield T = 1098-1335°C with an average of approximately 1200°C.

Malaga Province, Spain

Oen et al. (1973)

The harzburgite ultramafic massif contains ore lenses and veinlets of chromite + Ni-sulfides which are closely associated with pyroxenitic dikes. The spinels are distinctive in being Fe^{2+}-rich aluminous chromites with exceptionally high V_2O_3 contents (1.8-2.9 wt%) and high ZnO contents (0.5-1.0 wt%). Mg-cordierite (11 wt% MgO) + brown biotite are associated with some of the oxide-bearing dikes and the assemblage is considered to have formed at magmatic temperatures (T=1200°C), at low values of fO_2, and at P=5 Kb. Chromite textures are of two forms: (1) polygonally-textured; and (2) skeletal; these show differences in the range of spinel mineral chemistry which differ also from those chromites which are associated with cordierite.

Table Hg-14 (5)

Oen et al. (1973) cont.

Oxide	Polygonal		Skeletal		Cordierite-bearing	
Cr_2O_3	*28.6 -	43.2	*40.4 -	43.6	*25.8 -	33.8
Al_2O_3	33.8	19.0	20.7	18.0	34.3	25.5
Fe_2O_3	1.3	2.1	3.2	1.8	1.9	3.0
V_2O_3	1.9	2.1	1.8	1.9	2.0	2.4
TiO_2	0.2	0.2	0.1	0.6	-	-
FeO	26.2	28.0	26.9	28.7	30.0	30.4
MgO	6.5	3.6	4.4	3.5	4.0	2.9
MnO	0.3	0.5	0.5	0.5	0.4	0.4
ZnO	0.9	1.0	1.0	0.9	0.7	0.5
NiO	0.2	0.4	0.1	0.2	-	-
CoO	0.2	0.2	0.2	0.2	-	-
Total	100.3	100.3	99.5	100.0	99.3	99.1

* Choice of analyses based on min. and max. values for Cr_2O_3.

Table Hg-15

KIMBERLITES

Standard Reference Sources

The petrology of kimberlites in the USSR; ultramafic and related rocks, worldwide distribution; Lesotho kimberlites; Advances in kimberlite petrogenesis; Petrology, petrogenesis and distribution of African kimberlites; Kimberlites - a critical but constructive reappraisal.

Frantsesson (1970); Wyllie (1967); Nixon et al. (1963); Nixon (1973); Dawson (1962); Dawson (1971); Williams (1932); Ahrens et al. (1975); Mitchell (1970)

KIMBERLITE GROUNDMASS STUDIES

USSR

Ilmenite$_{ss}$ + spinel$_{ss}$

An appraisal of groundmass (sensu stricto) ilmenite compositions, in distinction to possible xenocrystic, phenocrystic or megacrystic ilmenites, is difficult but the range in MgO contents = 5.6-14.13 wt%. Cr_2O_3 is uniformly low 1.37 wt% max. Spinels are chromian picotites and a typical composition from the Mir pipe is as follows: TiO_2 = 1.01; Al_2O_3 = 19.61; Cr_2O_3 = 45.19; Fe_2O_3 = 22.92; MgO = 13.04 wt%.

Frantsesson (1970)

Spinel$_{ss}$

Detailed comparison of spinel compositions in kimberlites and spinels present as diamond inclusions. The major distinctions are in Cr_2O_3, Al_2O_3, and TiO_2 contents: groundmass spinels are diverse, having: Cr_2O_3 = 10-64 wt%, Al_2O_3 = 1-40 wt% and TiO_2 = 1-10 wt%. Diamond included spinels have: Cr_2O_3 = 62-68 wt%; Al_2O_3 = 2-7 wt%; and TiO_2 = <1-0.75 wt%. It is concluded that analogues comparable to these ranges in composition exist neither in other ultrabasic suites nor in lunar basalts.

Sobolev et al. (1975)

Table Hg-15 (2)

MAFIC AND ULTRAMAFIC SUITES

KIMBERLITES

Locality / Review	References
Peuyuk Island, Canadian Arctic	
$Spinel_{ss}$ + Ilm_{ss}	
Phenocrysts or xenocrysts of magnesian-ilmenite, magnetite, chromian spinel and chromite in association with olivine + garnet + phlogopite + bronzite; groundmass assemblage consists of chromian spinel + Mt-Sp-Usp_{ss} + rutile + perovskite + pyrite + heazelwoodite + chalcopyrite in dolomite + calcite silicate matrix. Complex epitaxial zonations around oxides resulted from multiple immiscible liquids (silicate + carbonate + sulfide) with late stage liquids being highly enriched in TiO_2. Epitaxial zones of spinels of unusual composition on chromites have the following compositions: (1) $Mt_{36}Sp_{18}Usp_{43}Chr_3$; and (2) $Mt_{47}Sp_{19}Usp_{34}Chr_{0.1}$. Ilmenites exhibit core-mantle zonation with early crystallization of high Mg-Ilm_{ss} (13.5 wt%) and later formation of low Mg-Ilm_{ss} (10.7 wt%). Early chromites, magnetite, and Mg-ilmenites are considered to be pre-fluidization constituents; chromian-spinels and Usp-Mt_{ss}-bearing spinels are largely post-fluidization crystallization products terminating in rutile + perovskite.	Clarke and Mitchell (1975)
A follow-up study of the above and representing the most detailed characterization of spinel petrogenesis in kimberlites to date. Peuyuk intrusion is divisible into three distinct phases of crystallization: Phase A: Spinel + perovskite; epitaxial reaction mantles are absent; no evidence of an immiscible carbonate liquid. Phase B: complexly zoned oxides but no evidence of an immiscible carbonate liquid. Phase C: complexly zoned spinels; immiscible carbonate liquid evolved during crystallization. Spinel compositions for each of these three phases are as follows: Phase A = Ti-Mg-Al-chromite; Phase B = zonation from Ti-Mg-Al-chromite to Mg-Usp-magnetite; and Phase C = discrete crystals of Mg-Usp_{ss} to Usp-Mt_{ss} or mantles of this composition on cores of Ti-Mg-Al-chromite. (Oxide compositional data listed below). Spinel compositional trends may reflect decreasing fO_2 (10^{-19} to 10^{-22} bars) in response to falling T = 800-600°C. Post-fluidization yield Ti-Mg-Al-chromite in Phases A, B and C but Ti and Fe enrichment trends (in Phases B and C) were inhibited in Phase A possibly by rapid cooling.	Mitchell and Clarke (1976)
Moses Rock Kimberlite dike, San Juan Co., Utah	
$Spinel_{ss}$ + $Ilmenite_{ss}$	
Groundmass spinels yield compositions between $(Mg_{1.2}Fe_{0.8})(Al_{2.4}Cr_{1.6})O_8$ and $(Mg_{1.4}Fe_{0.6})(Al_{3.5}Cr_{0.5})O_8$; this variation is systematic for discrete crystals and Al and Mg, (36.2-52.3 wt%) and (14.4-17.8 wt%) are similar to those of fragments of spinel lherzolite; coexisting ilmenite contains 10.1 wt% MgO and other minor elements constitute <0.5 wt%.	McGetchin and Silver (1970)
Norris Lake, Eastern Tennessee	
Mt_{ss} + Ilm_{ss}	
The opaque mineral assemblage consists of Mg-Ilm (8.2-14.2 wt% MgO), magnetite (0.29 wt% MgO), and rutile in association with CaO-poor and Cr_2O_3-rich garnet; the matrix is largely serpentine and the assemblage is referred to as dewelite.	Meyer (1975)

Table Hg-15 (3)

MAFIC AND ULTRAMAFIC SUITES

KIMBERLITES

Locality / Review	References
Benfontein Kimberlite Sills, Kimberley, S. Africa $Spinel_{ss} + Ilm_{ss}$ In each of three kimberlite sills there is evidence for: multiple injections; magmatic sedimentation; cumulus horizons; in-situ differentiation; and pre-injection differentiation. The transporting intercumulus liquid was "warm" and carbonatitic in composition which has resulted in part in cumulus calcite and the suggestions that: (a) the cold (plastic paste) hypothesis may be invalid for kimberlite intrusions; and (b) there is a genetic link between kimberlites and carbonatites. Ilmenite compositions are characteristic of kimberlite associations with MgO=13.8-14.4 wt%. Spinel compositions may be classified as Mg-Al-ulvőspinels ($Usp_{38.5}$-$Usp_{42.3}$ $Sp_{13.8-15.0}$ with significant $MgFe_2^{3+}O_4$ = 5.4-13.0 mole %). Details of the oxide compositions, which like the Benfontein Sills are unique in many respects, are given below (Table Hg-20).	Dawson and Hawthorne (1973)
The Igwisi Hills extrusive "kimberlites", Tanzania The "kimberlites" are described as igneous conglomerates, characterized by the presence of ellipsoids of olivine. The matrix contains major amounts of carbonate, serpentine and complexly zoned spinels which range from cores of Mg-Al-titanomagnetites to mantles of titan-pleonaste ($FeAl_2O_4$-$MgAl_2O_4$); although these phases are Cr-poor, chromite cores with Mt_{ss} rims are also present. These latter spinels are more in keeping with spinel compositions which are included in ellipsoidal olivine, and all sets differ from those which form mantles on these olivines (see details of oxide compositions below). The association of Al-enstatite, Al-Cr-diopside, Cr-pyrope and Mg-phlogopite suggest an affinity to garnet peridotite xenoliths (e.g. Leshaine); rapid intrusion in a fluidal carbonatitic matrix perhaps prevented interaction of fragmented xenolithic material which would normally have resulted in a kimberlite (sensu stricto), rather than in the observed "kimberlitic igneous conglomerate".	Reid et al. (1975)
Premier Mine, S. Africa $Ilm_{ss} + Spinel_{ss}$ Exsolution of Mg-Cr-ulvöspinel (TiO_2=23.0; Al_2O_3=6.0; Cr_2O_3=10.0; FeO=51.0; MgO=7.0 wt%) from picroilmenite (MgO=15.5 wt%) along {0001} planes. Spinel compositions are comparable to primary spinels in Apollo 11 Lunar basalts.	Danchin and d'Orey (1972)

Table Hg-15 (4)

MAFIC AND ULTRAMAFIC SUITES

KIMBERLITES

Locality / Review	References
Kao and Pipe 200, Lesotho Ilm_{ss} A study of the petrochemistry of kimberlitic autoliths (spherical bodies, ~50 mm diam., containing a central core of a mineral or rock fragment encased by fine grained equigranular kimberlite) which are regarded as being more closely related , or representative of the "true" groundmass (matrix) compositions of kimberlites (i.e. in contrast to the chaotic array of brecciated fragments normally present) shows: (1) that ilmenite nuclei are less magnesian than ilmenite in the encased surrounding kimberlite (4.49-6.63 vs 8.70-15.57 wt% respectively); and (2) no systematic variations exist, either in major or minor elements, in ilmenites from zones proceeding outwards from the nucleus to the autolith boundary. The compositions of the relatively ferriferous ilmenite nuclei compare favorably with those in olivine-ilmenite nodules.	Ferguson et al. (1973)
Wesselton Mine, S. Africa, and Lesotho Kimberlites An extension of the above autolith study extended to include the compositions of 5 ilmenite nuclei and 70 autolith groundmass ilmenites. These data show conclusively that nucleus ilmenite (~6 wt% MgO) is decidedly less magnesian and less Fe^{3+}-rich than groundmass ilmenite which reaches a max. of 20 wt% MgO. The data suggest furthermore that the compositional differences reflect primary ilmenite nuclei of high pressure origin and late stage magmatic ilmenite of somewhat lower pressure origin.	Danchin et al. (1975)
The Upper Canada Mine; and Kimberley area, S. Africa A study of the effects and influences of zoning on the distribution coefficients of major and minor elements for the groundmass assemblage phlogopite + $spinel_{ss}$, shows that more meaningful (coherent) patterns emerge if adjacent mineral zones are considered (e.g. in spinels containing Chr_{ss} cores and $Cr\text{-}Usp_{ss}$ mantles, or in phlogopites which are zoned from Cr-phl to Ti-biotite to Cr-poor phl) rather than cogenetic mineral pairs.	Rimsaite (1971)

Table Hg-15 (5)

MAFIC AND ULTRAMAFIC SUITES

KIMBERLITES

Locality / Review	References
Liqhobong Pipe, Lesotho Ilm_{ss} + $spinel_{ss}$ Xenocrystic picroilmenites have optically well defined but geochemically disordered reaction mantles which are serially zoned in the following sequence from cores of unaltered ilmenite towards the silicate groundmass: zoned primary picroilmenite (cores = 11 wt%, margins = 16 wt% MgO); magnesioferrite-Usp_{ss} (15 wt% MgO); secondary picroilmenite (20 wt% MgO); titanian-ferrianpleonaste (22 wt% MgO); and perovskite (0.2 wt% MgO). The spinels are of an unusual composition and compare with those from the Benfontein Sills (see Dawson and Hawthorne, 1973 above). The alternating sequence of ilmenite and spinel, and terminal perovskite suggests episodic oxidation and partial reduction, high CO-CO_2 and Ti activity and an absence of equilibrium between and among phases in each of the respective mineral zones. Note:These mantles, in common with the autolith geochemistry, confirm the notion that late stage (and by inference, groundmass) ilmenites are likely to be more magnesian than their deep seated high pressure counterparts (Danchin et al., 1975).	Haggerty (1973)
Nqechane, Lipelaneng and Kao, Lesotho; Monastery Mine, S. Africa Ilm_{ss} + $Spinel_{ss}$ + Rutile + Armalcolite A study of the chemistry of mineral oxides in kimberlite groundmass with special reference to zonal trends in the spinel mineral group; these data are summarized in Fig. Hg-37, and zonation is considered to have resulted by solid-liquid or solid-gas interaction. Ilmenite mineral chemistry is varied as shown in Fig. Hg-41, and the first terrestrial occurrence of armalcolite, (FeMg)Ti_2O_5 is reported.	Haggerty (1975a)
S. Africa, West Africa, Lesotho and Labrador An analysis of correlative and non-correlative oxide parameters for spinels (439 analyses) and ilmenites (350 analyses) in groundmass kimberlite shows that within the spinel mineral group the expected divalent and trivalent substitutional parameters are satisfied within experimental limits. Unexpected results show in addition that R^{3+} vs Ti^{4+}, R^{2+} vs Ti^{4+} and R^{2+} vs R^{3+} are also correlative. These relationships are in general linear or correspond to simple hyperbolic curves. A similar set of relationships hold true for the ilmenites, with the notable exceptions that inter-pipe variations show a lack of uniform correlations for Mn vs Mg and for Mn vs Cr. In addition to linear and hyperbolic relationships the ilmenites are also characterized by parabolic relationships for the oxides : Ti vs Cr; Mg vs Cr; Fe^{3+}vs Cr and Al vs Cr.	Haggerty (1975b)

Table Hg-16

MAFIC AND ULTRAMAFIC SUITES

CARBONATITES

Review	References
Standard References Sources Descriptions of carbonatite complexes; experimental studies; and economic aspects. Alkalic carbonatite complexes; carbonatites and kimberlites; geology, petrology and mineralogy; geochemistry and economic geology; origins and worldwide distribution. Alkaline pyroxenites; the Oka carbonatite complex; comparisons of Swedish, African, and USSR kimberlites and related carbonatites; experimental studies. Carbonatites - a general review.	Tuttle and Gittins (1966); Heinrich (1966); Wyllie (1967); Pecora (1956)
Locality **USSR, Europe, N. America** A comparison of magnetite compositions in veins in basalts ("trap rock") with compositions of Mt_{ss} from alkalic rocks. The vein deposits (hydrothermal) are characterized by Mt_{ss} ranging from near stoichiometric Fe_3O_4 to ferroan-magnesioferrites (Mt-magnesioferrite$_{ss}$); these Mt_{ss} are low in TiO_2 (<1 wt%), contain between 1 and 7 wt% Al_2O_3 and <0.5 wt% MnO. These contrast with Mt_{ss} in alkalic rocks and carbonatites which generally have higher MgO (16-26 wt%), extremely variable Al_2O_3 (0.07-15.14 wt%) and higher MnO (1.82-8.46 wt%) contents (Table Hg-20).	Fleischer (1965)
Oka Carbonatite Complex, Quebec, Canada The oxide minerals at Oka are: magnetite, maghemite, hercynite, hematite, ilmenite, rutile, pyrochlore, thorian-pyrochlore, perovskite, latrappite, and periclase. Compositional variations reported for magnetites are included in the comparative ranges discussed above by Fleischer (1965).	Gold (1966)
Experimental T^o and fO_2 determination for olivine + magnetite + latrappite yields T = $710^o \pm 15^oC$; $fO_2 = 10^{-17.1} \pm 10^{-0.5}$ atms. Full details are discussed in Table Hg-20.	Friel and Ulmer (1974)
Central and Southern Africa Magnetite compositions from 8 localities show that TiO_2 = 0.39-1.17; MnO = 0.18-1.65; MgO = 0.07-2.07; and Al_2O_3 = 0.07-0.96 wt%. Zonal trends indicate a decrease in TiO_2, MnO and MgO contents from core to grain boundaries; no general trends are exhibited for Al or V; in some cases the edges are enriched in Al + V whereas in others there are zones which are impoverished in these elements. Exsolution of hercynite is reported, and oxidation exsolution of ilmenite is considered to have taken place between T=540-575°C and $fO_2=10^{-23}$ to 10^{-24} atms. Oxidation induces a strongly fractionated distribution of minor elements for MnO and MgO towards Ilm_{ss}, but the expected enrichment of Al_2O_3 in Mt_{ss} is not borne out.	Prins (1972)

MAFIC AND ULTRAMAFIC SUITES

CARBONATITES

Locality / Review	References
Fen Complex, Norway The dominant rock type of the Fen carbonatite complex is damkjernite, a lamprophyric ultrabasic rock which has chemical affinities to alnoites, monchiquites and ouachitites. The major dilutant in groundmass magnetites is Al_2O_3 (1.5-4.6 wt%); $MgO + MnO + Cr_2O_3 = < 1$ wt%. Coexisting ilmenites on the other hand contain significant MnO (4.7-6.5 wt%) and minor MgO (0.10-0.15 wt%). $Mt\text{-}Usp_{ss}$ are present as large crystals and as skeletal crystals. The former yield T=970-880°C and $fO_2=10^{-11}$ to 10^{-12} bars; the latter yield T=580°C and $fO_2=10^{-21}$ bars. Both values of T°C are considered to be minima. Based on Fe/Mg of groundmass mica in association with K-feldspar + Mt a maximum of T=1050°C is suggested. If the bulk rock is assumed to have crystallized between 900-1100°C calculated fH_2O ranges from 60 bars at 900°C ($fO_2=10^{-12}$ bars) to 150 bars at 1000°C ($fO_2=10^{-10.5}$ bars). Calculated values of a_{SiO_2} suggest that the damkjernite magma could have been in equilibrium with peridotite at T=1200-1250°C and P=15-25 Kb.	Griffen and Taylor (1975)
Saguenay River Valley, Quebec, Canada Dikes of carbonatitic-kimberlites contain $Mt\text{-}Usp_{ss}$ which are zoned from $Usp_{14}Mt_{86}$ (cores) to $Usp_{61}Mt_{39}$ (margins). These magnetites are distinctly aluminous (max. 6 wt% Al_2O_3), magnesian (max. 8 wt% MgO) and manganoan (max. 5 wt% MnO) with trace amounts of Cr_2O_3. For the kimberlites in the suite Fe/Fe+Mg increases from core to rim; these rocks contain groundmass phlogopite which is considered to have crystallized throughout the period of Mt_{ss} crystallization. For the carbonatitic-kimberlites the reverse relationship exists with Al increasing sharply as Fe/Fe+Mg decreases from core to margin; groundmass phlog. is absent. The carbonatite suites show both trends and again phlog. is absent. The behavior of Ti shows that the Usp_{ss} component increases sharply as Fe/Fe+Mg increases from core to margin. Picroilmenite and Cr-pyrope are absent and there is some similarity to the Benfontein Sills (Dawson and Hawthorne, 1973).	Gittens et al. (1975)
Palabora, S. Africa Fe-Ti oxides reported from the Palabora Complex are $Mt\text{-}Usp_{ss}$, Ilm_{ss} and Pb_{ss}. The distribution of Mt_{ss} is clearly demarcated as follows: the central core of the complex is carbonatite; Mt_{ss} is concentrated in the range of 15-30% by wt., and TiO_2 <1 wt%; the margins of the complex are characterized by foskorite (also referred to as phoscorite), an apatite + olivine + magnetite rock; magnetite may be present in concentrations of up to 50% by wt., and $TiO_2 \approx 4$ wt%. Other apatite + Mt_{ss} associated are listed in Table Hg-19.	Palabora Mining Co. Staff (1976)

Table Hg-16 (3)

MAFIC AND ULTRAMAFIC SUITES

CARBONATITES

Locality / Review	References
Kodal, Vestfold, Norway Mt_{ss} + Ilm_{ss} + apatite in jacupirangite dikes; associated with pyroxene + amphibole + mica. Modal analyses show that Mt_{ss} vary between 11 and 60% (by vol.); ilmenite between 5 and 15% (by vol.); and apatite between 9 and 24% (by vol.). Ilmenite compositions are significant inasmuch as both discrete crystals and oxidation exsolution lamellae contain high MgO (3.8-5.1 wt%) and MnO (2.6-3.2 wt%) contents. Origin appears to be consistent with the liquid immiscible hypothesis (Philpotts, 1967 - see Table Hg-19 for other associations).	Bergstøl (1972)
Kaiserstuhl, Germany Magnetites have reported compositions with MgO=8.6 wt% max., and Al_2O_3=6.6 wt% max. Other analyses as summarized by Fleischer (1965) are given below (Table Hg-20).	Wimmenauer (1966)
Bukusu, Eastern Uganda Magnetites ("rubble magnetite") contain <1 % TiO_2 from apatite soil deposits, and 0.1 wt% V_2O_3 (c.f. Bushveld=0.8 wt% V_2O_3); other deposits exhibit a wide range in composition from <1 wt% TiO_2 to 6-20 wt% TiO_2 in titanomagnetite + perovskite assemblages.	Deans (1966)
Magnet Cove, Arkansas, U.S.A. Spectrographic analyses for magnetites show that the jacupirangites, and carbonatites contain Mt_{ss} with MgO>10 wt%; TiO_2=1.0-3.6 wt%; and MnO=0.8-2.7 wt%. It is noteworthy that the melteigites which contain as much MgO as the jacupirangites (~7 wt% MgO) contain Mt_{ss} with only 0.2 wt% MgO. Other analyses as summarized by Fleischer (1965) are given below (Table Hg-20).	Erickson and Blade (1963)

OXIDE-BEARING XENOLITHS IN IGNEOUS ROCKS

Locality	Review	References
Worldwide distribution	Geological setting: continental, coastal, intra-oceanic. Host rocks: basaltic and alkali-basaltic suites. Inclusion types: megacrysts, garnet peridotites, peridotite and eclogitic suites. Oxide components: spinel series. Origins: exotic vs. cognate cumulates.	Forbes and Kuno (1967)
Disko Island, Greenland	Basalts. Plagioclase + graphite + corundum + spinel ($Sp_{78}Her_{22}$) + metallic Ni-Fe. Origin: spinel may have resulted from decomposition of incorporated shale under low fO_2.	Melson and Switzer (1966)
Armidale, NSW, Australia	Bananites. Megacrysts: Al-augite, Al-bronzite, kaersutite, anorthoclase, zircon, ferrian pleonaste ($Sp_{68.5}$ $Her_{22.2}$ $Mt_{7.8}$ $Usp_{1.5}$) and ilmenite (Ilm_{60} Hem_9 $Geik_{21}$). Composite megacrysts: clinopyroxene-spinel; clinopyroxene-olivine-spinel; clinopyroxene-kaersutite-spinel-ilmenite. Origin: high pressure crystallization products probably cognate with enclosing lavas.	Binns (1969)
Northeastern NSW, Australia	Nepheline trachybasalt. Megacrysts: clinopyroxene, olivine, anorthoclase, spinel ($Sp_{61.1}$ $Her_{29.3}$ $Mt_{7.5}$ $Usp_{2.1}$). Olivine nephelinite. Megacrysts: clinopyroxene, olivine, spinel ($Sp_{80.5}$ $Her_{16.0}$ $Mt_{2.4}$ $Usp_{0.6}$). Analcimite. Megacrysts: kaersutite, titanbiotite, anorthoclase, ilmenite, Mg-Al-titanomagnetite ($Usp_{52.7}$ $Mt_{26.7}$ $Her_{11.6}$ $Sp_{9.0}$). Analcime trachybasalt. Megacrysts: kaersutite, anorthoclase, Mg-Al-titanomagnetite ($Usp_{46.0}$ $Mt_{28.9}$ $Her_{9.1}$ $Sp_{16.0}$) Origin: Most megacrysts differ in composition from equivalent groundmass constituents. Megacrysts are considered to be cognate precipitates formed at pressures broadly equivalent to the crust-mantle boundary.	Binns et al. (1970)
Northeastern NSW, Australia	Analcimite sill. Mafic-ultramafic inclusion suite. Spinel compositions in websterites ($Sp_{70.3}$ $Her_{25.5}$ $Chr_{2.1}$ $Mt_{2.1}$), ($Sp_{62.7}$ $Her_{32.6}$ $Chr_{2.1}$ $Mt_{2.6}$); olivine websterite ($Sp_{53.6}$ $Her_{32.8}$ $Chr_{2.6}$ $Mt_{10.2}$ $Usp_{0.8}$); mafic granulites ($Sp_{67.5}$ $Her_{30.0}$ $Chr_{0.9}$ $Mt_{1.6}$), ($Sp_{51.2}$ $Her_{45.3}$ $Chr_{0.6}$ $Mt_{2.9}$), ($Sp_{8.7}$ $Chr_{0.6}$ $Mt_{74.2}$ $Usp_{16.5}$); lherzolite ($Sp_{59.7}$ $Her_{13.9}$ $Chr_{19.9}$ $Mt_{5.7}$ $Usp_{0.8}$) and wehrlite ($Sp_{68.5}$ $Her_{27.1}$ $Chr_{2.5}$ $Mt_{1.9}$). Origin: Remnants of a layered ultramafic-mafic pluton which initially crystallized at ≃ 10kb and subsequently reequilibrated at subsolidus temperatures (≃ 950°C) and comparable pressures to the sill intrusion.	Wilkinson (1975)

Table Hg-17 (2)

OXIDE-BEARING XENOLITHS IN IGNEOUS ROCKS

Locality	Review	References
Itinome-gata, Japan	Alkali-basalts. Xenoliths of gabbro, hornblende gabbro, pyroxenite, peridotite, garnet peridotite. Spinels are $MgAl_2O_4$ and ferrian pleonaste.	Kuno (1967)
	Alkali-basalts. Xenoliths of harzburgite, lherzolite, and Fe-rich lherzolite. Harzburgitic spinels are characterized by higher contents of Cr_2O_3 than spinels in lherzolites (37.8 vs. 9.4 wt%); the former are picotites and the latter pleonastes. Garnet lherzolites contain pleonastes of comparable composition to garnet-free lherzolites; all spinels are TiO_2-poor (.06-.41 wt%). Origin: Cr-rich spinels are primary whereas Al-rich spinels result from garnet decomposition. Partial reequilibration of garnet lherzolite in spinel peridotite field with the formation of pyroxene + spinel.	Aoki and Prinz (1974)
Lanzarote, Canary Islands	Tuff. Dunite xenolith; olivine + spinel with spinel decomposition to chromite. Spinel $(Fe_{0.4}\ Mg_{0.6})^{2+}\ (Fe_{0.1}\ Cr_{0.8}\ Al_{1.1})^{3+}_{2}O_4$; chromite $(Fe_{0.5}\ Mg_{0.5})^{2+}\ (Fe_{0.5-0.6}\ Cr_{0.8}\ Al_{0.5-0.8})^{3+}_{2}O_4$. Origin: High T°C late magmatic and pre-eruption reaction.	Frisch (1971)
Kerguelen Archipelago, S. Indian Ocean	Olivine alkali basalts and trachybasalts. Xenoliths of spinel lherzolite, biotite-hornblende pyroxenite and spinel gabbro. Origin: lherzolitic spinels $(Fe_{0.64}\ Mg_{0.47})^{2+}(Fe_{.37}\ Cr_{.87}\ Al_{.53})^{3+}Ti_{.11}\ O_4$ and gabbroic spinels $(Mg_{.71}\ Fe_{.29})(Al_{1.98}\ Cr_{.02})_2O_4$ are probably unrelated although both are present in a single flow. Reaction of olivine + plag = Opx + spinel.	McBirney and Aoki (1973)
Canary Islands	Basalts. Inclusions are kaersutite-bearing clinopyroxenites with Fe-Ti oxides + pyrrhotite. Equilibrated coexisting oxides give T = 975°C; $fO_2 = 10^{-10.5}$ atms. Cumulate textures of hornblendites suggest crystal settling under high water pressures prior to incorporation into alkalic suite.	Frisch and Schmincke (1969)
Worldwide distribution	General comparison of dunite with olivine-rich inclusions in basalts. Compositions of Cr-bearing spinels in dunites (3 analyses) and in olivine-rich inclusions in basalts (10 analyses) vary in major oxides as follows: Oxide / Basalt-inclusions / Dunites Al_2O_3 / 19.29 - 48.06 / 8.81 - 26.93 Fe_2O_3 / 2.67 - 29.81 / 5.72 - 16.48	Ross et al. (1954)

Table Hg-17 (3)

	FeO 6.81 - 16.68 13.90 - 17.96 TiO_2 0.44 - 8.50 1.40 - 2.20 Cr_2O_3 11.00 - 34.87 24.54 - 55.17 MgO 12.14 - 21.30 10.02 - 13.66 MnO 0.11 - 0.18 0.07 - 0.16 Tectonized fabrics and close similarities in mineralogy and chemistry suggest a common origin for inclusions in basalts and dunites.	Ross et al. (1954) cont.
St. Vincent, West Indies	Basaltic. Magnetite in ejected plutonic blocks: plagioclase (An_{93}), olivine (Fo_{67}-Fo_{79}), salite (5-6 wt% Al_2O_3) and hastingsite. Magnetite contains 6 wt% Al_2O_3; 4 wt% MgO; 7 wt% TiO_2. Unit cell = 8.372; Reflectivity at 589 nm = 14.8 - 15.8 %; Hardness VHN = 782-824; Curie temp. = 400-480°C. Based on Al_2O_3 solubility in Fe_3O_4, temperature of equilibration = 700°C.	Lewis (1970)
Siberia, USSR	Basalts. Ultrabasic nodules of spinel lherzolites, websterites, clinopyroxenites, and plagioclase-clinopyroxenites. Spinel compositions show uniformly high Al_2O_3 contents (53-57 wt%), constant TiO_2 (0.3-0.4 wt%), but with widely variable concentrations of MgO (11.7-19.6 wt%); FeO (9.8-23.5 wt%); Fe_2O_3 (3.8-11.8 wt%); and Cr_2O_3 (0.33-13.50). Mantle origin for ultramafic suite; the suggestion that in transport to the surface, basaltic assimilation is more readily accomplished for pyroxene and spinel-bearing xenoliths and less corrosive on olivine-bearing suites - hence paucity of pyroxene and spinel-bearing inclusions.	Kutolin and Frolova (1970)
British Columbia, Canada	Basalts. Lherzolite nodules; spinel-olivine equilibration for Fe/Mg indicate temperatures of: 840°C for cumulate nodules; 1085°C and > 1600°C for two suites of tectonite nodules. Representative range of compositions for spinels are as follows:	Littlejohn and Greenwood (1974)

Oxide	840°C	1085°C	>1600°C
MgO	19.56 - 23.58	19.05	19.06 - 21.86
FeO	4.43 - 9.01	11.52	9.53 - 11.33
Fe_2O_3	2.98 - 8.42	-	0.46 - 0.97
Al_2O_3	35.19 - 54.19	58.71	57.35 - 64.59
Cr_2O_3	8.43 - 31.82	10.06	7.30 - 11.76

The higher temperature suites of tectonite nodules are considered to have formed in the upper mantle, whereas the lower temperature (840°C) lherzolites are crystal cumulates formed at "depth".

Table Hg-17 (4)

OXIDE-BEARING XENOLITHS IN IGNEOUS ROCKS

Locality	Review	References
Nunivak Island, Alaska	Alkali basalts. Originally troctolite cumulates; olivine + plagioclase reacted to form symplectites of spinel + Al clinopyroxene. Compositional comparisons of primary spinels and corona spinels are as follows: Oxide — Primary spinels — Corona spinels Al_2O_3 — 50.38 - 57.36 — 64.44 - 65.60 Cr_2O_3 — 9.72 - 15.32 — 0.37 - 0.82 FeO — 12.56 - 14.21 — 10.35 - 10.54 MnO — 0.24 - 0.33 — 0.06 - 0.07 MgO — 18.63 - 19.32 — 21.72 - 23.39 Application of pyroxene geothermometer indicates primary equilibrium at 950°C and 9 Kb with the formation of the coronites developing under isobaric cooling.	Francis (1976)
Nigeria	Alkali basalts. Megacrysts: clinopyroxene (7-10 wt% Al_2O_3); orthopyroxene (4 wt% Al_2O_3); garnet ($Py_{63}Alm_{23}Gr_{13}Sp_1$); Ti-pargasite; ilmenites (2.16-4.48 wt% MgO; and 0.48-1.37 wt% Al_2O_3); spinels: (a) Al-titanomagnetite ($Usp_{44.6}$ $Mt_{34.0}$ $Chr_{0.6}$ $Sp_{12.8}$ $Mgf_{8.0}$); (b) ferrian pleonaste ($Usp_{3.8}$ $Mt_{14.3}$ $Chr_{0.2}$ $Sp_{61.4}$ $Her_{20.3}$). Suggestion that Al_2O_3 contents of titanomagnetites may result from a high pressure origin. Comparison with other alkali basalt inclusions indicates a depth of origin of 30-60 km.	Frisch and Wright (1971)
Nigeria	Basanite. Pyrope megacryst ($Py_{71.0}$ $Alm_{13.9}$ $Gr_{12.4}$ $And_{2.1}$ $Sp_{0.6}$) containing tubes of magnesian trevorite ($Ni_{.63}$ $Mg_{.26}$ $Co_{.07}$ $Fe_{.02}$ $Mn_{.01}$)($Fe_{.99}$ $Al_{.005}$ $Cr_{.001}$)$_2$ O_4 + Po + Mt + Hem + ferroan trevorite ($Ni_{.59}$ $Fe_{.37}$ $Co_{.02}$ $Mg_{.01}$ $Mn_{.002}$)($Fe_{.99}$ $Al_{.002}$ $Cr_{.001}$)$_2$ O_4. Garnet crystallization at P>15 Kb with inclusions of an immiscible oxysulfide melt.	Irving and Watson (1976)
Calton Hill, Derbyshire, U.K.	Basalt. Xenoliths of olivine nodules: olivine (Fo_{90}-Fo_{91}); Opx ($Ca_{1.5}$ $Mg_{89.0}$ $Fe_{9.5}$ - $Ca_{0.8}$ $Mg_{88.6}$ $Fe_{10.6}$); Cpx ($Ca_{39.6}$ $Mg_{55.8}$ $Fe_{4.6}$); and chromian spinel (Al_2O_3 = 45.45; Fe_2O_3 = 5.84; Cr_2O_3 = 16.85; FeO = 11.69; MgO = 19.29; MnO = 0.05 wt%). Fragments of a deep-seated peridotite.	Hamad (1963)

Table Hg-17 (5)

OXIDE-BEARING XENOLITHS IN IGNEOUS ROCKS

Locality	Review	References
Western Victoria, S. E. Australia	Basinites. Lherzolite xenoliths with primary Opx, Cpx, olivine, and chromian spinels; secondary assemblage in pale to brown glass (53.2-61.2 wt% SiO_2) which develops at grain boundaries between Cpx and spinel or olivine and Opx consist of olivine + spinel $\pm$ Cpx $\pm$ plag. Primary spinels have the following range in wt%: TiO_2 = 0.13-0.45; Al_2O_3 = 42.63-52.26; Cr_2O_3 = 11.44-30.80; Fe_2O_3 = 2.6; FeO = 9.66-12.76; MgO = 16.60-20.32; MnO = 0.11-0.22; NiO = 0.22-0.25. Secondary spinels range as follows in wt% : TiO_2 = 0.4-1.4; Al_2O_3 = 21.1-52.9; FeO = 12.2-23.4; MgO = 13.2-20.6; Cr_2O_3 = 14.2-41.0; NiO = 0.3-0.4. Model,based on geochemical arguements,suggests that residual lherzolite is left in the lithosphere after partial fusion, and is later modified by a melt which has migrated to the top of the low velocity zone. Cooling, recrystallization and subsequent incorporation in explosive volcanism, accompanied by increasing T^oC and decreasing pressure during ascent, causes incongruent melting of hydrous phlogopite and amphibole.	Frey and Green (1974)
Black Rock Summit, Nevada	Basinoid flows and pyroclastics. Cpx megacrysts and olivine-rich spinel peridotites classified as wehrlites and lherzolites. Cr-spinels are zoned with respect to Cr_2O_3 and Al_2O_3 with the former higher in the cores of crystals and lower at the margins, and with the latter showing the reverse distribution. Al_2O_3 and Cr_2O_3 in lherzolite spinels vary respectively between 18.34-44.52 wt% Al_2O_3 and 19.96-50.10 wt% Cr_2O_3. Corresponding ranges for wehrlite spinels are 27.28-42.18 wt% Al_2O_3 and 16.47-26.70 wt% Cr_2O_3. Lherzolite suite yield T = 1166-1298^oC and P = 19.8-43.6 Kb; megacrysts range in T from 1050-1327^oC.	Pike (1976)

Table Hg-17 (6)

OXIDE-BEARING XENOLITHS IN IGNEOUS ROCKS

Locality	Review	References
Ndonyuo Olnchoro, Central Kenya	Olivine melanephelinite. Xenolith suite of peridotite nodules fall into two groups: (1) harzburgites and lherzolites; (2) websterites. Cr-spinels are present as discrete crystals, exsolved from Opx, and as symplectites in Opx, with no significant compositional variations among spinel types.	Suwa et al. (1975)

Oxide	Harz.	Lherz.	Web.
Al_2O_3	32.9 - 43.9	34.5 - 37.4	38.9 - 40.3
Cr_2O_3	26.9 - 37.8	32.8 - 35.8	26.5 - 27.9
Fe_2O_3	0.90- 2.82	0.49- 2.19	1.49- 1.82
FeO	5.75- 8.41	8.08- 10.10	14.6 - 14.9
MgO	18.6 - 20.4	17.7 - 18.9	14.7 - 14.8
MnO	0.11- 0.22	0.26- 0.27	0.22- 0.25
NiO	0.13- 0.22	0.15- 0.18	0.26

Harzburgites and lherzolites suggest T = 1200-1350°C, P = 36-38 Kb and are considered to have been garnet peridotites. Websterites are estimated to have formed at 1090°C and P = 16 Kb by crystallization from an alkaline magma.

Locality	Review	References
Fen Complex, Norway	Damkjernite (alkali ultrabasic): euhedral phenocrystic biotite, in groundmass of Ti-clinopyroxene, amphibole, mica, nepheline, K-spar, calcite; also Mt, Ilm, sphene, perovskite, apatite, pyrite. Lherzolite nodules with spinel partially replaced by amphibole. Compositions vary as follows in wt%: Al_2O_3 = 51.27-56.85; Cr_2O_3 = 8.07-14.58; FeO = 11.40-14.18; MnO = 0.16-0.30; and MgO = 18.49-21.10. Cr/Al ratios are related to the degree of amphibole resorption. Cpx compositions indicate P = 10-13 Kb and T = 1200-1250°C. Nodules are mantle fragments or cognate xenoliths. Rapid injection of damkjernite with no intracrustal differentiation.	Griffen (1973)

Table Hg-17 (7)

OXIDE-BEARING XENOLITHS IN IGNEOUS ROCKS

Locality	Review	References
Lashaine, Northern Tanzania	Ankaramitic scoria and carbonatite tuffs. Xenoliths: harzburgite containing Al-picrochromite (TiO_2 = 2.86; Al_2O_3 = 15.66; Cr_2O_3 = 43.65; Fe_2O_3 = 13.92; FeO = 6.88; MgO = 15.89; CaO = 1.42 wt%).	Dawson et al (1970)
	Garnet peridotites: reaction of olivine + garnet → aluminous orthopyroxene + spinel. Range in spinel compositions in wt% (3 analyses): Al_2O_3 (56.1-56.3); Cr_2O_3 (13.3-14.6); FeO_{total}(7.9-9.3); MgO (20.1-21.1). Reaction illustrates the transition from garnet peridotite to spinel peridotite facies.	Reid and Dawson (1972)
	Alkalic pyroxenites: Ilm with 5.87-8.18 wt% MgO; titanomagnetites with TiO_2 = 5.2-9.5 wt%; Al_2O_3 = 4.7-5.8 wt%; MgO = 4.1-4.9 wt%; and Cr_2O_3 = 5.7-6.8 wt% ; lherzolites and mica dunites with Mg-Cr-Al-titanomagnetite and Ti-Al-picrochromite respectively. Over-all assemblages are poorer in Al_2O_3 than those from other alkalic pyroxenite localities.	Dawson and Smith (1973)
	Peridotite xenoliths: primary minerals are Ni-rich (0.4 wt% NiO) Ca-poor olivine (Fo_{92}); low Ca, Al, Cr, Ti, Mn enstatite ($Wo_1En_{93}Fs_7$); low Al, Ti, Mn, chrome diopside ($Wo_{44}En_{52}Fs_4$); chrome pyrope; and Mg-Al-chromite. Range in chromite compositions in wt% (16 analyses): TiO_2 (0.19-5.50); Al_2O_3 (6.16-23.6); Cr_2O_3 (40.9-60.8); FeO_{total}(13.9-36.4); MnO (0.19-0.35); MgO (9.74-16.1). Comparisons with experimental data yield P = 50 Kb and T = 1050°C.	Reid et al. (1975)
Katanui, New Zealand	Ultramafic mineral breccia. Ilmenite xenocryst and ilmenite in melanephelinite. (see analyses below) Comparison based on MgO contents suggests depth of origin between 30-60 Km.	Reay and Wood (1974)

Oxide	Xenocryst	Inclusion
TiO_2	45.66	46.09
Fe_2O_3	17.44	17.76
FeO	31.72	32.19
Al_2O_3	1.06	0.96
Cr_2O_3	0.01	0.01
MgO	3.95	3.09
MnO	0.23	0.26

Table Hg-17 (8)

OXIDE-BEARING XENOLITHS IN IGNEOUS ROCKS

Locality	Review	References
San Quentin, Baha, California	Basanitoids and alkali basalts. Xenolith suite of spinel lherzolites and megacrysts of augite and andesine. Coequilibrated Fe-Ti oxides give T = 1005°C, $fO_2 = 10^{-11.4}$ atms. Thermodynamic expressions for stages in the P-T path of the ascending magma yield values for which the lavas could have been in equilibrium with the lherzolites: T = 1330-1410°C and P = 27.5-31.6 Kb, with the more SiO_2-poor liquid having the higher values (c.f. Basu and MacGregor, 1976 below). Corresponding values for the basanitoids and megacrysts, and the basanitoids and phenocrysts are T = 1130°C, and P = 10.5 Kb and 1.4 Kb respectively.	Bacon and Carmichael (1973)
Theoretical	Equilibration T°C and P (Kb) of various lava types with spinel - and garnet - peridotites. (see table below) Lavas with low $\underline{a}$ SiO_2 equilibrate at higher pressures than lavas with high $\underline{a}$ SiO_2.	Nicholls and Carmichael (1972)

Rock Type	Spinel-peridotite T°C	Spinel-peridotite P(Kb)	Garnet-peridotite T°C	Garnet-peridotite P(Kb)
Trachybasalt	1396	24.2	1321	22.7
High-Al basalt	1648	31.2	1588	30.0
Olivine tholeiite	1074	1.8	750	-5.1
Basaltic andesite	1065	2.2	767	-3.6
Ugandite	1340	45.2	1658	51.6

OXIDE-BEARING XENOLITHS IN KIMBERLITES

Locality	Review	References
San Quentin, Baja, California. Kimberley, South Africa	Alkali olivine basalt and kimberlites. Comparison of chromian spinels in ultramafic xenoliths. Textural types classified as : (a) euhedral spinels in kimberlites; (b) symplectites of spinels + clinopyroxene, orthopyroxene, amphibole, garnet in harzburgites and lherzolites in kimberlites; (c) exsolved spinels from Opx in xenoliths from basalts and kimberlites; (d) interstitial spinels in basaltic xenoliths; and (e) spinels in garnet kelyphites from kimberlites. Highest Cr/(Cr + Al), and lowest $Mg/(Mg + Fe^{2+})$ ratios in euhedral spinels, whereas exsolution and symplectic spinels are compositionally intermediate. Spinels from kimberlite xenoliths have higher $Fe^{3+}/(Cr + Al + Fe^{3+})$ than spinels from alkali olivine basalts. The ratio $Cr/(Cr + Al + Fe^{3+})$ increases with pressure. Al_2O_3 in spinel varies sympathetically with Al_2O_3 in Opx. Estimates of T indicate 850-1000°C for basaltic xenoliths and 940-1050°C for kimberlites. Values of P for these suites indicate 10-15 Kb and 35-55 Kb respectively.	Basu and MacGregor (1976)

Table Hg-18

OXIDE-BEARING XENOLITHS IN KIMBERLITES

Locality	Review	References
ILMENITE		
USSR	MgO contents vary between 5.6 and 14.13 wt% MgO. Cr_2O_3 is uniformly low with a maximum of 1.37 wt% recorded. Groundmass ilmenites are not distinguished. Curie points vary between -170°C and 260°C and Curie isotherms for the system $FeTiO_3$-Fe_2O_3-$MgTiO_3$ are provided.	Frantsesson (1970)
Bultfontein and Wesselton, S. Africa. Ison Creek, U.S.A.	Compositions for magnesian ilmenites vary as follows in wt%: TiO_2 = 49.48-55.90; FeO = 15.00-28.81; Fe_2O_3 = 6.27-15.71; MgO = 8.72-19.26; Al_2O_3 = 0.18-0.73; Cr_2O_3 = 0.20-2.39; NiO = 0.20-0.36. Range is limited to a maximum of Hem_{20} and to Geik contents between $Geik_{30}$-$Geik_{70}$.	Mitchell (1973a)
Worldwide distribution	Significance of magnesian-rich ilmenites and host rock compositions. Comparison of igneous, metamorphic and kimberlitic ilmenites. Review concludes that the range in MgO contents in kimberlites = 4.86-12.10 wt%; in basic igneous rocks is 0.46-3.27 wt%; and in kimberlitic xenoliths (granulites and eclogites) is 4.12-4.63 wt%.	Lovering and Widdowson (1968)
Monastery Mine, S. Africa	Critique of Lovering and Widdowson (1968) on the basis of Ilm compositions in single hand specs. Compositions vary as follows: MgO = 7.63-11.1 wt%; Cr_2O_3 = 0.21-0.82 wt%; Al_2O_3 = 0.08-0.54 wt %.	Frisch (1970)
Lesotho	Trace element variations in magnesian ilmenites. Zr/Hf = 29-40; Nb/Ta = 8.2-9.2; Zr/Nb = 0.6-1.36. Cr correlates with MgO contents, and the coupled substitution of $2Ti^{4+} \rightarrow Sc^{3+} + Nb^{5+}$ is suggested. Co correlates negatively with Cr (range in Co = 134-151 ppm; cf. Skaergaard 30-100 ppm Co).	Mitchell et al. (1973)
	Cr-Ni variations show a good positive correlation with Cr:Ni ratios clustering between 5 and 10.	Nixon and Kresten (1973)
	Discrete nodules. Ilm varies in MgO content between 9.0 and 11.1 wt%; Cr_2O_3 = 0.59-0.71 wt%. Carbonated ultramafic nodules (garnet + Cpx) contain Mg-Al-chromite (MgO = 13.39; Al_2O_3 = 12.32 wt%).	Nixon and Boyd (1973a)

Table Hg-18 (2)

OXIDE-BEARING XENOLITHS IN KIMBERLITES

Locality	Review	References

ILMENITE

Lesotho

Discrete silicate nodules with associated Ilm: Cpx + Ilm and garnet + Ilm; ilmenite nodules with associated garnet, Opx and olivine. Ilm compositions as follows:

Oxide	Cpx+Ilm	Gt+Ilm	Ilm+Gt	Ilm+Opx	Ilm+Ol
TiO_2	49.59	50.71	50.86	54.84	48.03
Al_2O_3	0.55	0.80	0.57	0.78	0.66
Cr_2O_3	0.53	0.18	0.79	0.85	0.56
Fe_2O_3	12.27	11.53	10.79	5.50	14.66
FeO	29.90	27.64	28.98	28.57	26.51
MnO	0.25	0.19	0.24	0.22	0.24
MgO	8.11	10.00	9.28	11.53	9.18
CaO	0.05	0.02	0.04	0.04	0.04

P-T conditions suggest that the discrete nodule suite formed under comparable conditions to the sheared lherzolite suite (see below); some megacrysts originated by disaggregation of unusually coarse grained rocks but the majority are phenocrysts in crystal-mush magmas in the low-velocity zone.

References: Nixon and Boyd (1973b)

Detailed discussion of Ilm-silicate nodules with comparisons from S. Africa. Garnets, enstatites, and diopsides which co-crystallize with Ilm have distinctive and restricted compositional ranges regardless of textures and intergrowths. A positive correlation is demonstrated between Mg/Mg + Fe in Ilm as a function of Mg/Mg + Fe in coexisting garnet. Average geikielite contents for Ilm from 5 pipes show a range in Geik (mole %) from 33.0 to 43.9. Ilm-bearing nodules fall midway between the ranges of sheared lherzolites and granular lherzolites (160 Km; T = 1100°C) i.e. in the range of the pyroxene geotherm inflexion. Cr-bearing ilmenites reported from Kentucky (5.04 wt% Cr_2O_3; MgO = 15.74), and from Monastery (2.27 wt% Cr_2O_3; 13.19 wt% MgO).

References: Boyd and Nixon (1973)

Ilmenite-zircon nodules from two localities: Mothae ($Ilm_{54.2}$ $Geik_{36.5}$ $Hem_{9.3}$); and Monastery ($Ilm_{62.1}$ $Geik_{27.7}$ $Hem_{10.2}$). These compositions are equivalent to 10.28 and 7.71 wt% MgO respectively. Cr_2O_3 values are 0.92 and 0.73 wt% respectively.

References: Nixon (1973); Whitelock (1973)

Colorado and Wyoming, U.S.A.

Ilmenite megacrysts. MgO = 7.5-13.8 wt%; Cr_2O_3 = 0.3-3.1 wt%. Range is equivalent to 45-65 mole % Geik.

References: McCallum et al. (1975)

Table Hg-18 (3)

OXIDE-BEARING XENOLITHS IN KIMBERLITES

Locality	Review	References
ILMENITE-CLINOPYROXENE LAMELLAR INTERGROWTHS		
Monastery Mine, S. Africa	Description and experimental phase equilibria study which suggested that the Ilm-Cpx intergrowth resulted by reequilibration and decomposition of a high pressure Ti-garnet.	Ringwood and Lovering (1970)
	Intergrowth is described as eutectic and textural association is suggested to have formed by exsolution of an ilmenite-structured pyroxene.	Dawson and Reid (1970)
	Test and evaluation of the hypotheses proposed above and a comparison of intergrowths from the Monastery mine with samples from Riley Co., Kentucky. Conclusion, based on extensive geochemical study suggests that the garnet origin is unlikely (noting that Ringwood and Lovering were unable to achieve reverse equilibrium); that exsolution of Ilm from an original Ilm-Cpx solid solution is doubtful; and that eutectic crystallization of Ilm + Cpx is the most probable mechanism as previously suggested by MacGregor (1970), MacGregor and Wittkop (1970) and Boyd (1971) and subsequently demonstrated experimentally by Wyatt, MacCallister and Boyd (1975).	Gurney et al. (1973)
	Mineral chemistry and REE data on Ilm-Cpx lamellar intergrowths show that magnesian-ilmenites react with late stage magmas to form Mn-rich ilmenite; a TiO_2-rich but Al_2O_3-poor phlogopite also forms from the primary assemblage. REE distribution patterns suggest that clinopyroxene could have been in equilibrium with the kimberlite magma, and is unlikely to have formed by the decomposition of a high pressure garnet.	Mitchell et al. (1973)

Table Hg-18 (4)

OXIDE-BEARING XENOLITHS IN KIMBERLITES

Locality	Review	References
SPINEL-SILICATE INTERGROWTHS IN KIMBERLITES		
Wesselton mine, S. Africa	Harzburgite. Symplectic chromite-paragasite intergrowth considered to have formed by reaction of garnet + chrome-diopside.	Boyd (1971)
Newlands, Monastery and Bultfontein mines, S. Africa and Lashaine, Tanzania	Lherzolite from Bultfontein, remainder are harzburgites. Assemblages are as follows:: (1) Newlands. Ol +Opx + Sp; (2) Monastery. Ol + Opx + Amph + Sp + Phl; (3) Bultfontein. Opx + Chr and Cpx + Chr; and (4) Leshaine. Ol + Opx + Chr. Origins for assemblages: reaction between pre-existing phases; metasomatic replacement of silicate by oxide; subsolidus exsolution; simultaneous (cotectic) crystallization; pressure induced sub-solidus recrystallization. Origin is enigmatic but tentatively suggests decomposition of an AB_2O_4 phase of spinel composition.	Dawson and Smith (1975)

Spinel Compositions

Oxide	1	2	3	4	5
SiO_2	0.15	0.09	0.16	0.16	0.00
TiO_2	<0.02	0.00	0.03	0.00	1.75
Al_2O_3	12.4	31.7	16.3	26.8	17.7
Cr_2O_3	48.8	38.4	49.8	42.4	49.1
Fe_2O_3	10.3	0.8	6.3	1.1	6.7
FeO	15.7	13.4	13.4	15.2	8.7
MnO	0.2	0.22	0.29	0.27	0.22
CaO	0.05	0.00	0.00	0.02	0.00
MgO	11.3	15.3	13.5	13.7	17.1
NiO	0.12	0.07	0.13	0.07	0.00
Total	99.5	99.9	100.0	99.62	99.4

References for analyses listed above: 1. Boyd (1971); 2 to 5. Dawson and Smith (1975).
Localities: 1. Wesselton; 2. Newlands mine; 3. Monastery mine; 4. Bultfontein mine; 5. Lashaine.

Table Hg-18 (5)

OXIDE-BEARING XENOLITHS IN KIMBERLITES

Locality	Review	References
ULTRAMAFIC NODULES IN KIMBERLITES		
Thaba Putsoa and Mothae, Lesotho	Comparison of granular and sheared lherzolites, with the observation that the granular suite contains equilibrated chromian-spinels (Al_2O_3 = 11.71-14.47 wt%; Fe_2O_3 = 3.33-6.74 wt% and MgO = 12.16-14.15) of a restricted compositional arnge; primary chromite is absent in the sheared suite and is restricted to kelyphitic reaction rims on garnet. Compositions of granular Ilm-bearing ultrabasic nodules are also included (MgO = 10.49-11.68 wt%).	Nixon and Boyd (1973)
Lesotho and Monastery mine, S. Africa	Detailed evaluation of the origin of ultramafic nodules in kimberlites, with particular reference to the distribution of chromite and to the association of Mg-Ilm in Cpx lamellar intergrowths. Harzburgites and spinel peridotites have granular textures and are considered to have formed in the range of granular lherzolites (i.e. lower P-T conditions). Cr/Al ratios in spinels ought to be indicative of depth of origin, yet garnet-free harzburgites contain either chromian-spinel (~ 14 wt% Al_2O_3; ~ 14 wt% MgO) or Mg-Al-chromite (~ 40 wt% Al_2O_3; ~ 18 wt% MgO). Ilmenite + enstatite assemblages fall largely within the sheared lherzolite field and are therefore of high P and T.	Boyd and Nixon (1975)
Lesotho, Kimberley, S. Africa and Lashaine, Tanzania	Lherzolites and harzburgite nodules. Detailed analyses of Ti-poor spinels which show that compositions fall into a narrow and restricted band which is similar to the trends exhibited for podiform chromite deposits in alpine-type peridotites and for ultramafic inclusions in basalts: $(Mg_2Fe^{2+}_6)(Al_{14}Cr_2)O_{32}$ to $(Mg_3Fe^{2+}_5)(Fe^{3+}AlCr_{14})O_{32}$. This trend is distinct from the trend for Archaen rocks and for layered basic intrusions. Cr-spinels in kimberlites are considered to be derived largely from lherzolites rather than harzburgites, but compositional data for both suites show extraordinary variations and overlapping fields for Al_2O_3 (2.55-49.0 wt%), and Cr_2O_3 (15.0-68.7 wt%); but with a narrow range for MgO (11.5-20.7 wt%).	Smith and Dawson (1975)

Table Hg-18 (6)

OXIDE-BEARING XENOLITHS IN KIMBERLITES

Locality	Review	References
ULTRAMAFIC NODULES IN KIMBERLITES		
De Beers mine, S. Africa	Lherzolite. Globules of magnetite + pentlandite intergrowths coexisting with massive pentlandite. Magnetite contains 1.0 wt% Ni and 0.1 wt% S. Origin results from cotectic crystallization of oxide-sulfide melt followed by partial decomposition of pyrrhotite → magnetite.	Bishop et al. (1975)
Premier mine, S. Africa	Harzburgites fall into three groups: (1) fluidal textured which are equivalent to sheared lherzolites, and conform to high P-T trend; (2) intermediately deformed with intermediate Mg/(Mg + Fe) and Ca/(Ca + Mg); and (3) granular harzburgites with equivalent P-T regime as granular lherzolites. Chromian spinels are restricted to the granular suite with Al_2O_3 (4.07-8.58 wt%), MgO (11.7-15.6 wt%) and particularly Cr_2O_3 (62.2-66.9 wt%) contents which are within the range of chromite inclusions in diamonds (see below). Groups 2 and 3 appear to have equilibrated in the range T = 1000-1200°C and between 140-180 Km. The sheared suite (spinel absent) are in the range T = ≃ 1300°C and depths of ~ 200 Km.	Danchin and Boyd (1976)
Matsoku Pipe, Lesotho	Ultramafic nodules and granulites. Significant difference with respect to other Lesotho and S. Africa xenolith suites inasmuch as the recrystallized or flaser textured nodules are enriched in spinel + phlogopite. Mg-Al-chromite (Al_2O_3 = 14.88; TiO_2 = 3.32; Cr_2O_3 = 46.19; FeO = 17.56; MgO = 14.76 wt%) coexisting with rutile (FeO = 1.68; Cr_2O_3 = 2.74 wt%). Ilmenite in Ilm-bearing ultrabasic nodule contains 12.44 wt% MgO and 2.27 wt% Cr_2O_3. Mantle origin at high P but moderate T = 1050°C is proposed.	Cox et al. (1973)
Stockdale, Kansas	Eclogite. Detailed consideration of oxide-sulfide associations. MgO contents of Ilm range between 9.56-10.0 wt%; coexisting rutiles have the following ranges in minor elements (Al_2O_3 = 0.10-0.18; Cr_2O_3 = 0.10-0.13; FeO = 0.33-0.45; MnO = 0.01-0.40; MgO = 0.02-0.03 wt%). Rutile contains exsolution lamellae of Ilm + Sp. Sulfides are Po + Pent + Cpy + Cub. Incorporation of the xenolith into the kimberlite is inferred to be in the range of T = 1100-1200°C.	Meyer and Boctor (1975)

Table Hg-18 (7)

OXIDE-BEARING XENOLITHS IN KIMBERLITES

Locality	Review	References
ULTRAMAFIC NODULES IN KIMBERLITES		
Holsteinborg-Ivigtut region, Greenland	Spinel-, garnet-, and phlogopite-peridotite xenoliths in dike and sheet intrusions of kimberlites. Spinel mineral chemistry exhibits a broad range among xenolith types particularly with respect to Al_2O_3, and between xenoliths and the host kimberlite.	Emeleus and Andrews (1975)

Oxide	Sp. Perid.	Gt. Perid.	Phlog. Perid.			Kim.	
			Di.	Phlog.	Exsoln.	Core	Euh.
Al_2O_3	48.8	11.8	1.53	7.67	9.86	4.82	7.23

Locality	Review	References
	Spinels in the phlogopite peridotite are respectively: (Di) enclosed in diopside; (Phlog) in equilibrium with phlogopite; (Exsoln) exsolved from Opx. Kimberlitic groundmass spinels are for the core of a Chr mantled by titanomagnetite; and for an euhedral (Euh) discrete grain. Uniform nature of high pressure mineral chemistry of garnet and spinel peridotites match closely similar inclusions from other inferred upper mantle localities.	
Montana, U.S.A.	Garnet peridotite xenoliths in sub-silicic-alkalic diatremes. Xenolith suite yields T = 920-1315°C and 106-148 Km (32-47 Kb). Ilmenite megacryst has MgO = 11.3 wt% and a composition of Ilm_{48} $Geik_{41}$ Hem_{11}.	Hearn and Boyd (1975)
OXIDE INCLUSIONS IN DIAMONDS		
Chromite Ghana, Venezuela, Sierra Leone	Relative abundances of inclusions: Ol > Gt > Chr > En > Di. Chr + Ol and Chr + Gt have been observed in the same inclusion, but Chr + Opx or Chr + Cpx have not been observed. Chromite compositions are characterized by high Cr_2O_3 contents (61-67 wt%) and with restricted ranges in MgO (14.2-16.4 wt%) and Al_2O_3 (5.12-6.74 wt%) contents. Exceptional Zn-bearing chromites are present in Sierra Leone diamonds (Zn = 2.38-2.60 wt%) and these have lower Al_2O_3 (3.23-3.29 wt%) and MgO (0.49-0.54 wt%) contents. Hypothesized that diamonds formed in igneous events and that the inclusions they contain crystallized in equilibrium with a liquid.	Meyer and Boyd (1972)

Table Hg-18 (8)

OXIDE-BEARING XENOLITHS IN KIMBERLITES

Locality	Review	References
OXIDE INCLUSIONS IN DIAMONDS		
Chromite		
West Africa	Mg-Al-chromite (Al_2O_3 = 5.9 wt%; MgO = 10.0 wt%; Cr_2O_3 = 66.2 wt%). Comparable to other inclusions except for small variation in MgO. Chromite is considered to be comparable to those in garnet-lherzolite suites.	Prinz et al. (1975)
Magnetite		
West Africa	Porous euhedral magnetite crystals which may have had associated sulfides. Some diamonds contain abundant magnetite crystals (max. 17). Magnetites are generally very pure with minor element concentrations of TiO_2 (max. = 0.99 wt%), Al_2O_3 (max. = 2.44 wt%), Cr_2O_3, MgO, MnO and V_2O_3. Magnetite (+ sulfides Po + Pent + Cpy) is considered to form part of an original eclogite suite.	Prinz et al. (1975)
Ilmenite		
Brazil	Ilmenite compositions differ markedly from those in kimberlites or in xenolith suites, particularly with respect to MgO (0.11-0.14 wt%). Ilmenites approach stoichiometric $FeTiO_3$; major dilutant is MnO (max. = 0.73 wt%).	Meyer and Svisero (1975)
Rutile		
Brazil	Analyses show TiO_2 = 99.6-99.8 wt%; total dilutants = 0.4 wt% with a max. for FeO = 0.24 wt%.	Meyer and Svisero (1975)
West Africa	Contrasting associations and compositions of rutiles show that in R + phlogopite + omphacite inclusions: Al_2O_3 = 1.35 wt% and Fe_2O_3 = 1.14 wt%. This contrasts with rutile-only inclusions where Fe_2O_3 is 7.2 wt%, Al_2O_3 = 1.91 wt%, and NiO = 0.21 wt%. Rutile, in common with magnetite, is considered to be typical of eclogite suites.	Prinz et al. (1975)

MAGMATIC ORE DEPOSITS

ANORTHOSITE ASSOCIATIONS

Locality	Review	T^oC	log 10 fO_2	References
Adirondacks	Titanomagnetite-ilmenite ore body in anorthosite. Sanford Hill.	750-710	-16.1 to -16.7	Buddington and Lindsley (1964)
	Oxide-rich norite in anorthosite. Sanford Hill.	650	-18.2	Buddington and Lindsley (1964)
	Oxide-rich noritic pyroxenite layer in anorthosite. Derrick.	810	-13.6	Buddington and Lindsley (1964)
	Oxide-rich feldspathic pyroxenite. Calamity Mill.	710	-16.0	Buddington and Lindsley (1964)
Newfoundland	Zoned oxide vein in anorthosite. Hayes prospect.	650	-15.8	Buddington and Lindsley (1964)
Quebec	La Blache titanomagnetite ore deposit in labradorite anorthosite. Mt-Usp_{ss} exsolution; granule and trellis oxidation exsolution; spinel exsolution from Mt-Usp_{ss}; discrete spinel $(MgFe)Al_2O_4$ with exsolved Mt-Usp_{ss}.	500-840	-21.0 to -14.0	Anderson (1968)
Quebec	Degrosbois deposit: titanomagnetite-ilmenite anorthosite intruded into gabbroic anorthosite; irregular intrusion with associated oxide-rich veinlets; pleonaste lamellae and ilmenite in Mt-Usp_{ss}.	570-630	-20.0 to -17.2	Lister (1966)
Norway	Egersund-Sogndal district. Ilm-Hem_{ss}; extensive exsolution of Ilm_{ss} and Hem_{ss}; magnetite, titanomagnetite, exsolved pleonaste and ilmenite oxidation lamellae. In associated anorthosites.			Gjelsvik (1957); Haggerty (unpublished)

Table Hg-19 (2)

MAGMATIC ORE DEPOSITS

Locality / Review	T^oC	log 10 fO_2	References
GABBROIC ASSOCIATIONS			
Norway			
Oxide-rich ferrogabbro, Øvre Røddal.	710	-18.1	Buddington and Lindsley (1964)
USSR			
Titanomagnetite gabbro in Tsaginsk gabbroic-anorthosite massif.	790	-14.6	Buddington and Lindsley (1964)
Minnesota			
Duluth gabbro: multiple intrusive complex with extensive continuous and discontinuous segregations of titanomagnetite-ilmenite lenses. Exsolution of Usp_{ss} from $Usp\text{-}Mt_{ss}$ and minor oxidation lamellae of trellis ilmenite.			Lister (1966)
North Range	760-915	-14.7 to -12.2	
South Range	820-850	-13.8 to -12.4	
APATITE ASSOCIATIONS			
Quebec			
Allard Lake: oxide and apatite-rich norite sheet in anorthosite.	600	-17.0	Buddington and Lindsley (1964); Hargraves (1966)
Ellen Lake: $Usp\text{-}Mt_{ss}$ and $Ilm\text{-}Hem_{ss}$ in apatite-rich (10%) norite sheet in anorthosite.	550	-17.0	Lister (1966)
St. Charles: lenses and dikes of titanomagnetite-ilmenite and apatite-rich (30%) associations with intrusive contacts in anorthosite.	700-950	-15.0 to -11.0	Lister (1966)
St. Charles: magnetite-ilmenite-apatite (30%) associations.	730-1000	-17.0 to -11.0	Philpotts (1967)

MAGMATIC ORE DEPOSITS

APATITE ASSOCIATIONS

Locality	Review	T^oC	log 10 fO_2	References
Quebec				
	Port Cartier: oxide-apatite-rich anorthosite.	850	-14.0	Philpotts (1967)
	Labrieville: magnetite (61%) - apatite (39%) in anorthosite.			Philpotts (1967)
	Mt. Johnson: magnetite (70%) - apatite (30%) in anorthosite.			Philpotts (1967)
Ontario				
	Pusey deposits: concordant lenses of pyroxenitic titano-magnetite-ilmenite and associated apatite in gabbro.	600-925	-19.1 to -12.0	Lister (1966)
	Nemegos: magnetite (66%) - apatite (34%) in anorthosite.			Philpotts (1967)
Northwest Territories				
	Magnetite-apatite intrusions associated with plutons of intermediate composition. Initial crystallization of Ti-poor Mt_{ss} + apatite followed by alteration stage (pneumatolitic) to form hematite, rutile and sphene.			Badham and Morton (1976)
Virginia				
	Nelsonite (rutile 35% + apatite 65%) dikes in anorthosite and hypersthene granodiorite.			Moore (1940); Ross (1941); Philpotts (1967)
South Africa				
	Upper Zone Bushveld Complex: olivine (37%) - apatite (35%) - magnetites (18%); apatite (10-36%) - magnetites (56-87%); plagioclase-apatite-magnetites; and feldspathic magnetites. These 4 types are all characterized by titanomagnetite + ilmenite. Oxide segregations are concordant with layering but are also present in discordant sheets.			Grobler and Whitfield (1970)
Sweden				
	Kiruna magmatic magnetite-apatite-rich ore deposits: main ores segregated between syenite porphyry and quartz porphyry; magnetite-hematite phosphorous-rich ores between quartz porphyry and pyroclastic sequences, or entirely in pyroclastic sequences; hematite-rich ores have developed by oxidation in volcanic pyroclastic series.			Geiger (1931); Geiger (1967); Freitsch (1973)

Table Hg-19 (4)

MAGMATIC ORE DEPOSITS

CHROMIAN SPINELS

MAFIC AND ULTRAMAFIC ASSOCIATIONS

Locality	Review	References
Chile	Laco magnetite-phosphate lava flows.	Park (1961); Rodgers (1968); Haggerty (1970)
Pacific Basio	Detailed review of magnetite-apatite dike associations.	Park (1972)
Western Hemisphere	Compilation of chromian spinel analyses from a wide variety of geological settings. Detailed Discussion of mutual solid-solubility among spinel types and development of the diagramatic multicomponent spinel prism. The predominant zone of isomorphism is a field extending from $MgAl_2O_4$ - $MgCr_2O_3$ towards $FeAl_2O_4$ - $FeCr_2O_4$ and sweeping to the terminal end member Fe_3O_4. Compositions along the following joins are essentially absent: Fe_3O_4 - $MgFe_2O_4$; $FeAl_2O_4$ - $FeCr_2O_4$; $MgFe_2O_4$ - $MgCr_2O_4$ and $MgFe_2O_4$ - $MgAl_2O_4$.	Stevens (1944)
Worldwide distribution	Distinctive characterization of podiform (alpine type) and stratiform (layered intrusive suites) chromite deposits.	Thayer (1970); Dickey and Yoder (1972); Dickey (1975)

	Stratiform	Podiform
Ages	Precambrian - shield areas.	Paleozoic or younger; island arcs or active mobile mountain belts.
Textures	Cumulus.	Nodular.
Chemistry	Fe increases rapidly as Cr_2O_3 decreases below 55 wt%. Most variable constituent is Fe.	Al_2O_3 increases reciprocally and total Fe remains about constant. Most variable constituent is Cr followed by Al.
Ratio	Fe_{total} vs Cr_2O_3	
	Ratio shows a positive correlation.	Ratio shows no correlation.
Ratio	Cr_2O_3 vs $Al_2O_3/(Al_2O_3 + Cr_2O_3 + Fe_2O_3)$	
	Ratio shows a good negative correlation.	Ratio shows an equally good negative correlation.

Positive correlation. | Weak to negative correlation.

Oxide	Av	Max	Min	S*	Av	Max	Min	S*
TiO_2	0.6	1.8	0.2	0.3	0.2	0.6	0.1	0.1
Al_2O_3	16.6	21.6	11.8	2.5	15.9	31.6	8.8	7.8
Cr_2O_3	46.2	58.1	34.9	3.7	53.1	61.4	35.5	9.0
FeO	25.0	38.7	11.6	4.9	16.9	21.9	13.6	2.0
MnO	0.3	0.4	0.2	0.04	0.3	0.5	0.2	0.1
MgO	11.1	16.9	6.0	1.8	13.8	16.7	11.0	1.9
Total	99.8	101.5	98.0	0.8	100.2	100.9	98.4	0.7
		(45 analyses)				23 analyses)		

Dickey and Yoder (1972)

Compositional Data: Bushveld; Selukwe; and Stillwater. | Day Brook (N. Carolina); Canyon Mt., Oregon; Mt. Albert, Quebec; McGuffy Creek, Calif. Little Castle Creek, Calif.; Cayoguan, Cuba; Camaguey, Cuba. — Dickey and Yoder (1972) cont.

S*: Standard deviation.

Host rocks: Norite, gabbroic anorthosite, harzburgite, bronzitite. | Dunites typically; less commonly in harzburgites; rarely in feldspathic associations.

Note: For further details consult Tables covering Layered Intrusions. | The paper by Thayer contains an extensive bibliography of studies and ideas prior to 1970. Dickey (1975) discusses a model for podifiorm deposits based on modern plate tectonic concepts.

Andizlik-Zimparalik region, Turkey — Engin and Hirst (1970)

Alpine type podiform and genetically associated stratiform type. Contrasts in mineral chemistry reveals a progressive decrease in Cr_2O_3 and Cr/Fe ratios from nodular ores, through massive and disseminated types, to the limited development of stratiform cumulus chromite which has low Cr_2O_3 and high Al_2O_3, FeO, ZnO, MnO and V_2O_3. Host rocks are peridotites and associated harzburgites.

Oxide	Nodular	Massive	Disseminated	Stratiform
Cr_2O_3	56.87-58.39	49.78-58.43	41.74-56.63	37.09-43.28
Cr:Fe	3.70-3.93	2.07-4.03	2.31-3.60	1.82-2.04

Stratiform ores are considered to have formed at a high level in the crust and at a late stage in the crystallization of the ore body.

Kemi, Northern Finland — Veltheim (1962)

Chromite deposit sandwiched between anorthosite gabbro and peridotite. Major element variations show Cr_2O_3 (42.4-45.7 wt%); Al_2O_3 (16.3-18.2 wt%); MgO (6.6-8.9 wt%); FeO (22.3-25.8 wt%); Fe_2O_3 (4.3-5.2 wt%). Considered to be partially stratiform (cumulus), partially intrusive.

Eastern Ghats, India — Mall and Rao (1970)

Regionally charnockitic; host rocks are chromitites and bronzitites. Analysis of Mg/Fe^{2+} relationships between Opx and coexisting chromite indicate a tight interdependency; $K_D(Mg/Fe)$ = 8.36-18.66 and is independent of trivalent cations. Chromite compositions vary as follows: Cr_2O_3 = 17.23-47.39; Al_2O_3 = 10.37-24.25; MgO = 3.97-9.99; FeO = 16.20-24.98; Fe_2O_3 = 14.56-38.04.

Table Hg-19 (6)

MAGMATIC ORE DEPOSITS

CHROMIAN SPINELS

MAFIC AND ULTRAMAFIC ASSOCIATIONS

Locality	Review	References
Cuttack District, Orissa, India	Bedded chromite deposits are confined to the tops of dunite-peridotite sheets. Vertical sections through layers show: (1) a linear size distribution variation; (2) decrease in Cr_2O_3 (from 58.42 to 50.67), and MgO (from 15.99 to 13.56); and (3) an increase in FeO (from 10.71 to 14.28), and Al_2O_3 (from 11.45 to 14.96); all suggestive of a single magmatic cumulative cycle.	Chakraborty (1973)
Zhob Valley, West Pakistan	An igneous complex where the dominant rock type is serpentinized harzburgite, though extensive areas are also occupied by dunite, gabbro and peridotite. Chromite deposits are classified as: (1) stringers and bands of massive chromite; (2) podiform or globular; and (3) disseminated type. The globular chromites contain the lowest Cr_2O_3 (44.7 wt%) and highest Al_2O_3 (21.0 wt%) contents; the massive varieties are characterized by high Cr_2O_3 (max 59.1 wt%) and low Al_2O_3 (min 10.2 wt%) contents; and the disseminated type are approximately intermediate between these two extremes. MgO contents for all types are approximately constant ~ 14 wt%. Estimates of the Cr content of the Zhob Valley complex suggest that if the magma originally contained ~ 0.05 wt% Cr_2O_3, chromite would have been precipitated; late differentiated dolerites vary between 79-276 ppm (c.f. Skaergaard = 170 ppm) and thus the original magma was somewhat higher than the expected average for basaltic magmas.	Bilgrami (1969)
Southern Appalachians, U.S.A	A comparison of disseminated chromite grains and massive chromite pods in olivine-rich ultramafics show that these two forms fall into distinct compositional fields when plotted as a function of MgO/RO vs Cr_2O_3/R_2O_3. The disseminated grains have a negative slope and parallel the 1000°C isotherm as determined from coexisting olivine. Massive chromite has a constant Cr_2O_3/R_2O_3 ratio and a variable MgO/RO ratio; these differences suggest variable depositional temperatures, comparable to that of the Stillwater chromite, and a lack of subsolidus reequilibration with olivine or with chromite in the host rock.	Fletcher and Carpenter (1972)

Table Hg-19 (7)

MAGMATIC ORE DEPOSITS

CHROMIAN SPINELS

MAFIC AND ULTRAMAFIC ASSOCIATIONS

Locality	Review	References
Massif du Sud, Southern New Caledonia	Chromite ore bodies are present in dunites and harzburgites and display a variety of textural forms from massive to disseminated and orbicular, which are in part consistent with a magmatic cumulative origin. The harzburgites contain < 1% of a chromian-spinel (28-45% Cr_2O_3) in association with olivine (Fo_{87-92}) and Opx (En_{89-92}); dunites contain 1-3% picrochromite (39-50% Cr_2O_3) in association with Fo_{87-93}; the chromitites also contain picrochromites (42-59% Cr_2O_3), and serpentinized olivine (Fo_{92-96}). Temperatures of crystallization are considered to have been ~ 1200°C, which contrasts with picrochromite-bronzite symplectites which are considered to have formed at ~ 790°C. Although classically an alpine type association, a cumulate origin is favored.	Rodgers (1973)
Darvel Bay, North Borneo	Chromite layers and pods in dunite and serpentinite. Review of earlier data and new analyses for the region show that Cr_2O_3 varies between 31.4-55.76 wt%; Al_2O_3 between 8.9-27.4 wt%; MgO between 7.99-19.05 wt%; Fe_2O_3 between 0.22-39.28 wt% and TiO_2 between 0.03-0.75 wt%. All ultramafic bodies in the region contain significant concentrations of Cr (2770 ppm) and Ni (1530 ppm) and yet the chromite ore bodies have a restricted distribution. A mantle origin is proposed for the ore bodies with subsequent emplacement as tabular masses into the crust.	Hutchison (1972)
Coolac District, N.S.W. Australia	Segregated pods of chromite in an alpine type peridotite complex. Chromite compositions display a bimodal distribution with maxima in Cr_2O_3 contents at ~ 38%. Pods of Cr-rich chromite (> 50 wt% Cr_2O_3, < 20 wt% Al_2O_3) are randomly distributed along the strike of the complex and are restricted to dunites. Pods of Al-rich chromite (~ 20 wt% Al_2O_3) are interspersed and are characteristically present in association with other mafic rocks (harzburgites) in the peridotite. The Al-rich pods are considered to be younger than the Cr-rich pods but the disposition of these pods in the protocomplex is conjectural. The complex is modelled on the early formation of Cr-rich chromite in dunite and later accumulation of Al-rich chromite in a pyroxenitic-mush.	Golding and Johnson (1971)

Compositions of Selected Coolac Chromites

	1	2	3	4	5	6	7	8	9	10	11	12
Cr_2O_3	33.3	36.3	42.5	45.6	49.5	50.8	53.2	56.5	58.2	60.0	62.6	58.0
Al_2O_3	35.3	30.7	27.4	22.1	19.8	17.4	17.0	12.2	11.5	8.5	5.6	5.4
Fe_2O_3	5.2	5.8	1.9	4.9	3.2	9.5	2.5	5.6	5.3	2.9	5.9	7.3
FeO	8.6	8.8	12.2	12.1	15.8	6.9	13.2	14.0	11.0	19.5	12.6	23.2
MgO	17.6	18.4	16.0	15.3	11.7	15.4	14.1	11.7	14.0	9.1	13.3	6.1

Table Hg-19 (8)

MAGMATIC ORE DEPOSITS

CHROMIAN SPINELS

MAFIC AND ULTRAMAFIC ASSOCIATIONS

Locality	Review	References
Hartley Complex, Great Dike, Rhodesia	The great dike consists of 4 ultrabasic lopolithic complexes. Each complex is synclinal and is divided into the following units: Unit 1 (top) is gabbroic; Unit 2 is pyroxenite, olivine in basal part, picrite and harzburgite, chromite seam #1, harzburgite, chromite seam #2; Unit 3 (bottom) is pyroxenite. Disseminated chromites in horizons adjacent to, above and below seams #1 and #2 are as follows in ascending order:	Bichan (1969)

Lower ⟶ Upper

Oxide	Harz	Seam #2		Harz	Harz Below Seam #1	Harz Above Seam #1	Picrite	
Cr_2O_3	50.30	*55.09	- 57.25	50.98	53.70	52.80	*50.40	- 48.00
Al_2O_3	15.45	11.80	11.05	13.94	12.70	14.35	14.96	15.58
Fe_2O_3	2.74	2.16	1.74	3.30	4.29	1.52	4.00	5.18
FeO	20.88	18.54	18.00	22.22	20.24	18.61	21.83	20.59
MgO	9.26	10.24	9.87	7.25	7.35	10.40	7.00	7.10
TiO_2	0.78	0.80	0.54	1.07	1.05	0.47	0.88	1.36
MnO	0.35	0.31	0.27	0.35	0.44	0.38	0.36	0.34
CaO	0.08	0.12	0.21	0.10	0.16	0.12	0.13	0.09
SiO_2	1.40	1.78	1.38	1.35	1.31	1.74	1.33	1.47
	101.24	100.84	100.31	100.56	101.24	100.39	100.89	99.71

*Choice of analysis based on min and max values for Cr_2O_3.

The major variations between the disseminated and the massive chromite bands are: massive chromite contains higher Cr_2O_3, lower Al_2O_3, and lower FeO, but comparable MgO and TiO_2 contents. A comparison of the great dike chromites with those in the Bushveld and Stillwater complexes are as follows:

Great Dike	Cr_2O_3	Cr/Fe
Disseminated Chr in Harz and Pic	48.0 - 53.7	1.67 - 2.32
Chr - seam #2	55.1 - 57.3	2.37 - 2.73
Chr - seam #1	51.0 - 51.68	1.93 - 2.06
Bushveld		
Disseminated Chr in pyroxenite	41.8 - 46.5	0.84 - 1.62
Chr bands in pyroxenite	35.6 - 47.3	0.87 - 1.59

Bichan (1969) cont.

Stillwater		
Disseminated Chr in dunite or Harz	40.5 - 47.82	1.23 - 1.57
Chromite bands in Harz	46.02 - 50.98	1.69 - 1.99

Origin and conclusion suggests that the chromite bands in Unit 2 of the Great Dike resulted from two discrete magmatic events and that convective overturning was minimal or absent.

Representative and Exotic Spinel Compositions in the Bushveld Complex

	Typical →		Atypical in V		Atypical in Ti+Cr+Al				Typical						Typical		Av
Oxide	1	2	3	4	5	6	7	8	9	10	11	12	13	14	15	16	17
SiO_2	0.34	0.88	-	-	-	-	-	-	0.50	-	-	-	-	-	0.34	0.23	-
TiO_2	15.38	11.30	7.10	10.37	7.54	3.92	7.94	3.21	1.90	1.28	2.07	0.59	0.43	1.03	1.60	0.91	0.6
Al_2O_3	3.55	2.99	1.79	6.20	12.53	16.55	13.30	20.49	14.62	12.18	14.47	14.77	13.27	15.87	13.0	14.4	16.6
Cr_2O_3	0.08	0.18	2.04	5.28	27.37	32.13	27.42	31.07	37.32	41.68	38.59	47.98	49.43	42.81	49.2	36.6	46.2
Fe_2O_3	39.11	45.66	47.04	27.35	13.81	11.15	10.07	7.93	16.37	12.65	12.77	7.10	6.82	8.22	3.1	16.0	-
FeO	39.15	36.64	37.51	41.75	33.25	31.61	36.86	31.69	21.19	27.19	25.37	18.87	20.55	19.93	25.3	24.0	25.0
MgO	1.31	0.55	0.29	0.54	5.08	4.49	2.81	4.02	8.00	5.20	7.12	10.42	9.03	9.68	6.5	7.2	11.1
CaO	-	-	-	-	-	-	-	-	-	-	-	-	-	-	0.3	0.01	-
MnO	0.27	0.23	-	-	-	-	-	-	0.42	-	-	-	-	-	0.31	0.19	0.3
NiO	-	-	-	-	-	-	-	-	0.14	-	-	-	-	-	0.09	0.15	-
V_2O_3	0.53	1.38	3.90	10.20	1.61	1.45	1.50	0.81	-	0.37	0.34	0.20	0.19	0.47	1.10	0.17	-
Total	99.72	99.81	99.67	101.69	101.19	101.30	99.90	99.22	100.65	100.55	100.73	99.33	99.72	98.02	100.9	99.9	99.8
	BUSHVELD														STILLWATER		AVERAGE STRATIFORM

References to analyses: (1-2) Molyneux (1972). Analyses 1 and 2 are from the eastern lobe of the Bushveld and are at 3084' (anorthosite) and 3610' (seam #11) respectively above the base of the Merensky Reef in the Upper Zone; (3-8) Cameron and Glover (1973). Samples are from "replacement" pegmatites in the Critical Zone and bridge the gap between Ti-poor chromites at the base of the Bushveld and titanomagnetites of the upper part. (9) Van Zyl (1969). Merensky Reef. (10-14) Cameron (1975), Critical Zone. Analysis #10 in plag.; #11 in bronzite; #12 intersertial chromite; #13 in olivine; and #14 is massive chromite. (15-16) Thayer (1969). Analysis #15 = G-zone; and #16 = B-zone in layered chromites. (17) Dickey and Yoder (1972). This analysis is the average stratiform chromite composition based on 3 complexes and 45 analyses.

Table Hg-20

Representative Coexisting Oxides in Igneous Suites

TITANOMAGNETITES

	Granites			Granodiorites		Pegmatites			Rhyolites		Obsidians		Pitchstones		Pumice	Rhyodacites		
Oxide	1	2	3	4	5	6	7	8	9	10	11	12	13	14	15	16	17	18
SiO_2	1.14	0.34	0.1	0.95	0.32	0.42	0.51	0.69	0.11	0.02	0.11	0.05	0.08	0.07	0.11	-	0.91	0.88
TiO_2	5.92	0.17	7.1	2.13	0.37	2.29	1.52	2.51	15.1	7.0	22.1	15.7	19.8	19.8	13.5	18.2	9.8	9.9
Al_2O_3	0.59	0.00	-	0.40	0.10	0.67	0.98	0.84	0.86	1.5	0.64	1.97	0.78	0.99	0.91	1.86	3.3	3.1
Cr_2O_3	0.00	0.00	-	0.03	0.00	-	-	-	0.01	0.05	0.01	0.05	0.02	0.03	0.01	-	0.18	0.19
Fe_2O_3	61.33	68.0	53.1	64.59	68.4	65.50	66.44	64.58	38.1	59.7	25.1	35.6	30.0	29.6	41.4	32.0	44.8	45.6
FeO	29.55	30.6	36.2	29.89	30.7	30.29	28.93	29.62	43.8	30.4	50.4	42.9	48.5	47.4	42.4	46.1	36.4	35.6
MgO	0.10	0.00	<0.05	1.85	0.03	0.09	0.30	0.40	0.13	1.05	0.06	1.81	0.34	0.40	0.18	0.84	1.2	1.9
CaO	0.74	0.01	0.02	0.17	0.01	0.01	0.04	0.03	0.01	-	0.04	0.01	0.01	0.01	0.01	-	0.13	0.13
MnO	0.50	0.44	0.5	0.17	0.29	0.20	0.44	0.61	0.96	0.61	1.00	0.42	0.64	0.76	0.85	0.70	0.52	0.58
NiO	-	-	-	-	-	-	-	-	-	0.02	-	-	-	-	-	-	-	-
V_2O_3	0.04	0.04	-	0.20	0.05	0.19	0.15	0.07	<0.01	0.50	<0.01	1.80	0.05	0.30	<0.01	-	0.26	0.26
ZnO	-	-	-	-	-	-	-	-	0.19	0.14	0.27	0.08	0.21	0.22	0.16	-	0.60	0.43
Total	99.91	99.60	97.0	100.35	100.27	99.66	99.31	99.35	100.3	100.7	99.7	100.4	100.6	99.6	99.5	99.7	98.10	98.57

ILMENITES

Oxide	1	2	3	4	5	6	7	8	9	10	11	12	13	14	15	16	17	18
SiO_2	1.49	0.26	0.1	-	0.24	1.07	1.08	0.65	0.02	0.02	0.03	0.03	0.04	0.03	0.02	-	0.81	0.80
TiO_2	46.05	49.1	48.9	47.31	49.0	46.38	44.18	43.25	49.8	41.0	49.9	46.7	49.6	48.9	48.5	49.8	42.8	42.2
Al_2O_3	0.68	0.06	-	-	0.11	0.94	0.55	1.25	0.04	0.21	0.03	0.15	0.06	0.06	0.04	0.30	0.57	0.52
Cr_2O_3	0.00	0.01	-	0.00	0.01	-	-	-	<0.01	<0.01	<0.01	0.04	<0.01	0.02	<0.01	-	<0.01	<0.01
Fe_2O_3	10.17	-	5.7	7.74	-	9.76	15.72	15.65	5.6	22.4	5.0	13.0	6.2	7.3	8.3	6.7	17.8	18.3
FeO	36.62	18.2	41.2	40.01	34.0	36.76	31.22	32.25	42.5	33.2	43.4	35.8	42.7	42.1	41.3	41.4	34.2	34.2
MgO	0.41	0.00	<0.05	-	0.03	0.48	0.43	0.19	0.31	1.93	0.10	3.15	0.63	0.47	0.40	1.33	2.5	2.2
CaO	1.15	0.14	0.02	-	0.01	0.31	0.10	0.10	0.02	-	0.04	0.02	0.02	<0.01	0.02	-	0.10	0.11
MnO	2.48	29.6	2.7	3.16	15.2	3.60	5.63	5.60	1.71	0.72	1.23	0.55	0.82	1.02	1.53	0.99	0.70	0.64
NiO	-	-	Nb_2O_3	-	-	-	-	-	-	<0.01	-	-	-	-	-	-	-	-
V_2O_3	0.00	0.11	=1.4	0.00	0.09	0.00	0.00	0.00	<0.01	0.15	<0.01	0.20	<0.01	<0.01	<0.01	-	0.06	0.07
ZnO	-	-	-	-	-	-	-	-	<0.01	0.05	<0.01	<0.01	<0.01	0.05	<0.01	-	<0.01	<0.01
Total	99.05	97.48	100.0	98.22	98.69	99.30	98.91	98.74	100.0	99.8	99.7	99.6	100.1	99.9	99.7	100.5	99.54	99.04
T^oC	685	-	680	580	-	570	560	600	810	860	915	930	900	925	820	890	855	865
$\log_{10}fO_2$	-16.3	-	-17.7	-19.0	-	-19.8	-19.4	-17.3	-14.7	-11.5	-12.9	-11.4	-12.9	-12.2	-13.9	-13.0	-11.7	-11.7

References to analyses listed above: (1,4) Puffer (1972); (2,5) Czamanske and Mihálik (1972); (3) Whitney and Stormer (1976); (6-8) Puffer (1975); (9,11-15) Carmichael (1976b); (10) Lowder (1970); (16) Nicholls (1971); (17-18) Bauer et al. (1973).

Table Hg-20 (2)

Representative Coexisting Oxides in Igneous Suites

TITANOMAGNETITES

	Dacites		Syenites			Monzonites		Foyaite	Diorite	Trachytes		Trachyandesites		Trachybasalts		Tholeiites		
Oxide	19	20	21	22	23	24	25	26	27	28	29	30	31	32	33	34	35	36
SiO_2	-	0.05	0.46	-	0.05	0.32	-	-	-	0.12	-	-	-	0.25	-	0.10	0.09	-
TiO_2	18.07	7.33	4.65	17.66	10.3	0.48	11.32	21.58	24.88	11.9	18.51	13.34	20.65	21.6	23.52	17.5	28.8	26.98
Al_2O_3	2.86	1.65	2.26	0.07	-	0.03	0.56	0.00	2.40	3.08	0.57	4.16	5.69	0.62	2.65	2.9	1.18	3.92
Cr_2O_3	0.18	0.09	-	-	-	0.02	-	-	-	0.01	-	-	-	<0.01	-	0.02	0.02	0.03
Fe_2O_3	41.43	52.5	62.12	34.90	49.0	67.8	46.51	27.15	17.48	42.7	33.87	39.84	23.31	24.5	23.35	31.4	10.7	13.88
FeO	32.93	35.0	29.26	42.28	40.3	30.5	40.46	47.25	50.53	37.1	44.14	38.64	48.12	49.3	46.39	46.0	56.1	53.40
MgO	1.54	1.60	0.35	0.12	0.05	0.03	0.09	0.08	1.20	2.73	1.66	3.48	0.63	0.04	4.31	0.9	0.50	1.83
CaO	1.08	0.01	-	0.00	0.01	0.04	-	0.03	0.00	0.03	-	-	-	0.11	-	0.11	0.14	-
MnO	0.40	0.49	0.10	4.96	0.4	0.08	1.06	3.53	1.61	1.18	1.81	0.00	0.68	1.29	0.59	0.34	0.82	0.87
NiO	0.04	-	-	-	-	-	-	-	-	-	-	-	-	-	-	-	-	-
V_2O_3	0.94	0.81	-	-	-	0.24	-	-	-	0.45	-	-	-	0.22	-	0.6	0.97	-
ZnO	0.12	0.10	-	-	-	-	-	-	-	0.15	-	-	-	0.15	-	0.10	0.12	-
Total	99.62	99.6	99.55	99.99	100.1	99.59	100.00	99.62	98.10	99.5	100.56	99.46	99.07	98.1	100.82	99.97	99.4	100.91

ILMENITES

Oxide	19	20	21	22	23	24	25	26	27	28	29	30	31	32	33	34	35	36
SiO_2	-	0.06	0.28	-	0.5	0.19	-	-	-	0.04	-	-	-	0.13	-	0.05	0.06	-
TiO_2	44.40	37.6	48.36	50.85	49.4	48.9	45.43	50.22	48.79	40.9	48.04	42.03	52.26	48.8	49.21	47.4	50.3	50.48
Al_2O_3	0.49	0.28	1.06			0.09	0.00	0.00	0.00	0.36	0.00	0.57	0.41	0.04	0.28	0.15	0.02	0.09
Cr_2O_3	0.04	0.03	-	-	-	0.01	-	-	-	0.01	-	-	-	-	-	0.00	-	0.02
Fe_2O_3	15.95	29.7	8.91	4.15	4.1	-	13.60	6.48	7.94	24.6	8.79	20.97	1.48	6.9	10.65	9.3	4.4	4.03
FeO	34.52	29.6	39.55	37.65	43.3	45.9	38.64	37.03	38.83	29.3	34.94	30.40	44.37	41.4	33.64	40.0	43.8	42.87
MgO	2.58	2.12	0.66	0.14	0.05	0.07	0.08	0.08	1.60	3.79	3.32	4.15	1.01	0.88	5.64	1.3	0.49	0.84
CaO	0.55	0.02	-	0.03	0.01	0.06	-	0.02	-	0.04	-	-	-	0.21	-	0.12	0.07	-
MnO	0.41	0.43	0.70	7.69	1.8	3.43	2.04	7.86	2.16	0.72	2.32	0.00	0.81	0.79	0.55	0.44	0.50	1.01
NiO	0.02	-	-	-	Nb_2O_3	-	-	-	-	-	-	-	-	-	-	-	-	-
V_2O_3	0.09	0.20	-	-	=0.3	0.12	-	-	-	0.07	-	-	-	<0.01	-	0.30	0.11	-
ZnO	0.04	<0.01	-	-	-	-	-	-	-	<0.01	-	-	-	-	-	0.04	-	-
Total	99.11	100.0	99.66	100.51	99.4	98.77	99.79	101.69	99.33	99.80	97.40	98.12	100.35	99.3	99.97	99.10	99.8	99.34
T^oC	1050	980	640	-	690	-	830	850	1080	1010	930	1110	705	975	1090	1070	1100	1020
$\log_{10}fO_2$	-9.6	-9.1	-18.8	-	-18.0	-	-13.0	-14.4	-10.0	-9.2	-11.7	-7.8	-19.1	-11.6	-9.5	-10.0	-10.1	-11.5

References to analyses listed above: (19) Wilkinson (1971); (20,28) Carmichael (1967b); (21) Anderson (1966); (22,25-27) Neumann (1974, 1976); (23) Whitney and Stormer (1976); (24) Czamanske and Mihálik (1972); (29-30,33) Anderson (1968); (31) Ridley and Baker (1973); (32,35) Carmichael (1967a); (34) Evans and Moore (1968); (36) Steinthórsson (1972).

Table Hg-20 (3)

Representative Coexisting Oxides in Igneous Suites

TITANOMAGNETITES

	Basalts				Gabbros				Pegmatite	Troctolites		Norites		Monzonorites		Perid.	Pyrox.	Mang.
Oxide	37	38	39	40	41	42	43	44	45	46	47	48	49	50	51	52	53	54
SiO_2	0.14	-	0.22	-	0.10	-	-	-	0.7	-	-	0.5	1.8	1.9	1.3	1.2	1.2	1.6
TiO_2	26.1	23.8	23.3	24.29	3.81	8.57	8.16	4.59	6.4	7.76	5.56	3.3	5.0	8.5	6.7	10.9	7.7	8.3
Al_2O_3	0.41	3.3	1.78	1.80	2.28	1.92	1.57	2.47	1.1	1.96	3.52	2.2	3.2	2.7	1.2	2.1	2.7	0.9
Cr_2O_3	0.02	<0.03	0.70	-	-	0.18	0.01	0.05	0.3	1.22	3.54	0.05	-	-	-	-	-	-
Fe_2O_3	17.3	20.8	20.2	20.27	62.39	54.48	56.93	59.30	53.8	53.34	52.19	59.2	57.8	52.2	57.3	49.7	53.1	54.2
FeO	53.2	47.6	51.0	51.80	30.46	32.18	31.21	30.52	35.0	32.78	32.14	31.5	31.7	32.9	32.7	33.7	32.9	33.7
MgO	0.87	3.6	1.24	0.19	0.40	1.38	1.07	1.77	0.5	1.52	2.10	0.81	0.47	0.57	0.11	0.90	1.1	0.15
CaO	0.10	-	0.12	-	-	-	-	-	0.1	-	-	0.07	0.11	0.14	0.09	0.16	0.14	0.19
MnO	0.67	0.7	0.46	0.53	0.13	0.36	0.24	0.26	0.3	0.41	0.25	0.06	0.08	0.21	0.22	0.22	0.12	0.22
NiO	-	-	-	-	-	-	-	-	-	-	-	0.01	0.06	-	-	-	-	-
V_2O_3	0.55	-	1.70	-	0.22	0.35	0.39	0.26	1.0	0.35	0.47	0.93	0.44	0.12	0.03	0.21	0.80	0.04
ZnO	0.09	-	0.07	-	-	-	-	-	-	-	-	0.11	0.13	0.18	0.13	0.21	0.13	0.14
Total	99.5	99.8	100.8	98.88	99.79	99.42	99.58	99.22	99.2	99.24	99.77	98.9	100.8	100.5	100.2	99.4	100.1	99.5

ILMENITES

Oxide	37	38	39	40	41	42	43	44	45	46	47	48	49	50	51	52	53	54
SiO_2	0.14	-	0.05	-	0.08	-	-	-	0.5	-	-	0.3	0.5	0.2	0.6	0.4	0.2	0.7
TiO_2	49.5	49.8	47.7	50.53	49.71	49.35	49.58	48.93	44.2	47.92	48.55	50.1	50.1	51.1	50.1	52.2	51.2	49.9
Al_2O_3	0.05	0.8	0.15	0.12	0.37	0.42	0.03	0.23	1.7	0.60	0.88	0.2	0.4	0.1	0.2	0.1	0.2	0.3
Cr_2O_3	<0.01	<0.03	-	-	-	0.05	0.00	0.02	0.2	0.11	0.12	-	-	-	-	-	-	-
Fe_2O_3	5.7	7.9	9.4	3.70	9.89	6.37	7.31	12.17	14.7	8.33	7.13	5.9	3.9	3.1	1.8	2.6	3.1	2.1
FeO	42.6	36.8	40.2	44.50	38.36	41.08	40.17	33.78	35.3	39.76	39.47	39.9	42.8	43.8	45.1	42.2	40.9	45.3
MgO	0.69	4.1	1.24	0.21	1.00	2.07	2.07	2.31	1.1	2.00	2.50	2.2	0.56	0.66	0.09	1.3	2.9	0.11
CaO	0.20	-	0.14	-	-	-	-	-	0.1	-	-	0.10	0.10	0.05	0.07	0.08	0.05	0.13
MnO	0.70	0.8	0.38	0.54	0.85	1.19	0.92	2.88	2.5	1.13	0.94	0.46	0.53	0.53	0.93	0.53	0.43	0.70
NiO	-	-	-	-	-	-	-	-	0.1	-	-	-	-	-	-	-	-	-
V_2O_3	0.21	-	0.54	-	<0.01	0.02	0.02	0.01	0.1	0.04	0.07	0.05	0.01	0.02	-	0.01	0.04	-
ZnO	<0.01	-	-	-	-	-	-	-	-	-	-	-	-	-	-	-	-	-
	99.8	100.2	99.8	99.60	100.27	100.55	100.10	100.33	100.5	99.89	99.66	99.4	99.0	99.7	99.0	99.5	99.3	99.4
T°C	1050	1000	1050	950	620	715	700	640	720	715	680	600	620	660	~600	670	660	610
$\log_{10} fO_2$	-10.6	-11.5	-10.2	-12.5	-19.5	-16.1	-16.6	-17.9	-15.0	-15.5	-16.7	-20.0	-20.2	-19.2	>-22.0	-19.2	-19.0	-21.5

References to analyses listed above: (37) Smith and Carmichael (1968); (38) Jakobsson et al. (1973); (39) Carmichael (1967a); (40) Ridley et al. (1974); (41) Anderson (1966); (42-44, 46-47) Mathison (1975); (45) Elsdon (1972); (48-54) Duchesne (1972).

Table Hg-20 (4)

Representative Compositions of Spinels in Basalts

	*C	M							P	O	P	P	P	P	C	Px →	M	O
Oxide	1	2	3	4	5	6	7	8	9	10	11	12	13	14	15	16	17	18
SiO_2	0.05	0.22	0.29	-	-	-	0.09	0.08	-	-	-	-	-	-	0.23	0.31	0.30	0.23
TiO_2	0.71	23.3	0.67	4.3	18.4	0.42	2.30	2.32	11.80	4.73	2.98	15.40	1.87	24.80	0.63	2.65	2.33	0.64
Al_2O_3	24.3	1.78	39.76	10.7	3.6	6.4	13.5	13.7	2.68	16.60	18.80	4.55	11.70	3.04	51.50	24.03	20.77	20.57
Cr_2O_3	37.0	0.70	25.06	36.6	15.5	35.3	43.3	43.0	7.35	27.70	34.40	17.50	48.70	7.10	0.00	10.39	16.83	35.93
Fe_2O_3	15.3	33.8	3.60	14.9	14.1	26.9	10.81	10.35	-	-	-	-	-	-	17.12	31.64	30.11	12.71
FeO	9.7	38.8	15.10	26.3	43.9	27.2	15.27	17.46	74.70	41.10	35.50	58.80	32.00	61.00	9.48	17.06	17.29	14.99
MgO	13.5	1.24	15.74	7.0	2.8	3.5	13.2	12.0	3.90	10.30	8.50	2.84	7.41	3.61	20.23	13.59	13.17	13.19
CaO	-	0.12	0.04	-	-	-	0.01	0.02	-	-	-	-	-	-	0.06	0.04	0.03	0.00
MnO	0.20	0.46	0.17	0.27	0.27	0.34	0.17	0.19	-	-	-	-	-	-	0.05	0.24	0.24	0.39
NiO	-	-	-	-	-	-	0.16	0.15	-	-	-	-	-	-	-	-	-	-
V_2O_3	0.18	1.70	-	-	-	-	0.19	0.22	-	-	-	-	-	-	-	-	-	-
ZnO	-	0.07	-	0.20	0.17	0.30	0.10	0.12	-	-	-	-	-	-	-	-	-	-
Total	100.9	101.2	100.43	100.27	98.74	100.36	99.10	99.61	100.43	100.43	100.18	99.09	101.68	99.75	99.30	99.95	101.07	98.65

	O	O	O	Gl	O/G	C	M		M	C	M	C	Px	Gm	Gm	P	P	P
Oxide	19	20	21	22	23	24	25	26	27	28	29	30	31	32	33	34	35	36
SiO_2	0.26	0.30	0.21	0.37	0.19	0.30	0.27	-	-	-	-	-	-	-	-	0.23	0.51	0.03
TiO_2	7.43	7.84	11.3	13.8	14.6	15.0	15.1	0.44	0.63	0.33	0.40	0.37	12.0	17.4	27.9	12.38	13.09	17.63
Al_2O_3	7.33	12.4	9.91	7.85	7.63	7.32	7.20	35.82	29.20	33.19	15.69	19.02	9.5	6.6	2.3	11.52	6.56	6.85
Cr_2O_3	4.62	22.9	19.3	17.8	19.1	16.9	17.0	29.37	35.30	35.23	46.20	45.09	1.8	0.6	0.6	0.00	-	-
Fe_2O_3	43.56	19.6	19.8	17.4	16.3	17.1	17.7	-	-	-	-	-	34.5	31.1	31.1	41.10	45.30	43.47
FeO	29.23	29.0	32.8	35.4	36.2	36.2	36.4	17.91	20.56	14.90	26.82	19.93	35.2	38.5	33.9	30.03	29.92	29.13
MgO	6.16	7.68	7.07	6.43	6.55	6.52	6.64	17.61	14.56	16.98	11.31	15.13	5.1	6.1	2.2	3.90	3.77	2.33
CaO	0.05	0.00	0.00	0.02	0.02	0.23	0.18	-	-	-	-	-	-	-	-	0.03	0.15	0.04
MnO	0.65	0.36	0.38	0.42	0.38	0.40	0.43	0.16	0.23	0.19	0.41	0.29	0.4	0.6	0.9	0.28	0.44	0.56
NiO	-	-	-	-	-	-	-	-	-	-	-	-	-	-	-	-	-	-
V_2O_3	-	-	-	-	-	-	-	-	-	-	-	-	-	-	-	-	-	-
ZnO	-	-	-	-	-	-	-	-	-	-	-	-	-	-	-	-	-	-
Total	99.29	100.0	100.7	99.4	100.9	100.0	100.9	101.31	100.47	101.81	100.93	99.83	98.5	100.9	98.9	100.12	99.99	100.04

References: Analyses (1-2) Carmichael (1967a), Thingmuli, Iceland; (3) Steinthórsson (1972), Surtsey, Iceland; (4-6) Evans and Moore (1968), Makaopuhi lava lake, Hawaii; (7-8) Evans and Wright (1972), Kilauea Iki, Hawaii; (9-14) Beeson (1976), East Molokai, Hawaii; (15-19) Arculus (1974), Granada, Lesser Antilles; (20-25) Thompson (1973), Snake River, Idaho; (26-30) Liou (1974), Taiwan, basalts in coastal ophiolite suite. Analyses are for a glassy basalt, an olivine basalt, and a dolerite. (31-33) Prévot and Mergoil (1973), Saint-Clément (Massif Central), France; analyses are for 3 generations of spinels, the earliest are in pyroxene and two later generations in the groundmass are distinguished on grain size and the degree of maghemitization; (34-36) Aoki (1966), Iki Island, Japan. Analyses 34-36 are phenocrysts in trachyandesites.

Table Hg-20 (5)

Representative Compositions of Spinels in Basalts

Oxide	P 37	P 38	P 39	P 40	Gm 41	Gm 42	C ⟶ 43	44	45	46	⟶ M 47	Av 48	O 49	50	51	O 52	Gm 53	P 54
SiO_2	0.28	0.32	0.26	1.16	1.31	1.91	-	-	-	-	-	-	-	0.03	0.04	-	-	-
TiO_2	17.62	11.46	19.94	26.48	21.17	19.97	1.22	3.00	9.27	15.00	17.43	22.8	2.61	0.5	0.8	1.70	1.30	22.10
Al_2O_3	7.50	11.13	4.65	1.56	5.60	6.56	13.89	11.83	5.67	5.35	4.02	1.3	13.11	9.4	8.9	35.00	35.50	3.90
Cr_2O_3	0.00	-	0.34	-	-	-	41.81	41.24	28.30	13.05	16.03	7.8	44.07	46.5	42.1	19.18	27.00	-
Fe_2O_3	30.71	42.25	26.77	36.21	30.66	26.74	-	-	-	-	-	-	-	17.1	20.5	13.30	4.00	21.60
FeO	39.55	27.97	44.89	31.08	39.44	37.95	33.20	42.75	55.20	65.69	55.92	66.1	32.20	12.3	16.3	12.90	21.60	45.40
MgO	3.02	4.52	2.18	1.31	1.56	4.69	9.95	1.00	1.00	1.00	4.46	1.2	9.43	14.2	11.8	16.90	11.00	4.30
CaO	0.15	0.11	0.03	1.07	-	1.55	-	-	-	-	-	-	-	0.00	0.00	-	-	-
MnO	0.52	0.34	0.76	1.04	0.58	0.70	-	-	-	-	-	-	-	0.2	0.2	-	-	-
NiO	-	-	-	-	-	-	-	-	-	-	-	-	-	0.1	0.1	-	-	-
V_2O_3	-	-	-	-	-	-	-	-	-	-	-	-	-	-	-	-	-	-
ZnO	-	-	-	-	-	-	-	-	-	-	-	-	-	-	-	-	-	-
Total	99.66	101.40	100.39	99.91	100.32	100.07	100.07	99.82	99.44	100.09	97.86	99.20	101.42	100.6	101.1	98.98	100.40	97.30

Deep Sea Basalts

Oxide	Plag 1	C ⟶ 2	3	⟶ M 4	C ⟶ 5	⟶ M 6	7	8	9	10	11	12	13	14	O/G 15	Gl 16	O 17	Gm 18
SiO_2	-	-	-	-	-	-	0.00	0.00	-	-	-	-	0.44	0.30	-	-	-	-
TiO_2	0.46	1.06	16.43	25.24	0.77	16.97	2.63	0.94	-	-	0.57	0.52	0.67	0.32	0.35	1.67	0.18	0.21
Al_2O_3	33.36	25.98	3.58	2.14	26.76	3.14	16.7	27.8	38.9	49.0	21.8	20.0	23.8	26.7	24.6	21.3	47.3	43.0
Cr_2O_3	29.89	30.17	13.11	0.07	33.97	18.69	28.9	27.8	28.7	20.0	41.5	43.1	39.1	38.2	42.3	38.6	19.5	23.5
Fe_2O_3	8.43	10.98	21.64	17.12	5.54	14.55	-	-	3.0	0.8	-	-	-	-	4.4	7.9	3.4	4.0
FeO	12.01	19.80	45.49	52.52	22.99	46.03	41.9	35.3	13.5	12.2	23.6	24.9	22.4	19.7	13.8	16.6	10.9	10.7
MgO	16.84	10.82	1.11	0.85	8.71	0.92	7.58	9.95	15.7	17.5	12.4	11.1	13.7	15.0	14.6	12.9	18.8	18.5
CaO	-	-	-	-	-	-	-	-	-	-	-	-	-	-	-	-	-	-
MnO	0.23	0.32	0.53	0.59	0.27	0.44	-	-	0.25	0.17	0.19	0.21	0.16	0.18	0.19	0.21	0.06	0.07
NiO	-	-	-	-	-	-	-	-	0.15	0.22	-	-	-	-	-	-	-	-
V_2O_3	-	-	-	-	-	-	-	-	-	-	-	-	-	-	-	-	-	-
ZnO	-	-	-	-	-	-	-	-	-	-	-	-	-	-	-	-	-	-
Total	101.22	99.13	101.89	98.53	99.01	100.75	97.7	101.8	100.47	99.95	100.0	99.9	100.7	100.4	100.2	99.2	100.3	100.0

References: (37-40) Aoki (1966), Iki Island, Japan. Analysis 37 is a phenocryst in a trachyandesite; 38 is in basalt, and 39-40 are spinels in mugearites; (41-42) Sasajima et al. (1975), Regional, Japan. (43-49) Gunn et al. (1970), East Island, Crozet Archipelago. Analysis #48 is an average of 10 grains <10 µm diam; (50-51) Irvine (1973), Bridet Cove, Alaska; (52-54) Babkine et al. (1965), Monistrol d'Allier, France.

DEEP SEA BASALTS. References:(1-6) Ridley et al. (1974), Leg 6, Philippine Sea. (7-8) Myers et al. (1976), Leg 17 Central Pacific Basin; (9-10) Prinz et al. (1976), MAR. Spinels are in spinel lherzolite; (11-12) Hodges and Papike (1976), Leg 37 MAR. Spinels are in mafic cumulates; (13-14) Ayuso (1976), Leg 11, Atlantic west of MAR. (15-18) Sigurdsson and Schilling (1976), MAR. Analyses are from the following rock types: #15 is magnesiochromite in olivine tholeiite; # 16 is Ti-magnesiochromite in olivine basalt; #17-18 are chromian spinels in picrites.

*C=core; M=margin; P=phenocryst; O=included in olivine; Px=included in pyroxene; Plag=included in plagioclase; Gl=glass; O/G=olivine-

Table Hg-20 (6)

Representative Spinel Compositions in Ultramafic Inclusions in Extrusive Rocks

Oxide	1	2	3	4	5	6	7	8	9	10	11	12	13	14	15	16	17	18
SiO_2	0.17	-	0.45	-	-	-	-	-	-	-	0.37	0.31	-	-	0.46	0.43	0.22	0.79
TiO_2	0.76	0.32	1.01	-	0.38	0.36	-	0.34	0.06	0.36	0.25	0.56	4.57	0.22	0.44	2.11	1.60	0.00
Al_2O_3	59.48	65.08	56.29	63.9	53.8	44.9	63.1	30.4	58.8	55.0	31.1	20.1	13.76	66.26	47.63	47.81	19.29	57.36
Cr_2O_3	0.02	0.52	0.01	2.08	2.40	18.0	2.50	37.8	9.4	9.1	31.0	31.1	33.46	-	19.24	14.05	34.87	9.72
Fe_2O_3	7.67	2.52	7.24	2.2	9.8	5.4	2.00	2.9	1.5	3.9	7.4	19.3	14.88	0.60	2.67	7.84	15.62	-
FeO	14.84	9.40	18.80	13.9	20.4	17.5	14.7	11.5	8.7	12.70	14.2	17.2	23.18	14.31	9.86	6.81	16.68	12.56
MnO	0.12	0.11	0.12	-	0.20	-	-	0.12	0.08	0.17	0.18	0.20	0.17	0.18	0.15	0.12	0.18	0.24
MgO	17.21	21.02	14.82	18.5	13.0	14.3	17.9	16.6	21.1	18.10	14.4	11.5	9.58	18.46	19.22	20.69	12.14	19.32
CaO	0.01	0.32	0.24	0.14	0.28	0.17	-	-	-	-	0.00	0.01	-	-	0.20	-	-	0.04
NiO	0.08	0.41	0.15	-	-	-	-	0.02	0.24	0.34	0.18	0.24	-	-	-	-	-	-
ZnO	0.12	0.09	0.14	-	-	-	-	0.03	0.02	0.15	0.22	0.00	-	-	-	-	-	-
V_2O_3	-	0.07	0.17	-	-	-	-	0.21	0.06	0.14	0.19	0.25	-	-	0.04	-	-	-
Total	100.48	99.86	99.44	100.7	100.3	100.6	100.2	99.92	99.96	99.96	99.5	100.8	99.60	100.02	99.91	99.90	100.68	100.03

Oxide	19	20	21	22	23	24	25	26	27	28	29	30	31	32	33	34	35	36
SiO_2	0.17	0.85	0.66	0.04	0.05	0.00	0.12	-	-	0.59	-	-	-	-	-	0.01	0.00	0.06
TiO_2	1.85	0.08	-	-	2.86	5.50	5.0	1.68	1.38	-	0.22	0.13	0.44	-	-	0.01	0.00	0.19
Al_2O_3	50.8	54.22	57.29	51.93	15.66	6.46	7.7	13.4	7.07	45.45	42.63	57.79	36.74	18.34	40.46	35.8	34.5	38.9
Cr_2O_3	0.16	11.50	8.07	14.58	43.65	25.4	53.0	53.5	60.8	16.85	27.76	10.98	30.80	50.01	26.54	34.3	35.8	27.9
Fe_2O_3	13.9	-	-	-	13.92	-	-	-	-	5.84	-	-	2.60	-	-	2.82	2.19	1.82
FeO	18.5	11.40	11.81	12.99	6.88	53.7	23.0	15.3	14.9	11.69	12.75	9.66	12.36	15.04	15.40	5.75	8.08	14.6
MnO	0.05	0.16	-	0.30	0.36	0.28	0.17	0.19	0.24	0.05	-	0.11	0.22	-	-	0.22	0.26	0.22
MgO	15.0	20.50	21.10	19.08	15.89	5.71	10.0	15.4	14.5	19.29	16.60	20.32	17.17	17.78	19.44	20.4	18.9	14.8
CaO	-	0.22	0.11	0.04	1.42	0.01	0.03	-	-	-	-	-	-	-	-	-	-	-
NiO	0.13	-	-	-	-	0.32	0.18	-	-	-	-	0.24	0.22	-	-	0.33	0.18	0.26
ZnO	0.11	-	-	-	-	-	-	-	-	-	-	-	-	-	-	-	-	-
V_2O_3	0.21	-	-	-	-	-	-	-	-	-	-	-	-	-	-	-	-	-
Total	100.9	98.93	99.04	98.96	100.86	97.4	99.7	99.4	98.8	99.80	99.96	99.23	100.55	101.17	101.84	99.6	99.9	98.8

References and rock type: 1. Binns (1969) megacryst in analcime basanite; 2-3. Binns et al. (1970) megacrysts in olivine nephelinite and nepheline trachybasalt; 4-7. Wilkinson (1975), websterite, olivine websterite, spinel lherzolite and wehrlite cumulate in analcimite; 8-10 Aoki and Prinz (1974), harzburgite, lherzolite and Fe-rich lherzolite in alkali-basalts; 11-12 Frisch (1971), dunite in tuff, core-mantle relationship; 13-14. McBirney and Aoki (1973), lherzolite and spinel gabbro inclusion respectively in alkalic-basalts; 15-17. Ross et al. (1954), megacrysts in basalts; 18. Francis (1976), pyroxene granulite in alkalic-basalt; 19. Frisch and Wright (1971), megacryst in alkali-basalt; 20-22. Griffen (1973), lherzolites in damkjernite; 23. Dawson et al. (1970), harzburgite in carbonatite tuff; 24-25. Dawson and Smith (1973), lherzolite and mica dunite respectively in carbonatite tuff; 26-27. Reid et al. (1975), garnet-free peridotite; 26 is core-spinel and 27 is a spinel mantle in carbonatite tuff; 28. Hamad (1963), dunite in basalt; 29-31. Frey and Green (1974), 29 is harzburgite, remainder are lherzolites in basanites; 32-33. Pike (1976), lherzolites in basanites; 34-36. Suwa et al. (1975), herzburgite, lherzolite and websterite respectively in olivine melanephelinite.
Localities: see tabulated summaries.

Table Hg-20 (7)

Representative Spinel Compositions in Ultramafic Inclusions in Kimberlites

Oxide	1	2	3	4	5	6	7	8	9	10	11	12	13	14	15	16	17	18
SiO_2	0.05	0.07	0.21	0.16	0.11	0.07	0.10	0.29	0.07	0.16	0.20	0.24	-	0.00	0.25	0.09	0.20	0.23
TiO_2	0.09	0.05	0.03	0.36	0.09	0.03	0.27	2.63	0.02	0.00	0.05	0.03	-	0.44	0.91	3.20	3.32	0.03
Al_2O_3	14.46	13.97	10.5	11.9	14.1	30.4	2.55	6.03	21.6	26.8	8.58	4.07	48.8	11.80	1.53	7.67	14.88	5.94
Cr_2O_3	55.05	56.07	58.3	58.2	54.9	40.1	68.70	59.2	49.0	42.4	62.2	66.90	20.2	53.5	58.5	51.9	46.19	64.0
Fe_2O_3	3.59	3.33	3.59	3.57	3.14	1.62	3.7	5.9	0.9	1.1	3.60	1.44	0.20	6.54	9.91	8.19	-	-
FeO	12.02	13.08	11.4	9.97	12.7	7.80	13.4	8.3	13.3	15.2	9.41	14.30	11.4	15.40	18.40	14.40	17.56	15.1
MnO	0.24	0.30	0.26	0.27	0.29	0.19	0.30	0.28	0.47	0.27	0.28	0.46	-	0.87	0.56	0.29	0.35	-
MgO	14.50	13.86	14.3	15.9	13.9	18.6	12.50	15.4	14.0	13.7	15.60	11.70	18.3	11.7	9.04	14.0	14.76	13.80
CaO	0.00	0.01	0.03	0.03	0.03	0.03	0.00	0.02	0.00	0.02	0.03	0.04	-	0.11	0.25	-	0.00	0.01
NiO	-	-	0.04	0.03	0.03	0.06	0.12	0.22	0.06	0.07	0.14	0.10	-	0.16	0.06	-	0.28	-
Total	100.00	100.74	98.6	100.3	99.2	98.8	101.60	98.4	99.40	99.6	100.10	99.20	98.9	100.5	99.4	99.7	97.54	99.1

References to analyses listed above: 1-2. Nixon and Boyd (1973); 3-6. Boyd and Nixon (1975); 7-10. Smith and Dawson (1975); 11-12. Danchin and Boyd (1976); 13-16. Emeleus and Andrews (1975); 17. Cox et al. (1973); 18. Meyer and Boyd (1972).

Rock types and localities: 1-2. granular lherzolites, Thaba Putsoa, Lesotho; 3. granular, graphite-bearing garnet-spinel harzburgite, Thaba Putsoa; 4. granular, graphite-bearing spinel harzburgite, Mothae, Lesotho; 5. granular spinel lherzolite, Thaba Putsoa; 6. granular spinel harzburgite, Liqhobong, Lesotho; 7. spinel-lherzolite, De Beers Mine, S. Africa; 8. spinel-rutile lherzolite, Bultfontein, S.A.; 9. spinel harzburgite, Jagersfontein, S.A.; 10. spinel harzburgite, Bultfontein; 11 and 12. gernet-harzburgite and spinel-harzburgite respectively, Premier Mine, S.A.; 13-16 S. W. Greenland, analyses are for spinel-peridotite, garnet peridotite, and for two phlogopite-peridotites; analysis 14 is for spinel in diopside, and 16 is for spinel in equilibrium with phlogopite; 17. spinel in sheared ultrabasic nodule, Matsoku pipe, Lesotho; 18. chrome-spinel inclusion in diamond, Ghana.

Table Hg-20 (8)

Representative Compositions of Spinels in Kimberlites

Oxide	1	2	3	4	5	6	7	8	9	10	11	12	13	14	15	16	17	18
SiO_2	-	-	-	-	-	-	-	-	-	0.10	0.97	-	-	-	-	-	-	-
TiO_2	0.10	0.63	1.20	3.81	5.24	9.17	12.90	16.48	0.05	18.0	22.6	2.40	1.34	1.98	3.45	18.2	12.0	4.08
Al_2O_3	36.82	17.31	15.34	7.67	8.90	10.69	9.24	10.69	0.05	7.8	8.0	9.63	52.0	47.1	47.2	7.68	13.8	9.12
Cr_2O_3	26.88	46.44	47.77	47.27	40.57	25.03	10.84	1.25	0.03	0.06	0.18	51.2	0.01	0.00	0.00	1.09	0.34	46.9
Fe_2O_3	4.76	7.18	6.39	11.01	13.45	19.78	28.10	30.61	67.38	-	-	-	-	-	-	-	-	-
FeO	10.55	15.36	15.44	15.23	17.94	20.52	23.28	26.61	28.71	58.6	51.5	22.3	23.2	29.1	29.1	58.9	57.8	26.1
MgO	17.46	12.80	12.71	13.40	12.62	13.33	13.24	13.57	0.85	13.4	15.2	13.5	21.9	20.7	20.4	13.5	15.0	12.9
CaO	-	-	-	-	-	-	-	-	-	0.13	0.20	-	-	-	-	-	-	-
MnO	0.15	0.34	0.33	0.44	0.49	0.47	0.50	0.47	0.26	0.73	0.60	0.41	0.17	0.19	0.24	0.45	0.62	0.31
NiO	-	-	-	-	-	-	-	-	-	-	-	-	-	-	-	-	-	-
V_2O_3	-	-	-	-	-	-	-	-	-	-	-	-	-	-	-	-	-	-
Total	98.51	100.36	99.18	98.82	99.21	98.99	98.21	99.0	97.33	98.82	99.25	99.4	98.6	99.1	100.4	99.7	99.5	99.4
	C	Int	M	C	Int	M	C	M	C	M	C	M	C	M	C	M		
Oxide	19	20	21	22	23	24	25	26	27	28	29	30	31	32	33	34	35	36
SiO_2	0.28	0.27	0.29	0.38	0.27	0.29	0.06	0.13	0.05	0.15	0.01	0.23	0.22	0.68	0.13	0.13	0.30	0.15
TiO_2	4.07	5.42	10.96	0.60	1.89	4.37	0.18	0.23	0.06	0.27	0.01	16.1	8.19	12.2	0.13	0.70	0.47	4.66
Al_2O_3	17.7	7.40	6.36	5.59	9.50	17.6	25.1	10.9	50.4	32.2	52.7	12.5	23.1	6.38	30.9	29.5	46.7	8.31
Cr_2O_3	39.2	25.3	2.50	60.1	53.5	37.9	46.0	59.0	15.7	35.6	7.51	2.95	17.4	11.0	35.6	29.5	16.3	48.8
Fe_2O_3	9.26	29.3	43.6	5.35	6.73	9.37	0.95	1.06	2.65	2.66	7.38	27.4	15.8	32.3	3.81	7.66	5.80	6.38
FeO	13.8	17.8	23.5	15.6	14.8	14.3	10.9	19.8	15.6	16.6	15.7	26.4	21.1	24.2	15.6	23.2	10.2	21.7
MgO	16.1	13.2	11.7	11.7	13.4	16.1	16.4	9.26	16.0	12.8	16.7	13.8	14.3	13.0	14.1	9.08	19.2	10.1
CaO	0.08	0.06	0.15	0.07	0.04	0.02	0.02	0.11	0.14	0.01	0.01	0.02	0.05	0.08	0.01	0.02	0.04	0.18
MnO	0.44	0.52	0.54	0.27	0.32	0.04	0.20	0.37	0.08	0.30	0.04	0.39	0.28	0.78	0.23	0.30	0.36	0.70
NiO	-	-	-	-	-	-	-	-	-	-	-	-	-	-	-	-	-	-
V_2O_3	-	-	-	-	-	-	-	-	-	-	-	-	-	-	-	-	-	-
Total	100.93	99.27	99.60	99.66	100.45	99.99	99.81	100.86	100.74	100.59	100.06	99.79	100.44	100.62	100.51	100.09	99.37	100.98

References to analyses: (1-9) Mitchell and Clarke (1976) Peuyuk, Somerset Island, N.W.T., Canada. Analyses #1-3 Al-Mg-chromite; #4-7 Ti-Mg-Al-chromite; #8 Magnesioferrian titanomagnetite; #9 Magnetite; (10-11) Dawson and Hawthorne (1973), Benfontein Sills, S. Africa. Cumulus Mg-Al-titanomagnetites. (12-18) Reid et al. (1975) Igwisi Hills, Tanzania. Analyses #12 Ti-Fe spinel enclosed in olivine; #13 Pleonaste mantle on ovoid olivine; #14-15 Ti-pleonaste mantles on titanomagnetite; #16-17 Matrix Mg-Al-titanomagnetites; #18 Mg-Al-Ti-chromite mantled by titanomagnetite; (19-36) Haggerty (1975). Analyses #19-26 Kao, Lesotho; #27-34 Lipelaneng, Lesotho; #35-36 Monastery, S. Africa. Data are plotted in Fig. Hg-37.

* C, Int and M are respectively core; intermediate core to margin; and margin of single grains.

Table Hg-20 (9)

Representative Compositions of Ilmenite and Ilmenite Reaction Mantles in Kimberlites

Oxide	C 1	M 2	3	4	B 5	W 6	W 7	A 8	N 9	N 10	N 11	I 12	Int 13	Out 14	PI 15	MFU 16	SI 17	TFP 18
SiO_2	-	-	-	-	-	-	-	-	-	-	-	-	-	-	0.05	0.49	0.08	0.15
TiO_2	58.45	57.45	53.9	56.3	49.48	49.57	55.15	54.35	45.55	44.90	42.83	52.29	51.74	52.88	42.63	26.23	56.40	20.61
Al_2O_3	0.06	0.06	0.42	0.37	0.18	0.57	0.43	0.35	0.27	0.22	0.19	0.50	0.45	0.40	0.17	1.06	0.30	10.99
Cr_2O_3	0.02	0.03	2.4	1.5	0.92	0.23	2.39	2.06	0.64	0.79	1.46	1.61	0.45	1.10	1.42	1.21	1.13	1.88
Fe_2O_3	-	-	-	-	11.68	12.50	6.27	6.72	-	-	-	-	-	-	0.33	21.78	1.91	24.23
FeO	26.17	30.88	26.3	27.9	28.72	26.05	15.00	21.76	44.67	48.62	49.64	33.49	34.37	31.68	32.17	32.40	19.90	18.02
MgO	13.46	10.66	14.4	13.8	8.72	10.28	19.26	15.05	6.63	5.14	4.49	11.37	11.11	12.72	12.02	15.47	20.23	22.80
CaO	0.29	0.30	-	-	-	-	-	-	-	-	-	-	-	-	0.05	0.24	0.10	0.18
MnO	0.65	0.58	0.33	-	-	-	-	-	0.30	0.20	0.20	0.10	0.20	0.20	0.39	0.64	0.52	0.72
NiO	0.05	0.04	0.19	-	0.24	0.22	0.27	0.30	-	-	-	-	-	-	-	-	-	-
Total	99.25	100.10	97.09	99.98	99.70	99.42	98.77	100.59	98.06	99.87	98.81	99.36	98.32	98.98	99.23	99.52	100.57	99.58

Representative Oxide Compositions in Carbonatite Suites

Oxide	→ 1	2	Mn-Mg-Al-Ti-magnetites 3	4	5	6	7	← 8	C.Mt 9	M.Mt 10	C.Mt 11	M.Mt 12	Ilm 13	Mt 14	Ilm 15	Ilm 16	Ilm 17	Ilm 18
SiO_2	-	0.13	0.38	1.10	0.34	0.65	0.16	0.83	-	-	-	-	-	1.3	-	0.0	-	-
TiO_2	2.40	5.01	6.98	1.31	4.70	1.76	5.32	3.71	1.29	1.39	0.91	0.78	51.85	12.9	51.4	46.9	49.3	50.3
Al_2O_3	10.37	15.14	3.62	6.57	6.80	0.07	0.62	0.15	0.92	0.34	0.09	0.18	0.07	3.2	0.0	0.04	-	-
Cr_2O_3	-	-	0.01	-	-	-	-	-	-	-	-	-	-	0.08	-	0.28	-	-
Fe_2O_3	59.01	46.95	57.11	61.95	62.39	67.94	59.71	64.28	63.82	65.04	65.30	65.72	2.97	36.9	1.9	11.1	-	-
FeO	16.82	26.58	21.83	18.72	23.24	26.95	22.70	24.52	31.82	32.04	32.17	32.38	40.69	45.2	41.5	35.6	41.8	43.1
MgO	9.47	5.83	7.18	6.74	2.59	0.15	3.24	0.60	2.05	1.25	0.20	0.20	0.52	0.33	0.10	0.15	5.1	4.2
CaO	-	0.11	0.94	-	-	0.30	-	1.58	0.05	0.05	0.08	-	-	0.18	-	0.13	-	-
MnO	2.10	-	1.82	-	-	2.19	8.46	3.71	0.67	0.64	1.04	0.36	4.95	0.18	4.7	6.5	2.6	3.2
V_2O_3	-	-	0.10	-	-	-	-	-	0.06	0.08	1.08	0.84	-	-	-	-	-	-
Total	100.17	99.75	99.97	99.79	100.06	100.39	100.21	99.69	100.27	100.86	100.87	100.46	101.05	100.27	99.6	100.7	98.8	100.8

KIMBERLITES. References to Analyses: (1-2) Clarke and Mitchell (1975), Peuyuk, Somerset Island, N.W.T., Canada; (3-4) Dawson and Hawthorne (1973), Benfontein Sills, S. Africa; (5-8) Mitchell (1973). B=Bultfontein, S. Africa; W=Wesselton, S. Africa; A=Ison Creek Pipe, U.S.A.; (9-14) Ferguson et al. (1973) Autolith study, Pipe 200, Lesotho. N=nucleus; I=inner; Int=intermediate; and Out=outer; these represent segments of the encasing kimberlite around an ilmenite nucleus; (15-18) Haggerty (1973), Liqhobong, Lesotho. Reaction mantles on primary picroilmenite (PI); MFU=magnesioferrite-ulvospinel; SI=secondary picroilmenite; and TFP=titanian-ferrian-pleonaste. This sequence proceeds outwards towards the kimberlitic groundmass.

CARBONATITES. References to Analyses: (1-8) As summarized by Fleischer (1965); analyses 1-3 Magnet Cove, Arkansas, U.S.A.; analyses (4-5) Kaiserstulh, Baden, Germany; (6) Mariupol (Zhdanov), U.S.S.R.; (7-8) Oka, Quebec, Canada; (9-13) Prins (1972). Regional East Africa analyses (9-12) are for core(C)-margin(M) pairs in magnetite; #13 is for oxidation exsolved ilmenite from a low Ti-magnetite; (14-16) Griffen and Taylor (1975), Fen Complex, Norway. Analyses are for magnetite and ilmenites in damkjerite. (17-18) Bergstøl (1972), Kodal, Vestfold, Norway. Analysis #17 is for discrete ilmenite and #18 is for oxidation exsolved ilmenite, both in jacupirangite and associated with Mt_{ss}+apatite.

Table Hg-20 (10)

ILMENITE COMPOSITIONS IN ILMENITE-PYROXENE LAMELLAR INTERGROWTHS

Oxide	1	2	3	4	5	6	7	8	9	10	11	12	13	14	15	16	17	18
SiO_2	0.04	0.04	0.10	0.08	0.05	0.00	0.10	0.09	0.10	0.18	0.08	0.14	0.14	0.12	0.00	0.21	0.08	0.20
TiO_2	48.77	50.22	49.50	51.46	48.30	53.32	52.40	51.90	54.00	47.68	49.96	48.70	47.05	53.90	54.48	48.60	50.92	54.90
Al_2O_3	0.50	0.55	0.25	0.55	0.61	0.77	0.70	0.60	0.40	0.97	0.75	0.77	0.75	0.95	0.14	0.01	1.08	0.90
Cr_2O_3	0.08	0.07	0.05	0.22	0.01	0.07	0.20	0.07	0.30	0.18	0.16	0.14	0.05	1.10	1.26	1.04	1.24	0.02
Fe_2O_3	9.63	8.38	12.0	7.93	14.0	0.73	8.6	9.2	7.2	14.84	10.84	15.3	15.38	15.26	-	-	12.80	-
FeO	28.65	30.12	28.0	29.19	28.1	39.35	27.8	27.5	26.5	25.40	26.73	25.7	26.44	16.00	31.29	44.75	18.28	30.8
MnO	0.23	0.23	0.23	0.22	0.20	5.35	0.2	0.3	0.2	0.21	0.2	0.23	0.23	0.26	0.31	4.31	0.20	0.25
MgO	11.82	8.31	9.16	9.43	8.52	0.81	10.7	10.6	12.3	9.67	10.07	10.1	8.80	11.80	12.34	0.41	15.30	12.80
CaO	0.08	-	0.03	0.09	0.01	0.06	0.05	0.05	0.05	0.06	0.02	0.03	0.05	0.08	0.06	0.10	0.07	0.04
Total	99.76	97.92	99.4	99.1	99.8	100.94	100.7	100.3	101.0	99.36	98.93	101.1	98.91	99.47	99.96	99.43	99.9	99.9

References for analyses as listed above: 1. Ringwood and Lovering (1970); 2. Dawson and Reid (1970); 3 and 12. Boyd and Nixon (1975); 4 and 17. Gurney et al. (1973); 5. Boyd (1971); 6. Mitchell et al. (1973); 7, 8 and 9. Boyd and Dawson (1972); 10. Nixon and Boyd (1973a); 11. Nixon (1973); 13. Boyd and Nixon (1973b); 14 and 16. Gurney and Haggerty (1975); 15. Frisk (1974, preprint); 18. McCallister et al. (1975). Analysis #5 is for Ilm associated with Opx; analysis 9 is for garnet + Ilm; 16 is for antigorite + Ilm. Mn-Ilm in analysis #6 is secondary, but in analysis #16 is primary. Data for NiO in excess of 0.1 wt% are reported for analyses 3, 7, 8, 9, 10, 11, 12 and 15. Localities: 1-6: Monastery mine, S.A.; 7: Uintjesburg, S.A.; 8: Sonop pipe, S.A.; 9: Excelsior pipe, S.A.; 10: Kao, Lesotho; 11: Mothae, Lesotho; 12: Thaba Putsoa, Lesotho; 13: Pipe 200, Lesotho; 14-16: Premier mine, S.A.; 17: Riley County kimberlite, Kentucky, U.S.A.; 18: Stockdale pipe, Kansas, U.S.A.

Table Hg-20 (11)

"Ferritchromit"

Cases	No Pattern	Enriched in Chromite	Enriched in "Ferritchromit"
1		Al, Mg	Fe, Cr
2		Al, Mg, Cr	Fe
3	Al, Mg, Cr		Fe
4		Al, Mg, Cr	Fe
5	Cr	Al, Mg	Fe
6		Al	Fe
7		Cr	Fe
8		Al, Mg, Cr, Zn	Fe, Ti, Ni
9		Al, Mg	Fe, Cr, Ni, Mn, Co, Ti, Zn, V
10		Al, Mg, Cr	Fe, Co, Zn, Ni
11		Al, Mg, Fe^{2+}	Cr, Fe^{3+}
12		Al, Mg	Fe, Cr
13		Al, Mg, Cr	Fe
14		Al, Mg, Cr	Fe
15		Cr, Ti, Fe^{2+}, Fe^{3+}	Al, Mg
16		Al, Mg, Cr	Fe^{2+}, Fe^{3+}, Ni

Cases 1-10 are as summarized by Onyeagocha (1974); 11. Golding and Bayliss (1968); 12. Engin and Aucott (1971); 13. Springer (1974); 14. Ulmer (1974); 15. Hamlyn (1975); 16. Bliss and Maclean (1975).

Oxide	Chr 1	"Ferrit" 2	Chr 3	Mt 4	Chr 5	"Ferrit" 6	"Ferrit" 7	"Ferrit" 8	Chr 9	"Ferrit" 10
TiO_2	0.08	0.13	0.05	0.02	0.29	0.21	0.78	0.14	2.14	0.00
Al_2O_3	17.5	9.9	18.1	0.56	15.27	9.25	0.08	0.15	15.1	63.6
Cr_2O_3	51.1	55.5	50.4	4.2	48.32	47.24	28.65	2.93	39.7	0.88
Fe_2O_3	4.1	7.9	3.9	64.9	7.47	15.14	41.58	66.02	9.2	1.2
FeO	12.4	14.1	12.6	27.2	17.57	18.21	23.57	29.46	30.7	18.9
MnO	0.44	0.55	0.4	0.45	-	-	-	-	-	-
MgO	14.3	12.4	14.2	1.7	11.07	9.89	4.88	0.96	2.14	15.1
CoO	0.07	0.10	0.05	0.21	-	-	-	-	-	-
ZnO	0.05	0.10	0.05	0.24	-	-	-	-	-	-
NiO	0.08	0.13	0.09	0.60	0.01	0.06	0.45	0.61	-	-
V_2O_3	0.16	0.19	0.12	0.13	-	-	-	-	-	-
Total	100.28	101.00	99.99	100.21	100.0	100.0	100.0	100.0	99.0	99.7

Analyses are for chromite + "ferritchromit", and chromite + magnetite pairs, or for chromite with successive (intermediate) zones of alteration. The examples quoted are representative of the three contrasting enrichment trends reported for "ferritchromit": (a) Fe(Fe^{2+}, Fe^{3+}) + Cr + Ni enrichment, Mg + Al depletion (# 1-2); (b) Fe (Fe^{2+}, Fe^{3+}) enrichment, Mg + Al + Cr depletion (# 3-4 and 5-8); (c) Mg + Al enrichment, Fe + Cr + Ti depletion (# 5-6). References: Analyses 1-4. Twin Sisters dunite, Washington. Onyeagocha (1974). Analyses 5-8. Serpentinized ultramafic, S.W. of the Manitoba Nickel Belt. Bliss and MacLean (1975). Analyses 9-10. Panton ultramafic sill, Western Australia. Hamlyn (1975

Table Hg-21

EXPERIMENTAL DETERMINATIONS OF T° AND fO_2 IN NATURAL SYSTEMS

Rock Type and Locality	T°C	log 10 fO_2	Initial FeO	Initial Fe_2O_3	References
Tholeiitic basalt. 1921 Kalaueu flow Hawaii	1200	-8.2	9.40	1.84	Fudali(1965)
Andesite basalt. Hawaii	1200	-7.2	9.07	3.20	
Hawaiite (andesine andesite). Hamakua series, Hawaii.	1200	-6.8	7.42	3.54	
Basalt. Picture Gorge, Oregon.	1200	-6.4	9.07	4.89	
Basalt. Late Yakima Group, Oregon.	1200	-8.0	9.95	2.10	
Olivine basalt. Hood River Quad., Oregon	1200	-8.5	8.70	1.38	
Olivine augite andesite. Lost Lake Butte,Mt. Hood area.	1200	-7.2*	5.78	2.34	
Hypersthene augite andesite. Parksdale District, N. base, Mt. Hood	1200	-8.2*	5.12	1.37	
Hypersthene augite andesite. Mt. Hood Volcano.	1200	-7.0*	4.03	1.63	

* A comparison of these samples by other methods is discussed hereunder by Ulmer et al. (1976).

Review

fO_2's in equilibrium with original FeO/Fe_2O_3 ratios. Experimental range in $fO_2 = 10^{-0.21}$ to $10^{-8.5}$ atm; gas-mixing at 1200°C.

Hypersthene augite andesites and an olivine augite andesite, Mt. Hood area. (see Fudali, 1965 above). — Ulmer et al. (1976)

A comparative study of geothermometric and geobarometric data by three independent techniques: (1) synthetic fO_2 reequilibration at bulk rock liquidus temperatures, see above (Fudali, 1965); (2) coexisting Fe-Ti oxide pairs, see Chapter 2, (Buddington and Lindsley, 1964); and (3) T-fO_2 determinations utilizing the ZrO_2 electrolytic cell technique, see below (Sato, 1970, 1971, 1972).

Rock Type	Method 1	Method 2	Method 3	
hypersthene augite andesite	-8.2	-8.0	-7.8	1200°C
hypersthene augite andesite	-7.0	-8.0	-7.5	1200°C
Olivine augite andesite	-7.2	-8.0	-8.2	1200°C
Error	±0.1	±0.7	±0.15	
	- log 10 fO_2			

Alkali basalt. Fayal Island, Azores. — Duke (1974)

Crystallization of bulk rock from liquidus temperatures. Range in T = 900-1200°C; $fO_2 = 10^{-7}$ to 10^{-14} atms. Gas mixing and fO_2 monitoring using the ZrO_2 electrolytic technique (Sato, 1970, 1971, 1972). Results: (1)Above 980°C ($fO_2 = 10^{-11.5}$) crystallization is above the FMQ buffer curve; Below 980°C crystallization is more reducing than the FMQ buffer; (2) the effects of oxidation are to: (a) increase the stabilities of spinel + Cpx, and to decrease the stabilities of Ol + Plag; (b) promote crystallization of a magnesioferrite-rich ($MgFe_2^{3+}O_4$) spinel and acmite; and (c) cause nepheline normative liquids to become saturated and oversaturated with respect to SiO_2.

Table Hg-21 (2)

EXPERIMENTAL DETERMINATIONS OF T^o AND fO_2 IN NATURAL SYSTEMS

Rock Type and Locality	Review	References
Stillwater Complex, G-Zone. Montana, U.S.A.	Coexisting cumulus chromite + olivine. Experimental range in T = 850-1200°C; fO_2 = 10^{-17} to 10^{-5} atms; gas-mixing and ZrO_2 electrolytic cell. Intrinsic fO_2 range for olivine and for chromite between 850 and 1200°C lie between the MW and WI buffer curves. Intrinsic curves for olivine and chromite intersect at T = 993°C; fO_2 =-13.6. The respective values at 850°C and 1200°C are: olivine = $10^{-15.9}$ and $10^{-9.9}$ atms; and chromite = $10^{-17.0}$ and $10^{-9.4}$ atms. Based on textural evidence that olivine and chromite coprecipitated (Jackson, 1961) and assuming T = 1200°C, then the magma probably had an initial fO_2 value = $10^{-9.6}$ atms.	Sato (1972)
Olivine phenocrysts in Hawaiian tholeiitic pillow basalt.	Experimental conditions as above. Intrinsic fO_2 curve falls close to the WI buffer curve and at 850 and 1200°C fO_2 values are 10^{-17} and 10^{-11} respectively. Assuming that phenocryst crystallization of T = 1250°C extrapolation yields $10^{-10.3}$ atms. which is considered low (c.f. Stillwater magma ~ $10^{-9.6}$, and Fudali's data listed above for the 1921 Kilaueu flow, ~ 10^{-8} atms.) and oxidation of the lava must have occured during intrusion, probably as a result of hydrogen loss.	Sato (1972)
Bushveld Complex, F-Horizon. S. Africa.	Experimental procedure as above. Mineral assemblages examined yield 3 separate interscetions for the following assemblages: (1) chromite-bronzite: T=1125°C; fO_2=$10^{-11.4}$ atms; (2) chromite-bronzite-plagioclase (An_{78}): T=1090°C; fO_2=10^{-12} atms; (3) chromite-plagioclase (An_{78}): T= 1050°C; fO_2=$10^{-12.5}$. A combined intersection among the assemblages yields T=1085$\pm$20°C; fO_2=$10^{-11.85}\pm10^{-0.40}$ atms. The temperature estimate is consistent with an independent experimental solidus determination which yields 1110°C$\pm$5°C (at 1 atm.) for the assemblage Chr+plag+bronzite.	Flynn et al. (1972); Ulmer et al. (1976)
Oka Carbonatite Complex, Quebec, Canada.	Experimental procedure as for Sato (1972). Range of T^{o}C and fO_2 = 600°-1000°C and 10^{-11} to 10^{-20} atms. respectively. Intrinsic fO_2 curves for olivine, Fo_{80}, magnetite and latrappite (Nb-perovskite) in carbonatite fall within a narrow envelope between the MH and WI buffer curves. The point of triple intersection yields T=710$^o\pm$15°C and fO_2 = $10^{-17.1}\pm10^{-0.5}$ atms. The estimate of T^{o}C is consistent with estimates derived from O^{18}/O^{16} studies, but subsequent data show that the magnetites are zoned and contain up to 10 wt% MnO (MacMahon and Haggerty, 1976).	Friel and Ulmer (1974)

Table Hg-21 (3)

Friel and Ulmer (1974) cont.

T^oC	$-\log fO_2$ Latrappite	Magnetite	Olivine
600 ± 12	19.5	-	18.5
650 ± 12	18.5	17.3	17.7
700 ± 12	17.2	17.0	17.2
750 ± 10	16.00	-	-
800 ± 8	14.6	15.3	15.7
850 ± 5	14.3	14.6	-
900 ± 5	13.1	14.0	14.6
950 ± 3	12.3	13.3	14.0
1000 ± 3	11.1	-	-

Kimberlite and a gem diamond from the Premier Mine, S. Africa.	Experimental procedure as for Sato (1972). Bulk kimberlite determination yields an intrinsic fO_2 curve with the following data points: (1) T=1000°, $fO_2=10^{-10}$; (2) T=850°, $fO_2=10^{-14}$; (3) T=725°C, $fO_2=10^{-17}$ atms. At T<850°C the FMQ buffer curve is intersected and the kimberlite is more reduced than FMQ; at T>850°C the kimberlite is more oxidized with respect to FMQ. Comparable experiments on gem diamond fall below the FMQ buffer and the oxidation activation temperature is inferred to be 960°C ± 15°C. The interpretation of these data suggests that diamond will oxidize in a cooling kimberlite matrix if T>960°C; below this temperature diamond is stable even although the matrix may have an oxidative capacity for oxidizing diamond. Diamond-containing kimberlites therefore are most likely to have groundmass intrusion temperatures between 725-850°C and fO_2 10^{-14} to 10^{-17} atms. or less.	Ulmer et al. (1976)
Basalt. Makaopuhi Lava Lake, Hawaii.	ZrO_2 electrolytic cell in-situ determinations of T°C and fO_2 in drill cores in an actively cooling lava lake. Range in T°C and fO_2 fall close to the FMQ buffer curve except for an anomalous zone in one drill hole where fO_2 reaches a max. of 10^{-5} atms. between 550-750°C (see chapter 5 for details). First appearance of opaque oxides is at 1070°C ± 10°C (ilmenite$_{ss}$ followed by titanomagnetite).	Sato and Wright (1966)
Basalt. Kilauea, Hawaii.	Similar experiment to that described above; first appearance of opaque mineral oxides is at 1070° ± 10°C, $fO_2 = 10^{-9.6} \pm 10^{-0.6}$ atms.	Peck and Wright (1966)

Table Hg-21 (4)

EXPERIMENTAL DETERMINATIONS OF T^o AND fO_2 IN NATURAL SYSTEMS

Rock Type and Locality	Review	References
Basalts. Picture Gorge, Oregon; the 1921 Kilauea olivine tholeiite and the 1801 Hualalai alkali basalt, Hawaii.	Phase relations of basalts in their melting range at P_{H_2O} = 5 Kb as a function of fO_2. Solid buffer technique (MH and FMQ buffers) in the range T=680-1000°C. The opaque mineral assemblages are as follows: FMQ=Ilm_{ss}+Usp_{ss}+sphene; MH=Pb_{ss}+Hem_{ss}+sphene. The 1921 olivine tholeiite contains sphene (FeO=1.7 wt%) in association with Ilm_{95} (FMQ 678-846°C), and in association with Pb_{91}+Hem_{ss} (MH 683-846°C); similarly in the 1801 lava, sphene is present in FMQ runs (725-823°C) in association with Ilm_{ss} but is absent in MH runs. A significant result shows that T^{o}C and fO_2 determinations using co-equilibrated oxide pairs (Buddington and Lindsley, 1964), which was possible only for the Picture Gorge basalts and FMQ runs, are well within the limits of experimental error. Exp T^{o}C / Oxide T^{o}C / Exp fO_2 / Oxide fO_2: 875 / 870 / -13.2 / -13.2 930 / 930 / -12.2 / -12.3 970 / 975 / -11.5 / -11.3 Coexisting Pb_{51-57}+Hem_{73-76} (MH and T=925-1000°C) have tie-line slopes comparable to those found in deuterically oxidized lavas (see chapter 5). Ilmenite compositions in FMQ runs between 875 and 1000°C from the 1921 lava contain significant MgO contents: 875°=1.4; 930°=3.9; 970°=5.5; and 1000°C=5.1 wt%. These high MgO values are considered to be due to initial compositional differences between this lava and the remaining two basalts.	Tuthill Helz (1973)
1921 Kilauea olivine tholeiite, Hawaii; Precambrian diabase dike, Kingston, Ontario.	Crystallization of spinels from basaltic liquids as a function of T= 1100-1325°C and fO_2=$10^{-0.68}$ to 10^{-14} atms. These data have a significant and direct bearing on the hiatus between chromite and Mt-Usp_{ss} commonly observed in basalts and in layered intrusions; and confirm also the earlier prediction by Irvine (1967) that the break in spinel solid solution (i.e. between the early crystallization of Chr_{ss} and the later precipitation of Mt-Usp_{ss}) is due to a peritectic-like reaction which involves the precipitation of pyroxene + plagioclase. Spinel crystallization is interrupted at values of fO_2 below approximately 10^{-8} atms. (~1120-1150°C) by the precipitation of Cpx; early spinels are Cr-enriched and later spinels are Fe+Ti enriched. Variations in oxide compositions at constant T=1200°C as a function of fO_2, and of constant fO_2=10^{-7} atms. but variable T=1145-1200°C are shown in Fig 18 . The maximum solubility of Cr in basaltic liquid is 200 ppm at T=1200°C, fO_2=10^{-8} atms; Cr is strongly partitioned into crystalline chromite and is approximately 1000 fold larger than that of the coexisting liquid.	Hill and Roeder (1974)

Table Hg-21 (5)

T-f_{O_2} Curves

ffer		Reference
	$fO_2 = 13.12 - \frac{32,730}{T} = -A$	Eugster and Wones (1962)
essure correction	$fO_2 = -A + .083 \frac{(P-1)}{T}$	
Q	$fO_2 = 9 - \frac{25,738}{T} = -A$	Wones and Gilbert (1969)
essure correction	$fO_2 = -A + .092 \frac{(P-1)}{T}$	Eugster and Wones (1962)
O	$fO_2 = 9.36 - \frac{24930}{T} = -A$	Huebner and Sato (1970)
essure correction	$fO_2 = -A + .046 \frac{(P-1)}{T}$	Eugster and Wones (1962)
$_{1-x}O-Mn_3O_4$	$fO_2 = 13.38 - \frac{25680}{T}$	Huebner and Sato (1970)
	$fO_2 = 14.41 - \frac{24912}{T} = -A$	Eugster and Wones (1962)
essure correction	$fO_2 = -A + .019 \frac{(P-1)}{T}$	Eugster and Wones (1962)
$_3O_4-Mn_2O_3$	$fO_2 = 7.34 - \frac{9265}{T}$	Huebner and Sato (1970)
kaopuhi # 9	$fO_2 = 6.44 - \frac{21900}{T}$	Sato and Wright (1966)

T = °K; P = bars.

Table Hg-22

ANALYTICAL TECHNIQUES

METHOD	COMMENTS	REFERENCES
Magnetic Property Measurements		
Techniques and Applications:		
	Sampling techniques; instrumentation for measurement of magnetic properties; procedures for magnetic stability tests; isotropic and anisotropic susceptibility measurements; applied field techniques; magnetic component measurements (optical and X-ray; chemical; magnetic; low temperature identification; magnetic domain structure).	Collinson et al. (1967)
	Review and update of techniques.	Collinson (1975)
Standard Texts:		
	Magnetism and the chemical bond.	Goodenough (1963)
	Rock magnetism.	Nagata (1961)
	Physical principles of rock magnetism.	Stacey and Banerjee (1974)
	Physics of the earth.	Stacey (1969)
	Paleomagnetism - application to geological and geophysical problems.	Irving (1964)
	Applications of paleomagnetism.	McElhinny (1973)
Reviews:		
	Magnetic properties of rocks and minerals.	Lindsley et al. (1966)
	Magnetic properties of minerals.	Akimoto (1966)
	Rock magnetism.	Runcorn (1966)
	Paleomagnetism and rock magnetism.	Wilson (1966)
	Magnetic properties of rocks.	Verhoogen (1969)
	Theory and nature of rock magnetism.	Hargraves and Banerjee (1973)
	Magnetic effects associated with chemical changes in igneous rocks.	Merrill (1975)
	Recent developments in magnetic materials.	Wernick (1972)
	Theory of magnetic viscosity of lunar and terrestrial rocks.	Dunlop (1973)
	Lunar magnetism.	Fuller (1974)
	Meteoritic magnetism.	Brecher and Arrhenius (1974); Larson et al. (1974); Stacey (1976).
Noteworthy Publications:		
	Superparamagnetic and single domain threshold sizes in Fe_3O_4; indices of multidomain behavior of Fe_3O_4 in igneous rocks; TRM of submicroscopic Fe_3O_4.	Dunlop (1973); Dunlop et al. (1973a); Dunlop et al. (1973b)
	Cationic distributions in titanomagnetites.	O'Reilly & Banerjee (1967); Readman and O'Reilly (1971); Stephenson (1969); Stephenson (1972); Bleil (1971)
Special Techniques:		
	Magnetic colloid; ionic bombardment and acid etching.	Petersen (1962); Soffel (1968); Soffel (1970); Soffel and Petersen (1971); Merrill and Kawai (1969)
Mossbauer Spectroscopy		
Standard Texts:		
	Principles and applications.	Frauenfelder (1962); Wertheim (1964); Goldanskii and Herber (1968)
Applications to Fe-Ti-Cr Oxides:		
	Fe and Ti spinels with local disorder.	Banerjee et al. (1967a)
	Behavior of Fe^{2+} in Fe-Ti spinels.	Banerjee et al. (1967b); Jensen and Shrive (1973)
	Cationic distribution in Fe_2TiO_4.	Rossiter and Clark (1965)
	Cationic distributions in pseudobrookites.	Virgo and Huggins (1975)
	Fe^{3+} : Fe^{2+} ratios in titanomagnetites.	Watt et al. (1973)

Table Hg-22 (2)

X-Ray Diffraction	
Standard Texts:	
Elements of X-ray crystallography and the powder method.	Azaroff (1962); Azaroff and Buerger (1963)
Tables for identification.	Berry and Thompson (1962)
International X-ray diffraction card index.	
Applications to Fe-Ti Oxides:	See Chapter I (this volume)
Electron, Ion and Laser Microprobe Analyses	
Standard Texts:	
Instrumentation, procedures and applications to microchemistry.	Heinrich (1966); Andersen (1973); Reed (1975); Goldstein and Yakowitz (1975); Smith (1976).
Noteworthy Publications:	
Ion microprobe mass analyzer.	Andersen and Hinthorne (1972)
Fe^{2+} : Fe^{3+} determinations and Lα X-ray emmission spectra.	Snetsinger (1969); Albee and Chodos (1970); O'Nions and White (1971)
Correction factors.	Ziebold and Ogilvie (1964); Bence and Albee (1968); Sweetman and Long (1969); Albee and Ray (1970)
Applications to Oxides:	See Chapters 3-8 (this volume)
Electron, Scanning and Transmission Microscopy	
Standard Texts:	
Principles, techniques and applications.	Hall (1953); Holt et al. (1974); Andersen (1973); Goldstein and Yakowitz (1975); Wenk (1976)
Applications to Oxides:	
Magnetite-ulvospinel exsolution.	Nickel (1968)
Titanomagnetite-ilmenite intergrowths.	Lewis (1963); Evans and Wayman (1970); Evans and Wayman (1972); McClay (1974); Hoblitt and Larson (1975); Davis and Evans (1976)
Ilmenite-hematite intergrowths.	Lally et al. (1976)
Reflection Microscopy	
Standard Texts:	
Physical and optical determinative methods: microhardness indentation techniques, reflectivity measurements, optical rotational properties.	Zussman (1967); Freund (1966); Cameron (1966); Hallimond (1970); Galopin and Henry (1972); Nakhla et al. (1973); Bowie et al.(1975)
Noteworthy Publications:	
Microhardness.	Millman and Young (1963)
Microhardness vs. composition in magnesian ilmenites.	Morton and Mitchell (1972)
Reflectivity vs. microhardness.	Millman and Kingston (1964); Chamberlain et al. (1967)
Textures and Genesis:	
Microscopic identification, textural intergrowths, comparative tables, and interpretation.	Ramdohr (1969); Oelsner (1966); Uytenbogaardt and Burke (1972); Galopin and Henry (1972); Cameron (1966); Edwards (1965); Bastin (1950); Shouten (1962); Deer et al. (1972)
Photomicrographic card index.	Maucher and Rehwald (1961-)

Noteworthy Publications:	
Textural interpretation of oxides.	Ramdohr (1969); Edwards (1965); Buddington and Lindsley (1964)
Ulvospinel-magnetite exsolution.	Ramdohr (1953); Vincent and Phillips (1954); Vincent (1960); Vaasjoki and Heikkinen (1962); Tsvetkov et al. (1965); Nickel (1968)
Ilmenite-hematite exsolution.	Ramdohr (1969); Carmichael (1961); Edwards (1965); Kretchsmar and McNutt (1971)
Ulvospinel-magnetite$_{ss}$ oxidation.	Ramdohr (1969); Vincent and Phillips (1954); Buddington and Lindsley (1964); Duchesne (1970)
Ilm_{ss} and Usp_{ss} oxidation.	See Chapter 4 (this volume)

<u>Petrographic Objectives</u>

Oxide Thermometry and Geobarometry:	
Experimental T°C and fO_2 determinative curves.	Buddington and Lindsley (1964)
Computational methods and procedures for calculating mole % Fe_2TiO_4 and mole % $FeTiO_3$ from microchemical and chemical data.	Buddington and Lindsley (1964); Carmichael (1967); Anderson (1968); Lindsley and Haggerty (1970); Rumble (1973).
Data Plotting Procedures:	
Ternary FeO-Fe_2O_3-TiO_2	See Fig. Hg-23 (this chapter)
Ternary Fe_2O_3-$FeTiO_3$-$MgTiO_3$	See Fig. Hg-42 (this chapter)
Spinel prisms	Stevens (1944); Jackson (1963); Irvine (1965); Haggerty (1971); Haggerty (1972); Busche et al. (1972). See Fig. Hg-24 (this chapter)

Abbott, D. (1962) The gabbro cumulates of the Kap Edvard Holm complex, east Greenland. Unpublished Ph.D. thesis, University of Manchester.

Ade-Hall, J. M. (1963) Electron probe microanalyser analyses of basaltic titanomagnetites and their significance to rock magnetism. *Geophys. J. Royal astr. Soc. 8*, 301.

________ (1964) A correlation between remanent magnetism and petrological and chemical properties of tertiary basalt lavas from Mull, Scotland. *Geophys. J. Royal astr. Soc. 8*, 403.

________ (1964) The magnetic properties of some submarine oceanic lavas. *Geophys. J. 9*, 85.

________ (1969) Opaque petrology and the stability of natural remanent magnetism in basaltic rocks. *Geophys. J. Royal astr. Soc. 18*, 93.

________ (1974) Strong field magnetic properties of basalts from DSDP Leg 26. In, T. A. Davies and B. P. Luyendyk, Eds., *Initial Reports of the Deep Sea Drilling Project, 26*, 529, U. S. Government Printing Office, Washington, D. C.

________ (1974) The opaque mineralogy of basalts from DSDP Leg 26. In, T. A. Davies and B. P. Luyendyk, Eds., *Initial Reports of the Deep Sea Drilling Project, 26*, 533, U. S. Government Printing Office, Washington, D. C.

________, and N. D. Watkins (1970) Absence of correlations between opaque petrology and natural remanence polarity in Canary Island lavas. *Geophys. J. Roy. astr. Soc. 19*, 351.

________, and R. L. Wilson (1963) Petrology and the natural remanence of the Mull lavas. *Nature 198*, 659.

________, and R. L. Wilson (1969) Opaque petrology and natural remanence polarity in Mull (Scotland) dykes. *Geophys. J. Roy. astr. Soc. 18*, 333.

________, L. K. Fink, and H. P. Johnson (1976) Petrography of opaque minerals, Leg 34. In, S. R. Hart and R. S. Yeats, Eds., *Initial Reports of the Deep Sea Drilling Project, 34*, U. S. Government Printing Office, Washington, D. C.

________, M. A. Khan, P. Dagley, and R. L. Wilson (1968) A detailed opaque petrological and magnetic investigation of a single tertiary lava flow from Skye, Scotland - I - Iron-titanium oxide petrology. *Geophys. J. Roy. astr. Soc. 16*, 375.

________, M. A. Khan, P. Dagley, and R. L. Wilson (1968) A detailed opaque petrological and magnetic investigation of a single tertiary lava flow from Skye, Scotland - II - Spatial variations of magnetic properties and selected relationships between magnetic and opaque petrological properties. *Geophys. J. Roy. astr. Soc. 16*, 389.

________, M. A. Khan, P. Dagley, and R. L. Wilson (1968) A detailed opaque petrological and magnetic investigation of a single tertiary lava flow from Skye, Scotland - III - Investigations into the possibility of obtaining the intensity of the ambient magnetic field (F_{ANC}) at the time of the cooling of the flow. *Geophys. J. Roy. astr. Soc. 16*, 401.

________, H. P. Johnson, and P. J. C. Ryall (1976) Rock magnetism of basalts, Leg 34. In, S. R. Hart and R. S. Yeats, Eds., *Initial Reports of the Deep Sea Drilling Project, 34*, U. S. Government Printing Office, Washington, D. C.

________, H. C. Palmer, and T. P. Hubbard (1971) The magnetic and opaque petrological response of basalts to regional hydrothermal alteration. *Geophys. J. Roy. astr. Soc. 24*, 137.

________, R. L. Wilson, and P. J. Smith (1965) The petrology, Curie points and natural magnetizations of basic lavas. *Geophys. J. Roy. astr. Soc. 9*, 323.

________, F. Aumento, P. J. C. Ryall, R. E. Gerstein, J. Brooke, and D. L. McKeown (1973) The mid-Atlantic ridge near 45°N, 21, magnetic results from basalt drill cores from the median valley. *Can. J. Earth Sci. 10*, 676.

________, P. Dagley, R. L. Wilson, A. Evans, A. Riding, P. J. Smith, R. Skelhorne, and T. Sloan (1972) A paleomagnetic study of the Mull regional dyke swarm. *Geophys. J. Roy. astr. Soc. 27*, 517.

Ahrens, L. H., J. B. Dawson, A. R. Duncan, and A. J. Erlank. *Physics and Chemistry of the Earth 9*, 940p.

Akimoto, S. (1966) Magnetic properties of minerals. In, S. K. Runcorn, Ed., *International Dictionary of Geophysics 2*, 861, Pergamon Press.

Akimoto, S-i. and I. Kushiro (1960) Natural occurrence of titanomaghemite and its relevance to the unstable magnetization of rocks. *J. Geomag. Geoelect. 11*, 94.

________, T. Nagata, and T. Katsura (1957) The $TiFe_2O_5$-Ti_2FeO_5 solid solution series. *Nature 179*, 37.

Albee, A. L. and A. A. Chodos (1970) Semiquantitative electron microprobe determination of Fe^{2+}/Fe^{3+} and Mn^{2+}/Mn^{3+} in oxides and silicates and its application to petrological problems. *Am. Mineral. 55*, 491.

________, and L. Ray (1970) Correction factors for electron probe microanalysis of silicates, oxides, carbonates, phosphates, and sulfates. *Anal. Chem. 42*, 1408.

Andersen, C. A. (1973) *Electron Microprobe Analysis.* John Wiley & Sons, 571p.

________, and J. R. Hinthorne (1972) Ion microprobe mass analyzer. *Science 175*, 853.

Anderson, A. T., Jr. (1966) Mineralogy of the Labrieville anorthosite, Quebec. *Am. Mineral. 51*, 1671.

________ (1968a) The oxygen fugacity of alkaline basalt and related magmas, Tristan da Cunha. *Am. J. Sci. 266*, 704.

________ (1968b) Oxidation of the LaBlache Lake titaniferous magnetite deposit, Quebec. *J. Geol. 76*, 528.

________ (1975) Some basaltic and andesitic gases. *Rev. Geophys. Space Phys. 13*, 37.

________, and M. Morin (1969) Two types of massif anorthosites and their implications regarding the thermal history of the crust. In, Y. N. Isachsen, Ed., *Origin of Anorthosites and Related Rocks*, N. Y. State Museum and Science Service, Memoir *18*, 57.

________, and T. L. Wright (1972) Phenocrysts and glass inclusions and their bearing on oxidation and mixing of basaltic magmas, Kilauea volcano, Hawaii. *Am. Mineral. 57*, 188.

Anderson, R. N., D. A. Clague, K. D. Klitgord, M. Marshall, and R. K. Nishimori. Magnetic and petrologic variations along the Galapagos spreading center and their relation to the Galapagos melting anomaly. *Geol. Soc. Am. Bull. 86*, 683.

Aoki, K. (1966) Phenocrystic spineliferous titanomagnetites from trachyandesites, Iki Island, Japan. *Am. Mineral. 51*, 1799.

________, and M. Prinz (1974) Chromian spinels in lherzolite inclusions from Itinomegata, Japan. *Contrib. Mineral. Petrol. 46*, 249.

Arculus, R. J. (1974) Solid solution characteristics of spinels: pleonaste-chromite-magnetite compositions in some island-arc basalts. *Ann. Rep. Dir. Geophys. Lab., Year Book 73*, 322.

Ayuso, R. A., A. E. Bence, and S. R. Taylor (1976) Upper Jurassic tholeiitic basalts from DSDP Leg 11. *J. Geophys. Res. 81*, 4305.

Aziaroff,L. V. (1962) *Elements of X-ray Crystallography.* McGraw Hill Company.

________, and M. J. Burger (1958) *The Powder Method in X-ray Crystallography.* McGraw Hill Company, 342p.

Babkine, J., F. Conquéré, J.-C. Vilminot, and K. D. Phan (1965) Les spinelles des basaltes de Monistrol d'Allier (chaîne du Devès, Haute-Loire). *Bull. Soc. franc. Minêral. Cristallogr. 88*, 447.

Bacon, C. R. and I. S. E. Carmichael (1973) Stages in the P-T path of ascending basalt magma: an example from San Quintin, Baja California. *Contrib. Mineral. Petrol. 41*, 1.

Badham, J. P. N. and R. D. Morton (1975) Magnetite-apatite intrusions and calc-alkaline magmatism, Camsell River, N.W.T. *Can. J. Earth Sci. 13*, 348.

Baker, I. (1969) Petrology of the volcanic rocks of Saint Helena, South Atlantic. *Geol. Soc. Am. Bull. 80*, 1283.

________, and S. E. Haggerty (1967) The alteration of olivine in basaltic and associated lavas. Part II: Intermediate and low temperature alteration. *Contrib. Mineral. Petrol. 16*, 258.

Banerjee, S. (1970) Decay of marine magnetic anomalies by ferrous ion diffusion. *Nature Phys. Sci. 229*, 181.

________, R. F. Butler, and J. H. Stout (1974) Magnetic properties and mineralogy of exposed oceanic crust on Macquarie Island. *Geophys. 40*, 537.

Banerjee, S., W. O'Reilly, and C. E. Johnson (1967) Mossbauer effect measurements in FeTi spinels with local disorder. *J. Appl. Phys. 38*, 1289.

________, W. O'Reilly, T. C. Gibb, and N. N. Greenwood (1967) The behavior of ferrous ions in iron-titanium spinels. *J. Phys. Chem. Solids 28*, 1323.

Basta, E. Z. (1959) Some mineralogical relationships in the system Fe_2O_3-Fe_3O_4 and the composition of titanomaghemite. *Econ. Geol. 54*, 698.

________ (1960) Natural and synthetic titanomagnetites (the system Fe_3O_4-Fe_2TiO_4-$FeTiO_3$). *M. Jahrb. Mineral., Abh. 94*, 1017.

Bastin, E. S. (1950) Interpretation of ore minerals. *Geol. Soc. Am. Memoir 45*.

Basu, A. R. and I. D. MacGregor (1975) Chromite spinels from ultramafic xenoliths. *Geochim. Cosmochim. Acta 39*, 937.

Bauer, G. R., R. V. Fodor, J. W. Husler, and K. Keil (1973) Contributions to the mineral chemistry of Hawaiian rocks. III. Composition and mineralogy of a new rhyodacite occurrence. *Contrib. Mineral. Petrol. 40*, 183.

Beck, M. E., Jr. (1966) The effect of magmatic differentiation on the magnetic properties of diabase sheets of southeastern Pennsylvania. *U. S. Geol. Survey Prof. Paper 550-D*, D109.

Beeson, M. H. (1976) Petrology, mineralogy, and geochemistry of the East Molokai volcanic series, Hawaii. *U. S. Geol. Survey Prof. Paper 961*, 1.

________, and E. D. Jackson (1969) Chemical composition of altered chromites from the Stillwater Complex, Montana. *Am. Mineral. 54*, 1084.

Bence, A. E. and A. L. Albee (1968) Empirical correction factors for the electron microanalysis of silicates and oxides. *J. Geol. 76*, 382.

________, J. J. Papike, and R. A. Ayuso (1975) Petrology of submarine basalts from the central Caribbean: DSDP Leg 15. *J. Geophys. Res. 80*, 4775.

Berg, J. H. (1976) Metamorphosed mafic and ultramafic rocks in the contact aureoles of the Nain complex, Labrador, and the miscibility gap between spinel and magnetite in natural Cr-Al-Fe^{3+}-Ti spinels. *Geol. Soc. Am. Abs. with Programs 8*, 773.

Bergstøl, S. (1972) The jacupirangite at Kodal, Vestfold, Norway. *Mineral. Deposita (Berl.) 7*, 233.

Bernal, J. D., D. R. Dasgupta, and A. L. Mackay (1957) Oriented transformations in iron oxides and hydroxides. *Nature 180*, 645.

Berry, L. G. and R. M. Thompson (1962) X-ray powder data for ore minerals. *Geol. Soc. Am. Bull. 85*.

Bichan, R. (1969) Chromite seams in the Hartley complex of the Great Dyke of Rhodesia. In, H. D. B. Wilson, Ed., Magmatic Ore Deposits, *Econ. Geol. Mono. 4*, 95.

Bilgrami, S. A. (1969) Geology and chemical mineralogy of the Zhob Valley chromite deposits, West Pakistan. *Am. Mineral. 54*, 134.

Binns, R. A. (1969) High-pressure megacrysts in basanatic lavas near Armidale, New South Wales. *Am. J. Sci. 267-A*, 33.

________, M. B. Duggan, and J. F. G. Wilkinson (1970) High pressure megacrysts in alkaline lavas from northeastern New South Wales. *Am. J. Sci. 269*, 132.

Bishop, F. C., J. V. Smith, and J. B. Dawson (1975) Pentlandite-magnetite intergrowth in De Beers spinel lherzolite: review of sulphides in nodules. *Phys. Chem. Earth 9*, 323.

Bliss, N. W. and W. H. MacLean (1975) The paragenesis of zoned chromite from central Manitoba. *Geochim. Cosmochim. Acta 39*, 973.

Bose, M. K. (1965) Iron-titanium oxide minerals in comagmatic rocks of Koraput, Orissa. *Am. J. Sci. 263*, 689.

Bowie, S. H. U., P. R. Simpson, and D. Atkin (1975) Reflectance measurements in monochromatic light on the Bowie-Taylor suite of 103 ore minerals. *Fortschr. Mineral. 52*, 567.

Boyd, F. R. (1971) Enstatite-ilmenite and diopside-ilmenite intergrowths from the Monastery Mine. *Ann. Rep. Dir. Geophys. Lab. Year Book 70*, 134.

________ (1971) Pargasite-spinel peridotite xenolith from the Wesselton Mine. *Ann. Rep. Dir. Geophys. Lab. Year Book 70*, 138.

Boyd, F. R. and P. H. Nixon (1973) Origin of the ilmenite-silicate nodules in kimberlites from Lesotho and South Africa. In, P. H. Nixon, Ed., *Lesotho Kimberlites*, 254, Lesotho National Development Corporation, Maseru.

________, and P. H. Nixon (1975) Origins of the ultramafic nodules from some kimberlites of northern Lesotho and the Monastery Mine, South Africa. *Phys. Chem. Earth 9*, 431.

Brecher, A. and G. Arrhenius (1974) The paleomagnetic record in carbonaceous chondrites: natural remanence and magnetic properties. *J. Geophys. Res. 79*, 2081.

Brett, P. R. (1962) Exsolution textures and rates in solid solutions involving bornite. *Carnegie Inst. Wash. Year Book 61*, 155.

________ (1964) Experimental data from the Cu-Fe-S system and their bearing on exsolution textures in ores. *Econ. Geol. 59*, 1241.

Brooke, J., E. Irving, and J. K. Park (1970) The mid-Atlantic ridge near 45°N. XIII. Magnetic properties of basalt bore-core. *Can. J. Earth Sci. 7*, 1515.

Brown, F. H. and I. Carmichael (1971) Quaternary volcanoes of the Lake Rudolf region: II. The lavas of North Island, South Island and the Barrier. *Lithos 4*, 305.

Brown, G. M. (1956) The layered ultrabasic rocks of Rhum, Inner Herbrides. *Phil. Trans. Roy. Soc. Lond., Ser. B, 240*, 1.

Bryan, W. B. (1971) Coral Sea drift pumice stranded on Eua Island, Tonga, in 1969. *Geol. Soc. Am. Bull. 82*, 2799.

________ (1972) Mineralogical studies of submarine basalts. *Ann. Rep. Dir. Geophys. Lab. Carnegie Inst. Year Book 71*, 396.

________, G. D. Stice, and A. Ewart (1972) Geology, petrography, and geochemistry of the volcanic islands of Tonga. *J. Geophys. Res. 77*, 1566.

Buchan, K. L. and D. J. Dunlop (1976) Paleomagnetism of the Haliburton intrusions: superimposed magnetizations, metamorphism, and tectonics in the late Precambrian. *J. Geophys. Res. 81*, 2951.

Buddington, A. F. and D. H. Lindsley (1964) Iron-titanium oxide minerals and synthetic equivalents. *J. Petrol. 5*, 310.

________, J. Fahey, and A. Vlisidis (1955) Thermometric and petrogenetic significance of titaniferous magnetite. *Am. J. Sci. 253*, 497.

________, J. Fahey, and A. Vlisidis (1963) Degree of oxidation of Adirondack iron-titanium oxide minerals in relation to petrogeny. *J. Petrol. 4*, 138.

Bunch, T. E. and R. LaBorde (1976) Mineralogy and compositions of selected basalts from DSDP Leg 34. In, S. R. Hart and R. S. Yeats, Eds., *Initial Reports of the Deep Sea Drilling Project, 34*, U. S. Government Printing Office, Washington, D. C.

Busche, F. D., M. Prinz, K. Keil, and T. E. Bunch (1972) Spinels and the petrogenesis of some Apollo 12 igneous rocks. *Am. Mineral. 57*, 1729.

Butler, R. F. (1973) Stable single-domain to superparamagnetic transition during low-temperature oxidation of oceanic basalts. *J. Geophys. Res. 78*, 6868.

________, and S. K. Banerjee (1973) Magnetic properties of exposed oceanic crust on Macquarie Island. *Nature Phys. Sci. 244*, 115.

Cameron, E. N. (1961) *Ore Microscopy*. John Wiley & Sons, Inc.

________ (1963) Structure and rock sequences of the critical zone of the eastern Bushveld complex. *Mineral. Soc. Am. Spec. Paper 1*, 93.

________ (1969) Postcumulus changes in the eastern Bushveld Complex. *Am. Mineral. 54*, 754.

________ (1970) Compositions of certain coexisting phases in the eastern part of the Bushveld complex, 46. In, D. J. L. Visser and G. von Gruenwaldt, Eds., Symposium on the Bushveld Igneous Complex and Other Layered Intrusions, *Geol. Soc. S. Africa Spec. Publ. No. 1*.

________ (1975) Postcumulus and subsolidus equilibration of chromite and coexisting silicates in the Eastern Bushveld complex. *Geochim. Cosmochim. Acta 39*, 1021.

________, and G. A. Desborough (1969) Occurrence and characteristics of chromite deposits - eastern Bushveld complex. In, H. D. B. Wilson, Ed., Magmatic Ore Deposits, *Econ. Geol. Mono. 4*, 23.

________, and M. E. Emerson (1959) The origin of certain chromite deposits in the eastern part of the Bushveld complex. *Econ. Geol. 54*, 1151.

Cameron, E. N. and E. D. Glover (1973) Unusual titanian-chromian spinels from the eastern Bushveld complex. *Am. Mineral. 58*, 172.

Cameron, K. L. and M. Cameron (1973) Mineralogy of ultramafic nodules from Knippa quarry, Uvalde, Texas. *Geol. Soc. Am. Abstr. with Programs 5*, 566.

Carmichael, C. M. (1961) The magnetic properties of ilmenite-haematite crystals. *Proc. Roy. Soc. A, 263*, 508.

_______ (1962) The magnetization of solid solutions of ilmenite in hematite. *J. Phys. Soc. Japan 17, Suppl. B-1*, 711.

_______ (1964) The magnetization of a rock containing magnetite and hemoilmenite. *Geophys. 29*, 87.

_______ (1970) The mid-Atlantic ridge near 45°N. VII. Magnetic properties and opaque mineralogy of dredged samples. *Can. J. Earth Sci. 7*, 239.

_______, and H. C. Palmer (1968) Paleomagnetism of the late Triassic, North Mountain basalt of Nova Scotia. *J. Geophys. Res. 73*, 2811.

Carmichael, I. S. E. (1967a) The mineralogy of Thingmuli, a tertiary volcano in eastern Iceland. *Am. Mineral. 52*, 1815.

_______ (1967b) The iron-titanium oxides of salic volcanic rocks and their associated ferromagnesian silicates. *Contrib. Mineral. Petrol. 14*, 36.

_______ (1967) The mineralogy and petrology of the volcanic rocks from the Leucite Hills, Wyoming. *Contrib. Mineral. Petrol. 15*, 24.

_______, and J. Nicholls (1967) Iron-titanium oxides and oxygen fugacities in volcanic rocks. *J. Geophys. Res. 72*, 4665.

_______, J. Nicholls, and A. L. Smith (1970) Silica activity in igneous rocks. *Am. Mineral. 55*, 246.

Chakraborty, K. L. (1973) Some characters of the bedded chromite deposits at Kalrangi, Cuttack district, Orissa, India. *Mineral. Deposita (Berl.) 8*, 73.

Challis, G. A. (1965) The origin of New Zealand ultramafic intrusions. *J. Petrol. 6*, 337.

Champness, P. E. (1970) Nucleation and growth of iron oxides in olivines, $(Mg,Fe)_2SiO_4$. *Mineral. Mag. 37*, 790.

_______, and P. Gay (1968) Oxidation of olivines. *Nature 218*, 157.

Chevallier, R., S. Mathieu, and E. A. Vincent (1954) Iron-titanium oxide minerals in layered gabbros of the Skaergaard intrusion, east Greenland, part 2, magnetic properties. *Geochim. Cosmochim. Acta 6*, 27.

Clarke, D. B. and R. H. Mitchell (1975) Mineralogy and petrology of the kimberlite from Somerset Island, N.W.T., Canada. *Phys. Chem. Earth 9*, 123.

Cockerham, R. S. and J. M. Hall (1976) Magnetic properties and paleomagnetism of some DSDP Leg 33 basalts and sediment and their tectonic implications. *J. Geophys. Res. 81*, 4207.

Coertze, F. J. (1970) The geology of the western part of the Bushveld igneous complex, 5. In, D. J. L. Visser and G. von Gruenwaldt, Eds., Symposium on the Bushveld Igneous Complex and Other Layered Intrusions. *Geol. Soc. S. Afr. Spec. Pub. No. 1.*

Collinson, D. W. (1975) Instruments and techniques in paleomagnetism and rock magnetism. *Rev. Geophys. and Space Sci. 13*, 659.

_______, K. M. Creer, and S. K. Runcorn, Eds. (1967) *Developments in Solid Earth Geophysics 3, Methods in Palaeomagnetism.* Elsevier Publishing Company.

Cornwall, H. R. (1951) Ilmenite, magnetite, hematite, and copper in lavas of the Keweenawen series. *Econ. Geol. 46*, 51.

Cox, A. and R. R. Doell (1962) Magnetic properties of the basalt in Hole EM 7, Mohole project. *J. Geophys. Res. 67*, 3997.

Cox, K. G., J. J. Gurney, and B. Harte (1973) Xenoliths from the Matsoku Pipe. In, P. H. Nixon, Ed., *Lesotho Kimberlites*, 76, Lesotho National Development Corporation, Maseru.

_______, and G. Hornung (1966) The petrology of the Karroo basalts of Basutoland. *Am. Mineral. 51*, 1414.

Cousins, C. A. (1969) The Merensky Reef of the Bushveld igneous complex. In, H. D. B. Wilson, Ed., Magmatic Ore Deposits, *Econ. Geol. Mono. 4*, 239.

Creer, K. M. (1971) Geophysical interpretation of remanent magnetization in oxidized basalts. *Z. Geophys. 37*, 383.

Creer, K. M. and J. D. Ibbetson (1970) Electron microprobe analyses and magnetic properties of non-stoichiometric titanomagnetites in basaltic rocks. *Geophys. J. R. astr. Soc. 21*, 485.

Cundari, A. (1975) Mineral chemistry and petrogenetic aspects of the Vico lavas, Roman volcanic region, Italy. *Contrib. Mineral. Petrol. 53*, 129.

Czamanske, G. K. and P. Mihálik (1972) Oxidation during magmatic differentiation. Finnmarka complex, Oslo area, Norway: Part I, The opaque oxides. *J. Petrol. 13*, 493.

________, and D. R. Wones (1973) Oxidation during magmatic differentiation, Finnmarka complex, Oslo area, Norway: Part 2, The mafic silicates. *J. Petrol. 14*, 349.

Dagley, P. R. L. Wilson, J. M. Ade-Hall, G. P. L. Walker, S. E. Haggerty, T. Sigurgeirsson N. D. Watkins, P. J. Smith, J. Edwards, and R. L. Grasty (1967) Geomagnetic polarity zones for Icelandic lavas. *Nature 216*, 25.

Danchin, R. V. and F. R. Boyd (1976, in press) Ultramafic nodules from the Premier kimberlite pipe, South Africa. *Ann. Rep. Dir. Geophys. Lab., Year Book 75*.

________, and F. d'Orey (1972) Chromian spinel exsolution in ilmenite from the Premier Mine, Transvaal, South Africa. *Contrib. Mineral. Petrol. 35*, 43.

________, J. Ferguson, J. R. McIver, and P. H. Nixon (1975) The composition of late stage kimberlite liquids as revealed by nucleated autholiths. *Phys. Chem. Earth 9*, 235.

Davidson, A. and P. J. Wyllie (1968) Opaque oxide minerals of some diabase-granophyre associations in Pennsylvania. *Econ. Geol. 63*, 950.

Davis, P. M. and M. E. Evans (1976) Interacting single-domain properties of magnetite intergrowths. *J. Geophys. Res.*, 989.

Dawson, J. B. (1962) Basutoland kimberlites. *Geol. Soc. Am. Bull. 73*, 545.

________ (1971) Advances in kimberlite geology. *Earth Sci. Rev. 7*, 187.

________, and J. B. Hawthorne (1973) Magmatic sedimentation and carbonatitic differentiation in kimberlite sills at Benfontein, South Africa. *J. Geol. Soc. Lond. 129*, 61.

________, and A. M. Reid (1970) A pyroxene-ilmenite intergrowth from the Monastery Mine, South Africa. *Contrib. Mineral. Petrol. 26*, 296.

________, and J. V. Smith (1973) Alkalic pyroxenite xenoliths from the Lashaine Volcano, Northern Tanzania. *J. Petrol. 14*, 113.

________, and J. V. Smith (1975) Chromite-silicate intergrowths in upper-mantle peridotites. *Phys. Chem. Earth 9*, 339.

________, D. G. Powell, and A. M. Reid (1970) Ultrabasic xenoliths and lava from the Lashaine Volcano, Northern Tanzania. *J. Petrol. 11*, 519.

Deans, T. (1966) Economic mineralogy of African carbonatites, 385. In, O. F. Tuttle and J. Gittins, Eds., *Carbonatites*, John Wiley & Sons, 591p.

De Boer, J. (1975) Variations of the magnetic properties of postglacial pillow lavas along the Reykjanes Ridge. *J. Geophys. Res. 26*, 3769.

________, J.-G. Schilling, and D. C. Krause (1969) Magnetic properties of pillow basalts from Reykjanes Ridge. *Science 166*, 996.

Deer, W. A. and D. Abbott (1965) Clinopyroxenes of the gabbro cumulates of the Kap Edvard Holm complex, east Greenland. *Mineral. Mag. 34*, 177.

________, R. A. Howie and J. Zussman (1972) *Rock-Forming Minerals*, Longman Press.

Desborough, G. A. (1963) Mobilization of iron by alteration of magnetite-ulvöspinel in basic rocks in Missouri. *Econ. Geol. 58*, 332.

Deutsch, E. R. and R. R. Pätzold (1976) Magnetism of basalt cores from the Nazca Plate and implications for magnetic anomaly interpretation. *J. Geophys. Res. 81*, 4188.

________, and C. Somayajulu (1970) Palaeomagnetism of Ordovician ignimbrites from Killary Harbour, Eire. *Earth Planet. Sci. Lett. 7*, 337.

de Villiers, J. S. (1970) The structure and the petrology of the mafic rocks of the Bushveld complex south of Potgietersrus, 23. In, D. J. L. Visser and G. von Gruenwaldt, Eds., Symposium on the Bushveld Igneous Complex and Other Layered Intrusions, *Geol. Soc. S. Afr. Spec. Publ No. 1*.

Dick, H. (1976) Spinels in fracture zone "B" and median valley basalts, Famous area, mid-Atlantic ridge. *Trans. Am. Geophys. U. 57*, 341.

Dickey, J. S., Jr. (1975) A hypothesis of origin for podiform chromite deposits. *Geochim. Cosmochim. Acta 39*, 1061.

________, and H. S. Yoder, Jr. (1972) Partitioning of chromian and aluminum between clinopyroxene and spinel. *Ann. Rep. Dir. Geophys. Lab., Year Book 71*, 384.

Donaldson, C. H. (1975) A petrogenetic study of harrisite in the Isle of Rhum pluton, Scotland. Unpublished Ph.D. thesis, St. Andrews Univ.

Duchesne, J.-C. (1972) Iron-titanium oxide minerals in the Bjerkrem-Sogndal Massif, south-western Norway. *J. Petrol. 13*, 57.

Duke, J. M. (1974) The effect of oxidation on the crystallization of an alkali basalt from the Azores. *J. Geol. 82*, 524.

Dunlop, D. J. (1972) Magnetite: Behavior near the single-domain threshold. *Science 176*, 41.

________ (1973) Theory of magnetic viscosity of lunar and terrestrial rocks. *Rev. Geophys. & Space Physics 11*, 855.

________ (1973) Thermoremanent magnetization in submicroscopic magnetite. *J. Geophys. Res. 78*, 7602.

________ (1976) Thermal fluctuation analysis: A new technique in rock magnetism. *J. Geophys. Res. 81*, 3511.

________, J. A. Hanes, and K. L. Buchan (1973) Indices of multidomain magnetic behavior in basic igneous rocks: Alternating-field demagnetization, hysteresis, and oxide petrology. *J. Geophys. Res. 78*, 1387.

Edwards, A. B. (1965) *Textures of the Ore Minerals and Their Significance.* The Australasian Institute of Mining and Metallurgy.

El Goresy, A. and E. C. T. Chao (1976) Identification and significance of armalcolite in the Ries glass. *Earth Planet. Sci. Lett. 30*, 200.

Elsdon, R. (1972) Iron-titanium oxide minerals in the upper layered series, Kap Edvard Holm, east Greenland. *Mineral. Mag. 38*, 946.

________ (1975) Manganoan ilmenite from the Leinster granite, Ireland. *Mineral. Mag. 40*, 419.

Emeleus, C. H. and J. R. Andrews (1975) Mineralogy and petrology of kimberlite dyke and sheet intrusions and included peridotite xenoliths from south-west Greenland. *Phys. Chem. Earth 9*. 179.

Engin, T. and J. W. Aucott (1971) A microprobe study of chromites from the Andizlik-Zimparalik area, south-west Turkey. *Mineral. Mag. 38*, 76.

________, and D. M. Hirst (1970) The Alpine chrome ores of the Andizlik-Zimparalik area, Fethiye, southwest Turkey. *Trans. Inst. Mineral. Metal., Sec. B, 79*, 16.

Erickson, R. L. and L. V. Blade (1963) Geochemistry and petrology of the alkalic igneous complex, Arkansas. *U. S. Geol. Surv. Prof. Paper 425*, 95p.

Eugster, H. P. and D. R. Wones (1962) Stability relations of the ferruginous biotite, annite. *J. Petrol. 3*, 82.

Evans, B. W. and J. G. Moore (1968) Mineralogy as a function of depth in the prehistoric Makaopuhi tholeiitic lava lake, Hawaii. *Contrib. Mineral. Petrol. 17*, 85.

________, and T. L. Wright (1972) Composition of liquidus chromite from the 1959 (Kilauea Iki) and 1965 (Makaopuhi) eruptions of Kilauea volcano, Hawaii. *Am. Mineral. 57*, 217.

Evans, M. E. and M. W. McElhinny (1970) An investigation of the origin of stable remanence in magnetite-bearing igneous rocks. *J. Geomag. Geoelec. 21*, 757.

________, and M. L. Wayman (1970) An investigation of small magnetic particles by means of electron microscopy. *Earth Planet. Sci. Lett. 9*, 365.

________, and M. L. Wayman (1972) The mid-Atlantic ridge near 45°N. XIX. An electron microscope investigation of the magnetic minerals in basalt samples. *Can. J. Earth Sci. 9*, 671.

Ewart, A. (1967) Pyroxene and magnetite phenocrysts from the Taupo quaternary rhyolitic pumice deposits, New Zealand. *Mineral. Mag. 36*, 180.

________, D. C. Green, I. Carmichael, and F. B. Brown (1971) Voluminous low temperature rhyolitic magmas in New Zealand. *Contrib. Mineral. Petrol. 33*, 128.

Ferguson, J., R. V. Danchin and P. H. Nixon (1973) Petrochemistry of kimberlite autoliths. In, P. H. Nixon, Ed., *Lesotho Kimberlites*, 285, Lesotho National Development Corporation, Maseru.

Findlay, D. C. and C. H. Smith (1965) The Muskox drilling project. *Geol. Survey Can. Paper 64*, 170.

Fleischer, M. (1965) Composition of magnetite as related to type of occurrence. *U. S. Geol. Surv., 525-D*, 82.

Fletcher, J. S. and J. R. Carpenter (1972) Chemical differences between massive and disseminated chromite from some ultramafics of the southern Appalachians and the petrogenetic implications. *Geol. Soc. Am. Abstr. with Programs 4*, 505.

Flynn, R. T., C. Sutphen, and G. C. Ulmer (1972) Petrogenesis of the Bushveld Complex: Pyroxene-spinel-feldspar peritectic? *Geol. Soc. Am. Abstr. with Programs 4*, 507.

Fodor, R. V. (1975) Petrology of basalt and andesite of the Black Range, New Mexico. *Geol. Soc. Am. Bull. 86*, 295.

________, K. Keil, and T. E. Bunch (1972) Mineral chemistry of volcanic rocks from Maui, Hawaii: Fe-Ti oxides. *Geol. Soc. Am. Abstr. with Programs 4*, 507.

Forbes, R. B. and H. Kuno (1967) Peridotite inclusions and basaltic host rocks, 328. In, P. J. Wyllie, Ed., *Ultramafic and Related Rocks*. John Wiley & Sons, 464p.

Fox, P. J. and N. D. Opdyke (1973) Geology of the oceanic crust: Magnetic properties of oceanic rocks. *J. Geophys. Res. 78*, 5139.

Francis, D. M. (1976) Corono-bearing pyroxene granulite xenoliths and the lower crust beneath Nuivak Island, Alaska. *Can. Mineral. 14*, 291.

Frankel, J. J. (1942) Chrome-bearing magnetic rocks from the eastern Bushveld. *S. Afr. J. Sci. 38*, 152.

Frantsesson, E. V. (1970) The petrology of the kimberlites. Trans. D. A. Brown, *Australian Natl. Univ. Pub. 150*, 194p.

Frauenfelder, H. (1962) *The Mössbauer Effect*. W. A. Benjamin, New York.

Frenzel, G. (1953) Die Erzparagenese des Katzenbuckels im Odenwald. *Heidelberger Beitrage zur Mineral. und Petrogr. 3*, 409.

Freund, H. (1966) *Applied Ore Microscopy: Theory and Techniques*. The MacMillan Company, 607p.

Frey, F. A. and D. H. Green (1974) The mineralogy, geochemistry and origin of lherzolite inclusions in Victorian basanites. *Geochim. Cosmochim. Acta 38*, 1023.

________, W. B. Bryan, and G. Thompson (1974) Atlantic ocean floor: Geochemistry and petrology of basalts from Legs 2 and 3 of the deep-sea drilling project. *J. Geophys. Res. 79*, 5507.

Frick, C. (1974, preprint) Ilmenite replacements in chrome diopsides from South African kimberlites.

Friel, J. J. and G. C. Ulmer (1974) Oxygen fugacity geothermometry of the Oka carbonatite. *Am. Mineral. 59*, 314.

Frietsch, R. (1973) The origin of the Kiruna iron ores. *Geol. Foeren. Stockholm Foerh. 95*, 375.

Frisch, T. (1970) A note on "the petrological environment of magnesium ilmenites" by J. F. Lovering and J. R. Widdowson. *Earth Planet. Sci. Lett. 8*, 329.

________ (1971) Alteration of chrome spinel in a dunite nodule from Lanzarote, Canary Islands. *Lithos 4*, 83.

________, and D. Bridgwater (1976) Iron- and maganese-rich minor intrusions emplaced under late-Orogenic conditions in the Proterozoic of south Greenland. *Contrib. Mineral. Petrol. 57*, 25.

________, and H. U. Schmincke (1969) Petrology of clinopyroxene-amphibole inclusions from the Roque Nublo Volcanics, Gran Canaria, Canary Islands. *Bull. Vol. 33-4*, 1073.

________, and J. B. Wright (1971) Chemical composition of high-pressure megacrysts from Nigerian Cenozoic lavas. *Jahrb. Mineral. Mh. 19*, 289.

Frodesen, S. (1968) Coronas around olivine in a small gabbro intrusion, Bamble Area, South Norway. *Norsk Geol. Tidss. 48*, 201.

Fudali, R. F. (1965) Oxygen fugacities of basaltic and andesitic magmas. *Geochim. Cosmochim. Acta 29*, 1063.

Fuller, M. (1974) Lunar magnetism. *Rev. Geophys. & Space Phys. 12*, 23.

Galopin, R. and N. F. M. Henry (1972) *Microscopic Study of Opaque Minerals.* W. Heffer and Sons, Ltd.

Gasparrini, E. and A. J. Naldrett (1972) Magnetite and ilmenite in the Sudbury nickel irruptive. *Econ. Geol. 67*, 605.

Geijer, P. (1931) The iron ores of the Kiruna type. *Sveriges geol. undersokning, Ser. C, No. 288*, 22.

_______ (1967) Internal features of the apatite-bearing magnetite ores. *Sveriges geol. undersokning, Ser. C., No. 624*, 32.

Gidskehaug, A., K. M. Creer, and J. G. Mitchell (1975) Palaeomagnetism and K-Ar ages of the South-west African basalts and their bearing on the time of initial rifting of the South Atlantic Ocean. *Geophys. J. R. astr. Soc. 42*, 1.

Gittins, J., R. H. Hewins, and A. F. Laurin (1975) Kimberlitic-carbonatitic dikes of the Saguenay River Valley, Quebec, Canada. *Phys. Chem. Earth 9*, 137.

Gjelsvik, T. (1957) Geochemical and mineralogical investigation of titaniferous iron ores, west coast of Norway. *Econ. Geol. 52*, 482-498.

Gold, D. P. (1966) The minerals of the Oka carbonatite and alkaline complex, Oka, Quebec. *Mineral. Soc. India, IMA*, 109.

Goldanskii, V. I. and R. H. Herber, Eds. (1968) *Chemical Applications of Mössbauer Spectroscopy.* Academic Press, New York.

Golding, H. G. and P. Bayliss (1968) Altered chrome ores from the Coolac serpentine belt, New South Wales, Australia. *Am. Mineral. 53*, 162.

________, and K. R. Johnson (1971) Variation in gross chemical composition and related physical properties of podiform chromite in the Coolac district, N.S.W., Australia. *Econ. Geol. 66*, 1017.

Goldstein, J. I. and H. Yakowitz, Eds. (1975) *Practical Scanning Electron Microscopy.* Plenum Press.

Goodenough, B. (1963) *Magnetism and the Chemical Bond.* John Wiley & Sons, New York.

Gray, I. M. and A. P. Millman (1962) Reflection characteristics of ore minerals. *Econ. Geol. 57*, 325.

Green, D. H. (1964) The petrogenesis of the high-temperature peridotite intrusion in the Lizard Area, Cornwall. *J. Petrol. 5*, 134.

Griffin, W. L. (1970) Genesis of coronas in anorthosites of the Upper Jotun Nappe, Indre Sogn, Norway. *J. Petrol. 12*, 219.

________ (1973) Lherzolite nodules from the Fen Alkaline Complex, Norway. *Contrib. Mineral. Petrol. 38*, 135.

________, and K. S. Heier (1973) Petrological implications of some corona structures. *Lithos 6*, 315.

________, and P. N. Taylor (1975) The fen damkjernite: Petrology of a "central-complex" kimberlite. *Phys. Chem. Earth 9*, 169.

Grobler, N. J. and G. G. Whitfield (1970) The olivine-apatite magnetites and related rocks in the Villa Nora occurrence of the Bushveld igneous complex, 208. In, D. J. L. Visser and G. von Gruenwaldt, Eds., Symposium on the Bushveld Igneous Complex and Other Layered Intrusions, *Geol. Soc. S. Afr. Spec. Pub. 1.*

Groeneveld, D. (1970) The structural features and the petrography of the Bushveld complex in the vicinity of Stoffberg, Eastern Transvaal, 36. In, D. J. L. Visser and G. von Gruenwaldt, Symposium on the Bushveld Igneous Complex and Other Layered Intrusions, *Geol. Soc. S. Afr. Spec. Pub. 1.*

Grommé, S. and E. A. Mankinen (1976) Natural remanent magnetization, magnetic properties, and oxidation of titanomagnetite in basaltic rocks from DSDP Leg 34. In, S. R. Hart and R. S. Yeats, Eds., *Initial Reports of the Deep Sea Drilling Project, 34*, U. S. Government Printing Office, Washington, D. C.

________, T. L. Wright, and D. L. Peck (1969) Magnetic properties and oxidation of iron-titanium oxide minerals in Alae and Makaopuhi lava lakes, Hawaii. *J. Geophys. Res. 74*, 22.

Gunn, B. M., R. Coy-Yll, N. D. Watkins, C. E. Abranson, and J. Nougier (1970) Geochemistry of an oceanite-ankaramite-basalt suite from East Island, Crozet Archipelago. *Contrib. Mineral. Petrol. 28*, 319.

Gurney, J. J., H. W. Fesq, and E. J. D. Kable (1973) Clinopyroxene-ilmenite intergrowths from kimberlite: a re-appraisal. In, P. H. Nixon, Ed., *Lesotho Kimberlites*, 238, Lesotho National Development Corporation, Maseru.

Haspala, I. and P. Ojanpera (1972) Magnetite and ilmenite from some Finnish rocks. *Bull. Geol. Soc. Finland 44*, 13.

Haggerty, S. E. (1968) Fe-Ti oxides in Icelandic basic rocks and their significance in rock magnetism. Ph.D. thesis, University of London.

________ (1970) Mid-Atlantic ridge near 45°N: Magnetic mineralogy. *Trans. Am. Geophys. U. 51*, 273.

________ (1970) The Laco magnetite lava flow, Chile. *Ann. Rep. Dir. Geophys. Lab. Year Book 68*, 329.

________ (1971) High-temperature oxidation of ilmenite in basalts. *Ann. Rep. Dir. Geophys. Lab. Year Book 70*, 165.

________ (1972) Luna 16: An opaque mineral study and a systematic examination of compositional variations of spinels from Mare Fecunditatis. *Earth Planet. Sci. Lett. 13*, 328.

________ (1973) Spinels of unique composition associated with ilmenite reactions in the Liqhobong kimberlite pipe, Lesotho. In, P. H. Nixon, Ed., *Lesotho Kimberlites*, 149, Lesotho National Development Corporation, Maseru.

________ (1975) The chemistry and genesis of opaque minerals in kimberlites. *Phys. Chem. Earth 9*, 295.

________ (1976) Kennedyite: A comparison with lunar and terrestrial armalcolites. *Geol. Soc. Am. Abstr. with Programs 8*, 898.

________, and I. Baker (1967) The alteration of olivine in basaltic and associated lavas. Part I: High temperature alteration. *Contrib. Mineral. Petrol. 16*, 233.

________, and E. Irving (1970) On the origin of the natural remanence (NRM) of the mid-Atlantic ridge at 45°N. *Trans. Am. Geophys. Union 51*, 273.

________, and D. H. Lindsley (1970) Stability of pseudobrookite (Fe_2TiO_5)-ferropseudobrookite ($FeTi_2O_5$) series. *Carnegie Inst. Washington Year Book 68*, 247.

________, and N. D. Watkins (1966) Iron-titanium oxide variation in an Icelandic composite dike. *Trans. Am. Geophys. Union 47*, 210.

________, G. D. Borley, and M. J. Abbott (1966) Iron-titanium oxides in a suite of alkaline volcanic rocks from Tenerife (abstr.). *Int. Mineral. Assoc. Meet.*, Cambridge University.

Hall, A. (1955) *Introduction to Electron Microscopy*. McGraw Hill and Company.

Hallimond, A. F. (1970) *The Polarizing Microscope, 3rd ed.*, Vickers Ltd., 302p.

Hamad, El. D. (1963) The chemistry and mineralogy of the olivine nodules of Calton Hill, Derbyshire. *Mineral. Mag. 33*, 491.

Hamlyn, P. R. (1975) Chromite alteration in the Panton sill, East Kimberley region, Western Australia. *Mineral. Mag. 40*, 181.

Hargraves, R. B. (1959) Magnetic anisotropy and remanent magnetism in hemo-ilmenite from ore-deposits at Allard Lake, Quebec. *J. Geophys. Res. 64*, 1565.

________ (1962) Petrology of the Allard lake anorthosite suite, Quebec. In, "Petrological Studies," *Geol. Soc. Am., Buddington Vol.*, 163.

________, and J. M. Ade-Hall (1975) Magnetic properties of separated mineral phases in unoxidized and oxidized Icelandic basalts. *Am. Mineral. 60*, 29.

________, and S. K. Banerjee (1973) Theory and nature of magnetism in rocks. In, F. A. Donath, Ed., *Annual Review of Earth and Planetary Sciences*, Annual Reviews, Inc.

________, and D. M. Burt (1967) Paleomagnetism of the Allard Lake anorthosite suite. *Can. J. Earth Sci. 4*, 357.

________, and N. Petersen (1971) Notes on the correlation between petrology and magnetic properties of basaltic rocks. *Z. Geophys. 37*, 367.

________, and W. M. Young (1969) Source of stable remanent magnetism in Lamberville diabase. *Am. J. Sci. 267*, 1161.

Haselton, J. D. and W. P. Nash (1975) Ilmenite-orthopyroxene intergrowths from the Moon and the Skaergaard intrusion. *Earth Planet. Sci. Lett. 26*, 287.

Hearn, B. C., Jr. and F. R. Boyd (1975) Garnet peridotite xenoliths in a Montana, U.S.A., kimberlite. *Phys. Chem. Earth 9*, 247.

Heinrich, E. W. (1966) *The Geology of Carbonatites.* Rand McNally & Company, 555p.

Heinrich, K. F. J., Ed. (1966) *The Electron Microprobe.* John Wiley and Sons.

Heinrichs, D. F. (1973) Paleomagnetic studies of basalt core from DSDP 163. In, T. H. van Andel and G. R. Heath, *Initial Reports of the Deep Sea Drilling Project, 19*, U. S. Government Printing Office, Washington, D. C.

Heller, F. (1971) Remanent magnetization of the Bergell granite. *Z. Geophys. 37*, 557.

Helsley, C. E. and H. Spall (1972) Paleomagnetism of 1140 to 1150 million-year diabase sills from Gila County, Arizona. *J. Geophys. Res. 77*, 2115.

Heming, R. F. and I. Carmichael (1973) High-temperature pumice flows from the Rabaul Caldera Papua, New Guinea. *Contrib. Mineral. Petrol. 38*, 1.

Henderson, P. (1975) Reaction trends shown by chrome-spinels of the Rhum layered intrusion. *Geochim. Cosmochim. Acta 39*, 1035.

________, and P. Suddaby (1971) The nature and origin of the chrome-spinel of the Rhum layered intrusion. *Contrib. Mineral. Petrol. 33*, 21.

Hess, H. H. (1950) Vertical mineral variation in the Great Dyke of Southern Rhodesia. *Trans. Geol. Soc. S. Afr. 53*, 159.

________ (1960) Stillwater igneous complex--a quantitative, mineralogical study, Montana. *Geol. Soc. Am. Mem. 80*, 230.

Hill, R. and P. Roeder (1974) The crystallization of spinel from basaltic liquid as a function of oxygen fugacity. *J. Geol. 82*, 709.

Himmelberg, G. R. and R. G. Coleman (1968) Chemistry of primary minerals and rocks from the Red Mountain-Del Puerto ultramafic mass, California. *U. S. Geol. Surv. Prof. Pap. 600*, C18.

________, and R. A. Loney (1975) Petrology of the Vulcan Peak alpine-type peridotite, southwestern Oregon. *Geol. Soc. Am. Bull. 84*, 1585.

Hoblitt, R. P. and E. E. Larson (1975) New combination of techniques for determination of the ultrafine structure of magnetic minerals. *Geology 3*, 723.

Hodges, F. N. and J. J. Papike (1976) DSDP Site 334: Magmatic cumulates from oceanic layer 3. *J. Geophys. Res. 81*, 4135.

Holt, D. B., M. D. Muir, P. R. Grant, and I. M. Boswarva, Eds. (1974) *Quantitative Scanning Electron Microscopy.* Academic Press, New York.

Hor, A. K., D. K. Hutt, J. Wakefield, and B. F. Windley (1975) Petrochemistry and mineralogy of early Precambrian anorthositic rocks of the Limpopo belt, southern Africa. *Lithos 8*, 297.

Hoye, G. S. and W. O'Reilly (1972) A magnetic study of the ferro-magnesian olivines $(Fe_{\chi}Mg_{1-\chi})SiO_4$, $0 < \chi < 1$. *J. Phys. Chem. Solids 33*, 1827.

________, and W. O'Reilly (1973) Low temperature oxidation of ferro-magnesian olivines - a gravimetric and magnetic study. *Geophys. J. R. astr. Soc. 33*, 81.

Huebner, J. S. and M. Sato (1970) The oxygen fugacity-temperature relationships of manganese oxide and nickel oxide buffers. *Am. Mineral. 55*, 934.

Hughes, C. J. (1970) Major rhythmic layering in ultramafic rocks of the Great Dyke of Rhodesia, with particular reference to the Sebakwe area, 594. In, D. J. L. Visser and G. von Gruenvaldt, Eds., Symposium on the Bushveld Igneous Complex and Other Layered Intrusions, *Geol. Soc. S. Afr. Spec. Pub. 1.*

Hutchinson, C. S. (1972) Alpine-type chromite in North Borneo, with special reference to Darvel Bay. *Am. Mineral. 57*, 835.

Irvine, T. N. (1963) Origin of the ultramafic complex at Duke Island, southeastern Alaska. *Mineral. Soc. Am. Spec. Pap. 1*, 36.

________ (1965) Chromian spinel as a petrogenetic indicator. Part 1. Theory. *Can. J. Earth Sci. 2*, 648.

________ (1967) Chromian spinel as a petrogenetic indicator. Part 2. Petrologic applications. *Can. J. Earth Sci. 4*, 71.

Irvine, T. N. (1970) Crystallization sequences in the Muskox intrusion and other layered intrusions. I, olivine-pyroxene-plagioclase relations, 441. In, D. J. L. Visser and G. von Gruenwaldt, Eds., Symposium on the Bushveld Igneous Complex and other Layered Intrusions, *Geol. Soc. S. Afr. Spec. Pub. 1*.

________ (1973) Bridget Cove volcanics, Juneau area, Alaska: Possible parental magma of Alaskan-type ultramafic complexes. *Ann. Rep. Dir. Geophys. Lab. Year Book 72*, 478.

________ (1975) Crystallization sequences in the Muskox intrusion and other layered intrusions. II. Origin of chromitite layers and similar deposits of other magmatic ores. *Geochim. Cosmochim. Acta 39*, 991.

________, and C. H. Smith (1969) Primary oxide minerals in the layered series of the Muskox intrusion. In, H. D. B. Wilson, Ed., Magmatic Ore Deposits, *Econ. Geol. Mono. 4*, 76.

Irving, A. J. and E. B. Watson (1976) Trevorite-bearing Fe-Ni-oxide-sulfide inclusions in a high-pressure pyrope megacryst from a Nigerian basanite. *Geol. Soc. Am. Abstr. with Programs 8*, 935.

Irving, E. (1964) *Paleomagnetism - Application to Geological and Geophysical Problems*. John Wiley and Sons, New York.

________ (1970) The mid-Atlantic ridge at 45°N. XIV. Oxidation and magnetic properties of basalt; review and discussion. *Can. J. Earth Sci. 7*, 1528.

________, and J. C. McGlynn (1976) Polyphase magnetization of the Big Spruce Complex, Northwest Territories. *Can. J. Earth Sci. 13*, 476.

________, J. K. Park, and R. F. Emslie (1974) Paleomagnetism of the Morin Complex. *J. Geophys. Res. 79*, 5482.

________, W. A. Robertson, and F. Aumento (1970) The mid-Atlantic ridge near 45°N. VI. Remanent intensity, susceptibility, and iron content of dredged samples. *Can. J. Earth Sci. 7*, 226.

________, J. K. Park, S. E. Haggerty, F. Aumento, and B. Loncarevic (1970) Magnetism and opaque mineralogy of basalts from the mid-Atlantic ridge near 45°N. *Nature 228*, 974.

Jackson, E. D. (1961) Primary textures and mineral associations in ultramafic zone of the Stillwater complex, Montana. *U. S. Geol. Surv. Prof. Pap. 358*, 106.

________ (1963) Stratigraphic and lateral variation of chromite composition in the Stillwater complex. *Mineral. Soc. Am. Spec. Pap. 1*, 46.

________ (1967) Ultramafic cumulates in the Stillwater Great Dyke and Bushveld intrusions. In, P. J. Wyllie, Ed., *Ultramafic Rocks*, John Wiley and Sons, New York.

________ (1969) Chemical variation in coexisting chromite and olivine in chromite zones of the Stillwater complex. In, H. D. B. Wilson, Ed., Magmatic Ore Deposits, *Econ. Geol. Mono. 4*, 41.

________ (1970) The cyclic unit in layered intrusions - a comparison of repetitive stratigraphy in the ultramafic parts of the Stillwater, Muskox, Great Dyke and Bushveld complexes, 391. In, D. J. L. Visser and G. von Gruenwaldt, Eds., Symposium on the Bushveld Igneous Complex and Other Layered Intrusions, *Geol. Soc. S. Afr. Spec. Pub. 1*, 391.

________, and T. L. Wright (1970) Xenoliths in the Honolulu volcanic series, Hawaii. *J. Petrol. 11*, 405.

Jaeger, J. C. (1961) The cooling of irregularly shaped igneous bodies. *Am. J. Sci. 259*, 721.

Jakobsson, S. P., A. K. Pedersen, J. G. Ronsbo, and L. M. Larsen (1973) Petrology of mugearite-hawaiite: Early extrusives in the 1973 Heimaey eruption, Iceland. *Lithos 6*, 203.

Jensen, A. (1966) Mineralogical variations across two dolerite dykes from Bornholm. *Medd. fra Dansk Geol. Foren. 16*, 370.

Jensen, S. D. and P. N. Shrive (1973) Cation distribution in sintered titanomagnetites. *J. Geophys. Res. 78*, 8474.

Jones, W. R., J. W. Peoples, and A. L. Howland (1960) Igneous and tectonic structures of the Stillwater complex, Montana. *U. S. Geol. Soc. Bull. 1071-H*, 281.

Joplin, G. A., E. Kiss, N. G. Ware, and J. R. Widdowson (1972) Some chemical data on members of the shoshonite association. *Mineral. Mag. 38*, 936.

Joshima, M. (1973) Magnetization of oceanic basalts. *Rock Magnetism and Paleogeophysics 1*, 9.

Katsura, T. and I. Kushiro (1961) Titanomaghemite in igneous rocks. *Am. Mineral. 46*, 134.

Klootwijk, C. T. (1975) A note on the paleomagnetism of the late Precambrian Malani rhyolites near Jodhpur, India. *J. Geophys. 41*, 189.

Kobayashi, K. and K. Momose (1969) Thermomagnetic analysis of ferromagnetic minerals in pumice as a method in tephrochronology. *Etudes sue le quaternaire dans le monde* (Union internationale pour l'étude du quaternaire), VIIIe Congrès INQUA, Paris 1969, 959.

Kohlstedt, D. L. and J. B. Vander Sande (1975) An electron microscopy study of naturally occurring oxidation produced precipitates in iron-bearing olivines. *Contrib. Mineral. Petrol. 53*, 13.

________, C. Goetze, W. B. Durham, and J. Vander Sande (1976) New technique for decorating dislocations in olivine. *Science 191*, 1045.

Kono, M. (1971) Intensity of the earth's magnetic field during the Pliocene and Pleistocene in relation to the amplitude of mid-ocean ridge magnetic anomalies. *Earth Planet. Sci. Lett. 11*. 10.

Kretchsmar, U. H. and R. H. McNutt (1971) A study of the Fe-Ti oxides in the Whitestone anorthosite, Dunchruch, Ontario. *Can. J. Earth Sci. 8*, 947.

Kullerud, G., G. Donnay, and J. D. H. Donnay (1969) Omission solid solution in magnetite: Kenotetrahedral magnetite. *Z. Kristallogr. 128*, 1.

Kuno, H. (1967) Mafic and ultramafic nodules from Itinome-Gata, Japan, 337. In, P. J. Wyllie, Ed., *Ultramafic and Related Rocks*, John Wiley and Sons, New York.

________ (1969) Mafic and ultramafic inclusions in basaltic rocks and the nature of the upper mantle. In, P. Hart, Ed., *The Earth's Crust and Upper Mantle, Geophys. Mono. 13*, 507.

Kutolin, V. A. and V. M. Frolova (1970) Petrology of ultrabasic inclusions from basalts of Minusa and Transbaikalian Regions (Siberia, USSR). *Contrib. Mineral. Petrol. 29*, 163.

Lally, J. S., A. H. Heuer, and G. L. Nord, Jr. (1976) Precipitation in the ilmenite-hematite system. In, H.-R. Wenk, Ed., *Electron Microscopy in Mineralogy*, Springer Verlag.

Larson, E. E., D. E. Watson, and W. Jennings (1971) Regional comparisons of a Miocene geomagnetic transition in Oregon and Nevada. *Earth Planet. Sci. Lett. 11*, 391.

________, D. E. Watson, J. M. Hendon, and M. W. Rowe (1974) Thermomagnetic analysis of meteorites. I. Cl chondrites. *Earth Planet. Sci. Lett. 21*, 345.

Lawley, E. A. and J. M. Ade-Hall (1971) A detailed magnetic and opaque petrological study of a thick palaeogene tholeiite lava flow from northern Iceland. *Earth Planet. Sci. Lett. 11*, 113.

Legg, C. A. (1969) Some chromite-ilmenite associations in the Merensky Reef, Transvaal. *Am. Mineral. 54*, 1347.

Lerbekmo, J. F. and D. G. W. Smith (1972) Temporal distinction between lobes of the White River ash by composition of iron-titanium oxide minerals. *Geol. Soc. Am. Abstr. with Programs*, 190.

Lewis, J. F. (1970) Chemical composition and physical properties of magnetite from the ejected plutonic blocks of the Soufrière Volcano, St. Vincent, West Indies. *Am. Mineral. 55*, 793.

Lewis, M. (1963) Reversed partial thermo-magnetic remanence in natural and synthetic titanamagnetites. Unpubl. Ph.D. thesis, University of London.

Lidiak, E. G. (1974) Magnetic characteristics of some Precambrian basement rocks. *J. Geophys. 40*, 549.

Lightfoot, B. (1940) The Great Dyke of Southern Rhodesia. *Proc. Geol. Soc. S. Afr. 43*, 27.

Lindsley, D. H. (1960) Geology of the Spray quadrangle, Oregon: with special emphasis on the petrography and magnetic properties of the Picture Gorge basalt. Unpubl. Ph.D. thesis, The Johns Hopkins University.

________ (1962) Investigations in the system FeO-Fe_2O_3-TiO_2. *Carnegie Inst. Washington Year Book 61*, 100.

________ (1963) Equilibrium relations of coexisting pairs of Fe-Ti oxides. *Carnegie Inst. Washington Year Book 62*, 60.

Lindsley, D. H. (1965) Lower thermal stability of $FeTi_2O_5$-Fe_2TiO_5 (pseudobrookite) solid solution series (abstr.). *Progr., Ann. Meet., Geol. Soc. Am.*, 96.

________ (1971) Synthesis and preliminary results on the stability of aenigmatite ($Na_2Fe_5TiSi_6O_{20}$). *Ann. Rep. Dir. Geophys. Lab. Year Book 69*, 188.

________ (1973) Delimitation of the hematite-ilmenite miscibility gap. *Geol. Soc. Am. Bull. 84*, 657.

________, and S. E. Haggerty (1971) Phase relations of Fe-Ti oxides and aenigmatite; oxygen fugacity of the pegmatoid zones. *Ann. Rep. Dir. Geophys. Lab. Year Book 69*, 278.

________, G. E. Andreasen, and J. R. Balsley (1966) Magnetic properties of rocks and minerals. *Geol. Soc. Am. Mem. 97*, 544.

________, D. Smith, and S. E. Haggerty (1971) Petrography and mineral chemistry of a differentiated flow of Picture Gorge basalt near Spray, Oregon. *Ann. Rep. Dir. Geophys. Lab. Year Book 69*, 264.

________, S. E. Kesson, M. J. Hartzman, and M. K. Cushman (1974) The stability of armalcolite: Experimental studies in the system MgO-Fe-Ti-O. *Proc. Lunar Sci. Conf. 5th*, 521.

Liou, J. G. (1974) Mineralogy and chemistry of glassy basalts, coastal range ophiolites, Taiwan. *Geol. Soc. Am. Bull. 85*, 1.

Lipman, P. W. (1971) Iron-titanium oxide phenocrysts in compositionally zoned ash-flow sheets from southern Nevada. *J. Geol. 79*, 438.

________, and I. Friedman (1975) Interaction of meteoric water with magma: An oxygen-isotope study of ash-flow sheets from southern Nevada. *Geol. Soc. Am. Bull. 86*, 695.

Lister, G. F. (1966) The composition and origin of selected iron-titanium deposits. *Econ. Geol. 61*, 275.

Littlejohn, A. L. and H. J. Greenwood (1973) Lherzolite nodules in basalts from British Columbia, Canada. *Can. J. Earth Sci. 11*, 1288.

Loney, R. A., G. R. Himmelberg, and R. G. Coleman (1971) Structure and petrology of the Alpine-type peridotite at Burro Mountain, California, U.S.A. *J. Petrol. 12*, 245.

Lovering, J. F. and J. R. Widdowson (1968) The petrological environment of magnesium ilmenites. *Earth Planet. Sci. Lett. 4*, 310.

Løvlie, R. and N. D. Opdyke (1974) Rock magnetism and paleomagnetism of some intrusions from Virginia. *J. Geophys. Res. 79*, 343.

Lowder, G. G. (1970) The volcanoes and caldera of Talasca, New Britain: Mineralogy. *Contrib. Mineral. Petrol. 26*, 324.

Lowrie, W. (1973) Magnetic properties of deep-sea drilling project basalts from the north Pacific ocean. *J. Geophys. Res. 78*, 7647.

________ (1973) Viscous remanent magnetization in oceanic basalts. *Nature 243*, 27.

________, and D. V. Kent (1976) Viscous remanent magnetization in basalt samples. In, S. R. Hart and R. S. Yeats, Eds., *Initial Reports of the Deep Sea Drilling Project*, U. S. Government Printing Office, Washington, D. C.

________, R. Lovlie, and N. D. Opdyke (1973) Magnetic properties of deep-sea drilling project basalts from the North Pacific Ocean. *J. Geophys. Res. 78*, 7647.

Lufkin, J. L. (1976) Oxide minerals in miarolitic rhyolite, Black Range, New Mexico. *Am. Mineral. 61*, 425.

Luyendyk, B. P. and W. G. Melson (1967) Magnetic properties and petrology of rocks near the crest of the mid-Atlantic ridge. *Nature 215*, 147.

MacDonald, J. G. (1967) Variations within a Scottish lower carboniferous lava flow. *Scot. J. Geol. 3*, 34.

MacGregor, I. D. and A. R. Basu (1976) Geological problems in estimating mantle geothermal gradients. *Am. Mineral. 61*, 715.

________, and C. H. Smith (1963) The use of chrome spinels in petrographic studies of ultramafic intrusions. *Can. Mineral. 7*, 403.

Mall, A. P. and M. K. Rao (1970) Distribution of iron and magnesium between chromites and orthopyroxenes in ultrabasics from Ganginemi, India. *Lithos 3*, 113.

Malpas, J. and D. F. Strong (1975) A comparison of chrome-spinels in ophiolites and mantle diapirs of Newfoundland. *Geochim. Cosmochim. Acta 39*, 1045.

Marshall, M. and A. Cox (1971) Effect of oxidation on the natural remanent magnetization of titanomagnetite in suboceanic basalt. *Nature 230*, 28.

________, and A. Cox (1971) Magnetism of pillow basalts and their petrology. *Geol. Soc. Am. Bull. 82*, 537.

________, and A. Cox (1972) Magnetic changes in pillow basalt due to sea floor weathering. *J. Geophys. Res. 77*, 6459.

Mathison, C. I. (1967) The Somerset Dam layered basic intrusion. *J. Geol. Soc. Aust. 14*, 57.

________ (1975) Magnetites and ilmenites in the Somerset Dam layered basic intrusion, southeastern Queensland. *Lithos 8*, 93.

Matthews, D. H. (1961) Lavas from an abyssal hill on the floor of the North Atlantic Ocean. *Nature 190*, 158.

Maucher, A. and G. Rehwald (1961) *Bildkartei der Erzmikroskopie*. Umschau Verlag, Frankfurt.

Mazzullo, L. J. and A. E. Bence (1976) Abyssal tholeiites from DSDP Leg 34: the Nazca Plate. *J. Geophys. Res. 81*, 4327.

McBirney, A. R. and K. Aoki (1973) Factors governing the stability of plagioclase at high pressures as shown by spinel-gabbro xenoliths from the Kerguelen Archipelago. *Am. Mineral. 58*, 271.

McCallister, R. H. and H. O. A. Meyer (1975) "Pyroxene"-ilmenite xenoliths from the Stockdale Pipe, Kansas: Chemistry, crystallography, and origin. *Phys. Chem. Earth 9*, 287.

McCallum, M. E., D. H. Eggler, and L. K. Burns. Kimberlitic diatremes in northern Colorado and southern Wyoming. *Phys. Chem. Earth 9*, 149.

McClay, K. R. (1974) Single-domain magnetite in the Jimberlana norite, western Australia. *Earth Planet. Sci. Lett. 21*, 267.

McDonald, J. A. (1965) Liquid immiscibility as one factor in chromitite seam formation in the Bushveld igneous complex. *Econ. Geol. 60*, 1674.

McElhinny, M. W. (1973) *Paleomagnetism and Plate Tectonics*. Cambridge University Press.

McGetchin, T. R. and L. T. Silver (1970) Minerals from kimberlite. *Am. Mineral. 55*, 1738.

McLeod, C. R. and J. A. Chamberlain (1968) Reflectivity and Vickers microhardness of ore minerals. *Geol. Surv. Can., Paper 68-64*.

McMahon, B. M. and S. E. Haggerty (1976) Oka carbonatite complex: Oxide mineral zoning in mantle petrogenesis. *Geol. Soc. Am. Abstr. with Programs 8*, 1006.

McTaggart, K. C. (1971) On the origin of ultramafic rocks. *Geol. Soc. Am. Bull. 82*, 23.

Medaris, L. G., Jr. (1972) High-pressure peridotites in southwestern Oregon. *Geol. Soc. Am. Bull. 83*, 41.

________ (1975) Coexisting spinel and silicates in alpine peridotites of the granulite facies. *Geochim. Cosmochim. Acta 39*, 947.

Melson, W. G. and G. Switzer (1966) Plagioclase-spinel-graphite xenoliths in metallic iron-bearing basalts. *Am. Mineral. 51*, 664.

Merrill, R. T. (1975) Magnetic effects associated with chemical changes in igneous rocks. *Geophys. Surv. 2*, 277.

________, and R. E. Burns (1972) A detailed magnetic study of Cobb Seamount. *Earth Planet. Sci. Lett. 14*, 413.

________, and C. S. Gromme (1969) Nonreproducible self-reversal of magnetization in diorite. *J. Geophys. Res. 74*, 2014.

________, and N. Kawai (1969) A method for detecting self-reversals using etching. *J. Geomag. Geoelectr. 21*, 507.

Meyer, H. O. A. (1975) Kimberlite from Norris Lake, eastern Tennessee: Mineralogy and petrology. *J. Geol. 83*, 518.

________, and N. Z. Boctor (1975) Sulfide-oxide minerals in eclogite from Stockdale kimberlite, Kansas. *Contrib. Mineral. Petrol. 52*, 57.

________, and F. R. Boyd (1972) Composition and origin of crystalline inclusions in natural diamonds. *Geochim. Cosmochim. Acta 36*, 1255.

________, and D. P. Svisero (1975) Mineral inclusions in Brazilian diamonds. *Phys. Chem. Earth 9*, 785.

Michael, W. H., Jr. R. H. Tolson, and J. P. Gapcynski (1966) Oxygen fugacities directly measured in magmatic gases. *Science 153*, 1103.

Miller, C. (1974) Reaction rims between olivine and plagioclase in metaperidotites, Otztal Alps, Austria. *Contrib. Mineral. Petrol. 43*, 333.

Mitchell, R. H. (1970) Kimberlite and related rocks - a critical reappraisal. *J. Geol. 78*, 686.

________ (1973) Composition of olivine, silica activity and oxygen fugacity in kimberlite. *Lithos 6*, 65.

________ (1973) Magnesian ilmenite and its role in kimberlite petrogenesis. *J. Geol. 81*, 301.

________, and D. B. Clark (1976) Oxide and sulphide mineralogy of the Peuyuk Kimberlite, Somerset Island, N.W.T., Canada. *Contrib. Mineral. Petrol. 56*, 157.

________, A. O. Brunfelt, and P. H. Nixon (1973) Part II: Trace elements in magnesian ilmenites from Lesotho kimberlites. In, P. H. Nixon, Ed., *Lesotho Kimberlites*, 230, Lesotho National Development Corporation, Maseru.

________, D. A. Carswell, and A. O. Brunfelt (1973) Ilmenite association trace element studies. In, P. H. Nixon, Ed., *Lesotho Kimberlites*, 224, Lesotho National Development Corporation, Maseru.

Molyneux, T. G. (1970) The geology of the area in the vicinity of Magnet Heights, Eastern Transvaal, with special reference to the magnetic iron ore, 228. In, D. J. L. Visser and G. von Gruenwaldt, Eds., Symposium on the Bushveld Igneous Complex and Other Layered Intrusions, *Geol. Soc. S. Afr. Spec. Pub. 1.*

________ (1972) X-ray data and chemical analyses of some titanomagnetite and ilmenite samples from the Bushveld Complex, South Africa. *Mineral. Mag. 38*, 863.

Momose, K. and K. Kobayashi (1972) Thermomagnetic properties of ferromagnetic minerals extracted from the Pumice-Fall deposit "Pm-I" of the Ontake volcano. *J. Geomag. Geoelectr. 24*, 127.

________, K. Kobayashi, K. Minagawa, and M. Machida (1968) Identification of tephra by means of ferromagnetic minerals in pumice. *Bull. Earthquake Res. Inst. 46*, 1275.

Moore, O. H. (1940) Origin of the Nelsonite dikes of Amherst County, Virginia. *Econ. Geol. 35*, 629.

Morse, S. A. Layered intrusions and anorthosite genesis. In, Y. N. Isachsen, Ed., *Origin of Anorthosites and Related Rocks*, N. Y. State Museum and Science Service, Albany, *Memoir 18*, 175.

Morton, R. D. and R. H. Mitchell (1972) The relationship between micro-indentation hardness and chemical composition in magnesian ilmenites. *Neues Jahrb. Mineral. Monatsch. 7*, 289.

Muir, J. E. and A. J. Naldrett (1973) A natural occurrence of two-phase chromium-bearing spinels. *Can. Mineral. 11*, 930.

München, U. B. (1971) Cation distribution in titanomagnetites. *Z. Geophys. 37*, 305.

Murthy, G. S. and K. V. Rao (1976) Paleomagnetism of Steel Mountain and Indian Head anorthosites from western Newfoundland. *Can. J. Earth Sci. 13*, 75.

________, M. E. Evans, and D. I. Gough (1971) Evidence of single-domain magnetite in the Michikamau anorthosite. *Can. J. Earth Sci. 8*, 361.

Myers, C. W., A. E. Bence, J. J. Papike, and R. A. Ayuso (1975) Petrology of an alkali-olivine basalt sill from Site 169 of DSDP Leg 16: The central Pacific basin. *J. Geophys. Res. 80*, 807.

Nagata, T. (1961) *Rock Magnetism.* Maruzen Publishing Co., Plenum Press.

Nakhla, F. M., S. A. Saleh, and O. Y. Abd El-Aal (1973) Contributions to reflectivity characteristics of some ore minerals and metallic elements. *Chem. Erde 32*, 279.

Naldrett, A. J., R. H. Hewins, and L. Greenman (1972) The main irruptive and the sub-layer at Sudbury Ontario. *Proc. 24th Int. Geol. Congr. 4*, 206.

Nash, W. P. and J. F. G. Wilkinson. Shonkin Sag Laccolith, Montana. 1. Mafic minerals and estimates of temperature, pressure, oxygen fugacity and silica activity. *Contrib. Mineral. Petrol. 25*, 241.

________, I. S. E. Carmichael, and R. W. Johnson (1969) The mineralogy and petrology of Mount Suswa, Kenya. *J. Petrol. 10*, 409.

Navrotsky, A. and O. J. Kleppa (1967) The thermodynamics of cation distributions in simple spinels. *J. Inorg. Nucl. Chem. 29*, 2701.

Neumann, E.-R. (1974) The distribution of Mn^{2+} and Fe^{2+} between ilmenites and magnetites in igneous rocks. *Am. J. Sci. 274*, 1074.

_______ (1976) Compositional relations among pyroxenes, amphiboles and other mafic phases in the Oslo Region plutonic rocks. *Lithos 9*, 85.

Neumann, H. and S. Bergstøl (1964) Contributions to the mineralogy of Norway, No. 25. Pyrophanite in the southern part of the Oslo area. *Norsk Geol. Tidss. 44*, 39.

Nicholls, G. D. (1955) The mineralogy of rock magnetism. *Advances in Physics, Q. Suppl. Phil. Mag. 4*, 113.

Nicholls, I. A. (1971) Petrology of Santorini volcano, Cyclades, Greece. *J. Petrol. 12*, 67.

Nicholls, J. and I. S. E. Carmichael (1972) The equilibration temperature and pressure of various lava types with spinel- and garnet-peridotite. *Am. Mineral. 57*, 941.

_______, I. S. E. Carmichael, and J. C. Stormer, Jr. (1971) Silica activity and P_{total} in igneous rocks. *Contrib. Mineral. Petrol. 33*, 1.

Nickel, E. H. (1958) The composition and microtexture of an ulvöspinel-magnetite intergrowth. *Can. Mineral. 6*, 191.

Nishida, J. and S. Sasajima (1974) Examination of self-reversal due to N-type magnetization in basalt. *Geophys. J. R. astr. Soc. 37*, 453.

Nitsan, U. (1974) Stability field of olivine with respect to oxidation and reduction. *J. Geophys. Res. 79*, 706.

Nixon, P. H. (1973) *Lesotho Kimberlites.* Lesotho National Development Corporation, Maseru, 340p.

_______ (1973) The geology of Mothae, Solane, Thaba Putsoa and Blow 13. In, *Lesotho Kimberlites*, 39.

_______, and F. R. Boyd (1973) Carbonated ultrabasic nodules from Sekameng. In, *Lesotho Kimberlites*, 190.

_______, _______ (1973) Deep seated nodules. In, *Lesotho Kimberlites*, 106.

_______, _______ (1973) Petrogenesis of the granular and sheared ultrabasic nodule suite in kimberlites. In, *Lesotho Kimberlites*, 48.

_______, _______ (1973) The discrete nodule association in kimberlites from Northern Lesotho. In, *Lesotho Kimberlites*, 67.

_______, and P. Kresten (1973) Chromium and nickel in kimberlitic ilmenites. In, *Lesotho Kimberlites.*

_______, O. von Knorring, and J. M. Rooke (1963) Kimberlites and association inclusions of Basutoland: A mineralogical and geochemical study. *Am. Mineral. 48*, 1090.

Oelsner, O. W. (1966) *Atlas of the Most Important Ore Mineral Paragenesis Under the Microscope.* Pergamon Press.

Oen, I. S., C. Kieft, and A. B. Westerhof (1973) Composition of chromites in cordierite- and mica-bearing Cr-Ni ores from Malaga Province, Spain. *Mineral. Mag. 39*, 193.

O'Nions, R. K. and D. G. White (1971) Investigations of the L_{II-III} x-ray emission spectra by electron microprobe. *Am. Mineral. 56*, 1452.

Onyeagocha, A. C. (1974) Alteration of chromite from the Twin Sisters dunite, Washington. *Am. Mineral. 59*, 608.

Opdyke, N. D. and R. Hekinian (1967) Magnetic properties of some igneous rocks from the mid-Atlantic ridge. *J. Geophys. Res. 72*, 2257.

O'Reilly, W. and S. K. Banerjee (1967) The mechanism of oxidation in titanomagnetites: A magnetic study. *Mineral. Mag. 35*, 29.

Otteman, J. and G. Frenzel (1966) Der Chemismus der Pseudobrookite von Vulkaniten (Eine Untersuchung mit der Elektronen-Mikrosonde). *Schweiz. Mineral. Pet. Mitt. 45*, 819.

Ozima, M. and E. E. Larson (1967) Study on irreversible change of magnetic properties of some ferromagnetic minerals. *J. Geomag. Geoelectr. 19*, 117.

_______, _______ (1970) Low- and high-temperature oxidation of titanomagnetite in relation to irreversible changes in the magnetic properties of submarine basalts. *J. Geophys. Res. 75*, 1003.

Ozima, M. and M. Ozima (1967) Self-reversal of remanent magnetization in some dredged submarine basalts. *Earth Planet. Sci. Lett. 3*, 213.

________, ________ (1971) Characteristic thermomagnetic curve in submarine basalts. *J. Geophys. Res. 76*, 2051.

Palabora Mining Co. Ltd. Mine Geological and Mineralogical Staff (1976) The geology and the economic deposits of copper, iron, and vermiculite in the Palabora igneous complex: A brief review. *Econ. Geol. 71*, 177.

Parak, T. (1975) Kiruna iron ores are not "intrusive-magmatic ores of the Kiruna type." *Econ. Geol. 70*, 1242.

Park, C. F., Jr. (1961) A magnetite "flow" in Northern Chile. *Econ. Geol. 56*, 431.

________ (1972) The iron ore deposits of the Pacific Basin. *Econ. Geol. 67*, 339.

Park, J. K. and E. Irving (1970) The mid-Atlantic ridge near 45°N. XII. Coercivity, secondary magnetization, polarity, and thermal stability of dredge samples. *Can. J. Earth Sci. 7*, 1499.

________, ________, and J. A. Donaldson (1973) Paleomagnetism of the Precambrian Dubawnt group. *Geol. Soc. Am. Bull. 84*, 859.

Pavlov, N. V. (1949) Chemical composition of chrome spinels in relation to the composition of rocks of ultrabasic intrusives. *An SSSR Inst. Mineral Fuels Trudy, Ser. Geol. 103*.

Peck, D. L. and T. L. Wright (1966) Experimental studies of molten basalt *in situ*: A summary of physical and chemical measurements on recent lava lakes of Kilauea volcano, Hawaii. *Geol. Soc. Am. Abstr. 101*, 158.

Pecora, W. T. (1966) Carbonatites: A review. *Geol. Soc. Am. Bull. 67*, 1537.

Petersen, V. N. (1962) Untersuchungen magnetischer Eigenschaften von Titanomagnetiten in Basalt des Rauhen Kulm (Oberpfalz) in Verbindung mit elektronen-kroskopischer Beobachtung. *Z. Geophys. 28*, 79.

Philpotts, A. R. (1967) Origin of certain iron-titanium oxide and apatite rocks. *Econ. Geol. 62*, 303.

Pike, J. E. N. (1976) Pressures and temperatures calculated from chromium-rich pyroxene compositions of megacrysts and peridotite xenoliths, Black Rock Summit, Nevada. *Am. Mineral. 61*, 725.

Prévot, M. and S. Grommé (1975) Intensity of magnetization of subaerial and submarine basalts and its possible change with time. *Geophys. J. R. astr. Soc. 40*, 207.

________, and J. Mergoil (1973) Crystallization trend of titanomagnetites in an alkali basalt from Saint-Clément (Massif Central, France). *Mineral. Mag. 39*, 474.

________, G. Rémond, and R. Caye (1968) Étude de la transformation d'une titanomagnétite en titanomaghémite dans une roche volcanique. *Bull. Soc. franc. Mineral. Cristallogr. 91*, 65.

Prins, P. (1972) Composition of magnetite from carbonatites. *Lithos 5*, 227.

Prinz, M., D. V. Manson, P. F. Hlava, and K. Keil (1975) Inclusions in diamonds: Garnet lherzolite and eclogite assemblages. *Phys. Chem. Earth 9*, 797.

________, K. Keil, J. A. Green, A. M. Reid, E. Bonatti, and J. Honnorez (1976) Ultramafic and mafic dredge samples from the equatorial mid-Atlantic ridge and fracture zone. *J. Geophys. Res. 81*, 4087.

Puffer, J. H. (1972) Iron-bearing minerals as indicators of intensive variables pertaining to granitic rocks from the Pegmatite Points area, Colorado. *Am. J. Sci. 272*, 273.

________ (1975) Some North American iron-titanium oxide bearing pegmatites. *Am. J. Sci. 275*, 708.

________, and J. J. Peters (1974) Magnetite veins in diabase of Laurel Hill, New Jersey. *Econ. Geol. 69*, 1294.

Radhakrishnamurty, C. (1974) Magnetic techniques for ascertaining the nature of iron oxide grains in basalts. *J. Geophys. 40*, 453.

________, N. P. Sastry, and E. R. Deutsch (1973) Ferromagnetic behavior of interacting superparamagnetic particle aggregates in basaltic rocks. *Pramāna L*, 61.

________, S. D. Likhite, P. K. S. Raja, and P. W. Sahasrabudhe (1971) Magnetic grains in Bombay columnar basalts. *Nature Phys. Sci. 235*, 1.

Radhakrishnamurty, C., P. K. S. Raja, S. D. Likhite, and P. W. Sahasrabudhe (1972) Problems concerning the magnetic behavior and determination of Curie Points of certain basalts. *Pure & Appl. Geophys. 93*, 129.

Ramdohr, P. (1953) Ulvöspinel and its significance in titaniferous iron ores. *Econ. Geol. 48*, 677.

________ (1962) Erzmikroskopische Untersuchungen an Magnetiten der Exhalationen im Valley of the 10,000 Smokes. *Neues Jahrb. Mineral. Monatsch. 3/4*, 49.

________ (1969) *The Ore Minerals and Their Intergrowths.* Pergamon Press.

Readman, P. W. and W. O'Reilly (1971) Oxidation processes in titanomagnetites. *Z. Geophys. 37*, 329.

________, ________ (1972) Magnetic properties of oxidized (cation-deficient) titanomagnetites $(Fe,Ti,\square)_3O_4$. *J. Geomag. Geoelectr. 24*, 69.

Reay, A. and C. P. Wood (1974) Ilmenites from Kakanui, New Zealand. *Mineral. Mag. 39*, 721.

Reed, S. J. B. (1975) *Electron Microprobe Analysis.* Cambridge University Press.

Reid, A. M. and J. B. Dawson (1972) Olivine-garnet reaction in peridotites from Tanzania. *Lithos 5*, 115.

________, C. H. Donaldson, R. W. Brown, W. I. Ridley, and J. B. Dawson (1975) Mineral chemistry of peridotite xenoliths from the Lashaine Volcano, Tanzania. *Phys. Chem. Earth 9*, 525.

________, ________, J. B. Dawson, R. W. Brown, and W. I. Ridley (1975) The Igwisi Hills extrusive "kimberlites." *Phys. Chem. Earth 9*, 199.

Rice, J. M., J. S. Dickey, Jr., and J. B. Lyons (1971) Skeletal crystallization of pseudobrookite. *Am. Mineral. 56*, 158.

Riding, A. (1969) (A) Magnetic materials in oxidized olivine and their contribution to the natural remanent magnetization of rocks. (B) The natural remanent magnetization of some Triassic red sandstone. Unpubl. Ph.D. thesis, University of Liverpool.

Ridley, W. I. and I. Baker (1973) The petrochemistry of a unique cordierite-bearing lava from St. Helena Island, South Atlantic. *Am. Mineral. 58*, 813.

________, J. M. Rhodes, A. M. Reid, P. Jakes, C. Shih, and M. N. Bass (1974) Basalts from Leg 6 of the deep-sea drilling project. *J. Petrol. 15*, 140.

Rimsaite, J. (1971) Distribution of major and minor constituents between mica and host ultrabasic rocks, and between zoned mica and zoned spinel. *Contrib. Mineral. Petrol. 33*, 259.

Ringwood, A. E. and J. F. Lovering (1970) Significance of pyroxene-ilmenite intergrowths among kimberlite xenoliths. *Earth Planet. Sci. Lett. 7*, 371.

Robertson, W. A. (1963) Paleomagnetism of some Mesozoic intrusives and tuffs from eastern Australia. *J. Geophys. Res. 68*, 2299.

Rodgers, K. A. (1973) Chrome-spinels from the Massif du Sud, southern New Caledonia. *Mineral. Mag. 39*, 326.

Rogers, D. P. (1968) The extrusive iron oxide deposits, "El Laco," Chile. *Geol. Soc. Am. Abstr. with Programs 63*, 700.

Ross, C. S. (1941) Occurrence and origin of the titanium deposits of Nelson and Amherst Counties. *U. S. Geol. Surv. Prof. Pap. 198.*

________, M. D. Foster, and A. T. Myers (1954) Origin of dunites and of olivine-rich inclusions in basaltic rocks. *Am. Mineral. 39*, 693.

Rossiter, M. J. and P. T. Clark, 1965. Cation distribution in ulvöspinel Fe_2TiO_4. *Nature 207*, 402.

Rothstein, A. T. V. (1972) Spinels from the Dawros peridotite, Connemara, Ireland. *Mineral. Mag. 38*, 957.

Rumble, D., III (1973) Fe-Ti oxide minerals from regionally metamorphosed quartzites of western New Hampshire. *Contrib. Mineral. Petrol. 42*, 181.

Runcorn, S. K. (1966) Rock magnetism. *Sci. Progr., Oxford 54*, 467.

Sanderson, D. D. (1974) Spatial distribution and origin of magnetite in an intrusive igneous mass. *Geol. Soc. Am. Bull. 85*, 1183.

Sasajima, S., J. Nishida, and T. Katsura (1975) Impure titanomagnetites characteristic to some alkaline basalts in southwest Japan. *Rock Magnetism & Paleogeophysics 3*, 1.

Sato, M. (1970) An electrochemical method of oxygen fugacity control of furnace atmosphere for mineral syntheses. *Am. Mineral. 55*, 1424.

________ (1971) Electrochemical measurements and control of oxygen fugacity and other gaseous fugacities with solid electrolyte sensors. In, G. C. Ulmer, Ed., *Research Techniques for High Pressure and High Temperature*, Springer-Verlag, Chapter 3, 43.

________ (1972) Intrinsic oxygen fugacities for iron-bearing oxide and silicate minerals under low total pressure. *Geol. Soc. Am. Mem. 135*, 289.

Schaeffer, R. M. and E. J. Schwarz (1970) The mid-Atlantic ridge near 45°N. IX. Thermomagnetics of dredged samples of igneous rocks. *Can. J. Earth Sci. 7*, 268.

Schult, A. (1968) Self-reversal of magnetization and chemical composition of titanomagnetites in basalts. *Earth Planet. Sci. Lett. 4*, 57.

Shouten, E. (1962) *Determination Tables for Ore Microscopy.* Elsevier Publishing Co.

Sigurdsson, H. and J.-G. Schilling (1976) Spinels in mid-Atlantic ridge basalts: chemistry and occurrence. *Earth Planet. Sci. Lett. 29*, 7.

Smith, A. L. (1970) Sphene, perovskite and coexisting Fe-Ti oxide minerals. *Am. Mineral. 55*, 264.

________, and I. S. E. Carmichael (1968) Quaternary lavas from the southern Cascades, western U.S.A. *Contrib. Mineral. Petrol. 19*, 212.

________, ________ (1969) Quaternary trachybasalts from southeastern California. *Am. Mineral. 54*, 909.

Smith, C. H. and H. E. Kapp (1963) The Muskox intrusion, a recently discovered layered intrusion in the Coppermine River area, Northwest Territories, Canada. *Mineral. Soc. Am. Spec. Pap. 1*, 30.

Smith, D. (1970) Mineralogy and petrology of the diabase rocks at a differentiated olivine diabase sill complex. *Contrib. Mineral. Petrol. 27*, 95.

Smith, D. G. W., Ed. (1976) *Short Course Handbook, Vol. 1, Microbeam Techniques.* Co-op Press.

Smith, J. V. and J. B. Dawson (1975) Chemistry of Ti-poor spinels, ilmenites and rutiles from peridotite and eclogite xenoliths. *Phys. Chem. Earth 9*, 309.

Smith, P. J. (1967) Electron probe microanalysis of optically homogeneous titanomagnetites and ferrian ilmenites in basalts of paleomagnetic significance. *J. Geophys. Res. 72*, 5087.

________ (1967) On the suitability of igneous rocks for ancient geomagnetic field intensity determinations. *Earth Planet. Sci. Lett. 2*, 99.

________ (1968) Paleomagnetism and the compositions of highly-oxidized iron-titanium oxides in basalts. *Phys. Earth Planet. Interiors 1*, 88.

Snetsinger, K. G. (1969) Manganoan ilmenite from a Sierran adamellite. *Am. Mineral. 54*, 431.

Sobelov, N. V. (1968) Deep-seated inclusions in kimberlites, and the problem of the composition of the earth's mantle. Extended thesis, abs. transl. by D. A. Brown, *Aust. Natl. Univ. Publ. 210.*

________, N. P. Pokhilenko, Yu. G. Lavrent'ev, and L. V. Usova (1975) Peculiarities in composition of Yakutian chromspinelides of diamonds and kimberlite rocks. *Acad. Sci. USSR, Siberian Div., 11*, 7.

Soffel, H. C. (1968) The behavior of the domain structure of polycrystalline magnetite at the margins of the crystallites. *Earth Planet. Sci. Lett. 4*, 53.

________, and N. Petersen (1971) Ionic etching of titanomagnetite grains in basalts. *Earth Planet. Sci. Lett. 11*, 312.

Spall, H. (1968) Paleomagnetism of basement granites of southern Oklahoma and its implications; progress report. *Okla. Geol. Surv. Notes 28*, No. 2, 65.

________ (1971) Paleomagnetism and K-Ar age of mafic dikes from the Wind River range, Wyoming. *Geol. Soc. Am. Bull. 82*, 2457.

________ (1972) On the remanent magnetism in Precambrian llanite and Town Mountain granite from Llano County, Texas. *Texas J. Sci. 23*, 479.

Spall, H. and H. C. Noltimier (1972) Some curious magnetic results from a Precambrian granite. *Geophys. J. R. astr. Soc. 28*, 237.

Springer, R. K. (1973) Contact metamorphosed ultramafic rocks in the Western Sierra Nevada foothills, California. *J. Petrol. 15*, 160.

Stacey, F. D. (1969) *Physics of the Earth.* John Wiley & Sons.

________ (1976) Paleomagnetism of meteorites. In, *Annu. Rev. Earth Planet. Sci. 4*, 147.

________, and S. K. Banerjee (1974) The physical principles of rock magnetism. *Developments in Solid Earth Geophysics, 5*, Elsevier Publishing Co.

Steinthorsson, S. (1972) The opaque mineraloty of Surtsey. *Surtsey Prog. Rep. VI*, 152.

Stephenson, A. (1969) The temperature dependent cation distribution in titanomagnetites. *Geophys. J. R. astr. Soc. 18*, 199.

________ (1972) Spontaneous magnetization curves and Curie points of spinels containing two types of magnetic ion. *Phil. Mag. 25*, 1213.

Stevens, R. E. (1944) Composition of some chromites of the Western hemisphere. *Am. Mineral. 29*, 1.

Storeetvedt, K. M. and K. M. Petersen (1970) On chemical magnetization in some Permian lava flows of southern Norway. *Z. Geophys. 36*, 569.

Stormer, J. C., Jr. (1972) Mineralogy and petrology of the Raton-Clayton volcanic field, northeastern New Mexico. *Geol. Soc. Am. Bull. 83*, 3299.

Strangway, D. W. (1970) *History of the Earth's Magnetic Field.* McGraw-Hill Co.

________, E. E. Larson, and M. Goldstein (1968) A possible cause of high magnetic stability in volcanic rocks. *J. Geophys. Res. 73*, 3787.

Stroh, J. M. (1976) Solubility of alumina in orthopyroxene plus spinel as a geobarometer in complex systems. Applications to spinel-bearing alpine-type peridotites. *Contrib. Mineral. Petrol. 54*, 173.

Stumpfl, E. F. (1961) Contribution to the study of ore minerals in some igneous rocks from Assynt. *Mineral. Mag. 32*, 767.

Surdam, R. C. (1968) Origin of native copper and hematite in the Karmutsen group, Vancouver Island, B. C. *Econ. Geol. 63*, 961.

Suwa, K., Y. Yusa, and N. Kishida (1975) Petrology of peridotite nodules from Ndonyuo Olnchoro, Samburu District, Central Kenya. *Phys. Chem. Earth 9*, 273.

Sweetman, T. R. and J. V. P. Long (1969) Quantitative electron probe microanalysis of rock forming minerals. *J. Petrol. 10*, 332.

Symons, D. T. A. (1967) The magnetic and petrologic properties of a basalt column. *Geophys. J. R. astr. Soc. 12*, 473.

Thayer, T. P. (1969) Gravity differentiation and magmatic re-emplacement of podiform chromite deposits. In, H. D. B. Wilson, Ed., Magmatic Ore Deposits, *Econ. Geol. Mono. 4*, 132.

________ (1970) Chromite segregations as petrogenetic indicators, 380. In, D. J. L. Visser and G. von Gruenwaldt, Eds., Symposium on the Bushveld Igneous Complex and Other Layered Intrusions. *Geol. Soc. S. Afr. Spec. Pub. 1*, 380.

Thompson, R. N. (1973) Titanian chromite and chromian titanomagnetite from a Snake River plain basalt, a terrestrial analogue to lunar spinels. *Am. Mineral. 58*, 826.

Traub, I. (1975) Chemische Zusammensetzung Natürlicher Pseudobrookite und Koexistierender Hämatite. Diplom. Mineral. Petrol. Inst. der Univ. Heidelberg.

Tsusue, A. (1973) The distribution of manganese and iron between ilmenite and granite magma in the Osumi Peninsula, Japan. *Contrib. Mineral. Petrol. 40*, 305.

Tsvetkov, A. I., V. S. Myasnikov, N. I. Shchepochkina, and N. A. Matveyeva (1965) Tabular formations in titanomagnetite. *Int. Geol. Rev. 8*, 676.

Turnock, A. C. and H. P. Eugster (1962) Fe-Al oxides: Phase relationships below 1000°C. *J. Petrol. 3*, 533.

Tuthill Helz, R. (1972) Phase relations of basalts in their melting range at P_{H_2O} = 5 kb as a function of oxygen fugacity. *J. Petrol. 14*, 249.

Tuttle, O. F. and J. Gittins, Eds. (1966) *Carbonatites.* John Wiley & Sons, 591p.

Ulmer, G. C. (1969) Experimental investigations of chromite spinels. In, H. D. B. Wilson, Ed., Magmatic Ore Deposits. *Econ. Geol. Mono. 4*, 114.

Ulmer, G. C., Ed. (1971) *Research Techniques for High Pressure and High Temperature.* Springer-Verlag.

________ (1974) Alteration of chromite during serpentinization in the Pennsylvania-Maryland district. *Am. Mineral. 59*, 1236.

________, M. Rosenhauer, E. Woermann, J. Ginder, A. Drory-Wolff, and P. Wasilewski (1976) Applicability of electrochemical oxygen fugacity measurements to geothermometry. *Am. Mineral. 61*, 653.

Unan, C. (1970) The relation between chemical analysis and mineralogy of oxidized titaniferous magnetites in some Scottish dolerites. *Geophys. J. R. astr. Soc. 22*, 241.

Uytenbogaardt, W. (1953) On the opaque mineral constituents in a series of amphibolitic rocks from Norra Storfjallet, Vasterbotten, Sweden. *Ark. Mineral. Geol. 1*, 527.

________, and E. A. J. Burke (1971) *Tables for Microscopic Identification of Ore Minerals.* Elsevier Publishing Co.

Vaasjoki, O. and A. Heikkinen (1962) On the significance of some textural and compositional properties of the magnetites of titaniferous iron ore. *Comptes rendus de la société geologique de Finlande 34*, 141.

________, and K. Puustinen (1966) On the titaniferous ore oxides in some subsilicic dikes and sills. *Comptes rendus de la société geologique de Finlande 38*, 289.

van Zyl, J. P. (1970) The petrology of the Merensky Reef and associated rocks on Swartklip 988, Rustenburg, 80. In, D. J. L. Visser and G. von Gruenwaldt, Symposium on the Bushveld Igneous Complex and Other Layered Intrusions. *Geol. Soc. S. Afr. Spec. Pub. 1.*

Velde, D. (1975) Armalcolite-Ti-phlogopite-diopside-analcite-bearing lamproites from Smoky Butte, Garfield County, Montana. *Am. Mineral. 60*, 566.

Veltheim, V. (1962) On the geology of the chromite deposit at Kemi, north Finland. *Bull. de la Comm. geol. de Finlande 194*, 1.

Verhoogen, J. (1962) Distribution of titanium between silicates and oxides in igneous rocks. *Am. J. Sci. 260*, 211.

________ (1962a) Oxidation of iron-titanium oxides in igneous rocks. *J. Geol. 70*, 168.

________ (1962b) Distribution of titanium between silicates and oxides in igneous rocks. *Am. J. Sci. 260*, 211.

________ (1969) Magnetic properties of rocks. In, P. Hart, Ed., *The Earth's Crust and Upper Mantle.* American Geophysical Union, 627.

Vincent, E. A. (1960) Ulvöspinel in the Skaergaard intrusion, Greenland. *Neues Jahrb. Mineral., Abh., 94*, 993.

________, and R. Phillips (1954) Iron-titanium oxide minerals in layered gabbros of the Skaergaard intrusion, East Greenland, Part 1. Chemistry and ore-microscopy. *Geochim. Cosmochim. Acta 6*, 1.

________, J. B. Wright, R. Chevallier, and S. Mathieu (1957) Heating experiments on some natural titaniferous magnetites. *Mineral. Mag. 31*, 624.

Vincenz, S. A., K. Yaskawa, and J. M. Ade-Hall (1975) Origin of the magnetization of the Wichita Mountains granites, Oklahoma. *Geophys. J. R. astr. Soc. 42*, 21.

Virgo, D. and F. E. Huggins (1975) Cation distributions in some compounds with the pseudobrookite structure. *Ann. Rep. Dir. Geophys. Lab. Year Book 74*, 585.

Vogt, P. R. and N. A. Ostenso (1966) Magnetic survey over the mid-Atlantic ridge between 42°N and 46°N. *J. Geophys. Res. 71*, 4389.

Vollstaadt, H., K. Rother, and P. Nozharov (1967) The palaeomagnetic stability and the petrology of some caenozoic and cretaceous andesites of Bulgaria. *Earth Planet. Sci. Lett. 3*, 399.

von Knorring, O. and K. G. Cox (1961) Kennedyite, a new mineral of the pseudobrookite series. *Mineral. Mag. 32*, 676.

Wager, L. R. and G. M. Brown, Eds. (1968) *Layered Igneous Rocks.* Oliver and Boyd, 588p.

________, and W. A. Deer (1939) Geological investigations in east Greenland, pt. III. The petrology of the Skaergaard intrusion, Kangerdlugssuaq, east Greenland. *Medd. om Grønland 105*, 1.

Wasilewski, P. J. (1968) Magnetization of ocean basalts. *J. Geomag. Geoelectr. 20*, 129.

Watkins, N. D. and F. W. Cambray (1971) Paleomagnetism of cretaceous dikes from Jamaica. *Geophys. J. R. astr. Soc. 22*, 163.

________, and B. M. Gunn (1971) Petrology, geochemistry, and magnetic properties of some rocks dredged from the Macquarie ridge. *New Zealand J. Geol. Geophys. 14*, 153.

________, and S. E. Haggerty (1965) Some magnetic properties and the possible petrogenetic significance of oxidized zones in an Icelandic olivine basalt. *Nature 206*, 797.

________, ________ (1967) Primary oxidation variation and petrogenesis in a single lava. *Contrib. Mineral. Petrol. 15*, 251.

________, ________ (1968) Oxidation and magnetic polarity in single Icelandic lavas and dikes. *Geophys. J. R. astr. Soc. 15*, 305.

________, and T. P. Paster (1971) The magnetic properties of igneous rocks from the ocean floor. *Phil. Trans. Roy. Soc. Lond. A., 268*, 507.

________, ________, and J. Ade-Hall (1970) Variation of magnetic properties in a single deep-sea pillow basalt. *Earth Planet. Sci. Lett. 8*, 322.

Watt, J. P., W. Keiper, J. M. Ade-Hall, and D. F. Goble (1973) Measurement of Fe^{3+}/Fe^{2+} ratios in basaltic class I titanomagnetites using the Mössbauer effect. *J. Geophys. Res. 78*, 3301.

Webb, A. J. (1970) Electrochemical determination of crystallization oxygen fugacities in Monteregion lamprophyre dikes. Unpubl. M.S. thesis, McGill University, Montreal.

Webster, A. H. and N. F. H. Bright (1961) The system iron-titanium-oxygen at 1200°C and oxygen partial pressures between 1 atm and 2 x 10^{-14} atm. *J. Am. Ceram. Soc. 44*, 110.

Wenk, H.-R., Ed. (1976) *Electron Microscopy in Mineralogy.* Springer-Verlag.

Wernick, J. H. (1972) Recent developments in magnetic materials. *Annu. Rev. Mater. Sci. 2*, 607.

Wertheim, G. K. (1964) *The Mössbauer Effect: Principles and Applications.* Academic Press, New York.

Whitelock, T. K. (1973) The Monastery Mine kimberlite pipe. In, P. H. Nixon, Ed., *Lesotho Kimberlites*, 214, Lesotho National Development Corporation, Maseru.

Whitney, J., H. P. Johnson, S. Levi, and B. W. Evans (1971) Investigations of some magnetic and mineralogical properties of the Laschamp and Olby Flows, France. *Quaternary Res. 1*, 511.

Whitney, J. A. and J. C. Stormer, Jr. (1976) Geothermometry and geobarometry in epizonal granitic intrusions: A comparison of iron-titanium oxides and coexisting feldspars. *Am. Mineral. 61*, 751.

Whitney, P. R. (1972) Spinel inclusions in plagioclase of metagabbros from the Adirondack Highlands. *Am. Mineral. 57*, 1429.

________, and J. M. McLelland (1973) Origin of coronas in metagabbros of the Adirondack Mountains, N. Y. *Contrib. Mineral. Petrol. 39*, 81.

Whitten, E. H. T. (1959) Tuffisites and magnetite tuffisites from Tory Island, Ireland, and related products of gas action. *Am. J. Sci. 257*, 113.

Wilkinson, J. F. G. (1957) Titanomagnetites from a differentiated teschenite sill. *Mineral. Mag. 31*, 443.

________ (1965) Titanomagnetites from a differentiation sequence, analcime-olivine theralite to analcime-tinguaite. *Mineral. Mag. 34*, 528.

________ (1971) The petrology of some vitrophyric calc-alkaline volcanics from the carboniferous of New South Wales. *J. Petrol. 12*, 587.

________ (1975) An Al-spinel ultramafic-mafic inclusion suite and high pressure megacrysts in an analcimite and their bearing on basaltic magma fractionation at elevated pressures. *Contrib. Mineral. Petrol. 53*, 71.

Willemse, J. (1969) The geology of the Bushveld igneous complex, the largest repository of magmatic ore deposits in the world. In, H. D. B. Wilson, Ed., Magmatic Ore Deposits. *Econ. Geol. Mono. 4*, 1.

________ (1969) The vanadiferous magnetic iron ore of the Bushveld igneous complex. In, H. D. B. Wilson, Ed., Magmatic Ore Deposits. *Econ. Geol. Mono. 4*, 187.

Williams, A. F. (1932) *The Genesis of Diamond, Vols. 1 & 2*, E. Benn, London.

Williams, R. J. (1971) Reaction constants in the system Fe-MgO-SiO_2-O_2: Intensive parameters in the Skaergaard intrusion, east Greenland. *Am. J. Sci. 271*, 132.

Wilson, R. L. (1966) Palaeomagnetism and rock magnetism. *Earth Sci. Rev. 1*, 175.

________, and N. D. Watkins (1967) Correlation of petrology and natural magnetic polarity in Columbia Plateau basalts. *Geophys. J. R. astr. Soc. 12*, 405.

________, and S. E. Haggerty (1966) Reversals of the earth's magnetic field. *Endeavor 25*, 104.

________, P. Dagley, and J. M. Ade-Hall (1972) Palaeomagnetism of the British tertiary igneous province: the Skye lavas. *Geophys. J. R. astr. Soc. 28*, 285.

________, S. E. Haggerty, and N. D. Watkins (1968) Variation of palaeomagnetic stability and other parameters in a verticle traverse of a single Icelandic lava. *Geophys. J. R. astr. Soc. 16*, 79.

________, N. D. Watkins, Tr. Einarsson, Th. Sigurgeirsson, S. E. Haggerty, P. J. Smith, P. Dagley, and A. G. McCormack. Palaeomagnetism of ten lava sequences from South-western Iceland. *Geophys. J. R. astr. Soc. 29*, 459.

Wimmenauer, W. (1966) The eruptive rocks and carbonatites of the Kaisersthuhl, Germany, 183. In, O. F. Tuttle and J. Gittins, *Carbonatites*. John Wiley & Sons, 591p.

Wolejszo, J., R. Schlich, and J. Segoufin (1974) Paleomagnetic studies of basalt samples, deep-sea drilling project, Leg 25. In, E. S. W. Simpson and R. Schlich, *Initial Reports of the Deep Sea Drilling Project 25*, U. S. Government Printing Office, Washington, D. C.

Wones, D. R. and M. C. Gilbert (1969) The fayalite-magnetite-quartz assemblage between 600° and 800°C. *Am. J. Sci. 267-A*, 480.

Worst, B. G. (1960) The Great Dyke of Southern Rhodesia. *Southern Rhodesia Geol. Surv. Bull. 47*.

________ (1964) Chromite in the Great Dyke of Southern Rhodesia. *Geol. Soc. S. Afr., The Geology of Some Ore Deposits in Southern Africa, II*, 209.

Wright, J. B. (1961) Solid-solution relationships in some titaniferous iron oxide ores of basic igneous rocks. *Mineral. Mag. 32*, 778.

________ (1967) The iron-titanium oxides of some Dunedin (New Zealand) lavas, in relation to their palaeomagnetic and thermomagnetic character (with an appendix on associated chromiferous spinel). *Mineral. Mag. 36*, 425.

________, and J. F. Lovering (1965) Electron-probe micro-analysis of the iron-titanium oxides in some New Zealand ironsands. *Mineral. Mag. 35*, 604.

Wright, T. L. and P. W. Weiblen (1967) Mineral composition and paragenesis in tholeiitic basalt from Makaoupuhi Lava Lake, Hawaii. *Geol. Soc. Am. Abstr. 115*, 242.

Wu, Y. T., M. Fuller, and V. A. Schmidt (1974) Microanalysis of N.R.M. in a granodiorite intrusion. *Earth Planet. Sci. Lett. 23*, 275.

Wyllie, P. J., Ed. (1967) *Ultramafic and Related Rocks*. John Wiley & Sons, 464p.

Yagi, K. (1964) Pillow lavas of Kerflavik, Iceland and their genetic significance. *J. Fac. Sci. Hokkaido Univ. Ser. IV: Geol. and Mineral. 12*, 171.

Young, B. B. and A. P. Millman (1964) Microhardness and deformation of ore minerals. *Bull. Trans. Inst. Mineral. Met. 73*, 437.

Zeuch, D. H. and H. W. Green, II (1976, in press). Naturally decorated dislocations in olivine from peridotite xenoliths. *Geology*.

Ziebold, T. O. and R. E. Ogilvie (1964) An empirical method for electron microanalysis. *Anal. Chem. 36*, 322.

Zussman, J., Ed. (1967) *Physical Methods in Determinative Mineralogy*. Academic Press, New York.